HEATH

PRE-ALGEBRA

David W. Lowry
Earl G. Ockenga
Walter E. Rucker

D.C. Heath and Company
Lexington, Massachusetts / Toronto, Ontario

Heath Pre-Algebra Components

Pupil's Edition
Teacher's Edition
Teacher's Resource File
Practice Workbook
Practice Workbook, Teacher's Edition
Transparencies: Visual Aids
Computer Generated Practice and Tests

Authors

David W. Lowry
Experienced teacher of mathematics in junior and senior high school. More than thirty years in mathematics curriculum development. Coauthor of numerous mathematics textbooks.

Earl G. Ockenga
Teacher of mathematics at Price Laboratory School, University of Northern Iowa, Cedar Falls, Iowa. Recipient of the Presidential Award for Excellence in Mathematics Teaching, 1989. Coauthor of numerous mathematics textbooks.

Walter E. Rucker
Experienced teacher of mathematics in junior and senior high school. More than thirty-five years in mathematics curriculum development. Coauthor of numerous mathematics textbooks.

3 4 5 6 7 8 9 0

CONTENTS

1 Solving Equations Using Addition and Subtraction

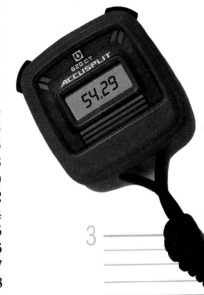

2 Solving Equations Using Multiplication and Division

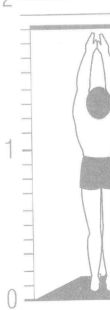

3　Solving Equations Using Number Properties

wet concrete roads

« dry concrete roads

10　20　30　40　50

4　Integers and Equations

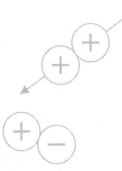

Algebra Using Exponents and Fractions

Algebra Using Sums and Differences of Fractions

v

Algebra Using Products and Quotients of Fractions

$$\begin{array}{r} BE \\ \times\ BE \\ \hline ARE \end{array}$$

8 Algebra Using Rational Numbers

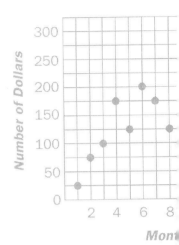

11 Geometry—Surface Area and Volume

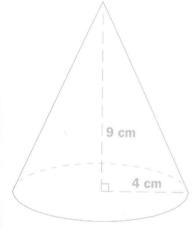

12 Graphing Equations and Inequalities

1 < H T

2 < H T

3 < H T H T

4 < H T H T

5 < H T

6 < H T

ix

15 Similar and Right Triangles

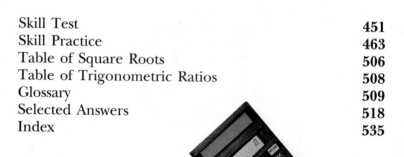

n

6 m

Solving Equations Using Addition and Subtraction

1

$$p + c = g$$

Paddling speed plus current speed equals ground speed.
(page 21)

(page 21)

Canoeists use information about river currents to plan trips and schedule pickup times.

1

1-1 Variables and Algebraic Expressions

OBJECTIVE: To write algebraic expressions that describe problem situations.

Financially successful movies usually have big box office sales during their opening weekend.

Movie	Opening weekend sales
Dick Tracy	n million dollars
Batman	$n + 19$ million dollars
Steel Magnolias	$n - 14$ million dollars
Driving Miss Daisy	22 million dollars less than *Dick Tracy*'s sales
Teenage Mutant Ninja Turtles	2 million dollars more than *Dick Tracy*'s sales

The letter n is a **variable**. It represents *Dick Tracy*'s total box office sales in millions of dollars for its opening weekend. Variables, numbers, and operation signs can be combined to form **algebraic expressions**. If *Dick Tracy*'s sales were n million dollars, then *Teenage Mutant Ninja Turtle*'s sales were $n + 2$ million dollars.

1. Which movie's sales were 19 million dollars more than *Dick Tracy*'s?

2. Which movie's sales were 14 million dollars less than *Dick Tracy*'s?

3. If *Dick Tracy*'s sales were n million dollars, which movie's sales were $n - 22$ million dollars?

EXAMPLES | **Here's how to write algebraic expressions for word expressions.**

Word expression	Algebraic expression		Word expression	Algebraic expression
A a number n plus 5	$n + 5$	B a number s increased by 4	$s + 4$
C a number p minus 2	$p - 2$	D a number t decreased by 1	$t - 1$
E a number m less 3	$m - 3$	F the sum of a number y and 0.5	$y + 0.5$
G $\frac{1}{2}$ less than a number b	$b - \frac{1}{2}$	H 9 more than a number c	$c + 9$

Notice that any letter can be used as a variable.

CHECK for Understanding

4. Look at Example D. To read $t - 1$, you can say "a number t _?_ by 1."

5. Look at Example H. To read the expression $c + 9$, you can say "9 _?_ than a number c.

6. Which of the movies listed above would you expect to have the greatest financial success?

EXERCISES

▶ **Write an algebraic expression for each word expression.**

7. 6 more than a number n

8. 9 less than a number w

9. the sum of a number a and 9

10. a number k minus 7

11. 5 increased by a number g

12. a number m decreased by 4

13. a number b less 8

14. the sum of a number m and 10

15. 32 decreased by a number n

16. 11 more than a number d

17. a number d plus $12\frac{1}{2}$

18. a number p minus $\frac{1}{3}$

19. a number t increased by itself

20. a number s decreased by itself

▶ **Let x be the cost in dollars for a _Batman_ video. Write an algebraic expression for the cost of**

21. a video that is 6 dollars more than _Batman_.

22. a video that is 2 dollars less than _Batman_.

23. a _Batman_ video whose price has been decreased by 2 dollars.

24. a _Batman_ video whose price has been increased by 1 dollar.

25. a _Batman_ video that is on sale for 2 dollars off.

26. a _Batman_ video and a 1 dollar box of Cracker Jack.

Problem Solving USING ALGEBRAIC EXPRESSIONS

▶ **Decide whether Expression A or B would be used to complete each sentence.**

> **Expression A:** $m + 3$
> **Expression B:** $m - 3$

27. Together two movie tickets cost m dollars. If one of the tickets costs 3 dollars, then the other ticket costs _?_ dollars.

28. The late movie is 3 dollars more than the afternoon movie. If the afternoon movie is m dollars, then the late movie is _?_ dollars.

29. _Driving Miss Daisy_ was nominated for m awards. It won all but 3 of them. _Driving Miss Daisy_ won _?_ awards.

30. Write a problem that has Expression A as its answer.

OBJECTIVE: To substitute numbers for variables and then evaluate the resulting numerical expression.

The algebraic expressions in this chart show the players' scores for each half of the basketball game. The variables f and s represent the number of points Jordan scored in the first and second halves of the game.

Player	Number of points scored First half	Second half
Jordan	f	$s = 22$
Pippen	$f + 4$ 17	$s - 2$? = 24
Paxson	$f - 6$ 7	$s + 5$ 14 = 21
King	$f - 2$ 12	$s - 18$ = 20
Nealy	$f - 3$ 10	$s = 19$
Armstrong	$f + 5$ 18	$s - 4$
Hodges	$f - 8$ 5	$s - 5$ = 4
Grant	$f - 5$ 9	$s - 8$

1. Look at the chart. In the first half, who scored 5 more points than Jordan?

2. In the second half, who scored 2 fewer points than Jordan?

3. Who scored more points in the first half, Pippen or Armstrong?

4. Who scored more points in the second half, Pippen or King?

EXAMPLES | Here's how to evaluate algebraic expressions.

In this game, Jordan scored 13 points in the first half and 9 points in the second half. To find the number of points each player scored, substitute 13 for f and 9 for s in the expressions and simplify.

A Points Paxson scored in the first half:
$f - 6$
$13 - 6 = 7$
Paxson scored 7 points in the first half.

B Points Paxson scored in the second half:
$s + 5$
$9 + 5 = 14$
Paxson scored 14 points in the second half.

C Total points Nealy scored in the game:
$(f - 3) + s$
$(13 - 3) + 9$
$10 + 9 = 19$
Nealy scored a total of 19 points.

Do the operation in parentheses first.

CHECK for Understanding

5. Look at Examples A and B. How many points did Paxson score in all?

6. Who scored more points in the game, Paxson or Nealy?

7. Complete the examples. Evaluate each algebraic expression for $x = 7$ and $y = 9$.

a. $x + y + 3$

$7 + 9 + 3 = \,?$

b. $(y - x) + 3$

$(9 - 7) + 3 = \,?$

c. $x + x + 8.5$

$7 + 7 + 8.5 = \,?$

d. $y + y + 2\frac{1}{2}$

$9 + 9 + 2\frac{1}{2} = \,?$

EXERCISES

▷ Evaluate each algebraic expression for $x = 5$, $y = 8$, and $z = 6$.

8. $x + 7$

9. $y - 5$

10. $x + y$

11. $y - z$

12. $y - 4$

13. $x + 8$

14. $19 - x$

15. $z - 6$

16. $z + 9$

17. $z - \frac{1}{2}$

18. $6.5 - z$

19. $y - 0.4$

20. $x + z - 2$

21. $x + y - 5$

22. $(y - z) + 6.29$

23. $y + y + 4$

24. $x + x - 4$

25. $(15 - y) - z$

26. $x + z - 9$

27. $y + y + y$

28. $(12 - x) + y$

29. $(y - z) + 2$

30. $y + z - x$

31. $(14 - x) - z$

32. $(y - z) + \frac{2}{3}$

33. $x + x - y$

34. $x + x + x + x$

35. $9.3 + x - y$

▷ Evaluate each algebraic expression for $a = 3$, $b = 7$, and $c = 4$.
Here are scrambled answers for the next two rows of exercises: 10   1   7   11

36. c decreased by a

37. the sum of b and c

38. b increased by a

39. a plus c

40. a less than b

41. b minus c

42. a decreased by itself

43. c more than 2

44. the sum of b and 6, decreased by 8

45. the sum of a and a, increased by b

Problem Solving USING ALGEBRAIC EXPRESSIONS

▷ Solve. Use the chart on page 4. *Remember: $f = 13$ and $s = 9$.*

46. How many points did Grant score in the first half?

47. How many points did Hodges score in the second half?

48. Who scored 18 points in the first half?

49. Who scored 7 points in the second half?

50. How many points did Jordan score in all?

51. How many points did Pippen score in all?

52. Who scored a total of 23 points in the game?

53. Who scored a total of 19 points?

★**54.** Who scored a total of $f + s + 1$ points? ★**55.** Who scored a total of $f + s - 3$ points?

OBJECTIVE: To round decimals to the nearest whole number, tenth, or hundredth.

The gasoline pump shows the total cost and the total number of gallons.

1. What is the total cost?

2. What is the total number of gallons?

EXAMPLES | **Here's how to round a decimal.**

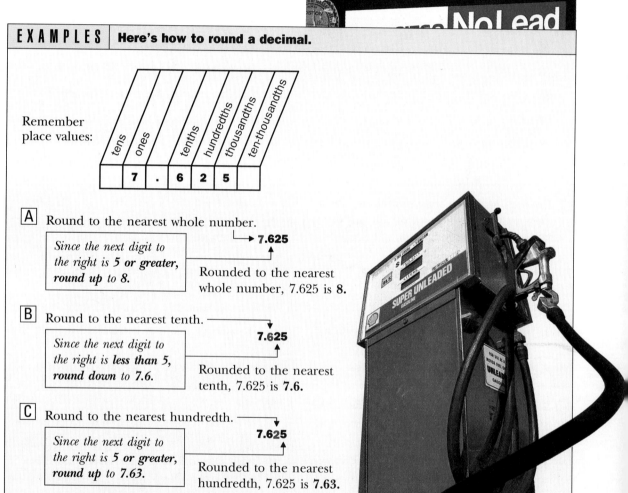

Remember place values:

tens	ones		tenths	hundredths	thousandths	ten-thousandths
	7	.	6	2	5	

A Round to the nearest whole number.

*Since the next digit to the right is **5 or greater**, round up to 8.*

7.**625**

Rounded to the nearest whole number, 7.625 is **8.**

B Round to the nearest tenth.

*Since the next digit to the right is **less than 5**, round down to 7.6.*

7.6**25**

Rounded to the nearest tenth, 7.625 is **7.6.**

C Round to the nearest hundredth.

*Since the next digit to the right is **5 or greater**, round up to 7.63.*

7.62**5**

Rounded to the nearest hundredth, 7.625 is **7.63.**

CHECK for Understanding

3. Look at the examples above. What is 7.625 rounded to the nearest
a. whole number?   **b.** tenth?   **c.** hundredth?

4. What is the total cost rounded to the nearest dollar?

EXERCISES

▶ **Round to the nearest whole number.**

5. 25.6	**6.** 36.3	**7.** 0.38	**8.** 84.44	**9.** 0.94
10. 43.17	**11.** 38.927	**12.** 89.821	**13.** 54.0361	**14.** 38.27

▶ **Round to the nearest tenth.**

15. 3.74	**16.** 36.205	**17.** 8.36	**18.** 0.554	**19.** 9.238
20. 3.470	**21.** 49.183	**22.** 41.08	**23.** 6.452	**24.** 7.1205
25. 605.38	**26.** 714.82	**27.** 0.057	**28.** 27.021	**29.** 64.398

▶ **Round to the nearest hundredth.**

30. 16.318	**31.** 42.106	**32.** 2.563	**33.** 72.835	**34.** 25.614
35. 72.573	**36.** 4.3059	**37.** 0.0351	**38.** 6.4745	**39.** 0.018
40. 8.4536	**41.** 0.005	**42.** 27.938	**43.** 165.789	**44.** 17.046

▶ **Round to the nearest dollar.**

45. $6.66	**46.** $15.39	**47.** $146.68	**48.** $58.51	**49.** $27.87
50. $42.48	**51.** $55.37	**52.** $138.50	**53.** $99.75	**54.** $14.09

Problem Solving NUMBER SENSE

▶ **Solve. Use the gasoline pump shown here.**

55. What is the total cost rounded to the nearest dollar?

56. What is the number of gallons rounded to the nearest
 a. whole number? **b.** tenth?
 c. hundredth?

57. What is the price per gallon rounded to the nearest cent?

Challenge! ESTIMATION

58. You're the driver! You look at your fuel gauge shown at the right. You know that you can travel about 30 miles on a gallon of gasoline. You also know that the tank has a capacity of 19.5 gallons. Do you think that you can drive 390 more miles without refueling?

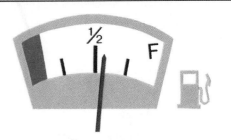

OBJECTIVE: To estimate and compute decimal sums.

1.28 g of sodium **0.098 g of sodium** **0.31 g of sodium**

1. How many grams (g) of sodium does a cheeseburger contain?

2. Which food contains about 0.1 g of sodium?

EXAMPLES	Here's how to estimate and compute decimal sums.

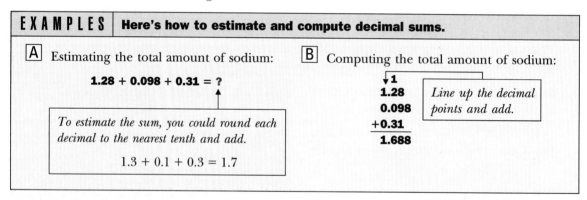

A Estimating the total amount of sodium:

1.28 + 0.098 + 0.31 = ?

To estimate the sum, you could round each decimal to the nearest tenth and add.

1.3 + 0.1 + 0.3 = 1.7

B Computing the total amount of sodium:

```
 ↓1
 1.28
 0.098
+0.31
 1.688
```
Line up the decimal points and add.

CHECK for Understanding

3. Look at Example A. To estimate the sum, each decimal was rounded to the nearest ？ .

4. Look at Example B. Was the computed sum close to the estimate?

EXERCISES

▶ **Three of the calculator answers are wrong. Find them by estimating.**

5. 6.244 ⊞ 0.3862 (*10.106*)

6. 9.8654 ⊞ 1.0737 (*10.9391*)

7. 80.099 ⊞ 20.377 (*90.476*)

8. 36.084 ⊞ 11.769 (*47.853*)

9. 13.521 ⊞ 2.7286 (*16.2496*)

10. 57.875 ⊞ 14.069 (*81.944*)

▶ **Use estimation to choose the sum.**

11. $15.29 + 1.68$
  a. 15.97
  b. 16.97
  c. 17.97

12. $3.845 + 0.216$
  a. 4.061
  b. 5.061
  c. 6.061

13. $8.14 + 2.683$
  a. 9.823
  b. 10.823
  c. 11.823

14. $6.8 + 3.02 + 0.61$
  a. 8.43
  b. 9.43
  c. 10.43

▶ **Evaluate each expression for $w = 2.4$, $x = 3.3$, $y = 5.1$, and $z = 2.68$.**

Here are scrambled answers for the next two rows of exercises:
5.7   8.9   8.4   10.4   12.1   7.5

Example:
$y + z$
$5.1 + 2.68 = 7.78$

15. $w + 8$

16. $w + 6.5$

17. $w + 9.7$

18. $w + x$

19. $x + y$

20. $w + y$

21. $y + y$

22. $x + z$

23. $w + z$

24. $w + x + 10.2$

25. $x + y + 20.5$

26. $w + y + 2.5$

27. $w + x + y$

28. $x + y + z$

29. $w + y + z$

30. $w + x + x$

31. $x + x + z$

32. $z + z + z$

Problem Solving   USING DATA

33. How many grams of sodium are contained in each breakfast menu?
  a. Menu #1
    2 large eggs
    4 strips of bacon
    2 slices of toast
    1 pat of butter
    1 cup of milk
  b. Menu #2
    2 pancakes
    2 pats of butter
    1 ounce of syrup
    1 cup of milk

Challenge!   NUMBER SENSE

34. The recommended daily limit of sodium is 2 grams. Using your answers to Exercise 33, do you think the average American consumes 1.5 g, 15 g, or 150 g of sodium a day?

NUMBER OF GRAMS OF SODIUM

1 pancake
1.2 g

4 strips of bacon
0.48 g

2 slices of toast
0.25 g

1 cup of milk
0.13 g

2 large eggs
0.12 g

1 ounce of syrup
0.065 g

1 pat of butter
0.04 g

Solving Equations Using Addition and Subtraction | **9**

OBJECTIVE: To estimate and compute decimal differences.

Several students had a contest to see who could come closest to estimating the length of a minute without looking at a clock. The results are shown in the table.

1. How many seconds did Rob guess?

2. From what number would you subtract 54.29 to compute how close Rob came to a minute?

Name	Time (seconds)
Rob C.	54.29
Angela G.	57.08
Susan Jo R.	63.41
Ben M.	65.63
Dean R.	56.79
Nara T.	64.80

EXAMPLES Here's how to estimate and compute decimal differences.

A Estimating how close Rob came to a minute:

$$60 - 54.29 = ?$$

To estimate the difference, you could round each decimal to the nearest whole number and subtract.

$$60 - 54 = 6$$

B Computing how close Rob came to a minute:

Line up the decimal points, annex the 0's, and subtract.

$$
\begin{array}{r}
5\ 9\quad 9 \\
\cancel{6}\cancel{0}.\cancel{1}0^{10} \\
-5\ 4.\ 2\ 9 \\
\hline
5.\ 7\ 1
\end{array}
$$

CHECK for Understanding

3. Look at Example A. To estimate the difference, each decimal was rounded to the nearest _?_ number.

4. Look at Example B. Was the computed difference close to the estimate?

EXERCISES

▶ Three of the calculator answers are wrong. Find them by estimating.

5. 87.42 ⊟ 4.25 (*83.17*)

6. 68.12 ⊟ 39.31 (*38.81*)

7. 6.01 ⊟ 2.99 (*5.02*)

8. 18.2 ⊟ 16.95 (*1.25*)

9. 243.8 ⊟ 22.12 (*122.6*)

10. 17.21 ⊟ 13.194 (*4.016*)

▶ **Use estimation to choose the difference.**

11. $39.48 - 7.294$
   a. 30.186
   b. 32.186
   c. 34.186

12. $50 - 13.683$
   a. 36.317
   b. 38.317
   c. 40.317

13. $6.1 - 3.374$
   a. 1.726
   b. 2.726
   c. 3.726

14. $8.304 - 6.91$
   a. 0.394
   b. 1.394
   c. 2.394

▶ **Give each difference.**
Here are scrambled answers for the next row of exercises: *28.2 2.45 6.25 8.2*

15. $6.43 - 0.18$

16. $39.6 - 11.4$

17. $15 - 6.8$

18. $2.93 - 0.48$

19. $1.742 - 0.815$

20. $21 - 3.75$

21. $78.4 - 26.95$

22. $62.49 - 35.71$

23. $5.74 - 3.8$

24. $27 - 10.4$

25. $80.03 - 4.38$

26. $6.83 - 2.7$

27. $6.2 - 3.851$

28. $75.25 - 18$

29. $9.006 - 4.44$

30. $100 - 26.7$

▶ **Use your mental math skills. Evaluate each expression for $x = 10$, $y = 5.5$, and $z = 12.25$.**
Here are scrambled answers for the next row of exercises: *15.5 2.25 22.25 4.5*

31. $x - y$

32. $z - x$

33. $x + y$

34. $z + x$

35. $x - x$

36. $y + z$

37. $y + z - x$

38. $y + x - z$

39. $x + z - y$

40. $x + y + z$

41. $(z - y) + x$

42. $(z - x) + y$

Problem Solving   USING DATA

▶ **Solve. Use the chart on page 10.**

43. How close did Nara come to guessing how long a minute is?

44. Who came closer to guessing a minute, Rob or Ben?

45. Who came closer to guessing a minute, Dean or Susan Jo?

46. Who came closest to guessing a minute?

Group Project   ANALYZING DATA   *Just One Minute!*

▶ **Work in a small group. The group will need a timepiece that can be read to the nearest second.**

47. Each group member, without looking at a timepiece, takes a turn guessing how long a minute is. Record each time to the nearest second.

48. a. List the times of all class members.
   b. Order the times for the class from least to greatest.

49. a. How many seconds from a minute was your guess?
   b. How many classmates came closer than you?
   c. How many came within 8 seconds of guessing how long a minute is?

50. Write a paragraph that summarizes the data.

Solving Equations Using Addition and Subtraction | **11**

PROBLEM SOLVING POWER

You will use a variety of problem-solving strategies on these two pages. For some problems, you may wish to use a calculator.

1 PATTERNS

With seven straight cuts across a pizza, what is the greatest number of pieces that can be formed?

✓ Problem Solving Tips

Look for a Pattern.* Put the data into a table and then look for a pattern.

Number of cuts	1	2	3	4	5	6	7
Number of pieces	2	4	7	11	?	?	?

*At least one more of the problems on these two pages can be solved by looking for a pattern.

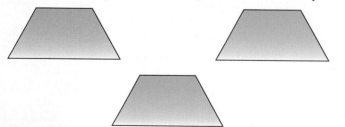

2 GEOMETRY—VISUAL THINKING

Think about rearranging these simple shapes to form a triangle. Then draw the triangle. *Hint: You may wish to trace the shapes.*

3 RELATIONSHIPS

Joe's grandmother on his father's side had two children, who each had two children.

His grandmother on his mother's side also had two children. They, too, each had two children.

How many cousins does Joe have?

4 PATTERNS

Which one of these tickets doesn't belong?
Why? *Hint: In each number, how are the digits related?*

6 PATTERNS

Look for the pattern. Find the missing numbers.

$$78 + 23 = 101$$
$$778 + 223 = 1001$$
$$7778 + 2223 = 10001$$
$$\vdots$$
$$a + b = 10000001$$

5 SEQUENCES

How many red triangles would there be in the 15th figure in this sequence?

1st   2nd   3rd   4th   15th

▲   ▲▽▲   ▲▽▲▽▲   ▲▽▲▽▲▽▲   … ?

7 WHO WENT WITH WHOM?

Dan, Bob, and Jerry went to the Bruce Springsteen concert with Mary Ellen, Katy, and Sue.

I'm not sure who went with whom, but I do know that

- Mary Ellen's sister went with Jerry.
- Dan did not go with Sue, who is an only child.

Match each girl with her date.

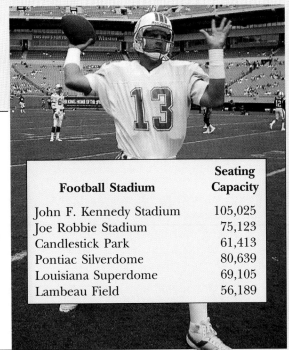

APPLICATIONS

Complete.
You may wish to use a calculator.

8 If the seating capacity of _?_ was increased by 13,710, it would hold as many fans as Joe Robbie Stadium.

9 _?_ can seat 18,934 more football fans than _?_ .

10 The total seating capacity of Candlestick Park, Lambeau Field, and the Louisiana Superdome is about the same as the total seating capacity of _?_ and _?_ .

Football Stadium	Seating Capacity
John F. Kennedy Stadium	105,025
Joe Robbie Stadium	75,123
Candlestick Park	61,413
Pontiac Silverdome	80,639
Louisiana Superdome	69,105
Lambeau Field	56,189

Rate Your Problem-Solving Power: *6–7 correct = Good;   8–9 correct = Excellent;   10 correct = Exceptional*

OBJECTIVE: To use map skills in solving problems.

▶ **Use the map to answer the questions about highway distances.**

1. How far is it from the Sea World exit on highway 528 to the exit for Orlando International Airport?

2. How many miles is it from the exit on highway 528 for Orlando International Airport to the exit for Walt Disney World's EPCOT Center on highway 4?

3. What is the distance between Orlando International Airport and Orlando Executive Airport, if you use highways 408 and 436?

4. How far is it from exit 80 on Florida's Turnpike to the Lake Buena Vista exit on highway 4?

5. You want to take the shortest route from Azalea Park to Disney World. Which of these routes should you use: 408 west to 4 south, *or* 436 south to 528 west to 4 south?

6. Going south on highway 4, you are 6 miles north of the highway 408 intersection. How far are you from the exit for Disney World's EPCOT Center?

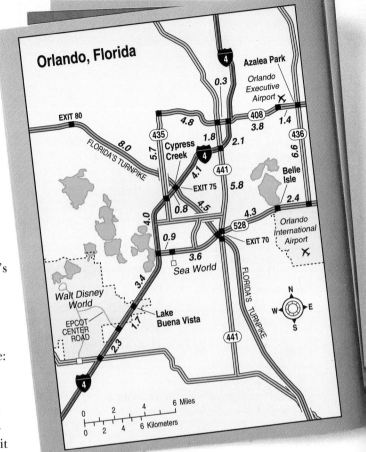

7. You are 2 miles west of Azalea Park on highway 408. How far are you from the intersection of highways 4 and 435?

▶ **Decide whether Expression A, B, or C would be used to complete each sentence.**

Expression A:	$n + 4.5$
Expression B:	$n - 4.5$
Expression C:	$4.5 - n$

8. You are driving east on highway 528. When you are n miles east of the highway 441 intersection, you are __?__ miles from the highway 4 intersection.

9. You are driving on Florida's Turnpike between exits 70 and 75. When you are n miles northwest of exit 70, you are __?__ miles from exit 75.

10. It takes 4.5 minutes to drive from Cypress Creek to highway 441 on highway 4. After driving n minutes, it will take __?__ minutes to complete the trip.

11. **Write a problem** that has Expression B for an answer.

▶ **Write an algebraic expression for each word expression.** *(page 2)*

1. 4 less than a number y
2. 4 more than a number w
3. 7 increased by a number d
4. a number x minus 17
5. the sum of a number b and 6
6. a number t decreased by 4
7. 5 decreased by s
8. the sum of a number f and 16
9. 11 more than a number x
10. a number h less 9
11. a number z minus $7\frac{1}{6}$
12. a number m increased by 1.8
13. a number r plus 0.3
14. 12.8 less than a number v
15. a number z minus a number y
16. a number w plus a number t
17. a number x decreased by itself
18. a number k increased by itself

▶ **Evaluate each algebraic expression for $x = 7$, $y = 10$, and $z = 8$.** *(page 4)*

19. $x + z$
20. $y - 10$
21. $x + 5$
22. $y - 8$
23. $x - 6$
24. $y + z$
25. $x + 9$
26. $y - x$
27. $z - 7$
28. $12 - z$
29. $x - 0$
30. $y + 14.5$
31. $x + x + 8.73$
32. $x + y - 1$
33. $z + y - x$
34. $(y - z) + 5.1$

▶ **Round these calculator answers to the nearest tenth.** *(page 6)*

35. $\boxed{8.24}$
36. $\boxed{36.48}$
37. $\boxed{63.807}$
38. $\boxed{8.259}$
39. $\boxed{78.095}$
40. $\boxed{356.209}$
41. $\boxed{263.74}$
42. $\boxed{750.06}$

▶ **Round these calculator answers to the nearest hundredth.** *(page 6)*

43. $\boxed{59.508}$
44. $\boxed{44.592}$
45. $\boxed{270.503}$
46. $\boxed{0.306}$
47. $\boxed{40.961}$
48. $\boxed{446.515}$
49. $\boxed{6.397}$
50. $\boxed{1.996}$

▶ **Evaluate each algebraic expression for $x = 5.7$, $y = 7.4$, and $z = 1.88$.** *(page 8)*

51. $x + y$
52. $x + z$
53. $x + y + z$
54. $y + 4 + z$
55. $z + x + 3.1$
56. $8.2 + y + x$
57. $6 + x + 9.08$
58. $2 + 4.9 + y$

▶ **Evaluate each algebraic expression for $a = 7.5$, $b = 2.4$, and $c = 8.39$.** *(page 10)*

59. $a - b$
60. $c - a$
61. $c - b$
62. $a - 2.46$
63. $(8 - b) + a$
64. $c + 8 + b$
65. $(19 - a) - c$
66. $9.1 + a - c$
67. $a + a - c$
68. $(c - b) - b$
69. $a + 2.5 - c$
70. $(10 - c) + b$

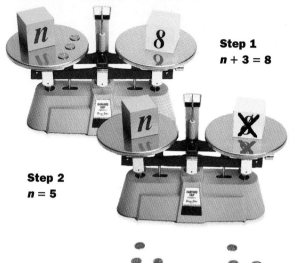

OBJECTIVE: To solve addition equations by subtracting the same number from both sides.

Step 1

$n + 3 = 8$

Look at the balance in Step 1. If we let n be the number of pennies in the blue box, we have the equation $n + 3 = 8$.

1. Look at the balance in Step 2.
   a. How many pennies were taken away from each side of the balance?
   b. Do the two sides still balance?
   c. What equation does the balance show now?
   d. How many pennies are in the blue box?

Step 2

$n = 5$

To **solve the equation** is to find the number that we can substitute for n to make the equation true.

EXAMPLES | Here's how to solve an addition equation.

A **Equation:**

$n + 3 = 8$

Addition and subtraction are **inverse operations.** Each operation undoes the other.

*To undo adding 3 and to get n by itself on one side of the equation, subtract 3 from both sides.

$n + 3 - 3 = 8 - 3$

Simplify both sides.

$n = 5$

✔ **CHECK**

To *check* the solution, substitute 5 for n in the equation:

$$n + 3 = 8$$
$$5 + 3 \stackrel{?}{=} 8$$
$$8 = 8$$

It checks!

> *The property used to subtract a number from both sides of an equation is called the **subtraction property of equality.**

B
$$d + 8 = 30$$
$$d + 8 - 8 = 30 - 8$$
$$d = 22$$

C
$$m + 32 = 41$$
$$m + 32 - 32 = 41 - 32$$
$$m = 9$$

D
$$t + 18.5 = 80.2$$
$$t + 18.5 - 18.5 = 80.2 - 18.5$$
$$t = 61.7$$

CHECK for Understanding

2. Look at Example A. To find n, what number was subtracted from both sides of the equation? How many pennies are in the blue box?

3. What number would you subtract from both sides to solve the equation?

   a. $a + 6 = 20$ 　　　　　 b. $m + 16 = 45$ 　　　　　 c. $f + 15.3 = 28$

4. Explain how to get the numbers that are printed in red in Examples B, C, and D.

EXERCISES

▶ **Solve and check.**
Here are scrambled answers for the next two rows of exercises:
  21   18   7   17   20   46   16   8

5. $n + 26 = 42$ **6.** $t + 10 = 27$ **7.** $x + 34 = 52$ **8.** $z + 18 = 38$

9. $n + 28 = 36$ **10.** $y + 27 = 48$ **11.** $c + 17 = 63$ **12.** $m + 24 = 31$

13. $y + 35 = 48$ **14.** $d + 32 = 59$ **15.** $w + 77 = 91$ **16.** $m + 4 = 39$

17. $x + 27 = 74$ **18.** $d + 15 = 28$ **19.** $x + 18 = 52$ **20.** $y + 55 = 62$

21. $g + 54 = 81$ **22.** $f + 45 = 83$ **23.** $n + 65 = 84$ **24.** $h + 43 = 98$

25. $x + 14 = 57$ **26.** $m + 11 = 34$ **27.** $t + 18 = 56$ **28.** $h + 33 = 42$

29. $g + 88 = 94$ **30.** $s + 43 = 59$ **31.** $r + 15 = 60$ **32.** $b + 23 = 23$

33. $t + 2.9 = 6.8$ **34.** $x + 1.6 = 1.9$ **35.** $g + 0.02 = 0.94$ **36.** $n + 0.66 = 0.84$

37. $t + 0.77 = 0.91$ **38.** $c + 0.46 = 0.62$ **39.** $x + 0.24 = 0.39$ **40.** $f + 2.1 = 6.6$

41. $a + 0.52 = 8.5$ **42.** $w + 0.55 = 7.8$ **43.** $f + 0.37 = 0.37$ **44.** $c + 0.34 = 5.2$

Problem Solving   USING EQUATIONS

▶ **Write an equation and solve the problem.**

45. A number x plus 45 equals 72. What is the number?

46. A number c increased by 14 equals 48. What is the number?

47. The sum of a number w and 15 is 43. What is the number?

48. Five more than a number d is 28. What is the number?

49. When a number a is increased by 18, the result is 41. What is the number?

50. The sum of a number k and 42 is 73. What is the number?

51. Nine more than a number b is 17.3. What is the number?

52. A number t plus 1.9 equals 3.1. What is the number?

Challenge!   LOGICAL REASONING

▶ **Solve.**

53. Suppose that you have 8 pennies that look alike. However, one of the pennies is counterfeit and is slightly heavier than each of the other 7. Tell how you could use a balance just **two** times to find the counterfeit penny.

OBJECTIVE: To solve subtraction equations by adding the same number to both sides.

1. Decide whether Equation A or B would be used to solve the problem.

> Equation A: $n + 3 = 11$
> Equation B: $n - 3 = 11$

Three dollars off the regular price of a CD is $11. What is the regular price?

CD SALE!

$3 Off Regular Price

EXAMPLES | **Here's how to solve a subtraction equation.**

A **Equation:** $\quad n - 3 = 11$

*To undo subtracting 3, $\quad n - 3 + 3 = 11 + 3$
add 3 to both sides.
Simplify. $\qquad\qquad\qquad n = 14$

✓ **CHECK**

$n - 3 = 11$
$14 - 3 \overset{?}{=} 11$
$11 = 11$
     It checks!

> *The property used to add a number to both sides of an equation is called the **addition property of equality**.

B $\qquad r - 28 = 12$
$r - 28 + 28 = 12 + 28$
$\qquad\qquad r = 40$

C $\qquad p - 8 = 13$
$p - 8 + 8 = 13 + 8$
$\qquad\qquad p = 21$

D $\qquad t - 17.2 = 60.8$
$t - 17.2 + 17.2 = 60.8 + 17.2$
$\qquad\qquad t = 78.0$

CHECK for Understanding

2. Look at Example A.

   a. To find n, what number was added to both sides of the equation? What does n equal?

   b. To check the solution, what number was substituted for n in the equation $n - 3 = 11$?

3. What number would you add to both sides to solve the equation?

   a. $h - 7 = 16$ $\qquad\qquad\qquad$ **b.** $m - 23.6 = 50$

4. Complete these examples.

   a. $\qquad x - 5 = 18$
   $x - 5 + \underline{\ ?\ } = 18 + 5$
   $\qquad\qquad x = 23$

   b. $\qquad z - 6 = 52$
   $z - 6 + 6 = 52 + \underline{\ ?\ }$
   $\qquad\qquad z = 58$

   c. $\qquad p - 8.2 = 13$
   $p - 8.2 + 8.2 = 13 + 8.2$
   $\qquad\qquad p = \underline{\ ?\ }$

5. Explain how to get the numbers that are printed in red in Examples B, C, and D.

EXERCISES

▶ **Solve and check.**
Here are scrambled answers for the next two rows of exercises:
33   41   35   20   39   40   15   32

6. $n + 10 = 50$

7. $p - 3 = 38$

8. $n - 15 = 24$

9. $x - 27 = 5$

10. $t - 12 = 8$

11. $r - 24 = 11$

12. $m + 23 = 38$

13. $y + 34 = 67$

14. $y - 20 = 17$

15. $n + 11 = 35$

16. $x - 7 = 23$

17. $z - 17 = 1$

18. $s - 5 = 26$

19. $t - 16 = 0$

20. $n - 15 = 27$

21. $a + 24 = 24$

22. $y + 17 = 20$

23. $t - 24 = 56$

24. $c - 61 = 62$

25. $c - 4 = 19$

26. $a - 22 = 9$

27. $z - 45 = 4$

28. $r + 33 = 33$

29. $d - 56 = 77$

30. $t - 11 = 5$

31. $y + 7 = 77$

32. $c - 7 = 10$

33. $h + 23 = 48$

34. $t + 1.4 = 10.8$

35. $s - 0.23 = 1.23$

36. $b - 1.2 = 3.4$

37. $k - 0.13 = 0.57$

38. $e + 1.3 = 80.7$

39. $b + 4.8 = 7.8$

40. $v - 2.05 = 0.84$

41. $m - 1.12 = 0.63$

42. $b - 1.1 = 1.23$

43. $s + 1.2 = 2.75$

44. $a - 5.6 = 3.47$

45. $n - 8.2 = 5.53$

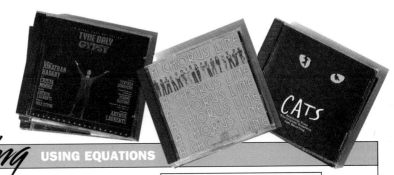

Problem Solving USING EQUATIONS

▶ **Decide whether Equation A or B would be used to solve each problem. Then solve the problem.**

Equation A:	$n + 12 = 60$
Equation B:	$n - 12 = 60$

46. Twelve less than a number n is 60. What is the number?

47. Twelve more than a number n is 60. What is the number?

48. The sum of a number n and 12 is 60. What is the number?

49. A number n decreased by 12 is 60. What is the number?

50. After 12 *Grease* CD's were sold, there were 60 left. How many *Grease* CD's were there in the beginning?

51. An *Annie* CD plays 12 minutes longer than a *Cats* CD. If the *Annie* CD plays for 60 minutes, how long does the *Cats* CD play?

52. **Write a problem** that can be solved by using Equation A.

53. **Write a problem** that can be solved by using Equation B.

Solving Equations Using Addition and Subtraction | **19**

1-9 More on Solving Equations

OBJECTIVE: To use the symmetric property of equality to solve equations that have the variable on the right side of the equal sign.

You have solved equations such as $x + 13 = 41$ and $y - 18 = 27$ by either subtracting the same number from both sides or adding the same number to both sides.

Notice that in both equations, the variable is on the left side of the equal sign. In this lesson, you will solve some equations that have the variable on the right side of the equal sign.

EXAMPLES | Here's how to use the symmetric property of equality to solve equations that have the variable on the right side of the equal sign.

A
$$86 = x + 63$$
$$x + 63 = 86$$
$$x + 63 - 63 = 86 - 63$$
$$x = 23$$

The **symmetric property of equality** tells you that if $86 = x + 63$, then $x + 63 = 86$.

✔ **CHECK**
$$86 = x + 63$$
$$86 \stackrel{?}{=} 23 + 63$$
$$86 = 86$$
It checks!

B
$$48 = y - 18.5$$
$$y - 18.5 = 48$$
$$y - 18.5 + 18.5 = 48 + 18.5$$
$$y = 66.5$$

✔ **CHECK**
$$48 = y - 18.5$$
$$48 \stackrel{?}{=} 66.5 - 18.5$$
$$48 = 48$$
It checks!

C
$$160 = k + 29$$
$$k + 29 = 160$$
$$k + 29 - 29 = 160 - 29$$
$$k = 131$$

D
$$83.4 = s - 36.7$$
$$s - 36.7 = 83.4$$
$$s - 36.7 + 36.7 = 83.4 + 36.7$$
$$s = 120.1$$

CHECK for Understanding

1. Look at the examples above.
   a. In each first equation, the expression containing the variable is on the __?__ side of the equation.
   b. In each second equation, the symmetric property of equality was used to get the expression containing the variable on the __?__ side of the equation.

2. Complete these examples.

   a.
$$59 = r + 42$$
$$r + 42 = \underline{?}$$
$$r + 42 = 59 - 42$$
$$r = 17$$

   b.
$$82 = n - 80$$
$$n - \underline{?} = 82$$
$$n - 80 + 80 = 82 + 80$$
$$n = 162$$

3. Explain how to get the numbers that are printed in red in Examples B, C, and D.

EXERCISES

Solve and check.
Here are scrambled answers for the next row of exercises: 58   19   37   27

4. $q - 26 = 32$ **5.** $56 = w + 19$ **6.** $a + 32 = 51$ **7.** $18 = a - 9$

8. $50 = y - 25$ **9.** $n - 18 = 35$ **10.** $65 = u + 59$ **11.** $b + 74 = 98$

12. $h + 18 = 100$ **13.** $97 = t + 30$ **14.** $25 = n - 33$ **15.** $c - 19 = 54$

16. $74 = y + 36$ **17.** $32 = m - 0$ **18.** $p - 29 = 77$ **19.** $d + 38 = 38$

20. $0 = z - 41$ **21.** $83 = s + 45$ **22.** $m + 53 = 110$ **23.** $h - 42 = 0$

24. $e + 6.5 = 9$ **25.** $c - 7.4 = 2.6$ **26.** $0.75 = q - 0.16$ **27.** $9.9 = k + 3.1$

28. $m - 30 = 3.6$ **29.** $7.3 = n + 0.25$ **30.** $j + 7.5 = 7.5$ **31.** $0 = m - 0.93$

Problem Solving   USING A FORMULA

Solve.

32. You're a canoeist. If you are paddling downstream with a river's current, will your ground speed be more or less than your paddling speed?

*Hint: The **formula** below shows how the paddling speed and current speed of a canoe are related to its ground speed.*

paddling speed	current speed	ground speed
↓	↓	↓

Formula: p + c = g

33. Use the **formula** in exercise 32. Complete the chart.

	p (miles per hour)	c (miles per hour)	g (miles per hour)
a.	6	2	?
b.	11	4	?
c.	15	?	21
d.	?	5	14

OBJECTIVE: To solve problems by using addition and subtraction equations.

Problems that can be solved with arithmetic can also be solved by writing and solving an equation.

EXAMPLES	Here's how to use arithmetic or algebra to solve a problem.

Problem: Jane's total score after three dives was 26. If she scored 8 points on her third dive, what was her point total after two dives?

A
Using arithmetic

$\quad$ 26 ← Total number of points
$\quad$ − 8 ← Number of points scored on third dive
$\quad$ 18 ← Number of points scored on first two dives

Eighteen was Jane's point total after two dives.

B
Using algebra

Step 1. Choose a variable. Use it and a fact from the problem to represent the numbers in the problem.

Let n = number of points scored on first two dives.

Then $n + 8$ = total number of points scored.

Step 2. Write an equation based on the facts.

$$n + 8 = 26$$

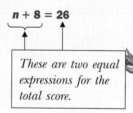

These are two equal expressions for the total score.

Step 3. Solve the equation.

$$n + 8 = 26$$
$$n + 8 - 8 = 26 - 8$$
$$n = 18$$

Eighteen was Jane's point total after two dives.

CHECK for Understanding

1. Look at Example A. The number of points scored on the __?__ dive was subtracted from the total number of points to get the number of points scored on the first two dives.

2. Look at Example B.
   a. In Step 1, the variable n equals the number of __?__ scored on the first two dives.

b. In Step 2, both the algebraic expression $n + 8$ and the number __?__ equal the total number of points scored.

c. In Step 3, to solve the equation, __?__ was subtracted from both sides.

3. Check the solution. Does 18 points on the first two dives and 8 points on the third dive give a total score of 26?

EXERCISES

▶ **Decide whether Equation A or B would be used to solve the problem. Then solve the problem.**

> **Equation A:** $n + 11 = 30$
> **Equation B:** $n - 11 = 30$

4. The sum of two numbers is 30. One of the numbers is 11. What is the other number?
(Let n = the other number.)

5. The difference between two numbers is 30. The lesser number is 11. What is the greater number?
(Let n = the greater number.)

6. There are 30 divers on the swim team. Eleven of the divers are boys. How many divers are girls?

7. The swim team finished first in 11 of the 30 meets it has entered. In how many of the meets did it not finish first?

8. **Write a problem** that can be solved using Equation A.

9. **Write a problem** that can be solved using Equation B.

▶ **Write an equation and solve the problem.**

10. Forty is 12 less than a number. What is the number?
(Let n = the number.)

11. The difference between two numbers is 5. The lesser number is 14. What is the greater number?
(Let g = the greater number.)

12. You had some money and then you spent 26 dollars. Now you have 9 dollars. How many dollars did you have before you spent some?
(Let d = the number of dollars you had before you spent some.)

13. You had some money and then you earned 12.75 dollars. Now you have 32 dollars. How many dollars did you have?
(Let d = the number of dollars you had.)

14. The regular price is 8 dollars more than the sale price. The regular price is 45 dollars. What is the sale price?
(Let s = the sale price in dollars.)

15. The sale price is 7.65 dollars less than the regular price. The sale price is 39.50 dollars. What is the regular price?
(Let r = the regular price in dollars.)

16. Ann scored 8 points less than Ted. Ann scored 27 points. How many points did Ted score?
(Let t = the number of points Ted scored.)

17. You scored 10 more points than your friend. You scored 34 points. How many points did your friend score?
(Let f = the number of points your friend scored.)

OBJECTIVE: To solve problems using a calculator when appropriate.

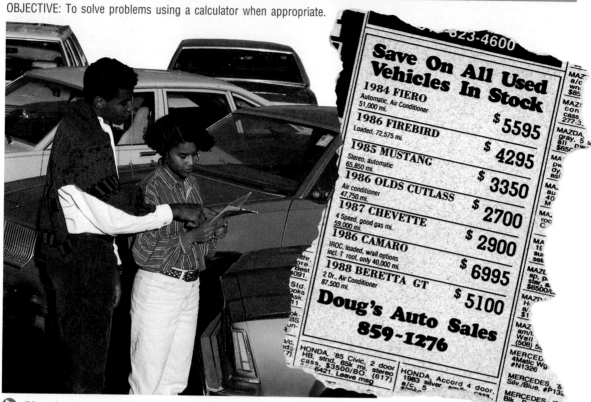

Save On All Used Vehicles In Stock

823-4600

1984 FIERO Automatic, Air Conditioner 51,000 mi.	**$5595**
1986 FIREBIRD Loaded, 72,575 mi.	**$4295**
1985 MUSTANG Stereo, automatic 65,850 mi.	**$3350**
1986 OLDS CUTLASS Air conditioner 47,750 mi.	**$2700**
1987 CHEVETTE 4 Speed, good gas mi. 59,000 mi.	**$2900**
1986 CAMARO IROC, loaded, w/all options incl. T roof, only 40,000 mi.	**$6995**
1988 BERETTA GT 2 Dr., Air Conditioner 87,500 mi.	**$5100**

Doug's Auto Sales
859-1276

▶ **Use the car ad. Fill in the facts that fit the calculator sequence.**

1. 72575 ⊟ 47750 ⊟ (24825)

The 1986 ___**a.**___ has been driven
⠀⠀⠀⠀⠀⠀⠀(car)

___**b.**___ more miles than the
(number)

___**c.**___ Olds.
(year)

2. 2239 ⊞ 2864 ⊟ 4295 ⊟ (808)

Beth has $2239 in one savings account

and $___**a.**___ in another. After she
⠀⠀⠀(number)

buys the 1986 ___**b.**___, she will have
⠀⠀⠀⠀⠀⠀⠀⠀⠀⠀(car)

$___**c.**___ left.
⠀(number)

▶ **Use the car ad. Solve each problem. Decide when a calculator would be useful.**

3. You have $3860. How much will you need to borrow to buy the 1988 Beretta?

4. Which car has been driven about 15,000 more miles than the 1984 Fiero?

5. Doug's Auto Sales guarantees its cars for 7500 miles after the purchase. How many miles will the Mustang have when the guarantee expires?

6. After buying the 1986 Camaro, Paul had $785 left in his savings account. How much money did he have in his savings account before he bought the car?

7. Write a problem that can be solved using the car ad. Then solve the problem.

8. Write an estimation problem that has an answer of about $1000. Use the car ad.

Cumulative Algebra Practice

▶ **Three of these calculator answers are wrong. Find them by estimating.** *(page 8)*

1. 38.4 + 1.63   (40.03)
2. 56.2 + 662.4   (718.6)
3. 4.65 + 2.81 + 7.82   (13.28)
4. 80.6 + 6.31 + 782   (868.91)
5. 481.7 + 87.21   (568.91)
6. 8.409 + 6.943   (16.352)
7. 23.107 + 31.479   (44.586)
8. 706.42 + 327.54   (1033.96)

▶ **Three of these calculator answers are wrong. Find them by estimating.** *(page 10)*

9. 703 − 276   (527)
10. 417 − 279   (98)
11. 90.31 − 8.46   (81.85)
12. 6.304 − 0.732   (5.572)
13. 708.1 − 239.2   (468.9)
14. 523.8 − 391.4   (924)
15. 837.03 − 35.619   (801.411)
16. 76.024 − 19.863   (56.161)

▶ **Solve and check.** *(page 16)*

17. $d + 11 = 25$
18. $g + 12 = 20$
19. $x + 9 = 17$
20. $b + 8 = 15$
21. $f + 19 = 33$
22. $y + 15 = 29$
23. $j + 12 = 12$
24. $r + 0 = 24$
25. $s + 2.6 = 5.8$
26. $y + 3.1 = 3.9$
27. $c + 4.2 = 5.7$
28. $h + 0.18 = 0.49$
29. $z + 0.65 = 0.84$
30. $e + 0.52 = 0.52$
31. $t + 0.37 = 6.5$
32. $a + 0.45 = 8.5$

▶ **Solve and check.** *(page 18)*

33. $b − 65 = 84$
34. $t − 9 = 7$
35. $m − 11 = 8$
36. $d − 10 = 7$
37. $r − 12 = 18$
38. $f − 15 = 41$
39. $s − 19 = 17$
40. $p − 20 = 36$
41. $h − 14 = 0$
42. $h − 21 = 38$
43. $c − 36 = 28$
44. $v − 39 = 41$
45. $e − 3.5 = 6.4$
46. $k − 2.3 = 2.3$
47. $q − 5.1 = 4.6$
48. $g − 0.52 = 0$
49. $m − 0.15 = 0.22$
50. $x − 0.14 = 7.2$
51. $r − 1.3 = 90$
52. $l − 1.3 = 0.13$

▶ **Solve and check.** *(page 20)*

53. $35 = a + 16$
54. $18 = x − 21$
55. $18 = h + 18$
56. $30 = r − 32$
57. $46 = b − 43$
58. $53 = y − 25$
59. $57 = k + 46$
60. $66 = s + 20$
61. $79 = c + 38$
62. $39 = z − 42$
63. $61 = m − 50$
64. $90 = t + 61$
65. $5.5 = d − 0$
66. $4.8 = u − 7.1$
67. $4.7 = p + 1.3$
68. $11.2 = u + 8.4$
69. $0.39 = e − 0.01$
70. $1.4 = v − 0.18$
71. $7.4 = q + 0.15$
72. $1 = m + 0.8$

Solving Equations Using Addition and Subtraction | **25**

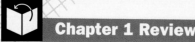

Here are scrambled answers for the review exercises:

1.9	12	75	expressions	inverse	right	symmetric
5	14.7	add	greater	left	simplify	up
6	65	down	hundredths	less	subtract	variable

1. Variables, numbers, and operation signs can be combined to form algebraic __?__. In the algebraic expression $x - 7$ the letter x is called a __?__. A word expression for the algebraic expression is "7 __?__ than a number x." *(page 2)*

2. If you evaluate the algebraic expression $x + y - z$ for $x = 5$, $y = 8$, and $z = 7$, you get __?__. If you evaluate the same expression for $x = 8$, $y = 10$, and $z = 6$, you get __?__. *(page 4)*

3. To round the number 67.439 to the nearest tenth, first look at the digit in the __?__ place. Since it is less than __?__, round the decimal __?__ to 67.4. *(page 6)*

4. To round the number 82.609 to the nearest hundredth, first look at the digit in the thousandths place. Since it is 5 or __?__, round the decimal __?__ to 82.61. *(page 6)*

5. If you estimate the sum $3.42 + 58.756 + 12.65$ by rounding each number to the nearest whole number, you get __?__. *(page 8)*

6. If you estimate the difference $17.86 - 3.216$ by rounding each number to the nearest tenth, you get __?__. *(page 10)*

7. Addition and subtraction are __?__ operations. Each operation undoes the other. To solve the equation $w + 8 = 73$, you would first __?__ 8 from both sides and then simplify both sides. To check the solution, you would substitute __?__ for w in the equation. *(page 16)*

8. To solve the equation $t - 0.5 = 1.4$, you would first __?__ 0.5 to both sides and then __?__ both sides. To check the solution, you would substitute __?__ for t in the equation. *(page 18)*

9. In the first equation at the right, the expression containing the variable is on the __?__ side of the equation. In the second equation, the __?__ property of equality was used to get the expression containing the variable on the __?__ side. *(page 20)*

$$27 = b + 11$$
$$b + 11 = 27$$

▶ **Write an algebraic expression for each word expression.** *(page 2)*

1. 5 more than a number x

2. 3 less than a number y

3. 14 decreased by a number a

4. 19 increased by a number d

5. a number z decreased by itself

6. a number b increased by itself

▶ **Round to the nearest hundredth.** *(page 6)*

7. 57.487 **8.** 0.462 **9.** 3.501 **10.** 9.055 **11.** 21.996

▶ **Evaluate each algebraic expression for $x = 1.4$, $y = 9.6$, and $z = 0.7$.** *(pages 4, 8, 10)*

12. $x - z$ **13.** $x + 3.17$ **14.** $x + z$ **15.** $20.3 + y$

16. $y + z$ **17.** $y - 1.8$ **18.** $y - 4$ **19.** $2.19 + z$

20. $(x - z) + z$ **21.** $7.96 + y - z$ **22.** $x + y - z$ **23.** $y + y - z$

24. $29 + x - y$ **25.** $(29 - x) - x$ **26.** $(29 - x) - y$ **27.** $29 + y - x$

▶ **Estimate each sum or difference by rounding. Round to the nearest tenth.** *(pages 6, 8, 10)*

28. $4.07 + 8.219$ **29.** $2.86 + 5.42 + 0.94$ **30.** $8.21 - 3.07$ **31.** $61.28 - 59.109$

32. $8.157 + 9.84$ **33.** $3.27 + 0.821 + 6.97$ **34.** $5.03 - 1.97$ **35.** $20.316 - 18.47$

▶ **Solve and check.** *(pages 16, 18, 20)*

36. $x + 18 = 39$ **37.** $z + 27 = 38$ **38.** $x + 23 = 23$ **39.** $y + 114 = 187$

40. $b - 14 = 19$ **41.** $a - 42 = 17$ **42.** $d - 46 = 0$ **43.** $c - 125 = 90$

44. $7.7 = r + 4.1$ **45.** $4.9 = t - 3.1$ **46.** $1.26 = s + 0.84$ **47.** $0 = u - 0.54$

48. $4 = w + 1.7$ **49.** $8.1 = r - 4$ **50.** $7.03 = n + 6$ **51.** $8 = p - 6.3$

▶ **Decide whether Equation A or B would be used to solve each problem. Then solve the problem.** *(page 22)*

| Equation A: | $n + 18 = 28$ |
| Equation B: | $n - 18 = 28$ |

52. The regular price decreased by \$18 is \$28. What is the regular price?

53. The cost was marked up \$18 to get the regular price \$28. What was the cost?

54. You gave 18 of your records to a friend. You then had 28 records. How many records did you have before you gave your friend the records?

55. Before your birthday, you had 18 silver dollars. After your birthday, you had 28 silver dollars. How many silver dollars did you get for your birthday?

▶ **Choose the correct letter.**

1. An algebraic expression for 17 less than a number x is

    A. $17 - x$

    B. $x + 17$

    C. $x - 17$

    D. none of these

2. An algebraic expression for a number a increased by a number b is

    A. $B - a$

    B. $a + b$

    C. $a - b$

    D. none of these

3. Evaluate $27 - z$ for $z = 12$.

    A. 39

    B. 12

    C. 15

    D. none of these

4. Evaluate $c + c - d$ for $c = 8$ and $d = 7$.

    A. 1

    B. 9

    C. 16

    D. none of these

5. 8.493 rounded to the nearest tenth is

    A. 8.4

    B. 8.5

    C. 9.0

    D. none of these

6. 45.815 rounded to the nearest hundredth is

    A. 45.80

    B. 45.81

    C. 45.82

    D. none of these

7. Give the sum.
Hint: Estimate.
$27.4 + 39.42 + 8.654$

    A. 12.870

    B. 65.474

    C. 75.474

    D. 128.70

8. Give the difference.
Hint: Estimate.
$50.06 - 17.38$

    A. 67.44

    B. 42.68

    C. 12.44

    D. 32.68

9. Solve.

$58 = x + 39$

    A. 19

    B. 97

    C. 29

    D. none of these

10. Solve.

$w - 27 = 42$

    A. 15

    B. 69

    C. 25

    D. none of these

11. Solve.

$10 = t - 6.2$

    A. 16.2

    B. 3.8

    C. 4.8

    D. none of these

12. Choose the equation.

Lisa scored 11 fewer points than Fernando. How many points did Fernando score if Lisa scored 14?

    A. $f + 11 = 14$

    B. $14 + f = 11$

    C. $f - 11 = 14$

    D. $14 - f = 11$

Solving Equations Using Multiplication and Division

$$\frac{d}{g} = m$$

Distance traveled in miles divided by the number of gallons used equals miles per gallon. (page 53)

In an automobile rally, contestants travel a well-defined course using as little fuel as possible. Because weight is one factor that affects gas mileage, the drivers remove as much weight as possible from their vehicles.

29

OBJECTIVE: To write algebraic expressions involving multiplication and division.

A group of high school students was asked to identify these photos.

Newsmakers in the Nineties	
Nelson Mandela (recognized by $3n$ students)	**Margaret Thatcher** (recognized by n students)
Mikhail Gorbachev (recognized by $3n + 10$ students)	**Corazon Aquino** (recognized by $n \div 2$ students)

1. Which face was recognized by n students?

2. Which face was recognized by one half as many students as Margaret Thatcher?

3. Which face was recognized by 3 times as many students as Margaret Thatcher?

4. Which face was recognized by 10 more students than Nelson Mandela?

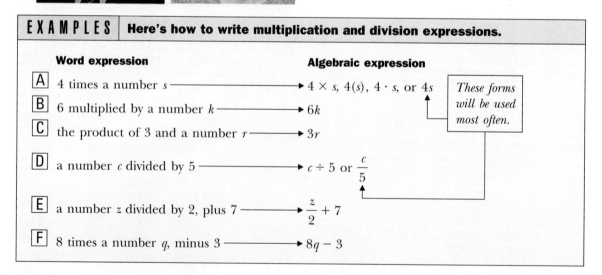

E X A M P L E S	**Here's how to write multiplication and division expressions.**

Word expression	Algebraic expression
A 4 times a number s ⟶	$4 \times s$, $4(s)$, $4 \cdot s$, or $4s$
B 6 multiplied by a number k ⟶	$6k$
C the product of 3 and a number r ⟶	$3r$
D a number c divided by 5 ⟶	$c \div 5$ or $\dfrac{c}{5}$
E a number z divided by 2, plus 7 ⟶	$\dfrac{z}{2} + 7$
F 8 times a number q, minus 3 ⟶	$8q - 3$

These forms will be used most often.

CHECK for Understanding

5. Look at Example A. To read the expression $4s$, you can say "4 _?_ a number s."

6. Look at Example D. To read $\dfrac{c}{5}$ you can say "a number c _?_ by 5."

EXERCISES

▶ **Write an algebraic expression for each word expression.**

7. a number p divided by 7

8. 9 times a number a

9. 18 divided by a number r

10. a number m multiplied by 7

11. the sum of a number z and 4

12. a number e minus 11

13. a number a divided by a number b

14. the product of a number d and a number x

15. 6 times a number c, minus 5

16. 5 times a number h, plus 12

17. a number d divided by k, minus 6

18. a number m divided by m, plus 9

19. The product of 0.8 and a number y

20. $\frac{1}{2}$ more than a number b

21. a number x decreased by $\frac{3}{4}$

22. 1.5 divided by a number c

▶ **Suppose that p is the number of presidents you can name. Write an algebraic expression for the number of people that is**

23. 5 times as many.

24. your number increased by 9.

25. your number divided by 5.

26. 4 times your number, minus 1.

27. the product of your number and 7, plus 4.

28. your number increased by 3.

29. 29 decreased by your number.

30. your number divided by 6, minus 4.

Problem Solving USING EXPRESSIONS

▶ **Decide whether expression A, B, C, or D would be used to complete each sentence.**

Expression A:	$4n$
Expression B:	$\dfrac{n}{4}$
Expression C:	$n + 4$
Expression D:	$n - 4$

31. President Reagan served n terms. If each term is 4 years, Reagan served __?__ years.

32. President Eisenhower served n years. The number of terms he served is __?__ .

33. You plan to memorize the names of n presidents. If you have memorized all but 4 of them, then you know __?__ presidents.

34. Carl knows the names of n presidents. You know 4 more presidents than Carl. Then you know the names of __?__ presidents.

35. Write a problem that has Expression A for an answer.

36. Write a problem that has Expression B for an answer.

2-2 Evaluating Algebraic Expressions

OBJECTIVE: To substitute numbers for variables and then evaluate the resulting multiplication or division expression.

Can you recognize the four common objects in these close-ups?

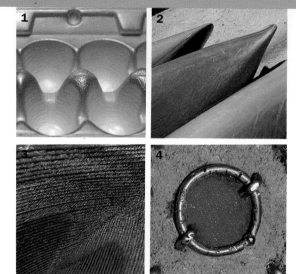

Survey Results		
Number of freshmen and sophomores that recognized each close-up.		
Close-up	Freshmen	Sophomores
1	f	$2s$
2	$3f$	s
3	$4f$	$5s$
4	$\dfrac{f}{3}$	$\dfrac{s}{2}$

1. Look at the chart. Which close-up was recognized by f freshmen?

2. Which close-up was recognized by more freshmen, close-up 2 or 3?

3. Which close-up was recognized by more sophomores, close-up 2 or 4?

EXAMPLES | Here's how to evaluate algebraic expressions.

12 freshmen recognized close-up 1; 8 sophomores recognized close-up 2. To find the number of people who recognized each close-up, substitute 12 for f and 8 for s in the expressions and simplify.

A Number of freshmen who recognized close-up 2:

$$3f$$
$$3 \cdot 12 = 36$$

Close-up 2 was recognized by 36 freshmen.

B Number of sophomores who recognized close-up 4:

$$\frac{s}{2}$$
$$\frac{8}{2} = 4$$

Close-up 4 was recognized by 4 sophomores.

C Number of students who recognized close-up 2:

$$(3f) + s$$
$$(3 \cdot 12) + 8$$
$$36 + 8 = 44$$

Close-up 2 was recognized by 44 students

Do the operation in parentheses first.

CHECK for Understanding

4. Look at Example A. How many freshman recognized close-up 2?

5. Look at Example C. Which close-up did 44 students recognize?

EXERCISES

▶ **Evaluate each expression for $a = 4$, $b = 12$, and $c = 6$.**
Here are scrambled answers for the next row of exercises: 3   1   24   48

6. $6a$

7. $\dfrac{b}{4}$

8. $8c$

9. $\dfrac{a}{4}$

10. ac

11. $\dfrac{b}{a}$

12. ab

13. $\dfrac{b}{c}$

14. $8 + c$

15. $b - 2$

16. $c - a$

17. $b - 6$

18. $(ac) - 4$

19. $(7a) - 8$

20. $\dfrac{8}{a} - 2$

21. $\dfrac{c}{2} - 2$

22. $\dfrac{18}{c} + 2$

23. $3ac$

24. $\dfrac{24}{c} - a$

25. $(3a) - 7$

26. $(ac) - b$

27. $(ab) - 40$

28. $(bc) - 50$

29. $(ac) - c$

30. $(3b) - 20$

31. $(2ac) - b$

32. $(2a) + c$

33. $(ac) \div b$

34. $\dfrac{36}{b} + c$

35. $\dfrac{16}{a} + b$

36. $\dfrac{12}{c} + a$

37. $\dfrac{60}{a} - b$

▶ **Evaluate each expression for $g = 4$, $k = 5$, and $p = 10$.**
Here are scrambled answers for the next two rows of exercises: 9   2   6   50

38. p divided by k

39. k times p

40. p decreased by g

41. g increased by k

42. the product of k and g

43. p minus k

44. five more than g

45. seven less than p

46. p divided by 2, minus g

47. g times k, plus 5

48. the product of g and k, minus p

49. twelve divided by g, plus p

50. g times g, decreased by p

51. k divided by k, increased by g

Problem Solving EVALUATING EXPRESSIONS

▶ **Solve. Use the chart on page 32.** *Remember:* $f = 12$ **and** $s = 8$

52. Which photograph was recognized by the same number of freshmen as sophomores?

53. Which photograph was recognized by more sophomores than freshmen?

54. Which photograph was recognized by 8 more freshmen than sophomores?

55. Write a problem that uses the data in the chart on page 32.

OBJECTIVE: To estimate and compute decimal products.

1. What is the cost of one pound of bananas?

2. To find the cost of 2.4 pounds of bananas, you would multiply 2.4 by what number?

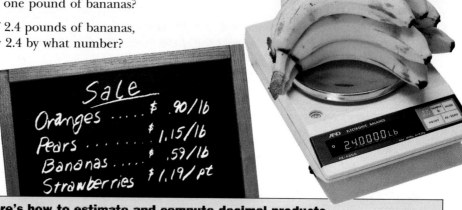

Sale
Oranges . . . . . $ .90/lb
Pears . . . . $ 1.15/lb
Bananas . . . . . $ .59/lb
Strawberries $ 1.19/pt

EXAMPLES | **Here's how to estimate and compute decimal products.**

A Estimating the cost of 2.4 pounds of bananas:

$$2.4 \times .59 = ?$$

> *To estimate the product, round each decimal to the nearest whole number and multiply.*
> **2 × 1 = 2.** *The estimated cost is $2.00.*

B Computing the cost of 2.4 pounds of bananas:

Multiply as whole numbers.	Count the digits to the right of the decimal points.	Count off the same number of digits in the product.
2.4 × .59 ——— 216 120 ——— 1416	2.4 ⟍ ⟍ 3 × .59 ⟋ ——— 216 120 ——— 1416	2.4 × .59 ——— 216 120 ——— 1.416

CHECK for Understanding

3. Look at the examples above. What is the cost of 2.4 pounds of bananas? Is $2 a good estimate?

4. Complete these examples.

a.
```
   9.2
  ×1.5
  ————
  4  0
    2
  ————
  3.  0
```

b.
```
   0.31
  ×0.27
  ————
   2 1
     2
  ————
  0. 8 7
```

> *You have to write a zero here to place the decimal point.*

EXERCISES

▶ **Three of the calculator answers are wrong. Find them by estimating.**

5. a. 7.1×0.98 `6.958`   **b.** 5.7×2.1 `119.7`

    c. 2.31×9.85 `22.7535`   **d.** 3.28×4.6 `15.088`

    e. 19×3.2 `6.08`   **f.** 8.74×2.1 `183.54`

▶ **Multiply.**
Here are scrambled answers for the next row of exercises: 8.4   33.12   2.604   9

6. 4.8×6.9	**7.** 7.5×1.2	**8.** 4.2×2	**9.** 0.31×8.4
10. 1.8×56	**11.** 7.4×20	**12.** 0.05×0.7	**13.** 0.75×16
14. 7.5×0.42	**15.** 3.5×6.3	**16.** 2.31×0.7	**17.** 8.76×2.6
18. 75.6×4.32	**19.** 0.89×8.76	**20.** 0.006×0.12	**21.** 64.2×9.5
22. 4.15×0.021	**23.** 97.1×0.0025	**24.** 6.4×9.05	**25.** 77×1.4

▶ **Use your mental math skills. Evaluate each expression for $r = 0.2$, $s = 0.6$, $t = 0.08$, $u = 9$, and $v = 0.1$.**

26. rs	**27.** st	**28.** tu	**29.** su
30. $r + s$	**31.** $u + v + s$	**32.** $s - r$	**33.** $(sv) + t$
34. $(rv) + u$	**35.** $(ru) - s$	**36.** $(uv) - r$	**37.** $(3s) + v$

Problem Solving   USING DATA

▶ **Solve. Use the prices on page 34.**

38. How much would 1.7 pounds of oranges cost?

39. A customer paid for 1.6 pounds of pears with a $5 bill. How much change should the customer receive?

40. You paid for 2 pints of strawberries with a $5 bill. You received 4 coins and 2 bills in change. What were the coins?

41. Write a problem that you could solve using this equation: $10 - (3 \times 1.15) = n$.

Challenge!   LOGICAL REASONING

▶ **Use the clues to complete the sentences.**

Clues:
- A lemon and a lime cost 61¢.
- A lime and a kiwifruit cost 54¢.
- A kiwifruit and a lemon cost 69¢.

42. Lemons cost _?_ ¢ each.

43. Limes cost _?_ ¢ each.

44. Kiwis cost _?_ ¢ each.

OBJECTIVE: To use mental math to multiply a decimal by 10, 100, or 1000.

A mail survey was taken to identify favorite outdoor recreation activities. The **bar graph** shows the results.

FAVORITE OUTDOOR RECREATIONS

1. What recreational activity was preferred by 1.8 thousand people?

2. What two numbers would you multiply to find how many people preferred bicycling?

EXAMPLES	Here's how to use mental math to multiply a decimal by 10, 100, or 1000.

When you multiply a number by 10, 100, or 1000, the product is greater than the number.

|A| **2.5 × 10 = ?**

Multiplying by 10 moves the decimal point 1 place to the right.

2.5 × 10 = 25.

|B| **2.5 × 100 = ?**

Multiplying by 100 moves the decimal point 2 places to the right.

2.5 × 100 = 250.

|C| **2.5 × 1000 = ?**

Multiplying by 1000 moves the decimal point 3 places to the right.

2.5 × 1000 = 2500.

CHECK for Understanding

3. Look at the examples above. How many people chose bicycling as their favorite outdoor recreation?

EXERCISES

▶ **Mental Math** **Write answers only.**
Here are scrambled answers for the next two rows of exercises.
7.2   856   85,600   72   860   7200   8600   8560

4. 7.2×10

5. 72×100

6. 0.72×10

7. 8.56×100

8. 856×100

9. 8.56×1000

10. 8.6×1000

11. 0.86×1000

12. 95×1000

13. 0.296×1000

14. 2.96×10

15. 2.96×100

16. 2.96×1000

17. 5.13×10

18. 5.13×100

19. 5.13×1000

20. 9400×10

21. 9400×100

22. 9400×1000

23. 940×100

24. 83×10

25. 83×100

26. 83×1000

27. 830×100

28. 0.76×1000

29. 0.76×100

30. 0.76×10

31. 7.6×1000

32. 0.09×1000

33. 0.09×100

34. 0.09×10

35. 0.9×10

36. 421×10

37. 421×1000

38. 421×100

39. 42.1×100

40. 64.2×100

41. 64.2×1000

42. 64.2×10

43. 0.642×1000

44. 4.76×1000

45. 4.76×10

46. 4.76×100

47. 476×100

Problem Solving READING GRAPHS

▶ **Solve. Use the bar graph on page 36.**

48. Which recreational activity was preferred by the most people?

49. Which activity was preferred by the fewest people?

50. Which activity was preferred by 1300 people?

51. How many people preferred fishing?

52. Which activities were preferred by more than 1500 people?

53. Which activities were preferred by fewer than 2000 people?

54. How many more people preferred bicycling to golf?

55. How many fewer people preferred tennis to fishing?

Challenge! MAKING GRAPHS

56. Using the data given at the right, make a bar graph like the one shown on page 36.

Students Trying Out for Springfield High School Teams			
Football	40	Girls' Gymnastics	15
Boys' Basketball	33	Boys' Track	43
Girls' Basketball	29	Girls' Track	37
Boys' Gymnastics	25	Baseball	22

Solving Equations Using Multiplication and Division | **37**

OBJECTIVE: To estimate and compute decimal quotients.

Time yourself! Can you identify, just from the shape, the content of each container?

A group of students timed themselves. Their results are shown in the table.

1. How many students timed themselves?

2. What was the total number of seconds?

3. To find the average time, you would divide 64.32 by what number?

Name	Time (seconds)
Ruella	10.26
Dave	9.78
Mike	9.92
Bev	12.09
Craig	10.75
Rita	11.52
Total	64.32

EXAMPLES | Here's how to estimate and compute decimal quotients.

A Estimating the average:

$$64.32 \div 6 = ?$$

To estimate the quotient, use nearby whole numbers that are easy to divide mentally.

$$60 \div 6 = 10$$

The quotient is near 10.

B Computing the average:

```
   10.72
6)64.32
  -6
   4
  -0
   43
  -42
   12
  -12
    0
```

Place the decimal point for the quotient. Then divide as you would whole numbers.

CHECK for Understanding

4. Look at the examples above.
   a. What was the average time for the students to identify the products?
   b. Why was the estimated quotient less than the computed quotient?

5. Complete these examples. Round each quotient to the nearest hundredth.

```
      0.0▮▮
a. 9)0.432
    -36
     72
    -72
      0
```

You have to write a zero here.

```
      0.▮▮▮
b. 26)6.280
     -5 2
      1 08
     -1 04
        40
       -26
        14
```

You can write a zero here and carry out the division to the next place.

EXERCISES

▶ **Use estimation to choose the quotient.**

6. 15.68 ÷ 2
  a. 0.784
  b. 7.84
  c. 78.4

7. 27.24 ÷ 4
  a. 0.681
  b. 6.81
  c. 68.1

8. 431.4 ÷ 6
  a. 0.719
  b. 7.19
  c. 71.9

9. 2.429 ÷ 7
  a. 0.347
  b. 3.47
  c. 34.7

▶ **Divide.**

Here are scrambled answers for the next row of exercises: 4.2   0.405   2.54   1.24

10. 8.68 ÷ 7

11. 33.02 ÷ 13

12. 2.025 ÷ 5

13. 134.4 ÷ 32

14. 10.32 ÷ 43

15. 38.36 ÷ 14

16. 24.08 ÷ 7

17. 358.4 ÷ 56

18. 107.01 ÷ 87

19. 300.51 ÷ 81

20. 16.5 ÷ 55

21. 362.25 ÷ 45

▶ **First carry out the division to the thousandths place. Then round the quotient to the nearest hundredth.**

Example:

$$0.078 \longrightarrow 0.08$$

$$
\begin{array}{r}
8\overline{)0.630} \\
-56 \\
\hline
70 \\
-64 \\
\hline
6
\end{array}
$$

22. 6.1 ÷ 7

23. 17.03 ÷ 9

24. 8.72 ÷ 3

25. 8.2 ÷ 12

26. 78.66 ÷ 15

27. 45 ÷ 64

28. 15 ÷ 8

29. 74.2 ÷ 60

30. 4.27 ÷ 23

31. 65.9 ÷ 9

32. 7.43 ÷ 14

33. 3.6 ÷ 7

▶ **Use your mental math skills. Evaluate each expression for $r = 2$, $s = 3$, $t = 0.6$, $u = 0.9$, and $v = 1.2$.**

34. $\dfrac{t}{r}$

35. $\dfrac{u}{s}$

36. $\dfrac{v}{6}$

37. su

38. tu

39. sv

40. $r + u$

41. $r + s + t$

42. $u - t$

43. $(v - u) + r$

44. $(rt) - v$

45. $\dfrac{v}{s} + t$

46. $\dfrac{8}{r} - u$

47. $\dfrac{v}{r} + s$

48. $(3t) + v$

Problem Solving   NUMBER SENSE

▶ **Solve. Use the time chart on page 38.**

49. Who identified the contents of the containers in the least amount of time?

50. Who identified the contents in a time that was 0.34 second faster than Ruella's time?

51. What was the average time for the three boys to identify the contents?

52. What was the average time for the three girls to identify the contents?

OBJECTIVE: To divide a decimal by a decimal.

Badid Captures Wheel Title

Special to the Star News

BOSTON - Moussetapha Badid of Pontoise, France broke the world wheelchair record in the Boston Marathon yesterday. Badid, 24, was an Olympic gold medalist in 1988. He completed the 26.2188-mile course in 1.5 hours, and received $25,000 in prize money.

Badid crosses finish line.

Manning
Grant
Vaught
Bannister
Garrick
Garland
Martin
Butler
Team
Opponents

Campbell
Corbin
Richards
Mitchell
Breuer
Murphy
Spencer
Glass
Brooks
West
Coffey
Thornt
Team
Oppoi

1. Read the newspaper article. How many miles is the Boston Marathon course?

2. To compute Badid's speed in miles per hour, you would divide 26.2188 by __?__ .

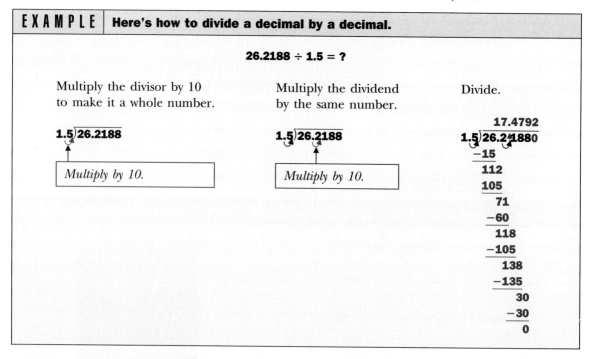

EXAMPLE | **Here's how to divide a decimal by a decimal.**

$$26.2188 \div 1.5 = ?$$

Multiply the divisor by 10 to make it a whole number.

1.5⟌26.2188

Multiply by 10.

Multiply the dividend by the same number.

1.5⟌26.2188

Multiply by 10.

Divide.

```
        17.4792
1.5⟌26.21880
   −15
    112
    105
     71
    −60
    118
   −105
    138
   −135
     30
    −30
      0
```

CHECK for Understanding

3. Look at the example above. To get a whole-number divisor, both decimal points were moved __?__ place(s) to the right.

4. What was Badid's average speed in miles per hour?

5. Complete these examples.

a.
$$\begin{array}{r} 32 \\ 0.54\overline{)17.28} \end{array}$$

Move both decimal points 2 places to the right.

$$\begin{array}{r} -\blacksquare\blacksquare \\ \overline{1\ 08} \\ -\blacksquare\blacksquare \end{array}$$

b.
$$\begin{array}{r} 46.7 \\ 0.006\overline{)0.280\ 2} \end{array}$$

Move both decimal points 3 places to the right.

$$\begin{array}{r} -\blacksquare \\ \overline{40} \\ -\blacksquare\blacksquare \\ \overline{4\ 2} \\ -4\ 2 \end{array}$$

EXERCISES

▶ **Place the decimal point in each quotient.**

6. $0.007\overline{)0.0098}$ quotient 14

7. $0.4\overline{)6.24}$ quotient 156

8. $1.2\overline{)31.2}$ quotient 26

9. $0.34\overline{)3.570}$ quotient 105

▶ **Divide.**

Here are scrambled answers for the next row of exercises: 910   38   6.5   12.7

10. $3.25 \div 0.5$ **11.** $0.76 \div 0.02$ **12.** $5.46 \div 0.006$ **13.** $1.016 \div 0.08$

14. $0.57 \div 0.15$ **15.** $3.91 \div 3.4$ **16.** $0.884 \div 0.26$ **17.** $0.126 \div 2.8$

18. $1.875 \div 0.25$ **19.** $26.22 \div 5.7$ **20.** $23.24 \div 2.8$ **21.** $40.95 \div 9.1$

▶ **Evaluate each expression for $a = 0.1$, $b = 0.3$, $c = 0.4$, and $d = 2.4$.**

22. $\dfrac{d}{b}$ **23.** $\dfrac{d}{c}$ **24.** $\dfrac{d}{a}$ **25.** $\dfrac{b}{a}$ **26.** $\dfrac{c}{a}$

27. $\dfrac{d}{b} - a$ **28.** $\dfrac{d}{c} - b$ **29.** $\dfrac{d}{a} - c$ **30.** $\dfrac{b}{a} + d$ **31.** $\dfrac{c}{a} + b$

Problem Solving   USING EXPRESSIONS

▶ **Decide which expression would be used to complete each sentence.**

32. The record for running the n-mile Boston Marathon is m hours. Therefore, the record average speed is __?__ miles per hour.

33. The longest recorded trampoline-bouncing marathon is m hours. The record was set by a team of n people. They averaged __?__ hours per person.

34. The longest distance recorded for walking backwards is __?__ miles at a speed of n miles per hour for m hours.

35. The record for swimming the English Channel by a relay team is m hours by n swimmers. The swimmers averaged __?__ hours apiece.

Expression A:	mn
Expression B:	$\dfrac{m}{n}$
Expression C:	$\dfrac{n}{m}$

OBJECTIVE: To use mental math to divide a decimal by 10, 100, or 1000.

Try your mental-math skills. Can you fill in the missing numbers in less than 15 seconds?

The vacation costs $3000.
You can buy it for __?__ $10 bills,
or __?__ 100 bills, or __?__ $1000 bills.

Remember: If you divide a number by 10, 100, or 1000, the quotient is less than the number.

7 Fun-Filled Days...

BERMUDA

Family vacation $3000
2 adults, 2 children,
other combinations available.

EXAMPLES | **Here's how to use mental math to divide by 10, 100, or 1000.**

A **3000 ÷ 10 = 300.0, or 300** Dividing by 10 moves the decimal point 1 place to the left.

B **3000 ÷ 100 = 30.00, or 30** Dividing by 100 moves the decimal point 2 places to the left.

C **3000 ÷ 1000 = 3.000, or 3** Dividing by 1000 moves the decimal point 3 places to the left.

D **47.125 ÷ 10 = 4.7125**

E **8.4 ÷ 100 = 0.084** ◀—— *Some zeros had to be written before the decimal point could be placed in the quotient.*

F **5.25 ÷ 1000 = 0.00525** ◀——

CHECK for Understanding

1. Look at Example A. Dividing by 10 moves the decimal point 1 place to the __?__ .

2. Look at Example B. Dividing by 100 moves the decimal point __?__ places to the left.

3. Look at Example C. Dividing by 1000 moves the decimal point __?__ places to the left.

4. Divide.
   a. $36.5 \div 10$ **b.** $6.3 \div 100$ **c.** $17.2 \div 1000$

EXERCISES

▷ **Mental Math**   Write answers only.

5. $76 \div 10$

6. $76 \div 100$

7. $76 \div 1000$

8. $7.6 \div 10$

9. $217.2 \div 10$

10. $217.2 \div 100$

11. $217.2 \div 1000$

12. $21.72 \div 1000$

13. $68.1 \div 1000$

14. $68.1 \div 100$

15. $68.1 \div 10$

16. $0.681 \div 10$

17. $80 \div 10$

18. $80 \div 100$

19. $80 \div 1000$

20. $800 \div 10$

21. $41.23 \div 10$

22. $41.23 \div 100$

23. $41.23 \div 1000$

24. $4.123 \div 1$

25. $846 \div 1000$

26. $846 \div 100$

27. $846 \div 10$

28. $84.6 \div 1000$

29. $214.7 \div 1000$

30. $214.7 \div 100$

31. $214.7 \div 10$

32. $21.47 \div 100$

33. $2.42 \div 10$

34. $2.42 \div 100$

35. $2.42 \div 1000$

36. $242 \div 1000$

Problem Solving USING DATA

▷ **Use the sales ad. Complete each sentence.**

37. A family vacation to Mexico costs __?__ hundred dollars.

38. You can buy the family vacation to Waikiki for __?__ hundred dollars.

39. The family vacation to the Bahamas costs __?__ thousand dollars.

40. You would pay __?__ thousand dollars for the family vacation to Maui.

41. The cost per day for the family vacation to Maui is __?__ hundred dollars.

42. **Write a problem** that can be solved using this equation: $5n = 2000$.

43. **Write a problem** using the data in the ad.

Family* Vacations		
Destination	**Days**	**Cost**
Florida	3	$1000
Mexico	4	$2500
Bahamas	5	$2000
Maui	8	$4000
Waikiki	8	$3900
*For 2 adults, 2 children		

Challenge! LOGICAL REASONING

▷ **Study the clues. Use the sales ad above to name the family vacation.**

44. *Clues:*
- This family vacation costs more than $1500.
- It lasts less than a week.
- It costs less than $600 per day.

45. *Clues:*
- This family vacation costs more than $2000.
- It lasts more than 4 days.
- The cost per person is less than $1000.

PROBLEM · SOLVING · POWER

You will use a variety of problem-solving strategies on these two pages. For some problems, you may wish to use a calculator.

I just figured it out! The product of our three consecutive locker numbers is 421,800.

1 GUESS AND CHECK

What are the three locker numbers?

✓ *Problem Solving Tips*

Guess and Check.* Choose three consecutive whole numbers, like 50, 51, 52. Multiply them.

50 ☒ 51 ☒ 52 ☐ (*132600*)   *Too small!*

Try 80, 81, 82.

80 ☒ 81 ☒ 82 ☐ (*531360*)   *Too big!*

Which three numbers would you try next?

*At least one more of the problems on these two pages can be solved by guessing and checking.

2 GEOMETRY—VISUAL THINKING

Think about drawing a path from each circle to its respective triangle. None of the paths should cross nor touch a side of the large rectangle. Then draw a picture to show your solution.

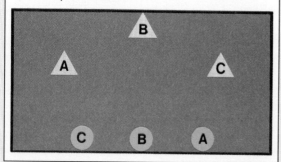

3 PATTERNS

Look for the pattern. Find the missing numbers.

$$1 \times 8 + 1 = 9$$
$$12 \times 8 + 2 = 98$$
$$123 \times 8 + 3 = 987$$
$$\vdots$$
$$a \times 8 + b = 98,765,432$$

4 WHAT'S YOUR GUESS?

The world's largest pumpkin was grown by Norman Gallagher of Chelan, Washington, in 1984. Guess the weight of the pumpkin and write it on a piece of paper. Then use the following information to see how close your guess was.

Here are three incorrect guesses:

| Mike 635 | Todd 511 | Kathy 487 |

One of these guesses is off by 125 pounds, one is off by 101 pounds, and one is off by 23 pounds. How many pounds did the pumpkin weigh? *Hint: If Mike's guess was off by 125 pounds, what would be the two possibilities for the correct weight?*

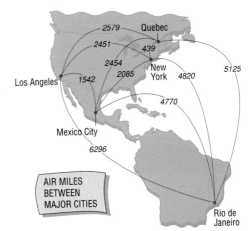

5 GEOMETRY—SPACE FIGURES

How many ways can the number cube be put into the box?

APPLICATIONS

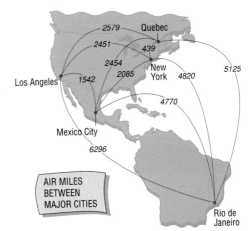

AIR MILES BETWEEN MAJOR CITIES

Complete.
You may wish to use a calculator.

8 Jane flew from Los Angeles to Mexico City to Rio de Janeiro. She flew a total of _?_ miles.

9 Rick flew a total of 8875 miles. He started in Rio de Janeiro, flew to Los Angeles, and ended his flight in _?_.

10 The total mileage in 3 round-trip flights between Rio de Janeiro and Quebec is about the same as the total mileage in 10 round-trip flights between Los Angeles and _?_.

6 SEQUENCES

How many blocks would there be in the 15th figure in this sequence?

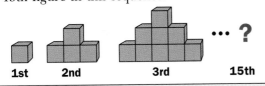

1st   2nd   3rd   15th

7 ALPHAMETICS

In this product, each different letter stands for a different digit. What digit can each letter stand for?

$$
\begin{array}{r}
BE \\
\times\ BE \\
\hline
ARE
\end{array}
$$

Hint: Since the product is a 3-digit number, BE is less than _?_.

Rate Your Problem-Solving Power: 6–7 correct = Good;   8–9 correct = Excellent;   10 correct = Exceptional

OBJECTIVE: To use a table of telephone rates to solve problems.

Here are rates for direct-dial telephone calls made from Dallas, Texas.

	M	T	W	T	F	S	S
8 A.M. to 5 P.M.	■	■	■	■	■	▨	▨
5 P.M. to 11 P.M.	▨	▨	▨	▨	▨	▨	▨
11 P.M. to 8 A.M.	▨	▨	▨	▨	▨	▨	▨

Dial direct	Weekday full rate ■		Evening ▨		Night & Weekend (except 5-11 P.M. Sun.) ▨	
Rates from Dallas to:	First Minute	Each Additional Minute	First Minute	Each Additional Minute	First Minute	Each Additional Minute
Tampa	$ .27	$ .23	$ .17	$ .15	$ .14	$ .13
Atlanta	.26	.24	.16	.14	.13	.12
Indianapolis	.28	.25	.18	.16	.15	.14
Seattle	.31	.28	.19	.17	.16	.14
Phoenix	.26	.24	.16	.14	.13	.12

▶ **Use the rates to solve these problems. Decide when a calculator would be helpful.**

1. Carla called her brother in Tampa. She called at 3 P.M. on Wednesday. They talked for 5 minutes.
   a. What was the cost for the first minute of their call?
   b. What was the cost for the next 4 minutes?
   c. What was the total cost of their 5-minute call?

2. Sam called his grandmother in Phoenix. He called at 5:45 P.M. on Tuesday. They talked for 8 minutes.
   a. What was the cost for the first minute?
   b. What was the cost for the next 7 minutes?
   c. What was the total cost?

3. Phil called his sister in Atlanta. He called at 9 A.M. on Saturday. They talked for 13 minutes. What was the cost?

4. Megan called her friend in Seattle on Sunday. They talked from 8:10 A.M. to 8:17 A.M. What was the cost?

5. A 20-minute call to Indianapolis on a weekday before 5 P.M. costs $5.03. How much money can be saved by waiting until 6 P.M. to make the call?

6. **Write a problem** about a call from Dallas to Tampa that you would solve using the equation $6 \times 0.13 + 0.14 = n$.

▶ **Let n be the number of additional minutes after the first minute. Write an expression for the cost (in cents) of a call to**

7. Indianapolis on a weekday before 5 P.M.

8. Phoenix at 10 A.M. on a Saturday.

9. Seattle on a weekday after 5 P.M.

10. Atlanta at 1:30 P.M. on a Sunday.

▶ **Solve and check.** *(page 16)*

1. $u + 14 = 32$ **2.** $x + 9 = 17$ **3.** $s + 11 = 27$ **4.** $a + 7 = 23$

5. $e + 23 = 36$ **6.** $p + 10 = 36$ **7.** $c + 41 = 41$ **8.** $y + 19 = 41$

9. $p + 9.3 = 47.2$ **10.** $c + 5.2 = 5.2$ **11.** $u + 14.7 = 32.3$ **12.** $a + 8.35 = 45.6$

13. $d + 4.9 = 7.1$ **14.** $g + 64.3 = 89.1$ **15.** $c + 6.12 = 9.93$ **16.** $w + 71.2 = 97.4$

▶ **Solve and check.** *(page 18)*

17. $a - 7 = 14$ **18.** $x - 5 = 16$ **19.** $u - 9 = 14$ **20.** $a - 8 = 37$

21. $b - 11 = 18$ **22.** $s - 17 = 32$ **23.** $b - 21 = 42$ **24.** $x - 14 = 45$

25. $b - 1.35 = 3.76$ **26.** $w - 9.3 = 12.7$ **27.** $y - 7.33 = 13.41$ **28.** $u - 9.25 = 25.7$

29. $x - 2.93 = 3.41$ **30.** $h - 5.7 = 7.1$ **31.** $w - 6.3 = 27.4$ **32.** $a - 22.3 = 29.4$

▶ **Write an algebraic expression for each word expression.** *(page 30)*

33. 3 times a number w

34. 9 decreased by a number a

35. a number z less 6

36. a number x decreased by 3

37. the product of a number c and 18

38. a number b minus 17

39. a number s plus 15

40. a number r divided by 24

41. a number s plus a number t

42. a number a minus a number b

43. a number r increased by itself

44. a number c decreased by itself

45. 4.7 less than a number y

46. $\frac{3}{4}$ more than a number p

47. 0.6 increased by a number t

48. a number k minus $\frac{1}{2}$

▶ **Evaluate each expression for $a = 1.2$, $b = 0.1$, $c = 2.4$, and $d = 0.4$.**
(pages 32, 34, 38, 40)

49. $a + c$ **50.** $a\,b$ **51.** $a \div d$ **52.** $a - d$

53. $\dfrac{c}{d}$ **54.** $(bc) + a$ **55.** $\dfrac{c}{a} + b$ **56.** $c + d$

57. $(6a) + c$ **58.** $(2c) - a$ **59.** $(ac) + 12$ **60.** $(c - d) + b$

61. $\dfrac{2.4}{c} + a$ **62.** $(a - b) - 0.7$ **63.** $\dfrac{3.6}{a} - b$ **64.** $4a + 3.2$

65. $(ad) + 1$ **66.** $(a + d) - b$ **67.** $\dfrac{a}{b} - d$ **68.** $6bd$

69. $\dfrac{1.5}{b} - c$ **70.** $\dfrac{d}{b} - 0.02$ **71.** $(bc) + a$ **72.** $\dfrac{12}{a} - 9$

73. $(10b) + a$ **74.** $\dfrac{4.8}{c} + 8$ **75.** $(cd) + 0.01$ **76.** $\dfrac{2}{d} + 6$

Solving Equations Using Multiplication and Division | **47**

2-9 Solving Multiplication Equations

OBJECTIVE: To solve multiplication equations by dividing both sides of the equation by the same number.

Look at the balance in Step 1. If we let n be the number of pennies in each box, we have the equation $3n = 12$.

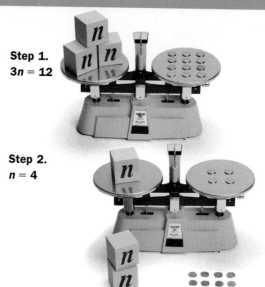

Step 1.
$3n = 12$

Step 2.
$n = 4$

1. Look at the balance in Step 2. The number of pennies on both sides was divided by 3.
   a. Do the two sides still balance?
   b. What equation does the balance show now?
   c. How many pennies are in the box?

E X A M P L E S | **Here's how to solve a multiplication equation.**

A **Equation:**

$3n = 12$

*Multiplication and division are inverse operations. To undo multiplying by 3 and to get n by itself on one side of the equation, divide both sides by 3.

$$\frac{3n}{3} = \frac{12}{3}$$

Simplify both sides.

$$n = 4$$

> *The property used to divide both sides of an equation by the same number is called the **division property of equality.**

✔ **CHECK**

Check the solution by substituting 4 for n in the equation $3n = 12$.

$$3n = 12$$
$$3(4) \stackrel{?}{=} 12$$
$$12 = 12$$

It checks!

B $7c = 56$
$$\frac{7c}{7} = \frac{56}{7}$$
$$c = 8$$

C $0.6b = 1.8$
$$\frac{0.6b}{0.6} = \frac{1.8}{0.6}$$
$$b = 3$$

CHECK for Understanding

2. Look at Example A. To find n, by what number were both sides of the equation divided?

3. By what number would you divide both sides to solve these equations?
   a. $5x = 75$ **b.** $0.9y = 9.9$ **c.** $3w = 2.4$

4. Explain how to get the numbers that are printed in red in Examples B and C.

EXERCISES

▶ **Solve and check.**
Here are scrambled answers for the next two rows of exercises:
42   4   12   9   7   17   6   5

5. $15k = 75$ **6.** $4a = 48$ **7.** $12f = 72$ **8.** $6k = 42$

9. $7d = 63$ **10.** $14j = 56$ **11.** $5n = 85$ **12.** $2r = 84$

13. $4h = 64$ **14.** $8y = 72$ **15.** $20c = 400$ **16.** $9v = 207$

17. $2x = 8.6$ **18.** $5y = 3.5$ **19.** $0.3b = 6$ **20.** $1.4q = 0$

21. $0.4s = 2.8$ **22.** $0.6e = 2.4$ **23.** $1.5q = 4.5$ **24.** $2.1c = 2.1$

▶ **Solve and check.**
Here are scrambled answers for the next two rows of exercises:
59   0   45   65   60   17   80   55

25. $c + 12 = 57$ **26.** $b + 10 = 75$ **27.** $6z = 102$ **28.** $a - 15 = 40$

29. $7j = 0$ **30.** $x - 19 = 61$ **31.** $4n = 240$ **32.** $t + 28 = 87$

33. $x - 28 = 79$ **34.** $5w = 235$ **35.** $d + 14 = 53$ **36.** $7k = 126$

37. $f + 0.9 = 2.4$ **38.** $10m = 4.3$ **39.** $y - 3.1 = 4.5$ **40.** $0.6c = 5.4$

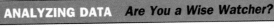

Challenge!   NUMBER SENSE

41. When completely filled with pennies, a bank held $9.68. Do you think that the bank could hold $100 in dimes? Why or why not?

Group Project   ANALYZING DATA   *Are You a Wise Watcher?*

▶ Work in a Small Group

42. Try to determine to the nearest hour how much time each group member spent watching television last week.
Hint: Try to recall the program(s) that you watched each day of the week.

43. a. List the number of hours that each class member spent watching television last week.
b. Order the numbers from least to greatest.

44. a. How many watched more television than you?
b. How many watched less television than you?

45. Write a paragraph that summarizes the data.

Solving Equations Using Multiplication and Division | **49**

OBJECTIVE: To solve division equations by multiplying both sides of the equation by the same number.

Decide whether Equation A or Equation B would be used to solve the problem.

> Equation A: $5n = 15$
>
> Equation B: $\dfrac{n}{5} = 15$

Jefferson Nickel

1963 1964 1975 1976 1977

1. Maria bought this set of 5 Jefferson nickels at a coin show. If the average value of a coin is $15, what is the value of the whole set?

EXAMPLES | Here's how to solve a division equation.

A **Equation:**

$$\dfrac{n}{5} = 15$$

To undo dividing by 5, multiply both sides by 5.

$$5 \times \dfrac{n}{5} = 5 \times 15$$

✔ CHECK

$$\dfrac{n}{5} = 15$$

$$\dfrac{75}{5} \stackrel{?}{=} 15$$

$$15 = 15$$

It checks!

Simplify.

$$n = 75$$

*The property used to multiply both sides of an equation by the same number is called the **multiplication property of equality**.

B

$$\dfrac{d}{9} = 7$$

$$9 \times \dfrac{d}{9} = 9 \times 7$$

$$d = 63$$

C

$$\dfrac{f}{0.4} = 20$$

$$0.4 \times \dfrac{f}{0.4} = 0.4 \times 20$$

$$f = 8.0$$

CHECK for Understanding

2. Look at Example A.
   a. To find n, what number were both sides of the equation multiplied by? What does n equal?
   b. To check the equation, what number was substituted for n in the equation $\dfrac{n}{5} = 15$?

3. By what number would you multiply both sides to solve these equations?

   a. $\dfrac{x}{3} = 11$ **b.** $\dfrac{z}{1.7} = 10$ **c.** $\dfrac{y}{5} = 9.2$

4. Explain how to get the numbers that are printed in red in Examples B and C.

EXERCISES

▶ **Solve and check.**
Here are scrambled answers for the next two rows of exercises:
32   70   36   78   56   21   63   72

5. $\dfrac{x}{5} = 14$ **6.** $\dfrac{n}{3} = 7$ **7.** $\dfrac{x}{2} = 16$ **8.** $\dfrac{b}{4} = 18$

9. $\dfrac{d}{2} = 28$ **10.** $\dfrac{r}{6} = 13$ **11.** $\dfrac{v}{7} = 9$ **12.** $\dfrac{e}{3} = 12$

13. $\dfrac{j}{6} = 14$ **14.** $\dfrac{d}{12} = 10$ **15.** $\dfrac{a}{4} = 2.3$ **16.** $\dfrac{y}{7} = 1.3$

17. $\dfrac{q}{1.5} = 10$ **18.** $\dfrac{w}{0.3} = 6$ **19.** $\dfrac{t}{0.5} = 1.7$ **20.** $\dfrac{c}{1.8} = 0.2$

▶ **Solve and check.**
Here are scrambled answers for the next two rows of exercises:
0   12   61   65   70   75   128   450

21. $10w = 1280$ **22.** $8k = 0$ **23.** $n + 19 = 94$ **24.** $x - 26 = 44$

25. $c - 16 = 49$ **26.** $11c = 132$ **27.** $\dfrac{b}{15} = 30$ **28.** $d + 37 = 98$

29. $y + 35 = 77$ **30.** $z - 32 = 0$ **31.** $10s = 3.7$ **32.** $\dfrac{z}{6} = 1.4$

33. $x - 7.2 = 11.3$ **34.** $1.6d = 16$ **35.** $\dfrac{x}{0.8} = 0$ **36.** $r + 12.7 = 18.4$

Problem Solving USING EQUATIONS

▶ **Decide whether Equation A, B, C, or D would be used to solve the problem. Then solve the problem.**

Equation A:	$n + 27 = 81$
Equation B:	$n - 27 = 81$
Equation C:	$27n = 81$
Equation D:	$\dfrac{n}{27} = 81$

37. Twenty-seven less than a number n is 81. What is the number?

38. The sum of a number n and 27 is 81. What is the number?

39. A number n divided by 27 is 81. What is the number?

40. Twenty-seven times a number n is 81. What is the number?

41. Maris bought 27 Mercury dimes for $81. How much did she pay for each Mercury dime?

42. George bought a silver dollar and an Indian head penny for $81. The Indian head penny cost $27. What did he pay for the silver dollar?

43. Write a problem that you would solve using Equation C.

44. Write a problem that you would solve using Equation D.

OBJECTIVE: To use the symmetric property of equality to solve multiplication and division equations that have the variable on the right side of the equal sign.

Variables can be on either side of the equation. In this lesson you will solve some equations that have the variable on the right side.

EXAMPLES | **Here's how to use the symmetric property of equality to solve equations that have the variable on the right side.**

A
$$56 = 7x$$
$$7x = 56$$
$$\frac{7x}{7} = \frac{56}{7}$$
$$x = 8$$

The symmetric property of equality tells you that if $56 = 7x$ then $7x = 56$.

✓ **CHECK**
$$56 = 7x$$
$$56 \overset{?}{=} 7(8)$$
$$56 = 56$$
It checks!

B
$$14 = \frac{n}{6}$$
$$\frac{n}{6} = 14$$
$$6 \times \frac{n}{6} = 6 \times 14$$
$$n = 84$$

✓ **CHECK**
$$14 = \frac{n}{6}$$
$$14 \overset{?}{=} \frac{84}{6}$$
$$14 = 14$$
It checks!

C
$$20 = \frac{w}{0.7}$$
$$\frac{w}{0.7} = 20$$
$$0.7 \times \frac{w}{0.7} = 0.7 \times 20$$
$$w = 14$$

D
$$9.6 = 1.2t$$
$$1.2t = 9.6$$
$$\frac{1.2t}{1.2} = \frac{9.6}{1.2}$$
$$t = 8$$

CHECK For Understanding

1. Look at the examples above.
   a. In each first equation, the expression containing the variable is on the ___?___ side of the equation.
   b. For each second equation, the symmetric property of equality was used to get the expression containing the variable on the ___?___ side of the equation.

2. Complete these examples.

   a.
   $$84 = v + 19$$
   $$v + 19 = \underline{\ ?\ }$$
   $$v + 19 - 19 = 84 - 19$$
   $$v = 65$$

   b.
   $$38 = r + 17$$
   $$r + 17 = 38$$
   $$r + 17 - \underline{\ ?\ } = 38 - 17$$
   $$r = 21$$

   c.
   $$5.3 = z - 4.1$$
   $$z - 4.1 = 5.3$$
   $$z - 4.1 + 4.1 = 5.3 + 4.1$$
   $$z = \underline{\ ?\ }$$

3. Explain how to get the numbers printed in red in Examples C and D.

EXERCISES

Solve and check.

Here are scrambled answers for the next row of exercises: 32   19   48   5

4. $12 = \dfrac{a}{4}$

5. $40 = 8j$

6. $21 = b - 11$

7. $36 = j + 17$

8. $0 = 6k$

9. $45 = m + 19$

10. $15 = \dfrac{d}{6}$

11. $30 = w - 14$

12. $42 = y - 9$

13. $11 = \dfrac{f}{6}$

14. $18 = g + 12$

15. $42 = 6p$

16. $59 = a + 31$

17. $45 = 5n$

18. $57 = y - 30$

19. $50 = \dfrac{g}{3}$

20. $80 = 16p$

21. $x - 18 = 33$

22. $\dfrac{m}{6} = 12$

23. $60 = a + 47$

24. $2 = \dfrac{c}{1.2}$

25. $x + 3.4 = 3.4$

26. $0.4q = 5.6$

27. $1.3 = b - 1.3$

28. $x + 2.9 = 10.3$

29. $19.5 = 1.5r$

30. $2.9 = m - 1.6$

31. $\dfrac{d}{0.7} = 1.4$

32. $12s = 16.8$

33. $0.14 = \dfrac{e}{10}$

34. $1.15 = z + 0.32$

35. $r - 0.28 = 0$

Problem Solving USING A FORMULA

Complete the table using the formula.

36. Gas mileage (miles per gallon) is the number of miles driven for each gallon of gasoline. You can use this formula to compute gas mileage:

Formula: *distance* ⟶ $\dfrac{d}{g}$ = m ⟵ *mileage*
number of gallons ⟶

	d (miles)	g (gallons)	m (miles per gallon)
a.	208	8	?
b.	330	15	?
c.	?	7	25
d.	?	9	24

Challenge! LOGICAL REASONING

37. How many months old are Ed's three baby cousins Nan, Julie, and Joey?

Clue 1: The product of their ages in whole months is 90.

Clue 2: The sum of their ages in whole months is the same as Ed's age in years.

Clue 3: Joey and Nan are twins.

Did you need to use all 3 clues? Why or why not?

OBJECTIVE: To use addition, subtraction, multiplication, and division equations to solve problems.

Remember that problems that can be solved with arithmetic can also be solved by writing and solving an equation.

EXAMPLE	**Here's how to use arithmetic or algebra to solve a problem.**

Problem: You bought 3 tickets to a Janet Jackson concert. The total cost for the tickets was $146.25. What was the cost of each ticket?

Using arithmetic

$$146.25 \div 3 = 48.75$$

↑ *total cost in dollars*   ↑ *number of tickets*   ↑ *cost in dollars of each ticket*

Each ticket cost $48.75.

Using algebra

Step 1. Choose a variable. Use it and the facts to represent the numbers in the problem.

Let t = cost in dollars of each ticket.
Then $3t$ = cost in dollars of three tickets.

Step 2. Write an equation based on the facts.

$3t = 146.25$

↑   ↑

These are equal expressions for the cost of three tickets.

Step 3. Solve the equation.

$$3t = 146.25$$
$$\frac{3t}{3} = \frac{146.25}{3}$$
$$t = 48.75$$

Each ticket cost $48.75.

CHECK for Understanding

1. Look at Example A. The total cost of the tickets was divided by the number of tickets to get the cost of _?_ ticket.

2. Look at Example B.
   a. In Step 1, the variable t equals the _?_ of each ticket.
   b. In Step 2, both the expression $3t$ and the number _?_ equal the total cost of the tickets.
   c. In Step 3, to solve the equation, each side of the equation was divided by _?_.

3. To check the solution, ask yourself whether the answer fits the facts in the problem. Do three tickets at $48.75 each cost a total of $146.25?

EXERCISES

▶ Decide whether Equation A, B, C, or D would be used to solve the problem. Then solve the problem.

> Equation A: $\dfrac{n}{1.5} = 9$
>
> Equation B: $1.5n = 9$
>
> Equation C: $n + 1.5 = 9$
>
> Equation D: $n - 1.5 = 9$

4. A number divided by 1.5 gives a result of 9. What is the number?
(Let n = the number.)

5. The product of a number and 1.5 is 9. What is the number?
(Let n = the number.)

6. When 1.5 is subtracted from a number the difference is 9. What is the number?
(Let n = the number.)

7. The sum of a number and 1.5 is 9. What is the number?
(Let n = the number.)

8. Sandy has 1.5 times as many CD's as Rick has. Sandy has 9 CD's. How many CD's does Rick have?
(Let n = the number of CD's that Rick has.)

9. Yesterday, The Sound Shop was open 1.5 fewer hours than Strawberries. Yesterday, The Sound Shop was open 9 hours. How long was Strawberries open yesterday?
(Let n = the number of hours Strawberries was open yesterday.)

10. A CD cost 1.5 times as much as an LP. The CD cost $9. How much did the LP cost?
(Let n = the cost in dollars of the LP.)

11. Write a problem that you would solve using Equation B.

▶ Write an equation and solve the problem.

12. The product of 8.25 and a number is 165. What is the number?
(Let k = the number.)

13. A number divided by 14 equals 6.5. What is the number?
(Let j = the number.)

14. The difference between two numbers is 23. The lesser number is 2.7. What is the greater number?
(Let m = the greater number.)

15. The sum of two numbers is 46. One of the numbers is 17.05. What is the other number?
(Let n = the other number.)

16. Bonnie scored 4.5 times as many points as Anna. Bonnie scored 36 points. How many points did Anna score?
(Let a = the number of points Anna scored.)

17. You scored 3 times as many points as your friend. You scored 27 points. How many points did your friend score?
(Let f = the number of points your friend scored.)

18. Kevin's age in years divided by 6 is 5. How old is Kevin?
(Let k = Kevin's age in years.)

19. Alicia's age in years minus 6 is 19. How old is Alicia?
(Let a = Alicia's age in years.)

20. Bonita is 36 years old. That is 3 times as old as Evan. How old is Evan?
(Let e = Evan's age in years.)

21. Brad is 6 years younger than Tracey. Brad is 17 years old. How old is Tracey?
(Let t = Tracey's age in years.)

OBJECTIVE: To solve problems using a calculator when appropriate.

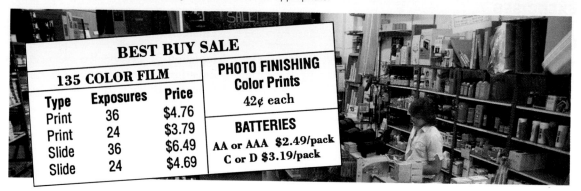

BEST BUY SALE

135 COLOR FILM			PHOTO FINISHING Color Prints 42¢ each
Type	**Exposures**	**Price**	
Print	36	$4.76	**BATTERIES**
Print	24	$3.79	AA or AAA $2.49/pack
Slide	36	$6.49	C or D $3.19/pack
Slide	24	$4.69	

You can use calculator memory keys to help find the total cost of the items on the list.

Scientific calculator

Press: 3 ⊠ 4.76 ⊟ [STO]   14.28 is stored in memory.

2 ⊠ 2.49 ⊟ [SUM]   4.98 is added to 14.28 in memory.

[RCL]   Memory is recalled and displayed.

Four-function calculator

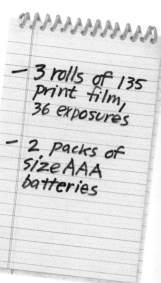

Press: [AC] or [MRC] [MRC]   The memory must be cleared (set to 0).

3 ⊠ 4.76 ⊟ [M+]   14.28 is added to 0 in memory.

2 ⊠ 2.49 ⊟ [M+]   4.98 is added to 14.28 in memory.

[MR] or [MRC]   Memory is recalled and displayed.

The display shows (19.26)   The total cost is $19.26.

Handwritten note:
— 3 rolls of 135 print film, 36 exposures
— 2 packs of size AAA batteries

▶ **Use the memory keys to find the total cost of the items on each list.**

1. 4 rolls of 36-exposure slide film
   3 rolls of 24-exposure print film

2. 3 packs of size AA batteries
   2 rolls of 24-exposure slide film
   2 rolls of 24-exposure print film

▶ **Use the sale prices. Decide when a calculator would be useful.**

3. Chad bought 3 rolls of identical film and 2 packs of C batteries. He spent $20.45. What kind of film did he buy?

4. Rachel bought 2 rolls of 36-exposure print film and picked up 36 prints from film she had left the day before. How much did she spend?

5. **Write a problem** that can be solved by using the sale ad and the memory keys. Then solve the problem.

6. **Write a problem** that can be solved by estimating the total cost. Use the sale ad.

Cumulative Algebra Practice

Solve and check. *(page 20)*

1. $10 = a - 7$ **2.** $14 = b + 9$ **3.** $17 = c - 8$ **4.** $3 = d + 2$

5. $9 = e + 9$ **6.** $15 = f - 8$ **7.** $0 = g - 6$ **8.** $13 = h + 1$

9. $21 = m + 6$ **10.** $16 = n - 7$ **11.** $19 = p - 4$ **12.** $20 = r - 20$

13. $3.1 = a - 2.7$ **14.** $0.7 = b + 0.6$ **15.** $4.7 = c - 1$ **16.** $8 = d + 2.3$

17. $0.8 = e + 0.8$ **18.** $1.9 = f - 1.9$ **19.** $0 = g - 7.1$ **20.** $3.26 = h - 1.6$

Solve and check. *(page 48)*

21. $3x = 21$ **22.** $4y = 0$ **23.** $6n = 42$ **24.** $8b = 72$

25. $4c = 80$ **26.** $10p = 130$ **27.** $8r = 8$ **28.** $5t = 40$

29. $2a = 26$ **30.** $7c = 28$ **31.** $6d = 0$ **32.** $9e = 90$

33. $5x = 3.5$ **34.** $0.3y = 6$ **35.** $0.6n = 4.2$ **36.** $7b = 0.7$

37. $0.9c = 9$ **38.** $4p = 0.12$ **39.** $1.9r = 0$ **40.** $2.8t = 2.8$

Solve and check. *(page 50)*

41. $\dfrac{d}{6} = 7$ **42.** $\dfrac{e}{3} = 6$ **43.** $\dfrac{f}{5} = 0$ **44.** $\dfrac{g}{4} = 8$

45. $\dfrac{n}{10} = 10$ **46.** $\dfrac{p}{6} = 12$ **47.** $\dfrac{r}{2} = 14$ **48.** $\dfrac{t}{8} = 2$

49. $\dfrac{a}{9} = 1$ **50.** $\dfrac{b}{1} = 5$ **51.** $\dfrac{c}{7} = 14$ **52.** $\dfrac{d}{12} = 4$

53. $\dfrac{d}{0.6} = 0.3$ **54.** $\dfrac{e}{4} = 0.5$ **55.** $\dfrac{f}{1.6} = 2$ **56.** $\dfrac{g}{1.5} = 0.4$

57. $\dfrac{n}{0.2} = 5$ **58.** $\dfrac{p}{8} = 0.4$ **59.** $\dfrac{r}{1.3} = 0$ **60.** $\dfrac{t}{1.1} = 1.1$

Solve and check. *(page 52)*

61. $15 = 3x$ **62.** $12 = \dfrac{y}{4}$ **63.** $40 = 5p$ **64.** $8 = \dfrac{b}{2}$

65. $9 = 9c$ **66.** $7 = \dfrac{w}{1}$ **67.** $100 = 2v$ **68.** $11 = \dfrac{d}{11}$

69. $0 = 16e$ **70.** $20 = \dfrac{f}{20}$ **71.** $52 = 52g$ **72.** $3 = \dfrac{a}{3}$

73. $4.8 = 0.6x$ **74.** $1.1 = \dfrac{y}{0.9}$ **75.** $1.2 = 8p$ **76.** $0.3 = \dfrac{b}{7}$

77. $23 = 0.1c$ **78.** $12 = \dfrac{w}{1.2}$ **79.** $29 = 2.9v$ **80.** $0.3 = \dfrac{d}{0.3}$

Solving Equations Using Multiplication and Division | **57**

Here are scrambled answers for the review exercises:

1	6	10	equality	left	right
2	7	12	estimate	multiply	simplify
3	8	divide	inverse	plus	times

1. A word expression for the algebraic expression **8z + 2** is "8 _?_ a number z, _?_ 2." *(page 30)*

2. If you evaluate the expression $\frac{a}{c} + b$ for $a = 24$, $b = 4$, and $c = 8$, you get _?_ . *(page 32)*

3. If you estimate the product **6.17 × 1.94** by rounding each decimal to the nearest whole number, you get _?_ . *(page 34)*

4. If you compute the product **3.6 × 7.1,** you multiply 36 and 71 as whole numbers and then count off _?_ decimal places in the product. *(page 34)*

5. Multiplying a decimal by 1000 moves the decimal point _?_ place(s) to the right. Dividing a decimal by 10 moves the decimal point _?_ place(s) to the left. *(pages 36, 42)*

6. To _?_ the quotient **372.8 ÷ 7,** you could use nearby whole numbers that are easy to divide mentally. *(page 38)*

7. If you compute the quotient **2.8)‾22.96,** you could first multiply both the divisor and the dividend by _?_ to get a whole number divisor. *(page 40)*

8. Multiplication and division are _?_ operations. To solve the equation **6x = 72,** you would first _?_ both sides by _?_ . Then you would simplify both sides. *(page 48)*

9. To solve the equation $\frac{a}{8} = 13$, you would first _?_ both sides by _?_ and then _?_ both sides. *(page 50)*

10. In the first equation at the right, the expression containing the variable is on the _?_ side of the equation. For the second equation, the symmetric property of _?_ was used to get the expression containing the variable on the _?_ side. *(page 52)*

$$56 = 4y$$
$$4y = 56$$

▶ **Write an algebraic expression for each word expression.** *(page 30)*

1. the product of 18 and b

2. 12 multiplied by a

3. a number c divided by 4, minus 12

4. 16 times a number r, plus 8

▶ **Evaluate each expression for $a = 9$, $b = 6$, $c = 18$, and $d = 30$.** *(page 32)*

5. bd

6. $\dfrac{c}{b}$

7. $(ac) - b$

8. $\dfrac{c}{a} + b$

9. $2ab$

10. $4c$

11. $(5a) - c$

12. $\dfrac{d}{5}$

▶ **Evaluate each expression for $e = 1.6$, $f = 0.2$, $g = 2.8$, and $h = 0.4$.**
(pages 34, 38, 40)

13. ef

14. $\dfrac{g}{h}$

15. $(fg) + e$

16. $\dfrac{e}{h} - f$

17. $(6h) + g$

18. $g - e$

19. $e + h$

20. $(g - h) + e$

21. $\dfrac{5}{f}$

22. $3fh$

23. $(eh) - f$

24. $\dfrac{g}{4} - h$

▶ **Mental Math** **Write answers only.** *(pages 36, 42)*

25. 0.62×10

26. 1.7×100

27. 49×100

28. 5.1×1000

29. $817.3 \div 100$

30. $298 \div 10$

31. $21.72 \div 1000$

32. $3 \div 100$

▶ **Solve and check.** *(pages 48, 50, 52)*

33. $a + 6 = 11$

34. $b - 4 = 9$

35. $7 = c + 1$

36. $14 = d - 9$

37. $7e = 42$

38. $\dfrac{f}{4} = 8$

39. $78 = 6g$

40. $15 = \dfrac{h}{5}$

41. $9 = p + 3.7$

42. $17.2 = q - 8.4$

43. $6.9 = 0.3r$

44. $1.2 = \dfrac{t}{9}$

▶ **First choose the equation and solve the problem. Then reread the problem to see if your answer makes sense.** *(page 54)*

> **Equation A:** $n + 5 = 47.5$
> **Equation B:** $n - 5 = 47.5$
> **Equation C:** $5n = 47.5$
> **Equation D:** $\dfrac{n}{5} = 47.5$

45. In one week Bart rode his bike 5 times over the same route. He rode 47.5 miles altogether. How many miles long is the route?

46. Carlin rode his bike 5 fewer miles than Katie rode hers. Carlin rode 47.5 miles. How many miles did Katie ride?

47. Marta biked an average of 47.5 miles daily. She found her average by dividing the number on her bike's odometer by 5. How many miles did she ride during the 5 days?

48. Tony rode his bike 5 more miles than Amy rode hers. Tony rode 47.5 miles. How many miles did Amy ride?

▶ **Choose the correct letter.**

1. An algebraic expression for 8 less than a number t is

   A. $t - 8$

   B. $8 - t$

   C. $t + 8$

   D. none of these

2. Evaluate $(20 - x) + y$ for $x = 6$ and $y = 9$.

   A. 5

   B. 17

   C. 23

   D. none of these

3. 6.4138 rounded to the nearest hundredth is

   A. 6.42

   B. 6.413

   C. 6.4

   D. none of these

4. Solve.
$$g + 0.5 = 1.3$$

   A. 0.8

   B. 1.8

   C. 18

   D. none of these

5. Solve.
$$9 = c - 4$$

   A. 5

   B. 13

   C. 49

   D. none of these

6. An algebraic expression for a number n divided by 3, plus 4 is

   A. $\dfrac{n}{3} + 4$

   B. $\dfrac{n}{7}$

   C. $\dfrac{3}{n} + 4$

   D. none of these

7. Evaluate $(2e) + f$ for $e = 7$ and $f = 3$.

   A. 11

   B. 17

   C. 20

   D. none of these

8. $61.3 \times 100 = \underline{\ ?\ }$

   A. 0.613

   B. 6.13

   C. 6130

   D. none of these

9. Evaluate $\dfrac{r}{t} - v$ for $r = 0.8$, $t = 0.2$, and $v = 0.1$.

   A. 8

   B. 3.5

   C. 3.9

   D. none of these

10. Solve.
$$\dfrac{p}{3} = 1.8$$

   A. 0.6

   B. 4.8

   C. 5.4

   D. none of these

11. Solve.
$$6 = 0.3t$$

   A. 0.2

   B. 1.8

   C. 5.7

   D. none of these

12. Choose the equation. Eileen earned $29.25 for working 9 hours. How much did she earn per hour?

   A. $n + 9 = 29.25$

   B. $n - 9 = 29.25$

   C. $9n = 29.25$

   D. $\dfrac{n}{9} = 29.25$

$c = 75d + 0.35m$

Cost equals $75 a day plus 35¢ a mile.
(page 86)

Vacation travelers sometimes fly to a major city near their final destination. There they rent a car, van, or truck to enjoy nearby points of interest.

61

OBJECTIVE: To apply the rules for order of operations in simplifying expressions.

1. How many baseball decals are there on the three sheets?

2. The number of baseball decals is given by the expression written in red.
   a. If you add first and then multiply, do you get the number of decals?
   b. If you multiply first and then add, do you get the number of decals?

3. Does the expression have a different value when you do the operations in a different order?

$3 + 2 \times 6$

EXAMPLES | **Here's how to simplify expressions having more than one kind of operation:**

So that an expression has only one value, we use these rules for the order of operations:

Rule 1. First, do the operation(s) within the grouping symbols, ( ).

Rule 2. Next, work from left to right doing any multiplication and division.

Rule 3. Last, work from left to right doing addition and subtraction.

A
$$
\begin{array}{c}
\text{3rd } \text{2nd } \text{1st} \\
8 + 3 \times (4 - 1) \\
8 + 3 \times \quad 3 \\
8 + 9 \\
17
\end{array}
$$

B
$$
\begin{array}{c}
\text{1st } \text{2nd } \text{3rd} \\
(8 + 3) \times 4 - 1 \\
11 \quad \times 4 - 1 \\
44 \quad - 1 \\
43
\end{array}
$$

C
$$
\begin{array}{c}
\text{2nd } \text{1st } \text{3rd} \\
8 + 3 \times 4 - 1 \\
8 + \quad 12 \quad - 1 \\
20 \quad - 1 \\
19
\end{array}
$$

CHECK for Understanding

4. In Example A, why is the subtraction done first?

5. Complete these examples.

a. $8 + 12 \div 4$
   $8 + \underline{\ ?\ }$
   $\underline{\ ?\ }$

b. $0.8 - 0.1 \times 6$
   $0.8 - \underline{\ ?\ }$
   $\underline{\ ?\ }$

c. $1.5 \times .7 + 2.6$
   $\underline{\ ?\ } + 2.6$
   $\underline{\ ?\ }$

EXERCISES

▶ **Simplify each expression.**
Hint: The order of operations is shown by the numbered arrows.

$$\overset{\text{2nd}}{\downarrow}\ \overset{\text{1st}}{\downarrow}$$
6. $9 + 3 \times 5$

$$\overset{\text{1st}}{\downarrow}\ \overset{\text{3rd}}{\downarrow}\ \overset{\text{2nd}}{\downarrow}$$
7. $32 \div 4 + 3 \times 10$

$$\overset{\text{2nd}}{\downarrow}\ \overset{\text{3rd}}{\downarrow}\ \overset{\text{1st}}{\downarrow}$$
8. $8 + 6 - 4 \times 3$

$$\overset{\text{2nd}}{\downarrow}\ \overset{\text{1st}}{\downarrow}$$
9. $0.7 \times (0.8 + 1.2)$

$$\overset{\text{1st}}{\downarrow}\ \overset{\text{2nd}}{\downarrow}$$
10. $(1.5 - 0.9) \div 0.5$

$$\overset{\text{2nd}}{\downarrow}\ \overset{\text{1st}}{\downarrow}$$
11. $3.6 - (1.2 - 0.8)$

$$\overset{\text{1st}}{\downarrow}\ \overset{\text{3rd}}{\downarrow}\ \overset{\text{2nd}}{\downarrow}$$
12. $(3 + 6) \times (12 - 9)$

$$\overset{\text{1st}}{\downarrow}\ \overset{\text{2nd}}{\downarrow}\ \overset{\text{3rd}}{\downarrow}$$
13. $(3 + 6) \times 12 - 9$

$$\overset{\text{2nd}}{\downarrow}\ \overset{\text{1st}}{\downarrow}\ \overset{\text{3rd}}{\downarrow}$$
14. $3 + 6 \times 12 - 9$

15. $33 - 8 - 6$

16. $16 + 30 \div 5$

17. $18 - 6 \times 3$

18. $(75 + 25) \div 10$

19. $96 \times (15 - 13)$

20. $20 \times (30 - 25)$

21. $48 + 8 \div 4 + 4$

22. $(24 + 8) \div 4 + 4$

23. $24 + 8 \div (4 + 4)$

24. $24 + 12 \times 6 - 1$

25. $(24 + 12) \times 6 - 1$

26. $(24 + 12) \times (6 - 1)$

▶ **Evaluate each algebraic expression for $m = 4$, $n = 6$, $r = 0.8$, and $t = 0.2$.**
Here are scrambled answers for the next row of exercises: 16 41 17.6

27. $4m + 2r$

28. $7n - 5t$

29. $5(m - r)$

30. $(t + r) \div (n + m)$

31. $9(n + 3)$

32. $(r - t) \div n$

33. $r \div (m + n) + t$

34. $r + t + m \div t$

35. $m + rt$

36. $5(n + m) + t$

37. $t(m + r)$

38. $m + n \div (r - t)$

Problem Solving CALCULATOR APPLICATION

Ron and Karen computed $9 + 3 \times 15$ by pressing these keys on their calculators:

$$9\ \boxed{+}\ 3\ \boxed{\times}\ 15\ \boxed{=}$$

Ron should have used this key sequence to get the right answer:

$$3\ \boxed{\times}\ 15\ \boxed{+}\ 9\ \boxed{=}$$

39. What key sequence should Ron use to simplify each of the following expressions?

   a. $16 + 8 \div 4$

   b. $20 + 4 \times 2 - 9$

   c. $22.1 - 5.3 \times 3.14$

Ron's simple
4-function
calculator

Karen's
scientific
calculator

3-2 Properties of Addition

OBJECTIVE: To recognize and use the basic properties of addition.

1. Choose values for *a* and *b*. Substitute your values in these expressions and simplify. Are your answers the same? $a + b$ $b + a$

2. Choose values for *a*, *b*, and *c*. Substitute your values in these expressions and simplify. Are your answers the same? $(a + b) + c$ $a + (b + c)$

3. Choose a value for *a*. Substitute your value in these expressions and simplify. Are your answers the same? $a + 0$ a

---

PROPERTIES OF ADDITION

COMMUTATIVE PROPERTY OF ADDITION
For any values of *a* and *b*,
$$a + b = b + a$$
    Changing the order of the addends does not change the sum.

ASSOCIATIVE PROPERTY OF ADDITION
For any values of *a*, *b*, and *c*,
$$(a + b) + c = a + (b + c)$$
    Changing the grouping of the addends does not change the sum.

ADDING 0 PROPERTY
For any value of *a*,
$$a + 0 = a$$
    The sum of any number and 0 is that number.

CHECK for Understanding

4. Complete these examples of the addition properties.

   a. $x + 0 = \underline{\ ?\ }$
   b. $r + 7 = \underline{\ ?\ } + r$
   c. $w + \underline{\ ?\ } = w$
   d. $(x + y) + z = x + (y + \underline{\ ?\ })$
   e. $(k + 4) + \underline{\ ?\ } = k + (4 + r)$

EXAMPLES	Here's how to use the properties of addition when doing mental math.

A $13 + 78 + 7$ ←	*13 and 7 are easy to add.*	**B** $3.7 + 2.2 + 5.4 + 6.3$ ←	*3.7 and 6.3 are an easy sum.*
$(13 + 7) + 78$ ←	*Rearrange the numbers using the commutative and associative properties.*	$(3.7 + 6.3) + (2.2 + 5.4)$ ←	*Rearrange the numbers using the commutative and associative properties.*
$20 + 78$		$10 + 7.6$	
98		17.6	

EXERCISES

▶ Use your mental math skills. The commutative and associative properties of addition allow you to add numbers in any order. Find these sums.

5. $12 + 39 + 8$

6. $15 + 52 + 35$

7. $11 + 29 + 37$

8. $59 + 47 + 23$

9. $25 + 68 + 75$

10. $46 + 70 + 30$

11. $19 + 9 + 11 + 6$

12. $28 + 9 + 12 + 7$

13. $50 + 16 + 50 + 6$

14. $11 + 36 + 14 + 23$

15. $29 + 56 + 44 + 12$

16. $50 + 30 + 70 + 36$

17. $45 + 27 + 55 + 7$

18. $85 + 53 + 15 + 20$

19. $29 + 26 + 18 + 74$

20. $60 + 38 + 22 + 40$

21. $25 + 34 + 75 + 30$

22. $10 + 65 + 90 + 25$

23. $3.6 + 1.8 + 8.2 + 2.1$

24. $4.8 + 2.9 + 7.1 + 3.5$

25. $7.7 + 5.2 + 2.3 + 1.9$

26. $0.70 + 0.25 + 0.80 + 0.20$

27. $0.64 + 0.75 + 0.36 + 0.10$

28. $0.56 + 0.25 + 0.75 + 0.18$

Problem Solving USING MENTAL MATH

▶ Use your mental math skills. Evaluate each algebraic expression to get the total cost.

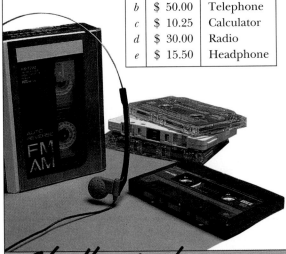

a	$150.00	Stereo
b	$ 50.00	Telephone
c	$ 10.25	Calculator
d	$ 30.00	Radio
e	$ 15.50	Headphone

29. $a + c$

30. $b + d$

31. $a + b$

32. $b + c$

33. $c + e$

34. $d + e$

35. $c + c$

36. $d + d$

37. $a + b + c$

38. $b + c + d$

39. $c + d + e$

40. $a + c + d$

41. $b + d + e$

42. $b + c + e$

43. $a + d + e$

44. $a + b + d$

45. $a + a + b$

46. $a + b + b$

47. $b + c + c$

48. $b + d + d$

49. $a + e + e$

50. $e + e + e$

Challenge! GUESS AND CHECK

▶ Solve.

51. Use the clues to find the cost of a camera.

Clues:
- The camera costs more than $80 but less than $100.
- You can buy it with the same number of $10 bills, $5 bills, and $1 bills.

OBJECTIVE: To recognize and use the basic properties of multiplication.

1. Choose values for x and y. Substitute your values in these expressions and simplify. Are your answers the same? xy yx

2. Choose values for x, y, and z. Substitute your values in these expressions and simplify. Are your answers the same? $(xy)z$ $x(yz)$

3. Choose a value for x. Substitute your value in these expressions and simplify. Are your answers the same? $x \times 1$ x

PROPERTIES OF MULTIPLICATION

COMMUTATIVE PROPERTY OF MULTIPLICATION
For any values of x and y,
$$xy = yx$$

Changing the order of the factors does not change the product.

ASSOCIATIVE PROPERTY OF MULTIPLICATION
For any values of x, y, and z,
$$(xy)z = x(yz)$$

Changing the grouping of the factors does not change the product.

MULTIPLYING BY 1 PROPERTY
For any value of x,
$$x \times 1 = x$$

The product of any number and 1 is that number.

CHECK for Understanding

4. Complete these examples of the multiplication properties.

 a. $n \times 1 = \underline{\ ?\ }$ **b.** $ab = b\underline{\ ?\ }$ **c.** $(rs)t = r(s\underline{\ ?\ })$ **d.** $(a\underline{\ ?\ })c = a(bc)$

 e. $\underline{\ ?\ }r = r(3)$ **f.** $\underline{\ ?\ } \times 1 = d$ **g.** $(\underline{\ ?\ }t)w = r(tw)$ **h.** $(5a)x = \underline{\ ?\ }(ax)$

EXAMPLES   **Here's how to use the properties of multiplication when doing mental math.**

A $5 \times 23 \times 6$ ←   *5×6 is an easy multiplication.*

$(5 \times 6) \times 23$ ←
$30 \quad \times 23$
690

Rearrange the numbers using the commutative and associative properties.

B $20 \times 7 \times 9 \times 5$ ←   *20×5 is an easy multiplication.*

$(20 \times 5) \times (7 \times 9)$ ←
$100 \quad \times \quad 63$
6300

Rearrange the numbers using the commutative and associative properties.

EXERCISES

▶ **Use your mental math skills. The commutative and associative properties of multiplication allow you to multiply numbers in any order. Find these products.**

5. $2 \times 16 \times 5$

6. $5 \times 22 \times 4$

7. $8 \times 7 \times 5$

8. $5 \times 18 \times 20$

9. $10 \times 17 \times 10$

10. $18 \times 25 \times 4$

11. $2 \times 8 \times 5 \times 9$

12. $4 \times 3 \times 6 \times 5$

13. $8 \times 10 \times 6 \times 10$

14. $9 \times 20 \times 8 \times 5$

15. $10 \times 7 \times 9 \times 10$

16. $4 \times 5 \times 25 \times 11$

17. $7 \times 10 \times 7 \times 10$

18. $11 \times 5 \times 8 \times 20$

19. $25 \times 9 \times 4 \times 9$

20. $2 \times 8 \times 0.5 \times 4$

21. $3 \times 0.25 \times 4 \times 12$

22. $0.2 \times 4 \times 15 \times 5$

▶ **Match each property with its example.**

23. Commutative property of addition

24. Associative property of addition

25. Adding 0 property

26. Commutative property of multiplication

27. Associative property of multiplication

28. Multiplying by 1 property

a. $26 + 0 = 26$

b. $18t = t(18)$

c. $67 \times 1 = 67$

d. $38 + 9 = 9 + 38$

e. $(y + 47) + 13 = y + (47 + 13)$

f. $(35 \times 18) \times 16 = 35 \times (18 \times 16)$

▶ **Complete these examples of the properties.**

29. $(9 + 18) + 27 = 9 + (\underline{\ ?\ } + 27)$

30. $26 \times 18 = \underline{\ ?\ } \times 26$

31. $73 \times 1 = \underline{\ ?\ }$

32. $(51 \times 8) \times 10 = \underline{\ ?\ } \times (8 \times 10)$

33. $y + 0 = \underline{\ ?\ }$

34. $12 + 74 = 74 + \underline{\ ?\ }$

35. $a(\underline{\ ?\ }) = 7a$

36. $(\underline{\ ?\ } + 16) + 37 = 45 + (16 + 37)$

37. $\underline{\ ?\ } + 84 = 84 + 27$

38. $\underline{\ ?\ } + 0 = 88$

39. $(6 \times \underline{\ ?\ })r = 6(4r)$

40. $\underline{\ ?\ } \times 1 = n$

Challenge! LOGICAL REASONING

▶ **Solve.**

41. Alexa, Brent, Sarah, and Dick collected a total of 14 souvenir hats. Each had more than one hat. Sarah had the most. Alexa had more than Dick. Brent had the fewest. How many souvenir hats did each person have?

3-4 Solving Equations Using the Commutative Properties

OBJECTIVE: To use the commutative properties to solve equations.

1. How many miles is it from Klamath Falls to Crater Lake?

2. Decide whether Equation A or Equation B would be used to solve the following problem.

> **Equation A:** $54 = n(18)$
> **Equation B:** $54 = 18 + n$

You are driving from Klamath Falls to Crater Lake. You have driven 18 miles. How many more miles is it to Crater Lake?

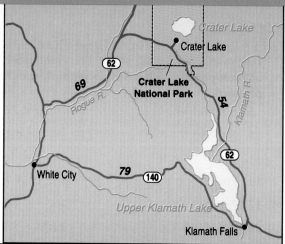

EXAMPLES | **Here's how to use the commutative properties to solve equations:**

A

$$54 = 18 + n$$
$$54 = n + 18$$
$$n + 18 = 54$$
$$n + 18 - 18 = 54 - 18$$
$$n = 36$$

Commutative property of addition

✔ **CHECK**

$$54 = 18 + n$$
$$54 \stackrel{?}{=} 18 + 36$$
$$54 = 54$$

It checks!

B

$$54 = n(18)$$
$$54 = 18n$$
$$18n = 54$$
$$\frac{18n}{18} = \frac{54}{18}$$
$$n = 3$$

Commutative property of multiplication

✔ **CHECK**

$$54 = n(18)$$
$$54 \stackrel{?}{=} 3(18)$$
$$54 = 54$$

It checks!

C

$$14 + r = 43$$
$$r + 14 = 43$$
$$r + 14 - 14 = 43 - 14$$
$$r = 29$$

D

$$t(11) = 132$$
$$11t = 132$$
$$\frac{11t}{11} = \frac{132}{11}$$
$$t = 12$$

E

$$5.4 = 1.7 + s$$
$$5.4 = s + 1.7$$
$$s + 1.7 = 5.4$$
$$s + 1.7 - 1.7 = 5.4 - 1.7$$
$$s = 3.7$$

CHECK for Understanding

3. Complete these examples.

a.
$$204 = 17v$$
$$\underline{?}\ v = 204$$
$$\frac{17v}{17} = \frac{204}{17}$$
$$v = 12$$

b.
$$26 + u = 75$$
$$u + 26 = 75$$
$$u + 26 - 26 = 75 - \underline{?}$$
$$u = 49$$

c.
$$w(0.2) = 2.8$$
$$0.2w = 2.8$$
$$\frac{0.2w}{0.2} = \frac{2.8}{?}$$
$$w = 14$$

4. Explain how to get the numbers that are printed in red in Examples C, D, and E.

EXERCISES

▶ **Solve and check.**
Here are scrambled answers for the next row of exercises: 8   37   26   11

5. $16 + f = 53$ **6.** $k(15) = 165$ **7.** $160 = z(20)$ **8.** $45 = 19 + a$

9. $220 = w(11)$ **10.** $25 + g = 74$ **11.** $63 = 27 + b$ **12.** $m(25) = 350$

13. $55 = 39 + x$ **14.** $9j = 108$ **15.** $w(9) = 144$ **16.** $19 = v - 12$

17. $288 = t(18)$ **18.** $8 = \dfrac{f}{16}$ **19.** $51 = t + 16$ **20.** $38 = j(4)$

21. $f - 38 = 38$ **22.** $52 = 48 + c$ **23.** $h(8) = 15$ **24.** $12k = 132$

25. $\dfrac{p}{6} = 20$ **26.** $12.3 + h = 41$ **27.** $37 = w - 29$ **28.** $20 = 6.4 + d$

29. $35 = t + 35$ **30.** $144 = 16b$ **31.** $b + 19 = 100$ **32.** $34 = 8c$

33. $17 = x - 6.8$ **34.** $5m = 42$ **35.** $22 = \dfrac{e}{8}$ **36.** $\dfrac{r}{3} = 4.8$

37. $c + 8.2 = 10$ **38.** $3.6 = \dfrac{h}{10}$ **39.** $g - 5.6 = 12$ **40.** $0.94 = y - 0.68$

41. $18.5 = 9.2 + e$ **42.** $8 = t + 6.5$ **43.** $p(5) = 67$ **44.** $1.75 + j = 2.65$

Problem Solving USING A FORMULA

▶ **Complete the chart using the formula.**
The distance (d) that you travel is equal to
the rate (r) multiplied by the time (t).

Formula: $d = rt$
distance   time   rate

★**45.**

	d (miles)	r (mph)	t (hours)
a.	?	43	4
b.	?	52	6
c.	270	54	?
d.	147	49	?
e.	204	?	4

Challenge! NUMBER SENSE

46. You are driving at the rate of 50 miles
per hour. You are 130 miles from
Klamath Falls. The time is shown on
the clock. Should you arrive at
Klamath Falls by 5 o'clock?

You will use a variety of problem-solving strategies on these two pages. For some problems, you may wish to use a calculator.

1 MAKE A TABLE

What are the ages of Jill and her grandfather?

> When I was 8, my grandfather was 68. He is now 5 times as old as I am.

✓ Problem Solving Tips

Make a Table.* Fill in the ages.

Jill's age	8	9	10	
Grandfather's age	68	69	70	71

*At least one more of the problems on these two pages can be solved by making a table.

2 GEOMETRY—VISUAL THINKING

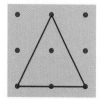

a. How many triangles like the one shown can be drawn on a 3-by-3 dot grid? The triangles must be the same size and shape.

b. How many different squares can you draw on a 3-by-3 dot grid?

3 GUESS AND CHECK

Which three gold nuggets have a combined weight of 239.9 grams?

77.76 grams

79.37 grams

82.77 grams

88.77 grams

4 PATTERNS

Look for the pattern. Then find three consecutive whole numbers whose sum is 1068.

$$14 + 15 + 16 = 45$$
$$21 + 22 + 23 = 66$$
$$29 + 30 + 31 = 90$$
$$\vdots$$
$$a + b + c = 1068$$

5 RELATIONSHIPS

Juan has twice as many brothers as sisters. His sister Maria has five times as many brothers as sisters. How many children are in the family?

6 RATE INCREASE

How many hours will a car traveling 45 mph take to catch up with another car traveling 30 mph, if the slower car starts one hour before the faster car?

7 ALPHAMETICS

In this division example, each different letter stands for a different digit. What digit does each letter stand for?

```
        G Y M
PE ) T I M E
        P E
        G G M
        G A A
        G M E
        G M E
```

$G \times PE = PE$
So $G = ?$

$M \times PE = GME$
What does that tell you about $M \times E$?

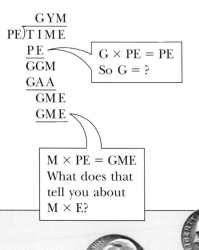

APPLICATIONS

The average number of coins minted by the U.S. Treasury during a 24-hour period:

25,616,438 pennies
1,917,808 nickels
3,479,452 dimes
3,561,644 quarters
2,192 fifty-cent pieces

Complete.
You may wish to use a calculator.

8 During a 5-day period, 17,397,260 __?__ are minted.

9 The value of the __?__ minted each day is $890,411.

10 The Treasury mints an average of __?__ coins per second.

Rate Your Problem-Solving Power: *6–7 correct = Good; 8–9 correct = Excellent; 10 correct = Exceptional*

OBJECTIVE: To use a graph for solving problems.

In an emergency, a driver might have to stop quickly. The braking distance of a car is the distance traveled after the brakes are applied and until the car comes to a stop. This distance depends on the speed of the car, the type of road surface, and the weather conditions.

The graph shows the braking distances for cars on wet and dry concrete roads at different speeds.

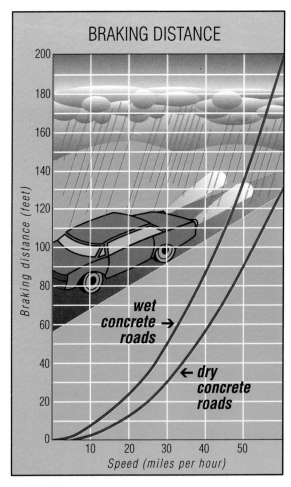

▶ **Use the graph to choose the correct answer.**

1. Suppose you are driving on dry concrete. About how many feet will your braking distance be if you are traveling 40 miles per hour?
   a. 30   **b.** 55   **c.** 90

2. About how many miles per hour was a car traveling on wet concrete if its braking distance was 110 feet?
   a. 20   **b.** 35   **c.** 45

3. What is the difference in braking distance between cars traveling 50 miles per hour on wet and dry concrete?
   a. about 10 feet
   b. about 45 feet
   c. about 100 feet

4. Compare the braking distances on dry concrete for speeds of 20 and 40 miles per hour. When the speed is doubled,
   a. the braking distance is the same.
   b. the braking distance is doubled.
   c. the braking distance is more than doubled.

▶ **Write an equation and solve the problem.**

5. The braking distance on wet asphalt is 1.32 times the braking distance on dry asphalt. When one is driving on wet asphalt at 40 miles per hour, the braking distance is 82.038 feet. What is the braking distance on dry asphalt at 40 miles per hour?
   (Let d = the braking distance in feet at 40 miles per hour on dry asphalt.)

6. The braking distance on dry concrete is half the braking distance on packed snow. When one is driving on dry concrete at 30 miles per hour, the braking distance is 30.3 feet. What is the braking distance at 30 miles per hour on packed snow?
   (Let s = the braking distance in feet at 30 miles per hour on packed snow.)

Cumulative Algebra Practice

▶ **Round to the nearest tenth.** *(page 6)*

1. `4.83` 2. `21.805` 3. `9.48` 4. `0.661` 5. `8.147`

6. `6.380` 7. `29.172` 8. `31.06` 9. `5.456` 10. `19.1308`

11. `406.29` 12. `823.74` 13. `0.059` 14. `46.032` 15. `34.296`

▶ **Evaluate each expression for $a = 13$, $b = 16.7$, $c = 1.4$, and $d = 20$.** *(page 10)*

16. $d + b$ 17. $a - c$ 18. $d + c$ 19. $b - c$

20. $b + 0.8$ 21. $2.1 - c$ 22. $c + b + a$ 23. $(b - a) - c$

24. $(d - c) + a$ 25. $d + a - b$ 26. $(a - 8.1) - c$ 27. $(d - 7) + a$

▶ **Solve and check.** *(page 20)*

28. $13 = q - 8$ 29. $b + 12 = 21$ 30. $r - 17 = 36$ 31. $5 = a + 5$

32. $t - 4.7 = 12$ 33. $9 = d + 1.8$ 34. $19.3 = v - 4$ 35. $n + 6 = 10.9$

36. $y + 2.6 = 3.01$ 37. $w - 9.7 = 16.8$ 38. $7.1 = c + 0.42$ 39. $6.4 = x - 6.4$

▶ **Four of the calculator answers are wrong. Find them by estimating.** *(page 38)*

40. a. $81.2 \div 4$ `2.3` b. $68.4 \div 9$ `7.6` c. $7.29 \div 3$ `2.43`

d. $815.6 \div 2$ `47.8` e. $32.7 \div 6$ `5.45` f. $250.6 \div 7$ `35.8`

g. $7.456 \div 8$ `9.32` h. $64.3 \div 5$ `128.6` i. $91.5 \div 6$ `15.25`

▶ **Solve and check.** *(page 52)*

41. $84 = 6x$ 42. $15 = \dfrac{w}{8}$ 43. $107 = y + 74$ 44. $z - 38 = 0$

45. $t + 83 = 100$ 46. $\dfrac{v}{16} = 3$ 47. $r - 46 = 47$ 48. $5u = 90$

49. $4 = \dfrac{k}{1.8}$ 50. $7p = 8.4$ 51. $3.7 = j - 5.4$ 52. $1.2y = 9$

▶ **Evaluate each expression for $r = 4.1$, $s = 7$, $t = 0.9$, and $u = 5$.** *(page 62)*

53. $u + st$ 54. $2tu$ 55. $t + ru$ 56. $s + us$

57. $s - (t + u)$ 58. $t + r \div u$ 59. $st - u$ 60. $s - ts$

61. $s \div u + r$ 62. $u - (t + r)$ 63. $s - (r - t)$ 64. $r \div (2u)$

65. $t \div u + s$ 66. $s - u \div 2$ 67. $s \div (r + t)$

OBJECTIVE: To solve two-step equations.

1. What is the cost of 1 book (not including shipping and handling)?

2. How much must be added for shipping and handling?

3. This equation can be used to find the number of books you can get for $50:

$$\overset{\text{cost of } 1 \text{ book}}{\underset{\text{number of books}}{\downarrow}} \overset{\text{shipping and handling}}{\underset{\text{total cost}}{\downarrow}}$$

$$16b + 2 = 50$$

Look at the equation. What does b represent?

Book Offer !

Get current books for only $16 each.
For each order, add $2 for shipping and handling.

E X A M P L E | **Here's how to solve a two-step equation.**

When solving an equation, you want to get the variable by itself on one side of the equation. You do this by undoing all the operations that were done on the variable. To solve the equation $16b + 2 = 50$, two steps are needed.

Equation: $16b + 2 = 50$

Step 1. To get $16b$ by itself, subtract 2 from both sides. Then simplify both sides.

$$16b + 2 - 2 = 50 - 2$$
$$16b = 48$$
$$\frac{16b}{16} = \frac{48}{16}$$

Step 2. To get b by itself, divide both sides by 16. Then simplify both sides.

$$b = 3$$

✔ **CHECK**

$$16b + 2 = 50$$
$$16(3) + 2 \overset{?}{=} 50$$
$$50 = 50$$

It checks!

CHECK for Understanding

4. Look at the example above. To find b first subtract __?__ from both sides of the equation and then divide both sides by __?__ .

5. Check the solution. How many books can you get for $50?

EXERCISES

▶ **Copy and finish solving each equation. Check your solution.**

6. $3y + 6 = 30$
$3y + 6 - 6 = 30 - \underline{\ ?\ }$
$3y = \underline{\ ?\ }$
$\dfrac{3y}{3} = \dfrac{24}{?}$
$y = \underline{\ ?\ }$

7. $5t - 4 = 26$
$5t - 4 + \underline{\ ?\ } = 26 + 4$
$5t = \underline{\ ?\ }$
$\dfrac{5t}{?} = \dfrac{30}{5}$
$t = ?$

8. $2.9 = 7r - 0.6$
$7r - 0.6 = \underline{\ ?\ }$
$7r - 0.6 + 0.6 = 2.9 + \underline{\ ?\ }$
$7r = \underline{\ ?\ }$
$\dfrac{7r}{7} = \dfrac{3.5}{?}$
$r = \underline{\ ?\ }$

▶ **Solve and check.**

Here are scrambled answers for the next two rows of exercises:

6   7   2   5   11   4   9   17

9. $4t + 8 = 28$

10. $2n + 5 = 39$

11. $3m - 2 = 25$

12. $27 = 4y + 3$

13. $6b - 3 = 39$

14. $20 = 6k - 4$

15. $32 = 2n + 10$

16. $8c - 2 = 14$

17. $5r - 6 = 29$

18. $22 = 3r + 10$

19. $9y + 10 = 19$

20. $4r + 13 = 41$

21. $6n - 19 = 53$

22. $77 = 10a + 7$

23. $4k - 42 = 18$

24. $5t - 35 = 0$

25. $50 = 5c - 10$

26. $3d + 14 = 35$

27. $19d - 1 = 56$

28. $85 = 15r + 10$

29. $3n + 6.2 = 9.2$

30. $5r + 6.3 = 8.8$

31. $4t - 1.8 = 10.6$

32. $9d - 3.1 = 14.9$

33. $10.2 = 6t - 1.2$

34. $22.3 = 4y + 1.5$

35. $6s - 5.1 = 18.9$

36. $12a + 6.2 = 42.2$

37. $3t + 9 = 21.6$

38. $5x - 2.2 = 10.3$

39. $2r + 23.5 = 53.7$

40. $7t + 6.1 = 9.6$

Problem Solving   USING A FORMULA

▶ **Substitute in the formula to complete this chart.**

You can use this formula to find the cost of buying books by mail.

Formula: $16\,b + s = c$

cost of 1 book   shipping and handling charge   total cost

number of books → $16\,b$

★41.

	b (books)	s (dollars)	c (dollars)
a.	3	1.85	?
b.	4	2.35	?
c.	?	3.45	99.45
d.	?	2.90	82.90
e.	8	?	132.20
f.	7	?	115.10
g.	9	4.80	?
h.	10	4.90	?
i.	11	?	181.15
j.	?	5.45	213.45

Solving Equations Using Number Properties | **75**

OBJECTIVE: To solve two-step equations.

If you divide Michael Jackson's age when he first sang in public by 7 and then add 19, you get 20. How old was he then?

You can find how old Michael Jackson was when he first sang in public by writing and solving an equation:

$$\frac{n}{7} + 19 = 20$$

E X A M P L E S | **Here's how to solve a two-step equation.**

A **Equation:** $\frac{n}{7} + 19 = 20$

Step 1. To get $\frac{n}{7}$ by itself, subtract 19 from both sides. Then simplify both sides.

$$\frac{n}{7} + 19 - 19 = 20 - 19$$

$$\frac{n}{7} = 1$$

Step 2. To get n by itself, multiply both sides by 7. Then simplify both sides.

$$7 \times \frac{n}{7} = 7 \times 1$$

$$n = 7$$

✔ **CHECK**

$$\frac{n}{7} + 19 = 20$$

$$\frac{7}{7} + 19 \overset{?}{=} 20$$

$$1 + 19 \overset{?}{=} 20$$

$$20 = 20$$

It checks!

B

$$2.5 = \frac{t}{2} - 0.9$$

$$\frac{t}{2} - 0.9 = 2.5$$

$$\frac{t}{2} - 0.9 + 0.9 = 2.5 + 0.9$$

$$\frac{t}{2} = 3.4$$

$$2 \times \frac{t}{2} = 2 \times 3.4$$

$$t = 6.8$$

C

$$4.5 = 0.3y + 1.5$$

$$0.3y + 1.5 = 4.5$$

$$0.3y + 1.5 - 1.5 = 4.5 - 1.5$$

$$0.3y = 3.0$$

$$\frac{0.3y}{0.3} = \frac{3.0}{0.3}$$

$$y = 10$$

CHECK for Understanding

1. a. Look at Example A. To find n, first subtract __?__ from both sides of the equation and then multiply both sides by __?__.

  b. How old was Michael Jackson when he sang for the first time in public? Is this answer reasonable?

2. Explain how to get the numbers that are printed in red in Examples B and C.

EXERCISES

> **Solve and check.**
> *Here are scrambled answers for the next two rows of exercises:*
> 2   0.4   7   21   10   27   6   3.2

3. $\dfrac{n}{7} + 5 = 8$ **4.** $\dfrac{t}{9} - 2 = 1$ **5.** $10 = \dfrac{c}{0.4} + 2$ **6.** $4 + \dfrac{r}{2} = 9$

7. $5n + 0.3 = 2.3$ **8.** $3n - 4 = 14$ **9.** $14 = 8y - 2$ **10.** $3m + 2 = 23$

11. $\dfrac{r}{7} - 50 = 20$ **12.** $\dfrac{k}{9} + 5 = 6$ **13.** $\dfrac{b}{3} - 0.2 = 1.1$ **14.** $24 = \dfrac{c}{3} + 4$

15. $0.6n - 1 = 17$ **16.** $31 = 5t + 6$ **17.** $12g - 3 = 33$ **18.** $4y + 7 = 51$

Problem Solving USING EQUATIONS

Equation A:	$\dfrac{n}{4} + 2 = 10$
Equation B:	$\dfrac{n}{4} - 2 = 10$
Equation C:	$4n + 2 = 10$
Equation D:	$4n - 2 = 10$

> **Decide which equation would be used to solve each problem. Then solve the problem.**

19. If you divide Paula's age by 4 and then add 2, you get 10. What is Paula's age?

20. If you multiply Jon's age by 4 and then subtract 2, you get 10. What is Jon's age?

21. Multiply Bruce's age by 4, then add 2, and you get 10. What is Bruce's age?

22. Write a problem that you would solve using Equation B.

Group Project ANALYZING DATA   *Happy Birthday!*

> **Work in a small group.**

23. a. Determine your age in months rounded to the nearest month.
b. Compute your age to the nearest tenth of a year.

24. a. List the ages of all class members.
b. Order the ages from least to greatest.

25. Compute the average age of the class members. Round to the nearest tenth of a year.

26. a. Are you older or younger than the average age?
b. Were more class members older or younger than the average age?

27. Write a paragraph that summarizes the data.

3-8 The Distributive Property

OBJECTIVE: To recognize and use the distributive property of multiplication over addition.

1. Choose values for *a*, *b*, and *c*. Substitute your values in these expressions and simplify.

$$a(b + c) \qquad ab + ac$$

Were your answers the same?

The next property links multiplication and division.

DISTRIBUTIVE PROPERTY

For any values of *a*, *b*, and *c*,
$$a(b + c) = ab + ac$$

CHECK for Understanding

2. Complete these examples of the distributive property.

   a. $4(2 + 6) = 4(2) + 4(\underline{\,?\,})$
   b. $3(\underline{\,?\,} + 8) = 3n + 3(8)$
   c. $9(7) + 9(2) = 9(7 + \underline{\,?\,})$
   d. $y(5) + \underline{\,?\,}(4) = y(5 + 4)$

EXAMPLES	Here's how to use the distributive property to compute mentally.

A Compute 6(17) mentally.

$$6(17) = 6(10 + 7)$$
$$= 6(10) + 6(7)$$
$$= 60 + 42$$
$$= 102$$

Since it is easier to multiply by 10, think of 17 as 10 + 7.

B Compute 8(75) + 8(25) mentally.

$$8(75) + 8(25) = 8(75 + 25)$$
$$= 8(100)$$
$$= 800$$

Notice that 8 is a common factor. Since it is easier to multiply by 100, add first and then multiply.

CHECK for Understanding

3. Look at the examples above.
   a. To compute 6(17) mentally, first think of 17 as 10 + _?_ , then multiply both 10 and 7 by _?_ . Next, add 60 and _?_ to get 102.
   b. In the second example, first use the distributive property to restate 8(75) + 8(25) as _?_ (75 + 25). Next, add 75 and _?_ to get _?_ , then multiply 100 by _?_ to get 800.

EXERCISES

▶ **Complete these examples of the distributive property.**

4. $2(5 + 4) = 2(5) + 2(\underline{\ ?\ })$

5. $8(t + 7) = 8(\underline{\ ?\ }) + 8(7)$

6. $12(3) + 12(7) = 12(3 + \underline{\ ?\ })$

7. $11(6) + 11(7) = 11(\underline{\ ?\ } + 7)$

8. $6(2 + d) = \underline{\ ?\ }(2) + 6d$

9. $\underline{\ ?\ }(1.7 + 4) = 6(1.7) + 6(4)$

10. $2(9 + 5) = 2(\underline{\ ?\ }) + 2(5)$

11. $15(6 + \underline{\ ?\ }) = 15(6) + 15y$

12. $4(3.5) + 4(6.1) = 4(3.5 + \underline{\ ?\ })$

13. $r(5) + r(4) = r(\underline{\ ?\ } + 4)$

▶ **Compute mentally.**

14. $7(23)$
Hint: $7(20) + 7(3)$

15. $5(16)$

16. $6(14)$

17. $8(31)$

18. $5(103)$

19. $7(202)$

20. $8(406)$

21. $9(204)$

22. $7(8) + 7(2)$
Hint: $7(8 + 2)$

23. $4(6) + 4(4)$

24. $6(2) + 6(8)$

25. $5(7) + 5(3)$

26. $3(80) + 3(20)$

27. $4(70) + 4(30)$

28. $8(90) + 8(10)$

29. $7(40) + 7(60)$

Problem Solving   USING EXPRESSIONS

▶ **Evaluate each expression to get the total cost. Use the distributive property whenever it makes your computation easier.**

30. $2a + 2c$
Hint: $2(\$3.75) + 2(\$2.25)$
$2(\$3.75 + \$2.25)$
$2(\$6.00)$

31. $3b + 3d$

32. $4c + 4d$

33. $2a + 2b$

34. $2a + 2d$

35. $2b + 2c$

36. $2a + b$

37. $a + 2c$

38. $2b + c$

Challenge!   LOGICAL REASONING

39. If you buy two adult tickets on Thursday, you get $2 off on a third adult ticket. How much will it cost a group of 4 adults to go to the aquarium on Thursday?

Aquarium Admissions		
a	$ 3.75	Adults
b	$ 3.25	Students
c	$ 2.25	Children
d	$ 1.75	Senior Citizens

OBJECTIVE: To simplify algebraic expressions by combining like terms.

The distributive property and the multiplying by 1 property can be used to simplify expressions such as

$5n + 4n$ $8m + m$

EXAMPLES | **Here's how to use the properties to simplify expressions.**

A Simplify **5n + 4n.**

$$5n + 4n = (5 + 4)n$$
$$= 9n$$

> *Use the distributive property.*

B Simplify **8m + m.**

$$8m + m = 8m + 1m$$
$$= (8 + 1)m$$
$$= 9m$$

> *Use the multiplying by 1 property.*

> *Use the distributive property.*

CHECK for Understanding

1. Complete these examples.

a. $8n + 3n = (8 + 3)n$
$\quad = \underline{?}$

b. $3r + r = (3 + \underline{?})r$
$\quad = 4r$

c. $6t + 4t = (6 + 4)\underline{?}$
$\quad = 10t$

d. $8k + k = (\underline{?} + 1)k$
$\quad = 9k$

e. $2m + m + 3m = (2 + \underline{?} + 3)m$
$\quad = 6m$

f. $3b + 4b + 5b = (3 + 4 + 5)b$
$\quad = \underline{?}\ b$

EXAMPLES | **Here's how to use a shortcut to simplify expressions.**

To simplify these expressions, combine **like terms** (terms that have the same variable) by adding.

C Simplify **5r + 2 + 4r + 8.**

Like terms

$$5r + 2 + 4r + 8 = 9r + 10$$

Like terms

D Simplify **6t + 5 + t + 1.**

Like terms

$$6t + 5 + t + 1 = 7t + 6$$

Like terms

EXERCISES

▶ **Simplify. Hint: Combine like terms.**

Here are scrambled answers for the next two rows of exercises:
14n   4n   8n   7n   9n   13n

2. $6n + 2n$

3. $3n + 4n + 7n$

4. $8n + 5n$

5. $5n + n + n$

6. $2n + 3n + 4n$

7. $n + n + n + n$

8. $5a + 2a$

9. $7b + 2b + 3b$

10. $8f + 2f + 3f$

11. $8m + 5m + 3m$

12. $3x + 4x + 2x$

13. $4c + 8c + 3c$

14. $7r + 2r + 5r$

15. $8y + 2y + 12$

16. $12c + 4c + 8$

17. $18g + 4 + 10g$

18. $36r + 16 + 5r$

19. $25t + 9t + 20t$

20. $d + d + 3d$

21. $5a + a + a$

22. $8z + 2z + 6 + 4$

23. $6r + 3r + 5r + 1$

24. $7x + 3 + 2 + 2x$

25. $14c + 10c + 2 + 1$

▶ **Use your mental math skills. Evaluate each expression for $v = 20$, $w = 60$, $y = 100$, $z = 200$.**

26. $5v + 2v$

27. $2w + 3w$

28. $7y + 3y$

29. $8z + 2z$

30. $6v + 4v$

31. $9w + w$

32. $3v + 2v + 4v$

33. $9w + w + w$

34. $8y + 2y + 3y$

35. $4z + 3z + 5z$

36. $v + 7v + 3v$

37. $7w + 2w + 5w$

38. $2v + 3v + 4v + v$

39. $w + 2w + 3w + w$

40. $2y + 2y + 3y + y$

41. $9z + 2z + 3z + 3z$

42. $10v + 3v + 2v + 4v$

43. $7y + 2y + 5y + 2y$

Challenge! USING EXPRESSIONS

▶ **Solve. Use the signpost.**

44. Which city is $2n + 10$ miles from Waukegan?

45. Which city is $5n + 13$ miles from Kenosha?

46. Which city is $5n + 13$ miles from Gary?

47. How many miles is it from Chicago to Kenosha?

★**48.** If $n = 12$, how many miles is it from Gary to Waukegan?

Chicago   n miles ▶

Gary   3n + 2 miles ▶

◀ Waukegan   n + 10 miles

◀ Kenosha   2n + 11 miles

OBJECTIVE: To solve equations that have the same variable in more than one term.

Diana bought 6 posters, and Frank bought 4 posters. Each poster cost the same. Together they spent $12.50. How much did each poster cost?

Let p be the cost of 1 poster. Then $6p$ is the cost of 6 posters, and $4p$ is the cost of 4 posters.

$$6p + 4p = 12.50$$

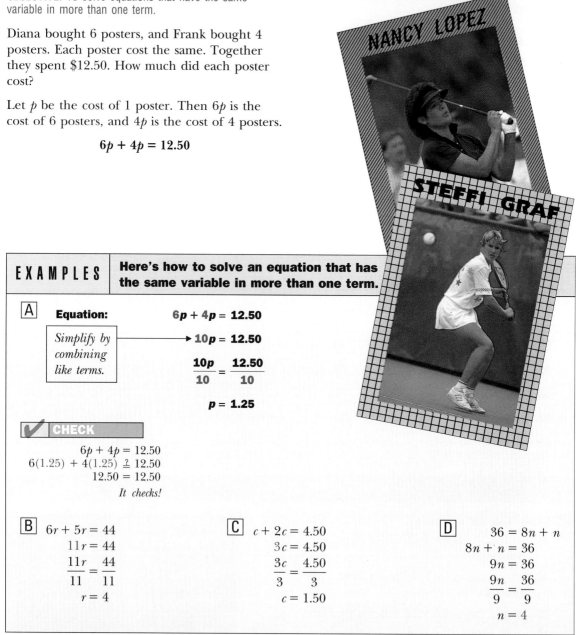

EXAMPLES Here's how to solve an equation that has the same variable in more than one term.

A Equation: $6p + 4p = 12.50$

Simplify by combining like terms. → $10p = 12.50$

$$\frac{10p}{10} = \frac{12.50}{10}$$

$$p = 1.25$$

✔ CHECK

$$6p + 4p = 12.50$$
$$6(1.25) + 4(1.25) \stackrel{?}{=} 12.50$$
$$12.50 = 12.50$$
It checks!

B
$$6r + 5r = 44$$
$$11r = 44$$
$$\frac{11r}{11} = \frac{44}{11}$$
$$r = 4$$

C
$$c + 2c = 4.50$$
$$3c = 4.50$$
$$\frac{3c}{3} = \frac{4.50}{3}$$
$$c = 1.50$$

D
$$36 = 8n + n$$
$$8n + n = 36$$
$$9n = 36$$
$$\frac{9n}{9} = \frac{36}{9}$$
$$n = 4$$

CHECK for Understanding

1. **a.** Look at Example A. To find p, first combine like terms to get __?__ $p = 12.50$ and then divide both sides by __?__ .
   b. How much does each poster cost? Is the answer reasonable?

2. Explain how to get the numbers that are printed in red in Examples B, C, and D.

EXERCISES

▶ **Solve and check.**
Here are scrambled answers for the next row of exercises: 5 16 4 7

3. $5n + 3n = 32$ **4.** $r + 4r = 25$ **5.** $49 = 6t + t$ **6.** $32 = c + c$

7. $6a + a = 140$ **8.** $t + 2t = 39$ **9.** $30 = 8y + 7y$ **10.** $120 = 9n + 6n$

11. $14n + 6n = 240$ **12.** $17w + 5w = 88$ **13.** $63 = 15b + 6b$ **14.** $30 = 12w + 3w$

15. $9n + 3n = 9.6$ **16.** $w + w = 24.6$ **17.** $8.8 = 3r + r$ **18.** $12.04 = 3s + s$

19. $6z + 2z = 1.92$ **20.** $3d + 4d = 2.45$ **21.** $12.6 = 2t + t$ **22.** $2.7 = 5c + c$

▶ **Solve and check.**
Here are scrambled answers for the next row of exercises: 9 2 50 6

23. $\dfrac{n}{5} + 2 = 12$ **24.** $4h + 5 = 29$ **25.** $2 = \dfrac{r}{3} - 1$ **26.** $10 = 7t - 4$

27. $15 = t + 6$ **28.** $\dfrac{d}{2} = 13$ **29.** $2n + 8n = 20$ **30.** $68 = 2f + 18$

31. $\dfrac{a}{7} + 6 = 7$ **32.** $2b - 6 = 8$ **33.** $8 = \dfrac{w}{2} + 2$ **34.** $\dfrac{b}{5} - 2 = 3$

35. $3h + h = 12$ **36.** $10 = \dfrac{c}{3} - 2$ **37.** $5 = 3n - 4$ **38.** $41 = 12n + 5$

39. $2n + 6.2 = 7.4$ **40.** $\dfrac{t}{5} - 3.2 = 4.3$ **41.** $n + 1.6 = 28$ **42.** $b - 6 = 9.3$

43. $4b - 2.2 = 10.3$ **44.** $9.6 = 8f + 6.1$ **45.** $9r + 3.1 = 5.8$ **46.** $8x + 2x = 6.6$

Problem Solving USING EQUATIONS

▶ **Decide which equation would be used to solve each problem. Then solve the problem.**

Equation A:	$4y + 6 = 7.40$
Equation B:	$4y + 6y = 7.40$
Equation C:	$4y = 7.40$
Equation D:	$y + 4 = 7.40$

47. Four posters cost $7.40. Each poster costs the same. How much does each poster cost?

48. Leo bought 4 pages for his photo album, and Jackie bought 6 pages. Each page cost the same. Together they spent $7.40. How much did one page cost?

49. Joanna bought a roll of film and some flashcubes. She spent $4 for the film. Altogether she spent $7.40. How much did she spend for the flashcubes?

50. Write a problem that you would solve using Equation A.

OBJECTIVE: To use equations to solve two-step problems.

In this lesson you will use what you learned about writing simple equations to write two-step equations.

EXAMPLE | **Here's how to use an equation to solve a problem.**

Problem: Arlana wants to buy a pair of roller blades for $98. She already has $30. If she earns $4 an hour as a cashier, how many hours must she work to earn the money she needs?

Step 1. Choose a variable. Use it and the facts to represent the numbers in the problem.

Let h = the number of hours Arlana must work.

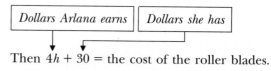

Dollars Arlana earns	Dollars she has

Then $4h + 30$ = the cost of the roller blades.

Step 2. Write an equation based on the facts.

$$4h + 30 = 98$$

These are two equal expressions for the cost of the roller blades.

Step 3. Solve the equation.

$$4h + 30 = 98$$
$$4h + 30 - 30 = 98 - 30$$
$$4h = 68$$
$$\frac{4h}{4} = \frac{68}{4}$$
$$h = 17$$

Arlana needs to work 17 hours to earn the money she needs.

CHECK for Understanding

1. Look at the example above.
   a. In Step 1, we let h equal the number of hours Arlana must work to earn the money she needs. Then the expression $4h + \underline{\ ?\ }$ equals the cost of the roller blades.
   b. In Step 2, the two equal expressions for the cost of the roller blades are $\underline{\ ?\ }$ and 98.
   c. An equation for the problem is $4h + \underline{\ ?\ } = \underline{\ ?\ }$.

2. To check the solution, ask yourself whether the answer fits the facts in the problem. Does Arlana need to work 17 hours to earn the money she needs?

EXERCISES

▶ **Read the facts. Then complete the steps to answer the question.**

3. *Facts:* The cost of a pair of roller blades is $169. That is $5 more than 4 times the cost of a helmet.

Question: What is the cost of a helmet?

a. If you let h equal the cost of the helmet, then $4h + \underline{\ ?\ }$ equals the cost of the roller blades.

b. The roller blades cost $\underline{\ ?\ }$ dollars.

c. Two equal expressions for the cost of the roller blades are $4h + \underline{\ ?\ }$ and 169.

d. To find the cost of the helmet, you can solve the equation $4h + 5 = \underline{\ ?\ }$.

e. Solve the equation in Part d. The cost of the helmet is $\underline{\ ?\ }$ dollars.

4. *Facts:* If you divide the cost of a pair of roller blades by 7 and subtract $15, you get the cost of a pair of knee pads. The knee pads cost $17.

Question: What is the cost of a pair of roller blades?

a. If you let r equal the cost of the roller blades, then $\frac{r}{?} - 15$ equals the cost of the knee pads.

b. The knee pads cost $\underline{\ ?\ }$ dollars.

c. Two equal expressions for the cost of the knee pads are $\frac{r}{?} - 15$ and 17.

d. To find the cost of the roller blades, you can solve the equation $\frac{r}{7} - 15 = \underline{\ ?\ }$.

e. Solve the equation in Part d. The cost of the roller blades is $\underline{\ ?\ }$ dollars.

▶ **Write an equation and solve the problem.**

5. If you multiply a number by 7 and subtract 9 from the result, you get 68. What is the number?
(Let n = the number.)

6. If you divide a number by 4 and add 3 to the result, you get 30. What is the number?
(Let k = the number.)

7. If a number is divided by 6 and then 5 is subtracted, the result is 4. What is the number?
(Let m = the number.)

8. If a number is multiplied by 8 and then 17 is added, the result is 65. What is the number?
(Let j = the number.)

9. Two more than the product of 7 and a number is 44. What is the number?
(Let p = the number.)

10. Five less than the product of 6 and a number is 61. What is the number?
(Let f = the number.)

11. The sum of 6 times a number and 10 is 64. What is the number?
(Let r = the number.)

12. If you divide a number by 8 and then add 7, you get 10. What is the number?
(Let d = the number.)

13. Martin earned $17 picking apples. He earned $2 more than 3 times what Lynn earned. How much money did Lynn earn?
(Let l = the money Lynn earned.)

14. Rose earned $26 washing cars. She earned $2 less than 4 times what Casey earned. How much money did Casey earn?
(Let c = the money Casey earned.)

OBJECTIVE: To solve problems using a calculator when appropriate.

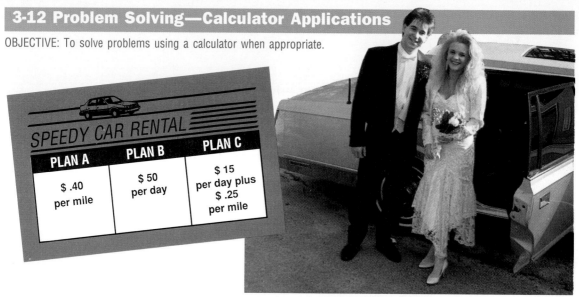

SPEEDY CAR RENTAL

PLAN A	PLAN B	PLAN C
$ .40 per mile	$ 50 per day	$ 15 per day plus $ .25 per mile

▶ **Use the car rental plans. Fill in the facts that fit the calculator sequence.**

1. 50 ⊠ 14 = (**700**)

   Using Plan ___**a.**___, it would cost
   (letter)

   $___**b.**___ to rent a car for ___**c.**___ days.
   (number)   (number)

2. 0.4 ⊠ 275 = (**110**)

   Using Plan ___**a.**___, it would cost
   (letter)

   $ ___**b.**___ to rent a car for
   (number)

   ___**c.**___ miles.
   (number)

▶ **Solve. Use the car rental formulas. Decide when a calculator would be useful.**

3. You rented a car using Plan A. It costs you $168. How many miles did you drive the car?

4. A salesperson used Plan B to rent a car. He paid $350. How many days did he have the car?

5. A corporation president rented a car using Plan C. She used the car for 4 days and was charged $200. How many miles did she drive the car?

6. A reporter rented a car using Plan C. He drove the car 660 miles and was charged $210. How many days did he have the car?

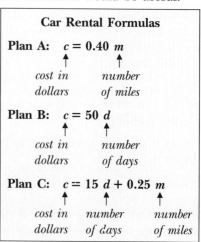

Car Rental Formulas

Plan A: $c = 0.40\ m$

↑ cost in dollars ↑ number of miles

Plan B: $c = 50\ d$

↑ cost in dollars ↑ number of days

Plan C: $c = 15\ d + 0.25\ m$

↑ cost in dollars ↑ number of days ↑ number of miles

7. You want to rent a car for 2 days and drive 500 miles. Which is the cheapest plan for you?

8. **Write a problem** that can be solved using the car rental formulas. Then solve the problem.

▶ **Give the product.** *(page 36)*

1. 0.08×1000 **2.** 0.08×100 **3.** 0.08×10 **4.** 0.8×10

5. 342×10 **6.** 23×1000 **7.** 136×100 **8.** 728×10

9. 56.7×100 **10.** 39.3×1000 **11.** 4.37×10 **12.** 6.593×100

▶ **Evaluate each expression for** $d = 9$, $e = 5$, $f = 7.2$, **and** $g = 2.4$. *(page 38)*

13. de **14.** eg **15.** $3f$ **16.** $7d$

17. $\dfrac{f}{d}$ **18.** $\dfrac{d}{e}$ **19.** $\dfrac{30}{e}$ **20.** $\dfrac{g}{3}$

21. $(df) + e$ **22.** $(ef) - g$ **23.** $(6g) + d$ **24.** $(dg) - 1$

▶ **Give the quotient.** *(page 42)*

25. $42 \div 10$ **26.** $356 \div 100$ **27.** $86 \div 1000$ **28.** $177 \div 100$

29. $38.51 \div 100$ **30.** $38.51 \div 10$ **31.** $38.51 \div 1000$ **32.** $3.851 \div 10$

33. $421 \div 10$ **34.** $53.6 \div 100$ **35.** $12.56 \div 1000$ **36.** $359.1 \div 100$

▶ **Solve and check.** *(pages 20, 52)*

37. $5x = 16.8$ **38.** $9.7 = w + 5.4$ **39.** $g - 8.25 = 0$ **40.** $7.1 = \dfrac{n}{6}$

41. $c + 0.88 = 1.6$ **42.** $\dfrac{p}{8} = 0.12$ **43.** $5.63 = 10y$ **44.** $h - 42.5 = 60$

45. $5.7 = j - 3.8$ **46.** $8k = 0.256$ **47.** $\dfrac{q}{6} = 4.08$ **48.** $42 = d + 6.55$

49. $\dfrac{r}{1.2} = 0.8$ **50.** $m - 2.85 = 4.75$ **51.** $f + 0.065 = 1.731$ **52.** $97.3 = 7z$

53. $7.42 = d + 3.58$ **54.** $2.6 = \dfrac{n}{8}$ **55.** $42 = m - 8.9$ **56.** $15.6 = 1.2r$

▶ **Evaluate each expression for** $a = 3$, $b = 5$, $c = 0.9$, **and** $d = 0.2$. *(page 62)*

57. $b - 1 + c$ **58.** $ac + 1$ **59.** $c - d + 1$ **60.** $c \div a + 1$

61. $4b + 2c$ **62.** $a(b + d)$ **63.** $(b - d) \div 2$ **64.** $2(d + a) - 6$

65. $(a - c) \div (c - d)$ **66.** $b + 2(a - d)$ **67.** $bc + ba$ **68.** $b(c + a)$

69. $b + c \div a$ **70.** $4(a + c) - b$ **71.** $d + 2c$ **72.** $ad + 9 - d$

73. $2(c + d) - 1$ **74.** $a(b - c)$ **75.** $ab - ac$ **76.** $2a + b \div d$

Here are scrambled answers for the review exercises:

4	18	addition	commutative	division	like	simplify
5	add	associative	distributive	grouping	multiply	subtract
17	adding	combine	divide	left	multiplying	

1. The rules for the order of operations tell you to first do the operations within the _?_ symbols. Next, work from left to right doing any multiplication and _?_ . Last, work from left to right doing any _?_ and subtraction. *(page 62)*

2. When you simplify $2 + 3 - (8 - 1) \div 7$, the result is _?_ . *(page 62)*

3. An example of the commutative property of addition is $a + b = b + a$. An example of the _?_ property of addition is $(a + b) + c = a + (b + c)$. An example of the _?_ 0 property is $a + 0 = a$. *(page 64)*

4. An example of the _?_ property of multiplication is $xy = yx$. An example of the associative property of multiplication is $(xy)z = x(yz)$. An example of the _?_ by 1 property is $x \times 1 = x$. *(page 66)*

5. To solve the equation $46 + n = 83$ you would first use the commutative property of addition. Then you would _?_ 46 from both sides and then _?_ both sides. *(page 68)*

6. To solve the equation $96 = n(4)$ you would first use the commutative property of multiplication. Next, you would use the symmetric property of equality to get the variable on the _?_ side of the equal sign. Finally, you would _?_ both sides by 4 and simplify both sides. *(page 68)*

7. To solve the two-step equation $5r + 18 = 78$ you would first subtract _?_ from both sides and simplify. Then you would divide both sides by _?_ and simplify. *(page 74)*

8. To solve the equation $\frac{t}{7} - 21 = 35$ you would first _?_ 21 to both sides and simplify. Then you would _?_ both sides by 7 and simplify. *(page 76)*

9. The property that links multiplication and addition is called the _?_ property. An example is $a(b + c) = ab + ac$. *(page 78)*

10. To simplify the expression $5x + 7 + 3x + 8$ you would combine _?_ terms by adding. *(page 80)*

11. To solve the equation $8y + 9y = 153$ you would first _?_ like terms. Then you would divide both sides by _?_ and simplify. *(page 82)*

▶ **Simplify each expression.** *(page 62)*

1. $18 + 6 \div 2$

2. $50 - 25 + 10$

3. $6 \times (13 + 7)$

4. $(6 + 24) \div 3 - 1$

5. $6 + 24 \div 3 - 1$

6. $(6 + 24) \div (3 - 1)$

7. $30 \div (2 + 4) - 3$

8. $30 \div 2 + 4 \times 4$

9. $2 + 4 \times (4 - 1)$

▶ **Match each property with its example.** *(pages 64, 66, 78)*

10. Commutative property of addition
11. Associative property of addition
12. Adding 0 property
13. Commutative property of multiplication
14. Associative property of multiplication
15. Multiplying by 1 property
16. Distributive property

a. $(15 \times y) \times 20 = 15 \times (y \times 20)$
b. $(12 + 18) + 30 = 12 + (18 + 30)$
c. $22 \times 10 = 10 \times 22$
d. $53 + n = n + 53$
e. $10 \times (13 + 27) = 10 \times 13 + 10 \times 27$
f. $112 + 0 = 112$
g. $t \times 1 = t$

▶ **Solve and check.** *(pages 68, 74, 76)*

17. $22 + x - 17$

18. $y(11) = 198$

19. $109 = 41 + z$

20. $476 = w(17)$

21. $5s - 3 = 57$

22. $40 = 2n + 18$

23. $62 = 6m - 10$

24. $3q + 19 = 79$

25. $\dfrac{d}{8} - 4 = 19$

26. $\dfrac{c}{6} + 14 = 23$

27. $10 = \dfrac{h}{7} - 5$

28. $32 = \dfrac{q}{5} + 25$

▶ **Simplify.** *(page 80)*

29. $5y + 3y$

30. $6x + x$

31. $z + z + z$

32. $3 + 2a + 5a$

33. $8w + w + 2w$

34. $4t + t + 9$

35. $4s + 2s + 3 + 8$

36. $8u + 6 + 5 + u$

▶ **Solve and check.** *(page 82)*

37. $6n + 2n = 96$

38. $8m + m = 90$

39. $120 = p + 7p$

40. $144 = 4q + 4q$

41. $3y + 5 = 29$

42. $\dfrac{w}{6} - 10 = 0$

43. $15.6 = 2x - 1.2$

44. $10 = \dfrac{z}{8} + 5.6$

▶ **Write an equation and solve the problem.** *(page 84)*

45. Six more than 4 times a number is 18. What is the number?

46. Six less than the product of 4 and a number is 18. What is the number?

47. If you divide a number by 4 and then add 6, you get 18. What is the number.

48. If 6 is added to 4 times a number, the result is 18. What is the number?

▷ **Choose the correct letter.**

---

1. 8.952 rounded to the nearest tenth is

A. 8.95

B. 8.9

C. 9.0

D. none of these

---

2. Solve.
$z + 5.4 = 9.75$

A. 9.21

B. 15.15

C. 4.35

D. none of these

---

3. Solve.
$139 = x - 21$

A. 139

B. 160

C. 118

D. none of these

---

4. Give the product.
12.7×1000

A. 127

B. 1270

C. 12,700

D. none of these

---

5. Give the quotient.
$29.74 \div 100$

A. 2974

B. 2.974

C. 0.02974

D. none of these

---

6. Solve.
$1.2m = 6.48$

A. 5.4

B. 7.776

C. 5.28

D. none of these

---

7. Solve.

$51 = \dfrac{y}{3}$

A. 17

B. 51

C. 153

D. none of these

---

8. Simplify.
$18 \div 2 + 4 \times 3$

A. 1

B. 21

C. 39

D. none of these

---

9. Evaluate.
$b + a(c - b)$
for $a = 2$, $b = 0.4$, and $c = 0.7$.

A. 1

B. 1.4

C. 0.72

D. none of these

---

10. $(y + 3) + 7 = y + (3 + 7)$ is an example of

A. a commutative property

B. an associative property

C. the distributive property

D. none of these

---

11. Solve.
$3n + n = 24$

A. 8

B. 4

C. 6

D. none of these

---

12. Choose the equation.

If you divide Maura's age by 2 and subtract 6 from the quotient, you get 8. How old is Maura?

A. $2n + 6 = 8$

B. $2n - 6 = 8$

C. $\dfrac{n}{2} + 6 = 8$

D. $\dfrac{n}{2} - 6 = 8$

Integers and Equations

$$a = \frac{t}{c}$$

The average passing yardage per completion of a football quarterback equals the total yards gained or lost on passes divided by the number of pass completions. (page 111)

One measure of a quarterback's effectiveness is the average number of yards gained for each completed pass. Joe Montana of the San Francisco 49ers has averaged over 11 yards per completion during his professional career.

91

OBJECTIVE: To compare integers. To state the absolute value of an integer.

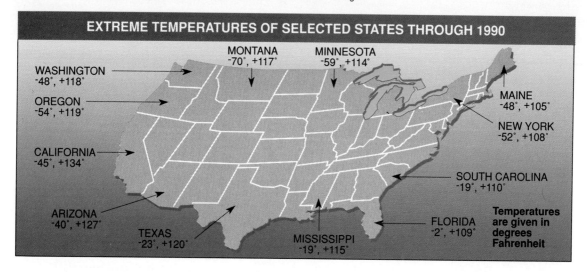

EXTREME TEMPERATURES OF SELECTED STATES THROUGH 1990

WASHINGTON
-48°, +118°

OREGON
-54°, +119°

CALIFORNIA
-45°, +134°

ARIZONA
-40°, +127°

TEXAS
-23°, +120°

MONTANA
-70°, +117°

MINNESOTA
-59°, +114°

MISSISSIPPI
-19°, +115°

MAINE
-48°, +105°

NEW YORK
-52°, +108°

SOUTH CAROLINA
-19°, +110°

FLORIDA
-2°, +109°

Temperatures are given in degrees Fahrenheit

1. The high temperature for Oregon is 119 degrees above zero, which can be written as $^+119°$ ("positive one hundred nineteen degrees"). What is the high temperature for Maine?

2. The low temperature for Florida is 2 degrees below zero, which can be written as $^-2°$ ("negative two degrees"). What is the low temperature for Mississippi?

E X A M P L E S | **Here's how to use the number line to compare two integers.**

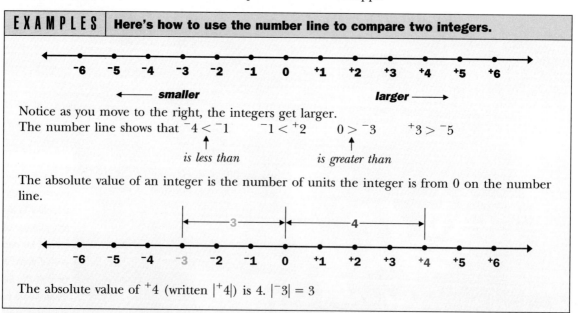

Notice as you move to the right, the integers get larger.

The number line shows that $^-4 < ^-1$ $^-1 < ^+2$ $0 > ^-3$ $^+3 > ^-5$

is less than   *is greater than*

The absolute value of an integer is the number of units the integer is from 0 on the number line.

The absolute value of $^+4$ (written $|^+4|$) is 4. $|^-3| = 3$

CHECK for Understanding

3. Look at the examples above.
   a. Is $^-4$ less than or greater than $^-1$?
   b. Is 0 less than or greater than $^-3$?

4. Look at the examples above. Give each absolute value.
   a. $|^+4|$ b. $|^-3|$

EXERCISES

▶ < or >?
5. $^+3 \Diamond ^+5$ 6. $^+3 \Diamond ^-5$ 7. $^-2 \Diamond ^+4$ 8. $^-2 \Diamond ^-4$ 9. $^-5 \Diamond ^+6$

10. $^+4 \Diamond ^-2$ 11. $^-4 \Diamond ^+2$ 12. $^-3 \Diamond ^-2$ 13. $^+3 \Diamond ^+2$ 14. $^-3 \Diamond 0$

15. $^-9 \Diamond ^+6$ 16. $0 \Diamond ^+8$ 17. $0 \Diamond ^-8$ 18. $^+10 \Diamond ^-7$ 19. $^-4 \Diamond ^-12$

20. $^-17 \Diamond ^+19$ 21. $^+17 \Diamond ^-19$ 22. $^+17 \Diamond ^+19$ 23. $^-17 \Diamond ^-19$ 24. $^+17 \Diamond 0$

▶ Give each absolute value.
25. $|^+6|$ 26. $|^-4|$ 27. $|^-5|$ 28. $|^+5|$ 29. $|0|$

30. $|^-11|$ 31. $|^+17|$ 32. $|^+20|$ 33. $|^-19|$ 34. $|^-22|$

▶ Complete.
35. All _?_ integers are greater than 0.

36. All _?_ integers are less than 0.

37. _?_ is neither positive nor negative.

38. Zero is _?_ than any negative integer.

39. A negative integer is _?_ than any positive integer.

40. A positive integer is _?_ than any negative integer.

Problem Solving USING DATA

▶ Solve. Use the information on page 92.
41. Which state has the highest temperature?

42. Which state has the lowest temperature?

43. Which state has the highest low temperature?

44. Which state has the lowest high temperature?

45. List the low temperatures in order from least to greatest. (If two or more states have the same low temperature, tell the number of states with that low temperature.)

46. List the high temperatures in order from least to greatest. (If two or more states have the same high temperature, tell the number of states with that high temperature.)

▶ Solve.
47. The relationship between a temperature in degrees Celsius (C) and the temperature in degrees Fahrenheit (F) is given by the following formula:

$$9C = 5F - 160$$

Substitute in the formula above to complete this chart.

	C (degrees Celsius)	F (degrees Fahrenheit)
a.	?	32
b.	?	50
c.	?	59
d.	25	?
e.	30	?
f.	100	?

OBJECTIVE: To think of integers as collections of charged particles.

Imagine some small particles, each of which has either a positive electrical charge or a negative electrical charge. The positive charges and negative charges are **opposites.** This means that when one positive charge and one negative charge are combined, the result is no charge, or a charge of 0.

2 positive charges

This collection has a charge of ⁺2.

1 positive charge and 1 negative charge

This collection has a charge of 0.

3 negative charges

This collection has a charge of ⁻3.

EXAMPLES | **Here's how to find the charge of a collection of charged particles.**

A

First look for pairs of opposites.

This collection has a charge of ⁺1.

B

A charge of 0

This collection has a charge of ⁻2.

C

This collection has a charge of ⁺2.

CHECK for Understanding

1. Look at the examples above. To find the charge of a collection of charged particles, first look for pairs of _?_ .

2. The charge of the collection in Example C is _?_ .

3. Give the charge of each collection of charged particles.

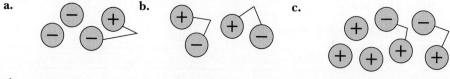

a. b. c.

EXERCISES

▶ Give the charge of each collection of charged particles.

4.

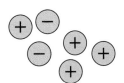

5.

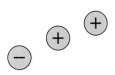

6.

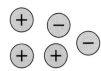

7.

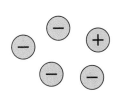

8.

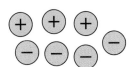

9.

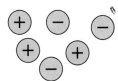

10.

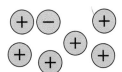

11.

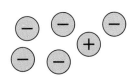

12.

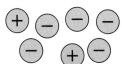

13.

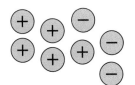

14.

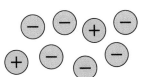

15.

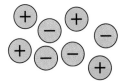

Problem Solving LOGICAL REASONING

▶ Solve. Draw charged particles if you need to.

16. What is the charge when three positive charges and two positive charges are combined?

17. What is the charge when four negative charges and five negative charges are combined?

18. What is the charge when two positive charges and two negative charges are combined?

19. If a collection has a charge of $^+4$ and you remove a charge of $^+2$, what will be the charge of the remaining collection?

20. If a collection has a charge of $^-6$ and you remove a charge of $^-2$, what will be the charge of the remaining collection?

★21. If a collection has a charge of $^+3$ and you remove a charge of $^-1$, what will be the charge of the remaining collection?

OBJECTIVE: To add integers.

1. Look at the first picture. What is the charge when four positive charges and one negative charge are combined?

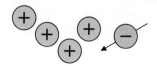

2. Look at the second picture. What is the charge when three positive charges and four negative charges are combined?

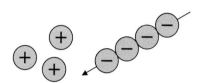

E X A M P L E S | **Here's how to add integers by thinking about combining charges.**

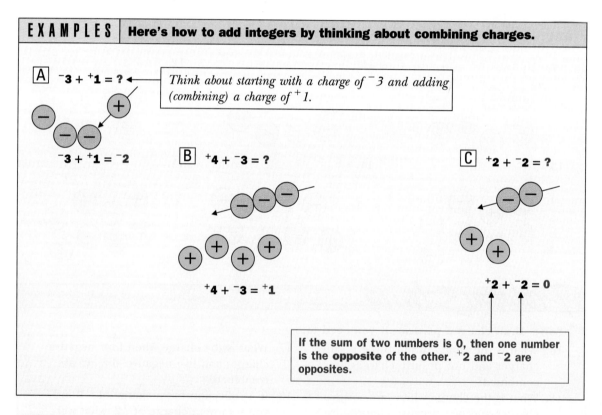

A $^-3 + {}^+1 = ?$ ◄—— *Think about starting with a charge of $^-3$ and adding (combining) a charge of $^+1$.*

$^-3 + {}^+1 = {}^-2$

B $^+4 + {}^-3 = ?$

$^+4 + {}^-3 = {}^+1$

C $^+2 + {}^-2 = ?$

$^+2 + {}^-2 = 0$

If the sum of two numbers is 0, then one number is the **opposite** of the other. $^+2$ and $^-2$ are opposites.

CHECK for Understanding

3. Look at the examples above.
   a. What is the sum of $^-3$ and $^+1$?
   b. What is the sum of $^+4$ and $^-3$?

4. What is the sum of $^-4$ and $^+4$? Are they opposites?

5. What is the opposite of $^+8$? of $^-6$? of 0?

You can use the following rules to add integers:

Adding an Integer and 0
- The sum of any integer and 0 is that integer.

Adding Two Positive Integers
- Add the absolute values of the numbers.
- The sign of the sum will be positive.

Adding Two Negative Integers
- Add the absolute values of the numbers.
- The sign of the sum will be negative.

Adding a Positive and a Negative Integer
- Subtract the smaller absolute value from the larger.
- Use the sign of the number with the larger absolute value. This will be the sign of the sum.

EXERCISES

▶ **Simplify.**
Here are scrambled answers for the next row of exercises: 0 $^+9$ $^+4$ $^+2$ $^-9$

6. $^+7 + ^+2$ **7.** $^+6 + ^-2$ **8.** $^-2 + ^+2$ **9.** $^-4 + ^-5$ **10.** $^+5 + ^-3$

11. $^+8 + ^-1$ **12.** $^-6 + ^+5$ **13.** $^-6 + ^-2$ **14.** $^+9 + ^-1$ **15.** $^+8 + ^-8$

16. $^-6 + ^+4$ **17.** $^+6 + ^+4$ **18.** $^-6 + ^-4$ **19.** $^+6 + ^-4$ **20.** $^+7 + ^-9$

21. $^-15 + ^-10$ **22.** $^-20 + ^+14$ **23.** $^+30 + ^-22$ **24.** $^-37 + 0$ **25.** $^-50 + ^-48$

26. $^+26 + ^-26$ **27.** $0 + ^-68$ **28.** $|^-4 + ^-5|$ **29.** $|^+6 + ^-10|$ **30.** $|^+7 + ^-4|$

▶ **Evaluate each expression for** $w = ^-20$, $x = ^+22$, $y = ^-22$ **and** $z = ^+10$.

31. $w + x$ **32.** $|w + x|$ **33.** $w + z$ **34.** $|w + z|$ **35.** $w + y$

36. $|w + y|$ **37.** $x + z$ **38.** $|x + z|$ **39.** $w + w$ **40.** $|w + w|$

41. $x + y + y$ **42.** $x + x + z$ **43.** $w + x + x$ **44.** $z + z + z$ **45.** $w + w + w$

▶ **True or false?**

46. The sum of two positive numbers is a positive number.

47. The sum of two negative numbers is a positive number.

48. The sum of a positive number and a negative number is a positive number.

49. The sum of two opposites is 0.

Challenge!   **NUMBER SENSE**

▶ **True or false?**   *Hint:* $^+3 + ^+5 = ^+8$

50. The sum of two positive integers is greater than either addend.   addends   sum

51. The sum of two negative integers is greater than either addend.

52. The sum of a positive and a negative integer is greater than the positive addend.

53. The sum of a positive and a negative integer is greater than the negative addend.

OBJECTIVE: To subtract integers.

Look at the picture at the right to answer the following questions.

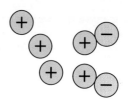

1. What is the charge of the collection of particles?

2. a. Suppose that you removed a charge of $^+1$. What would the charge be then?

   b. Suppose instead that you added a charge of $^-1$. What would the charge be then?

3. Compare the effect of removing a charge of $^-2$ with adding a charge of $^+2$.

EXAMPLES | **Here's how to subtract integers by thinking about removing charges.**

$\boxed{A}$ $^+2 - {}^-2 = ?$ ← *Think about starting with a charge of $^+2$ and subtracting (removing) a charge of $^-2$.*

$^+2 - {}^-2 = {}^+4$

Notice that *adding the opposite* of $^-2$ would have given the same result.

$^+2 + {}^+2 = {}^+4$

$\boxed{B}$ $^-5 - {}^+2 = ?$

$^-5 - {}^+2 = {}^-7$

Notice that *adding the opposite* of $^+2$ would have given the same result.

$^-5 + {}^-2 = {}^-7$

CHECK for Understanding

4. a. To subtract $^+7$, you could add ___?___.   **b.** To subtract $^-6$, you could add ___?___.

You can use the following rule to subtract integers:

• **To subtract an integer, add the opposite of the integer.**

EXAMPLES | **Here's how to subtract integers by using the rule.**

$\boxed{C}$ $^-8 - {}^+5 = {}^-8 + {}^-5$
$\quad\quad = {}^-13$

$\boxed{D}$ $^+10 - {}^-7 = {}^+10 + {}^+7$
$\quad\quad = {}^+17$

$\boxed{E}$ $^-7 - {}^-12 = {}^-7 + {}^+12$
$\quad\quad\quad = {}^+5$

CHECK for Understanding

5. Explain how to get the numbers that are printed in red in Examples C, D, and E.

EXERCISES

▶ **Give each difference.**
Here are scrambled answers for the next row of exercises: $^-9$ $^+9$ $^-5$ $^-3$

6. $^+7 - {}^-2$	**7.** $^-8 - {}^-5$	**8.** $^-3 - {}^+6$	**9.** $^+4 - {}^+9$
10. $^-10 - {}^+6$	**11.** $^-4 - {}^-8$	**12.** $^+6 - {}^+3$	**13.** $^-7 - {}^+6$
14. $^+9 - 0$	**15.** $0 - {}^+6$	**16.** $^+7 - {}^-3$	**17.** $0 - {}^-9$
18. $^+7 - {}^+9$	**19.** $^-5 - {}^+5$	**20.** $^-3 - {}^+5$	**21.** $^-8 - {}^+3$
22. $^+6 - {}^+7$	**23.** $^+4 - {}^-3$	**24.** $^-4 - {}^-3$	**25.** $^-9 - {}^+3$
26. $^+8 - 0$	**27.** $0 - {}^+9$	**28.** $^-7 - {}^+5$	**29.** $^+2 - {}^+9$
30. $^-11 - {}^-6$	**31.** $^-3 - {}^+9$	**32.** $^-9 - {}^+5$	**33.** $^+6 - {}^-3$
34. $^+3 - {}^+9$	**35.** $^-8 - {}^+8$	**36.** $^-10 - {}^+12$	**37.** $^+14 - {}^-12$
38. $^-14 - {}^-18$	**39.** $^+15 - {}^+15$	**40.** $^-21 - {}^-13$	**41.** $^+18 - {}^+18$
42. $^+18 - {}^+15$	**43.** $^-19 - {}^+11$	**44.** $^-17 - {}^-13$	**45.** $^-16 - {}^+18$

▶ **Simplify.**
Here are scrambled answers for the next row of exercises: $^+27$ $^-21$ $^+2$

46. $^-10 + {}^-11$	**47.** $^+12 - {}^-15$	**48.** $^+16 - {}^+14$
49. $^-18 + {}^+12$	**50.** $^-13 - {}^+12$	**51.** $^+14 + 0$
52. $^+11 + {}^+11$	**53.** $^+11 - {}^+11$	**54.** $|{}^+2 - {}^+6|$
55. $|{}^-8 + {}^-10|$	**56.** $|{}^+4 - {}^-3|$	**57.** $|{}^-6 - {}^+9|$
58. $(^+6 + {}^+3) - {}^-4$	**59.** $^+6 + (^+3 - {}^-4)$	**60.** $^+9 + (^+8 + {}^+6)$
61. $^+9 + (^+8 - {}^+6)$	**62.** $(^-5 - {}^+8) + {}^+2$	**63.** $^-5 - (^+8 + {}^+2)$
64. $(^-12 - {}^-10) - {}^-15$	**65.** $^-12 - (^-10 - {}^-15)$	**66.** $(^+18 - {}^-20) + {}^-12$
67. $^-23 + (^+19 + 0)$	**68.** $(^+27 - {}^+27) - {}^-31$	**69.** $^-36 + (^-29 + {}^+29)$

▶ **Evaluate each expression for** $n = {}^+5$, $p = {}^-8$, **and** $q = {}^-1$.

70. $n - p$	**71.** $|n - p|$	**72.** $p - q$	**73.** $|p - q|$	**74.** $q - n$
75. $|q - n|$	**76.** $p + n - q$	**77.** $n - p + q$	**78.** $p + n + q$	**79.** $q - n - p$

▶ **Write an algebraic expression for each word expression.**

80. the sum of y and $^-8$	**81.** a number k decreased by $^+8$
82. $^+6$ increased by a number n	**83.** a number x plus a number y
84. a number s less $^+7$	**85.** c subtracted from a number r
86. $^+3$ more than a number b	**87.** a number a added to itself
88. a number w minus $^-21$	**89.** $^-3$ less than a number d

OBJECTIVE: To multiply integers.

1. What is the charge of the collection of particles?

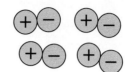

2. Suppose that you put in 2 sets of $^+2$ charges. What would the charge be then?

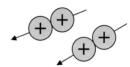

3. Suppose instead that you took out 2 sets of $^-2$. What would the charge be then?

E X A M P L E S	**Here's how to multiply integers by thinking about putting in and taking out sets of charges.**

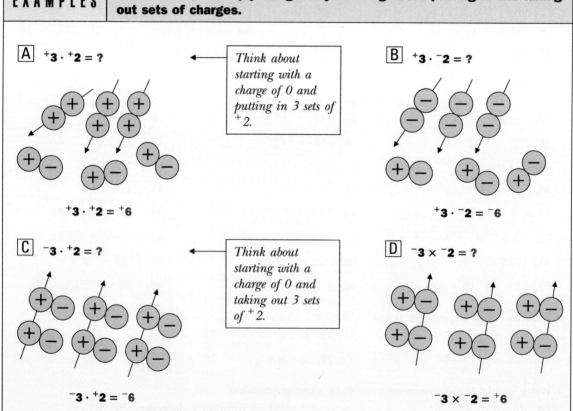

A $^+3 \cdot {}^+2 = ?$

Think about starting with a charge of 0 and putting in 3 sets of $^+2$.

$^+3 \cdot {}^+2 = {}^+6$

B $^+3 \cdot {}^-2 = ?$

$^+3 \cdot {}^-2 = {}^-6$

C $^-3 \cdot {}^+2 = ?$

Think about starting with a charge of 0 and taking out 3 sets of $^+2$.

$^-3 \cdot {}^+2 = {}^-6$

D $^-3 \times {}^-2 = ?$

$^-3 \times {}^-2 = {}^+6$

You can use the following rules to multiply integers:

- The product of two integers with the *same* sign is *positive*.
- The product of two integers with *different* signs is *negative*.
- The product of any integer and 0 is 0.

CHECK for Understanding

4. The product of a positive integer and a positive integer is a _?_ integer.

5. The product of a positive integer and a negative integer is a _?_ integer.

6. The product of a negative integer and a negative integer is a _?_ integer.

7. The product of an integer and 0 is _?_.

EXERCISES

▶ **Simplify.**
Here are scrambled answers for the next row of exercises: $^+15$ $^-8$ $^+28$ $^-24$ 0

8. $^-5 \cdot 0$	**9.** $^-6 \cdot {}^+4$	**10.** $^+3 \cdot {}^+5$	**11.** $^+1 \cdot {}^-8$	**12.** $^-4 \cdot {}^-7$
13. $^-4 \cdot {}^+2$	**14.** $^+9 \cdot {}^+5$	**15.** $^+8 \cdot {}^+8$	**16.** $^-7 \cdot {}^+7$	**17.** $0 \cdot {}^-8$
18. $^+9 \cdot {}^-3$	**19.** $^-8 \cdot {}^-5$	**20.** $^-2 \cdot {}^+5$	**21.** $^+8 \cdot {}^+9$	**22.** $^+3 \cdot {}^-5$
23. $^+6 \cdot {}^-4$	**24.** $^-7 \cdot {}^+8$	**25.** $0 \cdot 0$	**26.** $^-3 \cdot {}^-3$	**27.** $^+7 \cdot 0$
28. $^+9 \cdot {}^-9$	**29.** $^+2 \cdot {}^+6$	**30.** $^-6 \cdot {}^+4$	**31.** $^+4 \cdot {}^-7$	**32.** $^-9 \cdot {}^-8$
33. $\left| ^-3 \cdot {}^+5 \right|$	**34.** $\left| ^+7 \cdot {}^-6 \right|$	**35.** $\left| ^-8 \cdot {}^+4 \right|$	**36.** $\left| ^+9 \cdot {}^+6 \right|$	**37.** $\left| ^-6 \cdot {}^-4 \right|$
38. $^-12 \cdot {}^+12$	**39.** $^+18 \cdot {}^+11$	**40.** $^-15 \cdot {}^-10$	**41.** $^-16 \cdot {}^+14$	**42.** $^+18 \cdot {}^-18$

▶ **Evaluate each expression for** $a = {}^-6$, $b = {}^-3$, $c = 0$, **and** $d = {}^+8$.
Here are scrambled answers for the next row of exercises: $^+3$ $^-48$ $^-9$ $^+48$

43. $a + b$	**44.** $c - b$	**45.** ad	**46.** $\left| ad \right|$
47. cb	**48.** ab	**49.** $\left| ab \right|$	**50.** $d - b$
51. $a - b$	**52.** $\left| a - b \right|$	**53.** bb	**54.** $d - a$
55. $a(b + c)$	**56.** $ab + ac$	**57.** $a(b - c)$	**58.** $ab - ac$
59. $(a + b) + c$	**60.** $a + (b + c)$	**61.** $(a - b) - d$	**62.** $a - (b - d)$

Challenge! NUMBER SENSE

▶ **True or false?**

63. The product of three negative integers is negative.

64. The product of four negative integers is negative.   *Hint:* $^+3 \cdot {}^+5 = {}^+15$

65. The product of two positive integers is greater than either factor.   factors   product

66. The product of two negative integers is greater than either factor.

67. The product of a positive and a negative integer is greater than the positive factor.

68. The product of a positive and a negative integer is greater than the negative factor.

OBJECTIVE: To divide integers.

1. Look at the photo. What integer is the missing factor?

2. What would you multiply ⁻6 by to get ⁻54?

3. What would you multiply ⁻12 by to get 0?

$$^+7 \cdot \square = {}^+35$$

E X A M P L E S | **Here's how to divide integers by finding a missing factor.**

A $^+24 \div {}^+4 = ?$ ⟵ $\boxed{^+4 \cdot ? = {}^+24}$
$^+24 \div {}^+4 = {}^+6$
because
$^+4 \cdot {}^+6 = {}^+24$

B $^+24 \div {}^-4 = ?$ ⟵ $\boxed{^-4 \cdot ? = {}^+24}$
$^+24 \div {}^-4 = {}^-6$
because
$^-4 \cdot {}^-6 = {}^+24$

C $^-24 \div {}^+4 = ?$
$^-24 \div {}^+4 = {}^-6$
because
$^+4 \cdot {}^-6 = {}^-24$

D $^-24 \div {}^-4 = ?$
$^-24 \div {}^-4 = {}^+6$
because
$^-4 \cdot {}^+6 = {}^-24$

You can use the following rules to divide integers:

- The quotient of two integers with the *same* sign is *positive*.
- The quotient of two integers with *different* signs is *negative*.
- The quotient of 0 divided by any nonzero integer is 0.

CHECK for Understanding

4. If you divide a positive integer by a positive integer, the quotient is __?__.

5. If you divide a negative integer by a negative integer, the quotient is __?__.

6. If you divide a positive integer by a negative integer, the quotient is __?__.

7. If you divide a negative integer by a positive integer, the quotient is __?__.

EXERCISES

▶ Simplify.
Here are scrambled answers for the next row of exercises: ⁺8 ⁻4 ⁺10 ⁺6 ⁻8

8. $^+20 \div {}^+2$ **9.** $^+24 \div {}^-3$ **10.** $^-30 \div {}^-5$ **11.** $^-36 \div {}^+9$ **12.** $^-56 \div {}^-7$

13. $^+14 \div {}^-2$ **14.** $^+27 \div {}^+9$ **15.** $^-42 \div {}^-6$ **16.** $0 \div {}^+7$ **17.** $^+28 \div {}^-4$

18. $^-64 \div {}^-8$ **19.** $|^+45 \div {}^+9|$ **20.** $|^+18 \div {}^-9|$ **21.** $^-48 \div {}^+8$ **22.** $^+25 \div {}^-5$

▶ **Simplify.**

23. $\dfrac{^+36}{^-9}$ $\boxed{\dfrac{^+36}{^-9}\ \textit{means}\ ^+36\div{}^-9}$ **24.** $\dfrac{^+49}{^-7}$ **25.** $\dfrac{^+32}{^-8}$ **26.** $\dfrac{^-36}{^-6}$ **27.** $\dfrac{^-45}{^+9}$

28. $\dfrac{0}{^+7}$ **29.** $\dfrac{^+63}{^-9}$ **30.** $\dfrac{^-63}{^-7}$ **31.** $\dfrac{^-30}{^+5}$ **32.** $\left|\dfrac{^+64}{^+8}\right|$ **33.** $\left|\dfrac{^+56}{^-8}\right|$

> **NOTICE!!** From this point on we will no longer write the raised plus sign when writing a positive integer.

▶ **Simplify.**

Here are scrambled answers for the next row of exercises: 4 $^-7$ 21 $^-108$ $^-5$

34. $3 + 18$ **35.** $21 - 17$ **36.** $\dfrac{25}{^-5}$ **37.** $^-9 \cdot 12$ **38.** $20 - 27$

39. $^-45 + 20$ **40.** $\dfrac{^-72}{^-4}$ **41.** $48 - {}^-30$ **42.** $3 \cdot {}^-35$ **43.** $16 + {}^-28$

▶ **Evaluate each expression for $r = {}^-21$, $s = 12$, $t = {}^-6$, and $u = 3$.**

Here are scrambled answers for the next row of exercises: 2 15 $^-18$ $^-9$ $^-2$

44. $r + s$ **45.** $t - r$ **46.** tu **47.** $\dfrac{s}{t}$ **48.** $\left|\dfrac{s}{t}\right|$

49. $r - st$ **50.** $\dfrac{r}{u} - s$ **51.** $tu + ts$ **52.** $su - \dfrac{s}{t}$ **53.** $\dfrac{r}{u} - rt$

Group Project   ANALYZING DATA   *Your Pulse, Please*

▶ **Work in a small group.**

54. Have someone take your pulse and record how many times your heart beats in one minute.

55. a. Record the sex and pulse rate of each class member. For example, G74 would represent a girl having a heart rate of 74 beats per minute.
    b. Order the data from least to greatest for both boys and girls.

56. a. Compute the average pulse rates of both boys and girls.

    b. How do the averages compare?
    c. Was your pulse rate above or below the average for your sex?

57. Write a paragraph that summarizes the data.

BUILD YOUR
PROBLEM·SOLVING·POWER

You will use a variety of problem-solving strategies on these two pages. For some problems, you may wish to use a calculator.

1 USE LOGICAL REASONING

Amir, Lisa, Joel, and Mary are each active in a different sport. Their sports are golf, tennis, bowling, and running. Lisa is the sister of the tennis player. Mary's sport does not use a ball. Joel once made a "hole in one" in his sport. Which sport does each person play?

✓ Problem Solving Tips

Use Logical Reasoning.* Make a chart. Fill in the facts.

	Amir	Lisa	Joel	Mary
Golf	no	no	yes	no
Tennis			no	
Bowling			no	
Running			no	

*At least one more problem on these two pages can be solved by using logical reasoning.

2 GEOMETRY—VISUAL THINKING

Baron von Bucks, owner of a ranch shaped like this, put this clause in his will: "My children will each receive one quarter of my ranch if they can use just four lines to divide its map into four equal pieces, each shaped like the original ranch." Can you do it?

3 SALE TABLE

Raspberries Records is having a compact disk sale. Connie spent a total of $69.75 on CD's. How many did she get altogether? *Hint: Make a table.*

All CD's $7.75
BUY 3 GET 1 FREE

Number of CD's she got	1	2	3	4
Total cost ($)	7.75	15.50	?	?

104 | *Chapter 4*

4 IT'S LOGICAL

Karen, Jane, and Ann are all in my algebra class. They are going steady with three of my friends: Tom, Peter, and John. But I forget who's going with whom. However, I do know that

- Ann cannot bear the sight of Tom.
- Jane goes steady with Tom's brother.
- John goes steady with Jane's sister.

Match each girl with her steady.

5 GUESS AND CHECK

Arrange the digits 1 through 6 in the boxes so that the multiplication works out correctly.

6 COIN COUNT

How can a two-dollar bill be changed into the same number of nickels, dimes, and quarters?

7 ORDER OF OPERATIONS

Put in grouping symbols to make the equation true.

a. $28 + 12 \times 4 \div 2 = 80$
b. $28 + 12 \times 4 - 2 = 74$
c. $28 + 12 \times 4 \div 2 = 52$
d. $28 + 12 \times 4 - 2 = 158$

APPLICATIONS

8 Complete the chart using the formula. Round your answers to the nearest ten-thousandth.
You may wish to use a calculator.

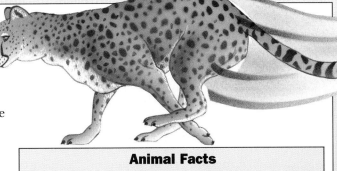

The distance (*d*) traveled is equal to the rate (*r*) multiplied by the time (*t*).

Formula: $d = rt$
distance = rate × time

Complete.

9 The fastest animal listed is the ___?___ .

10 For a short distance, an elephant can run about as fast as a ___?___ .

Animal Facts

Animal	d (miles)	r (mph)	t (hours)
Cheetah	0.0568	a.	0.0008
Elephant	b.	25	0.1666
Tortoise	4.08	0.17	c.
Greyhound	d.	41.72	0.0056
Snail	1.44	0.03	e.
Human	0.0084	f.	0.0003

Rate Your Problem-Solving Power: *6–7 correct = Good; 8–9 correct = Excellent; 10 correct = Exceptional*

OBJECTIVE: To use prices of items for solving problems.

▶ **Use the given information to solve the problems.**

1. What is the price per ounce of the Bosc pears?
Hint: 1 pound = 16 ounces

2. What is the price of 1 pound 7 ounces of Anjou pears?

3. How much do 6 pounds of pears cost if 2.25 pounds are Bartlett pears and the rest are Bosc pears?

PEARS
Bartlett $1.20 per pound
Bosc $1.28 per pound
Anjou $1.12 per pound

▶ **Read the facts. Then complete the steps to answer the question.**

4. *Facts:* You bought a $1.44 bag of apples and a 4-pound bag of peaches. You spent a total of $7.40.

Question: What is the cost of one pound of peaches?

a. Let c equal the cost in dollars of one pound of peaches. Then __?__ c equals the cost of four pounds of peaches. So __?__ $c + 1.44$ equals the total cost of the apples and peaches.

b. Two equal expressions for the total cost are __?__ $c + 1.44$ and 7.40.

c. To find the cost of one pound of peaches, you can solve the equation $4c + 1.44 =$ __?__ .

d. Solve the equation in Part c. What is the cost of one pound of peaches?

5. *Facts:* You bought a cantaloupe at $1.20 per pound and a $2.20 box of strawberries. You spent a total of $5.20.

Question: What is the weight of the cantaloupe?

a. Let p equal the weight of the cantaloupe in pounds. Then __?__ p equals the cost in dollars of the cantaloupe. So __?__ $p + 2.20$ equals the total cost of the cantaloupe and strawberries.

b. Two equal expressions for the total cost are __?__ $p + 2.20$ and 5.20.

c. To find the weight of the cantaloupe, you can solve the equation $1.20p + 2.20 =$ __?__ .

d. Solve the equation in Part c. What is the weight of the cantaloupe?

▶ **Write an equation and solve the problem.**

6. You bought a 15-pound watermelon and a $3.50 bag of oranges. You spent a total of $7.85. How much did the watermelon cost per pound?
(Let c equal the cost in dollars of one pound of watermelon.)

7. You paid $7.29 for a $1.89 melon and a bag of cherries. The cherries cost $1.80 a pound. How many pounds of cherries did you buy?
(Let p equal the number of pounds of cherries you bought.)

▶ **Solve and check.** *(pages 20, 52)*

1. $n + 2.3 = 4.5$
2. $x + 0.3 = 9.7$
3. $4.6 = m + 2$
4. $8 = y + 3.4$

5. $w - 3.5 = 6.4$
6. $10.8 = z - 5$
7. $c - 7.5 = 0$
8. $0.9 = k - 0.4$

9. $4t = 2.4$
10. $3b = 10.5$
11. $0.72 = 9s$
12. $1.44 = 8w$

13. $\dfrac{b}{1.2} = 5$
14. $\dfrac{m}{3} = 5.6$
15. $0.12 = \dfrac{f}{8}$
16. $15 = \dfrac{h}{0.04}$

▶ **Simplify each expression.** *(page 62)*

17. $12 \div 4 - 1$
18. $8 \times 5 - 3$
19. $24 - 4 \div 4$

20. $30 - 12 - 6$
21. $10 + 16 \div 4$
22. $18 + 6 \div 3$

23. $5 + 2 \times 5 - 1$
24. $5 \times 2 + 10 \div 2$
25. $5 + (3 + 9) \div 6$

26. $(4 + 5) \times 10 - 4$
27. $4 + 5 \times 10 - 4$
28. $4 + 5 \times (10 - 4)$

29. $16 + 8 \div 4 + 4$
30. $(16 + 8) \div 4 + 4$
31. $16 + 8 \div (4 + 4)$

32. $(20 + 12) \times 4 - 1$
33. $20 + 12 \times (4 - 1)$
34. $20 + 12 \times 4 - 1$

▶ **Solve and check.** *(page 68)*

35. $23 + g = 91$
36. $8d = 104$
37. $\dfrac{d}{6} = 13$
38. $x - 47 = 104$

39. $\dfrac{f}{16} = 10$
40. $129 = y - 36$
41. $117 = z + 83$
42. $153 = j(9)$

43. $35 + x = 71$
44. $k(20) = 340$
45. $\dfrac{g}{32} = 6$
46. $5 = \dfrac{f}{16}$

47. $5c = 11.5$
48. $r - 8.7 = 15.3$
49. $132 = r - 63$
50. $18.3 = y + 7.9$

▶ **Match each property with its example.** *(pages 64, 66, 78)*

51. The commutative property of addition

a. $134 + 0 = 134$

52. The associative property of addition

b. $(n \times 19) \times 54 = n \times (19 \times 54)$

53. The adding 0 property

c. $p + q = q + p$

54. The commutative property of multiplication

d. $26 \times (44 + 39) = (26 \times 44) + (26 \times 39)$

55. The associative property of multiplication

e. $27 \times 63 = 63 \times 27$

56. The multiplying by 1 property

f. $(71 + 45) + 39 = 71 + (45 + 39)$

57. The distributive property

g. $t \times 1 = t$

OBJECTIVE: To solve addition and subtraction equations involving integers.

You have already solved equations such as

$$x + 37 = 54 \qquad y - 24 = 16$$

In this lesson you will use what you have learned about solving addition and subtraction equations to solve equations involving integers.

E X A M P L E S	Here's how to solve addition and subtraction equations involving integers.

A Solve.

$$n + 14 = {}^-6$$
$$n + 14 - 14 = {}^-6 - 14$$
$$n = {}^-20$$

✓ CHECK

$$n + 14 = {}^-6$$
$${}^-20 + 14 \stackrel{?}{=} {}^-6$$
$${}^-6 = {}^-6$$

It checks!

B Solve.

$$m - {}^-8 = 2$$
$$m - {}^-8 + {}^-8 = 2 + {}^-8$$
$$m = {}^-6$$

✓ CHECK

$$m - {}^-8 = 2$$
$${}^-6 - {}^-8 \stackrel{?}{=} 2$$
$${}^-6 + {}^+8 \stackrel{?}{=} 2$$
$$2 = 2$$

It checks!

C

$${}^-28 = u + {}^-10 \longleftarrow$$
$$u + {}^-10 = {}^-28 \longleftarrow$$
$$u + {}^-10 - {}^-10 = {}^-28 - {}^-10$$
$$u = {}^-18$$

Symmetric property of equality

D

$$5 + w = {}^-10$$
$$w + 5 = {}^-10$$
$$w + 5 - 5 = {}^-10 - 5$$
$$w = {}^-15$$

CHECK for Understanding

1. Look at Examples A and B.
   a. To find n, what number was subtracted from both sides of the equation? What does n equal?
   b. To find m, what number was added to both sides of the equation? What does m equal?
   c. To check the solution, what number was substituted for m in the equation $m - {}^-8 = 2$?

2. Complete these examples.

   a.
   $$r + {}^-4 = 10$$
   $$r + {}^-4 - {}^-4 = 10 - \underline{\ ?\ }$$
   $$r = 14$$

   b.
   $$s - 7 = {}^-5$$
   $$s - 7 + \underline{\ ?\ } = {}^-5 + 7$$
   $$s = 2$$

   c.
   $$15 = v - {}^-18$$
   $$v - {}^-18 = 15$$
   $$v - {}^-18 + {}^-18 = 15 + \underline{\ ?\ }$$
   $$v = \underline{\ ?\ }$$

3. Explain how to get the numbers that are printed in red in Examples C and D.

EXERCISES

▶ **Solve and check.**

Here are scrambled answers for the next two rows of exercises:

$^-10 \quad 14 \quad 26 \quad ^-17 \quad 0 \quad ^-32 \quad 21 \quad ^-19$

4. $f + 12 = {}^-5$

5. $b + {}^-4 = 10$

6. $^-12 = y + 20$

7. $12 + z = 2$

8. $x - 6 = 15$

9. $w + {}^-6 = 20$

10. $^-23 = r - 4$

11. $p - {}^-2 = 2$

12. $r + 15 = {}^-25$

13. $k + 20 = 0$

14. $0 = c + {}^-46$

15. $5 + p = {}^-8$

16. $s - 19 = 18$

17. $g - 5 = {}^-6$

18. $^-40 = b - 35$

19. $n - 25 = 38$

20. $x + 16 = 9$

21. $d + {}^-18 = 4$

22. $a + 18 = {}^-15$

23. $n + {}^-4 = {}^-9$

24. $p - 8 = {}^-3$

25. $m - 8 = {}^-2$

26. $t - 1 = 5$

27. $x - 45 = 54$

28. $m + 6 = {}^-8$

29. $^-1 = q + 9$

30. $^-32 + b = 0$

31. $e + 10 = 3$

32. $c - {}^-7 = {}^-3$

33. $m - 100 = 0$

34. $a - {}^-2 = 9$

35. $^-11 = y - 6$

36. $w + {}^-3 = {}^-9$

37. $6 + y = {}^-2$

38. $10 = y + {}^-10$

39. $^-13 = s + 14$

Problem Solving USING EQUATIONS

▶ **Decide which equation would be used to solve each problem. Then solve the problem.**

Equation A:	$30 + n = 20$
Equation B:	$^-30 + n = 20$
Equation C:	$30 + n = {}^-20$
Equation D:	$^-30 + n = {}^-20$

40. In a card game, Chris made 30 points on his first hand. After the second hand, his total score was 20 points. What was his score on the second hand?

41. A checking account had a balance of $^-30$ dollars. (It was overdrawn.) After a deposit was made, the account had a balance of 20 dollars. How much money was deposited?

42. 30 increased by a number is 20. What is the number?

43. When a number is added to 30, the sum is $^-20$. What is the number?

Challenge! USING A CALCULATOR

Here's how to use a $+/-$ (change sign) key to operate with integers.

Example. Find $^-16 \div 2$.

$16 \boxed{+/-} \div 2 = \boxed{\qquad ^-8}$

Example. Find $18 - {}^-34$.

$18 - 34 \boxed{+/-} = \boxed{\qquad 52}$

44. Simplify.

a. $^-13 \times 17$

b. $252 \div {}^-14$

c. $402 - {}^-108$

d. $28 \times {}^-17$

e. $^-182 \div 26$

f. $^-8 \times {}^-33$

g. $^-106 - {}^-85$

h. $72 - {}^-112$

i. $^-306 \div {}^-9$

j. $12 \times {}^-26$

OBJECTIVE: To solve multiplication and division equations involving integers.

Remember that multiplication and division are inverse operations. That is, division undoes multiplication and multiplication undoes division.

EXAMPLES	Here's how to solve multiplication and division equations involving integers.

A Solve.

$$^-3n = {}^-15$$

$$\frac{^-3n}{^-3} = \frac{^-15}{^-3}$$

$$n = 5$$

 CHECK

$$^-3n = {}^-15$$
$$^-3 \cdot 5 \stackrel{?}{=} {}^-15$$
$$^-15 = {}^-15 \text{ \textit{It checks!}}$$

B Solve.

$$\frac{m}{^-4} = 20$$

$$^-4 \cdot \frac{m}{^-4} = {}^-4 \cdot 20$$

$$m = {}^-80$$

 CHECK

$$\frac{m}{^-4} = 20$$

$$\frac{^-80}{^-4} \stackrel{?}{=} 20$$

$$20 = 20 \quad \text{\textit{It checks!}}$$

C

$$\frac{s}{7} = {}^-3$$

$$7 \cdot \frac{s}{7} = 7 \cdot {}^-3$$

$$s = {}^-21$$

D

$$42 = {}^-6t$$

$$^-6t = 42$$

$$\frac{^-6t}{^-6} = \frac{42}{^-6}$$

$$t = {}^-7$$

CHECK for Understanding

1. Look at Examples A and B.

  a. To find n, both sides of the equation were divided by what number? What does n equal?

  b. To find m, both sides of the equation were multiplied by what number? What does m equal?

  c. To check the solution, what number was substituted for m?

2. Complete these examples.   **a.** $9r = {}^-72$ **b.** $^-5 = \dfrac{u}{^-3}$

$$\frac{9r}{9} = \frac{^-72}{9} \qquad\qquad \frac{u}{^-3} = {}^-5$$

$$r = \underline{\ ?\ } \qquad\qquad\quad ^-3 \cdot \frac{u}{^-3} = \underline{\ ?\ } \cdot {}^-5$$

$$u = \underline{\ ?\ }$$

3. Explain how to get the numbers that are printed in red in Examples C and D.

EXERCISES

▷ **Solve and check.**
Here are scrambled answers for the next row of exercises: $\quad 7 \quad {}^-1 \quad {}^-3 \quad 22 \quad {}^-8$

4. ${}^-2x = 6$
5. $30t = {}^-30$
6. ${}^-21 = {}^-3t$
7. $9y = {}^-72$
8. ${}^-330 = {}^-15y$

9. $\dfrac{b}{{}^-3} = 4$
10. $\dfrac{n}{6} = {}^-3$
11. $\dfrac{n}{{}^-2} = {}^-5$
12. $15 = \dfrac{a}{2}$
13. $0 = \dfrac{n}{6}$

14. ${}^-4s = {}^-40$
15. $57 = 3t$
16. ${}^-6c = {}^-66$
17. $7n = {}^-7$
18. ${}^-192 = {}^-24w$

19. $\dfrac{x}{4} = {}^-9$
20. $9 = \dfrac{n}{{}^-7}$
21. $\dfrac{y}{{}^-10} = 5$
22. $\dfrac{t}{{}^-8} = {}^-1$
23. ${}^-11 = \dfrac{y}{{}^-6}$

24. $7h = {}^-77$
25. ${}^-5 = {}^-5c$
26. $15t = 30$
27. ${}^-9w = 270$
28. $32j = 0$

▷ **Solve and check.**
Here are scrambled answers for the next row of exercises: $\quad 19 \quad {}^-3 \quad {}^-7$

29. $n + 7 = 4$
30. $t - 12 = 7$
31. ${}^-6n = 42$

32. $a + 6 = {}^-6$
33. $\dfrac{x}{3} = {}^-6$
34. $m - 9 = {}^-8$

35. ${}^-9x = {}^-81$
36. $\dfrac{r}{{}^-5} = {}^-2$
37. $5g = {}^-25$

38. $40 = {}^-10b$
39. $12 = {}^-6 + r$
40. $t - 15 = {}^-10$

41. $\dfrac{y}{8} = {}^-3$
42. ${}^-20 = \dfrac{y}{{}^-10}$
43. $3 + r = {}^-3$

44. $h - 11 = 3$
45. ${}^-55 = {}^-5g$
46. $\dfrac{t}{9} = 11$

47. $c + 7 = {}^-3$
48. $d - 4 = {}^-5$

Problem Solving USING A FORMULA

You can use this formula to find a quarter-back's average yardage per pass completion.

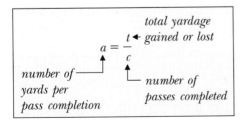

$$a = \frac{t}{c}$$

total yardage ← gained or lost

number of yards per pass completion

number of passes completed

49. *Substitute in the formula to complete this chart.*

	Total yardage (t)	Number of completions (c)	Number of yards per completion (a)
a.	56	7	?
b.	${}^-32$	8	?
c.	?	12	3
d.	?	9	${}^-2$

OBJECTIVE: To solve two-step equations involving integers.

Starting at 35 feet below the surface (⁻35 feet), a diver begins descending, moving at a rate of ⁻15 feet per minute. At that rate, how many minutes will it take him to reach a depth of ⁻140 feet?

1. At what depth was the diver before he started descending at a rate of ⁻15 feet per minute?

2. This equation can be used to find how many minutes it will take the diver to reach a depth of ⁻140 feet.

$$\overset{\underset{\text{Rate of descent}}{\downarrow}}{-15m} + \overset{\underset{\text{Starting depth}}{\downarrow}}{-35} = \overset{\underset{\text{Final depth}}{\uparrow}}{-140}$$
$$\underset{\underset{\text{Number of minutes of descent}}{\uparrow}}{}$$

Look at the equation. What does the variable m represent?

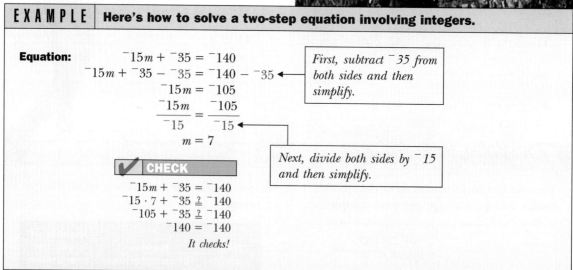

EXAMPLE Here's how to solve a two-step equation involving integers.

Equation:

$$-15m + -35 = -140$$
$$-15m + -35 - -35 = -140 - -35 \qquad \text{First, subtract } -35 \text{ from both sides and then simplify.}$$
$$-15m = -105$$
$$\frac{-15m}{-15} = \frac{-105}{-15}$$
$$m = 7$$

Next, divide both sides by ⁻15 and then simplify.

✔ CHECK

$$-15m + -35 = -140$$
$$-15 \cdot 7 + -35 \overset{?}{=} -140$$
$$-105 + -35 \overset{?}{=} -140$$
$$-140 = -140$$

It checks!

CHECK for Understanding

3. Look at the example above. To find m, first subtract ⟶?⟵ from both sides of the equation and then divide both sides by ⟶?⟵

4. How many minutes will it take the diver to descend to a depth of ⁻140 feet? Is the answer reasonable?

EXERCISES

▶ **Copy and finish solving each equation. Check your solution.**

5. $^-4c - 80 = ^-60$
$^-4c - 80 + 80 = ^-60 + \underline{\ ?\ }$
$^-4c = \underline{\ ?\ }$
$\dfrac{^-4c}{^-4} = \dfrac{20}{?}$
$c = \underline{\ ?\ }$

6. $\dfrac{d}{^-6} + 14 = 26$
$\dfrac{d}{^-6} + 14 - \underline{\ ?\ } = 26 - 14$
$\dfrac{d}{^-6} = \underline{\ ?\ }$
$\underline{\ ?\ } \cdot \dfrac{d}{^-6} = ^-6 \cdot 12$
$d = \underline{\ ?\ }$

7. $\dfrac{e}{4} - 8 = ^-40$
$\dfrac{e}{4} - 8 + 8 = ^-40 + \underline{\ ?\ }$
$\dfrac{e}{4} = \underline{\ ?\ }$
$4 \cdot \dfrac{e}{4} = \underline{\ ?\ } \cdot ^-32$
$e = \underline{\ ?\ }$

▶ **Solve and check.**
Here are scrambled answers for the next two rows of exercises:
$^-16 \quad 5 \quad 2 \quad ^-14 \quad ^-3 \quad ^-12 \quad ^-4 \quad ^-35$

8. $5n - 30 = ^-20$

9. $^-3m + 5 = ^-10$

10. $2d - 14 = ^-20$

11. $9e + 6 = ^-30$

12. $\dfrac{m}{2} + 5 = ^-2$

13. $\dfrac{f}{^-3} + 2 = 6$

14. $\dfrac{h}{4} + 6 = 2$

15. $\dfrac{j}{^-5} - 3 = 4$

16. $3r + 8 = 2$

17. $^-5t + 10 = ^-15$

18. $^-4u - 2 = ^-30$

19. $\dfrac{m}{3} + 7 = 5$

20. $\dfrac{n}{3} + 20 = 18$

21. $\dfrac{y}{^-4} - 2 = 10$

22. $\dfrac{s}{^-7} + 3 = 4$

23. $\dfrac{n}{7} + ^-5 = 0$

24. $3y - 5 = 7$

25. $^-5t + 7 = 27$

26. $6a - 12 = ^-36$

27. $^-4x + 8 = 0$

28. $\dfrac{d}{3} - 4 = 5$

29. $\dfrac{n}{^-2} + 7 = 8$

30. $\dfrac{m}{^-4} - 3 = ^-4$

31. $\dfrac{y}{^-6} - 3 = 4$

32. $^-3q - 4 = 5$

33. $2r - ^-6 = ^-4$

34. $^-7b + 10 = ^-11$

35. $5c - 2 = ^-37$

36. $\dfrac{t}{^-3} - 2 = ^-4$

37. $\dfrac{m}{5} - ^-6 = ^-1$

38. $\dfrac{n}{9} + ^-7 = 0$

39. $\dfrac{y}{^-4} + ^-8 = 1$

40. $^-3r + 9 = 0$

41. $^-7s + 10 = ^-4$

42. $6t - 8 = ^-20$

43. $^-9w + 8 = 8$

Challenge! LOGICAL REASONING

▶ **Use the clues to complete each sentence.**

44. The deepest-diving bird, the emperor penguin, can reach a depth of _?_ feet.

Clues:
- The number is between $^-810$ and $^-899$.
- One of its digits is a 0.
- Its digits add to 15.

45. The deepest-diving mammal, the bull sperm whale, can reach a depth of _?_ feet.

Clues:
- The number is between $^-3710$ and $^-3910$.
- One of its digits is a 2.
- Its digits add to 12.

OBJECTIVE: To use equations with integers to solve problems.

Earlier you wrote equations to solve problems involving whole numbers. In this lesson you will write equations involving integers.

| EXAMPLE | **Here's how to use an equation to solve a problem.** |

Problem: Montana's highest temperature was 187°F more than Montana's lowest temperature. The highest temperature was 117°F. What was Montana's lowest temperature?

Step 1. Choose a variable. Use it and the facts to represent the numbers in the problem.

Let n = Montana's lowest temperature.
Then $n + 187$ = Montana's highest temperature.

Step 2. Write an equation based on the facts.

$$n + 187 = 117$$

These are two equal expressions for Montana's highest temperature.

Step 3. Solve the equation.

$$n + 187 = 117$$
$$n + 187 - 187 = 117 - 187$$
$$n = {}^-70$$

CHECK for Understanding

1. Look at the example above.
   a. In Step 1, we let n equal Montana's _?_ temperature. Then Montana's highest temperature equals $n +$ _?_ .
   b. In Step 2, the two equal expressions for Montana's highest temperature are _?_ and _?_ .
   c. An equation for the problem is _?_ .
   d. To solve the equation, _?_ was subtracted from both sides of the equation.

2. To check the solution, ask yourself whether the answer fits the facts in the problem. Was Montana's lowest temperature ${}^-70$°F?

EXERCISES

▶ **Read the facts. Then complete the steps to answer the question.**

3. *Facts:* A record for extreme temperatures in a 24-hour period was set in Browning, Montana, in 1916. The temperature dropped 100 degrees. The low temperature was $^-56°F$.

Question: What was the high temperature?

a. If you let t equal the high temperature, then $t -$ _?_ equals the low temperature.

b. Two equal expressions for the low temperature are $t -$ _?_ and $^-56$.

c. To find the high temperature, you can solve the equation $t - 100 =$ _?_ .

d. Solve the equation in Part c. The high temperature was _?_ .

4. *Facts:* A most unusual rise in temperature occurred on January 22, 1943, in Spearfish, South Dakota. From 7:30 A.M. to 7:32 A.M. the temperature rose $49°F$. The temperature at 7:32 was $45°F$.

Question: What was the temperature at 7:30?

a. If you let n equal the temperature at 7:30 A.M., then $n +$ _?_ equals the temperature at 7:32 A.M.

b. Two equal expressions for the temperature at 7:32 A.M. are $n +$ _?_ and 45.

c. To find the temperature at 7:30 A.M., you can solve the equation $n + 49 =$ _?_ .

d. Solve the equation in Part c. The temperature at 7:30 A.M. was _?_ .

▶ **Write an equation and solve the problem.**

5. A number increased by 9 equals $^-7$. What is the number?
(Let $d =$ the number.)

6. A number decreased by 5 equals $^-9$. What is the number?
(Let $e =$ the number.)

7. A number times $^-12$ equals $^-72$. What is the number?
(Let $f =$ the number.)

8. A number divided by 8 equals $^-25$. What is the number?
(Let $g =$ the number.)

9. The sum of two numbers is $^-12$. One of the numbers is $^-9$. What is the other number?
(Let $j =$ the other number.)

10. The product of two numbers is $^-85$. One of the numbers is 5. What is the other number?
(Let $k =$ the other number.)

11. Multiply a number by 4, then subtract 8, and you get $^-32$. What is the number?
(Let $t =$ the number.)

12. Divide a number by $^-6$, then add 12, and you get 19. What is the number?
(Let $u =$ the number.)

13. From last night to this morning the temperature dropped $18°F$. The temperature this morning was $7°F$. What was the temperature last night?
(Let $n =$ the temperature last night.)

14. From this morning to this afternoon, the temperature rose $21°F$. The temperature this afternoon was $12°F$. What was the temperature this morning?
(Let $m =$ the temperature this morning.)

Integers and Equations | **115**

OBJECTIVE: To solve problems using a calculator when appropriate.

▶ **Use the mileage chart to solve the problems.**

1. It is 802 miles from New York to Chicago. How far is it from New York to Denver?

2. How far is it from New York to Los Angeles?

3. Fill in the facts that fit the calculator sequence.

996 ⊞ 1059 ⊟ (2055)

The total distance of a Chicago to ___a___ to ___b___ trip is ___c___ miles.
                                    (city)   (city)   (number)

MILEAGE CHART	Baltimore	Chicago	Cleveland	Denver	Houston	Los Angeles	New York	St. Louis
Baltimore		668	343	1621	1412	2636	196	817
Chicago	668		335	996	1067	2054	802	288
Cleveland	343	335		1321	1273	2367	473	546
Denver	1621	996	1321		1019	1059	1771	586
Houston	1412	1067	1273	1019		1538	1608	794
Los Angeles	2636	2054	2367	1059	1538		2786	1848
New York	196	802	473	1771	1608	2786		966
St. Louis	817	288	546	586	794	1848	966	

▶ **Solve. Use the mileage chart and the formulas. Decide when a calculator would be helpful.**

4. You and a friend want to drive from Houston to St. Louis to Chicago in less than 20 hours. Can you do it by traveling at a rate of 50 miles per hour?

5. You leave Houston and plan to drive 50 miles per hour. Will you be in St. Louis in less than 14 hours?

7. Your car's gas tank has 12.5 gallons of gas in it. If your gas mileage averages 25 miles per gallon, can you drive from St. Louis to Chicago without stopping for gas?

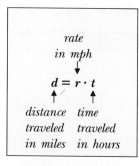

rate
in mph
↓
$d = r \cdot t$
↑   ↑
distance time
traveled traveled
in miles in hours

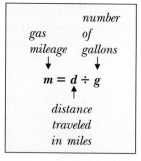

                number
                of
gas   gallons
mileage   ↓
↓
$m = d \div g$
↑
distance
traveled
in miles

6. You leave St. Louis at 1:00 P.M. and plan to drive 50 miles per hour on your trip to Chicago. Will you reach Chicago by 7:00 P.M.?

8. **Write a problem** that can be solved using the mileage chart. Then solve the problem.

▶ **Solve and check.** *(page 74)*

1. $3x + 6 = 21$
2. $2n + 10 = 34$
3. $5m - 6 = 49$
4. $4y - 11 = 89$

5. $39 = 7c + 18$
6. $75 = 6k - 15$
7. $18 = 12d + 18$
8. $0 = 5n - 60$

9. $3j + 25 = 100$
10. $20q - 10 = 130$
11. $136 = 10r + 16$
12. $100 = 5t + 5$

13. $2y + 1.4 = 5$
14. $4x - 3.4 = 5.8$
15. $8.1 = 5n - 6.4$
16. $1.7 = 6b - 4.3$

17. $8n - 0.5 = 0.3$
18. $18.4 = 3f - 0.2$
19. $15.9 = 5m + 7.4$
20. $12z - 8.5 = 1.1$

▶ **Solve and check.** *(page 76)*

21. $\dfrac{x}{4} + 9 = 17$
22. $5g + 3 = 88$
23. $39 = 2n - 5$
24. $4 = \dfrac{g}{3} - 5$

25. $9 = \dfrac{s}{7} + 8$
26. $78 = 4j + 18$
27. $\dfrac{y}{6} - 4 = 21$
28. $11m - 20 = 90$

29. $18 = \dfrac{r}{2} + 16$
30. $6n - 3 = 33$
31. $42 = 5j + 2$
32. $\dfrac{w}{3} - 10 = 15$

33. $12a + 0.6 = 12.6$
34. $\dfrac{z}{5} - 1.8 = 2.3$
35. $1.8 = \dfrac{b}{3} + 1.8$
36. $25.8 = 9d + 9.6$

37. $\dfrac{c}{7} + 0.9 = 5.2$
38. $2u + 1.7 = 4.5$
39. $8.5 = \dfrac{a}{4} + 8.5$
40. $0.86 = 12t - 0.58$

▶ **Simplify.** *(page 80)*

41. $5a + 3a$
42. $7n + 11n$
43. $3c + c$

44. $3f + 2f + 4f$
45. $9x + 3x + x$
46. $8y + y + y$

47. $13w + 4w + 6$
48. $12z + z - 8$
49. $r + r + 11$

50. $6k + 4k + 8 + 3$
51. $5n + n + 6 + 11$
52. $3j + j + 10 - 4$

53. $t + 3t + 4t + 3t$
54. $x + x + 3x + x$
55. $5n + 8 + n + 7$

▶ **Solve and check.** *(page 82)*

56. $4x + 2x = 66$
57. $5a + 3a = 120$
58. $2y + 8y = 160$
59. $3b + 2b = 65$

60. $7n + n = 136$
61. $m + 11m = 252$
62. $g + g = 46$
63. $c + 20c = 105$

64. $108 = 3j + 3j$
65. $130 = k + 9k$
66. $0 = d + d$
67. $125 = 13c + 12c$

68. $w + w = 4.8$
69. $9e + e = 7.9$
70. $12.5 = 3w + 2w$
71. $1.21 = 5z + 6z$

72. $9m + 8m = 8.5$
73. $j + j = 12.4$
74. $0.88 = 5c + 3c$
75. $12.45 = g + 4g$

Here are scrambled answers for the review exercises:

⁻15	⁻4	add	greater	negative	same
⁻10	⁻1	combining	larger	nonzero	simplify
⁻8	0	different	less	opposite	subtract
⁻6	15	divide	multiply	positive	

1. As you move to the right on the number line shown below, the integers get __?__ .

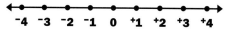

⁻2 is __?__ than ⁻1.   ⁺3 is __?__ than 0. *(page 92)*

2. The absolute value of an integer is the number of units the integer is from __?__ on the number line. *(page 92)*

3. To add integers you can think about __?__ charges.

         ⁻3 + ⁺2 = __?__

If the sum of two numbers is 0, then one number is the __?__ of the other. *(page 96)*

4. To subtract an integer, __?__ the opposite of the integer. To subtract ⁺6, you would add __?__ . ⁻4 − ⁺6 = __?__ *(page 98)*

5. The product of two integers with the same sign is __?__ . The product of two integers with different signs is __?__ . The product of any integer and 0 is __?__ . ⁻8 · ⁺1 = __?__ *(page 100)*

6. The quotient of two integers with the __?__ sign is positive. The quotient of two integers with __?__ signs is negative. The quotient of 0 divided by any __?__ integer is 0. ⁺24 ÷ ⁻6 = __?__ *(page 102)*

7. To solve the equation $x +$ ⁻12 = ⁻3, you would first __?__ ⁻12 from both sides and then __?__ both sides. *(page 108)*

8. To solve the equation ⁻15y = 60, you would first __?__ both sides by __?__ and then simplify both sides. *(page 110)*

9. To solve the equation $\frac{t}{-5} - 15 =$ ⁻6, you would first add __?__ to both sides and then simplify both sides. Next you would __?__ both sides by ⁻5 and simplify both sides. *(page 112)*

▶ **< or >?** (*page 92*)

1. $^+8 \diamond ^+11$

2. $^-8 \diamond ^-11$

3. $0 \diamond ^-9$

4. $^-6 \diamond ^+5$

▶ **Give the absolute value.** (*page 92*)

5. $|^+10|$

6. $|^-8|$

7. $|0|$

8. $|^-23|$

▶ **Give the sum or difference.** (*pages 96, 98*)

9. $^+8 + ^+9$

10. $^+11 + ^-4$

11. $^-16 + ^-10$

12. $^-12 + ^+12$

13. $^+11 - ^+7$

14. $^+15 - ^-12$

15. $^-20 - ^+6$

16. $^-14 - ^-22$

▶ **Give the product or quotient.** (*pages 100, 102*)

17. $^+6 \cdot ^-3$

18. $^-10 \cdot ^-6$

19. $^-8 \cdot ^+5$

20. $^+6 \cdot ^+9$

21. $^+21 \div ^+3$

22. $^+42 \div ^-6$

23. $0 \div ^-9$

24. $^-60 \div ^-12$

▶ **Evaluate each expression for** $e = ^-2, f = 4, g = ^-12,$ **and** $h = 8.$
(*pages 92, 96, 98, 100, 102*)

25. $e - h$

26. $e + f$

27. $\dfrac{g}{e}$

28. gh

29. $|e - f|$

30. $\left|\dfrac{g}{f}\right|$

31. $g + hf$

32. $f - (h - g)$

33. $f(g - e)$

34. $\dfrac{h}{e} - f$

▶ **Solve and check.** (*pages 108, 110, 112*)

35. $f + 11 = 8$

36. $13 + g = ^-4$

37. $s - ^-2 = 8$

38. $^-3 = g - ^-8$

39. $^-3k = 42$

40. $^-9m = ^-90$

41. $\dfrac{t}{^-2} = 16$

42. $^-21 = \dfrac{v}{^-4}$

43. $4n - 8 = ^-48$

44. $^-5m + 8 = ^-2$

45. $6k + 3 = 3$

46. $^-12j - ^-3 = ^-45$

47. $\dfrac{x}{6} + 3 = 3$

48. $\dfrac{w}{^-4} - 8 = 2$

49. $\dfrac{c}{^-3} - ^-4 = 10$

50. $\dfrac{f}{^-5} + ^-3 = 15$

▶ **Write an equation and solve the problem.** (*page 114*)

51. The product of two numbers is $^-10.$ One of the numbers is 5. What is the other number?

52. The sum of two numbers is $^-10.$ One of the numbers is 5. What is the other number?

53. A number divided by 5 equals $^-10.$ What is the number?

54. A number increased by 5 equals $^-10.$ What is the number?

▶ **Choose the correct letter.**

---

1. Solve.

$$6.3 = \frac{x}{3}$$

A. 3.2

B. 2.1

C. 9.3

D. none of these

---

2. An example of the associative property of addition is

A. $g + f = f + g$

B. $(q + r) + s = q + (r + s)$

C. $(de)f = d(ef)$

D. $m + 0 = m$

---

3. Solve.

$$83 = 21 + r$$

A. 62

B. 83

C. 104

D. none of these

---

4. Solve.

$$128 = j(4)$$

A. 124

B. 32

C. 512

D. none of these

---

5. Solve.

$$8b - 32 = 88$$

A. 7

B. 15

C. 120

D. none of these

---

6. Solve.

$$9.6 = \frac{b}{2} + 6.4$$

A. 32

B. 1.6

C. 6.4

D. none of these

---

7. An example of the distributive property is

A. $(ab)c = a(bc)$

B. $m + n = n + m$

C. $2(4 + 7) = 2 \cdot 4 + 2 \cdot 7$

D. $8 + (3 \cdot 0) = 8 \cdot 3 + 8 \cdot 0$

---

8. Simplify.

$$9x + 3x$$

A. $12x$

B. 12

C. $6x$

D. none of these

---

9. Simplify.

$$5y + 8 + y + 7$$

A. $21y$

B. $14y + 7$

C. $6y + 15$

D. none of these

---

10. Solve.

$$7.5 = g + 4g$$

A. 1.5

B. 2.5

C. 37.5

D. none of these

---

11. Evaluate $r - ty$ for $r = {}^-6$, $t = 4$ and $y = {}^-7$.

A. 22

B. 70

C. $^-34$

D. none of these

---

12. Choose the equation.

Al bought 3 red hats and 2 blue hats for $12.50. Each hat cost the same. How much did 1 hat cost?

A. $\dfrac{n}{3} + 2 = 12.50$

B. $\dfrac{n}{2} + 3 = 12.50$

C. $3n + 2n = 12.50$

D. $3n + 2 = 12.50$

Algebra Using Exponents and Fractions

$$h = \frac{d^2}{8000}$$

The height in miles at which you can see for d miles is equal to d^2 divided by 8000. (page 131)

M any tourists in Alaska enjoy *flight-seeing* trips. These flights allow visitors to view huge glaciers, ice fields, lakes, wildlife, and scenic wonders that are not accessible in any other way.

OBJECTIVE: To use exponents to simplify algebraic expressions.

Given below are two ways to write the number of blocks. The dot (·) means multiplication.

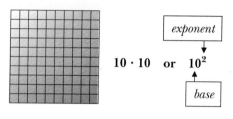

$10 \cdot 10$ or 10^2

exponent

base

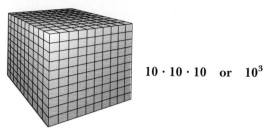

$10 \cdot 10 \cdot 10$ or 10^3

Read "10^2" as "10 squared" or "10 to the second power."

Read "10^3" as "10 cubed" or "10 to the third power."

Notice that instead of writing the same factor several times, you can write the factor once and use an **exponent**. The exponent tells you how many times the **base** is used as a factor.

EXAMPLES | Here's how to write algebraic expressions using exponents.

In arithmetic

A $2 \cdot 2 \cdot 2 \cdot 2 = 2^4$ *2 to the fourth power*

B $3 \cdot 3 \cdot 5 \cdot 5 \cdot 5 = 3^2 \cdot 5^3$

Remember that the commutative and associative properties of multiplication allow you to multiply numbers in any order.

When simplifying an algebraic expression having more than one variable, always list the variables in alphabetical order.

In algebra

C $n \cdot n \cdot n = n^3$ *n to the third power*

D $r \cdot r \cdot r \cdot s \cdot s = r^3 s^2$

E $3 \cdot a \cdot 2 \cdot a \cdot a = 6a^3$

F $2 \cdot x \cdot x \cdot y \cdot 5 \cdot x \cdot y = 10x^3 y^2$

CHECK for Understanding

1. Complete each example.

a. $5 \cdot 5 \cdot 5 = 5^{\underline{?}}$

b. $x \cdot x = \underline{?}^2$

c. $3 \cdot 3 \cdot 4 \cdot 4 \cdot 4 = 3^2 4^{\underline{?}}$

d. $a \cdot a \cdot a \cdot b \cdot b = a^3 \underline{?}^2$

e. $4 \cdot a \cdot 3 \cdot a = 12a^{\underline{?}}$

f. $2 \cdot a \cdot 7 \cdot a \cdot a = \underline{?} a^3$

g. $3 \cdot y \cdot x \cdot 3 \cdot y = 9x \underline{?}^2$

h. $5 \cdot x \cdot y \cdot y \cdot 3 \cdot y = 15x \underline{?}^3$

i. $2 \cdot x \cdot 3 \cdot x \cdot x \cdot y = 6x^{\underline{?}} y$

EXERCISES

▶ **Write using exponents.**

    2. $9 \cdot 9 \cdot 9 \cdot 9$ **3.** $4 \cdot 4 \cdot 4$

    4. $10 \cdot 10$ **5.** $2 \cdot 2 \cdot 2 \cdot 3 \cdot 3$

    6. $7 \cdot 7 \cdot 10 \cdot 10$ **7.** $8 \cdot 8 \cdot 10 \cdot 10 \cdot 10$

    8. $a \cdot a \cdot b$ **9.** $a \cdot b \cdot b \cdot b$

    10. $a \cdot a \cdot a \cdot b \cdot b$ **11.** $x \cdot x \cdot x \cdot y$

    12. $x \cdot x \cdot y \cdot y \cdot y$ **13.** $x \cdot x \cdot x \cdot y \cdot y$

    14. $m \cdot m \cdot n \cdot n \cdot n \cdot n$

    15. $m \cdot m \cdot m \cdot n \cdot n \cdot n$

    16. $m \cdot m \cdot m \cdot m \cdot n \cdot n$

▶ **Simplify using exponents with the variables.**

Here are scrambled answers for the next two rows of exercises:

    $21a^2$ $20a^3b$ $6a^2$ $27a^2b$ $30a^3$ $12ab^2$

    17. $3 \cdot a \cdot 2 \cdot a$ **18.** $5 \cdot a \cdot a \cdot 6 \cdot a$ **19.** $7 \cdot a \cdot a \cdot 3$

    20. $9 \cdot a \cdot 3 \cdot a \cdot b$ **21.** $6 \cdot 2 \cdot a \cdot b \cdot b$ **22.** $a \cdot 5 \cdot a \cdot 4 \cdot a \cdot b$

    23. $a \cdot c \cdot c \cdot d \cdot d$ **24.** $2 \cdot c \cdot d \cdot d \cdot d$ **25.** $a \cdot c \cdot c \cdot c \cdot d$

    26. $8 \cdot m \cdot m \cdot 2 \cdot n$ **27.** $6 \cdot m \cdot 3 \cdot m \cdot n \cdot n$ **28.** $m \cdot 4 \cdot m \cdot 3 \cdot m \cdot n$

    29. $4 \cdot a \cdot b \cdot 3 \cdot a \cdot b$ **30.** $5 \cdot a \cdot a \cdot b \cdot 2 \cdot a$ **31.** $6 \cdot a \cdot b \cdot 4 \cdot a \cdot a \cdot b$

    32. $8 \cdot r \cdot s \cdot 9 \cdot r \cdot r$ **33.** $4 \cdot r \cdot r \cdot s \cdot 6 \cdot r \cdot s$ **34.** $7 \cdot r \cdot s \cdot s \cdot 8 \cdot r \cdot r \cdot s$

    35. $3 \cdot x \cdot 2 \cdot y \cdot x \cdot y$ **36.** $6 \cdot x \cdot y \cdot 3 \cdot x \cdot y \cdot x$

Problem Solving  **USING A FORMULA**

Time in seconds *(t)*	Distance in feet *(d)*
a. 1	?
b. 2	?
c. 3	?
d. 4	?
e. 5	?

▶ **Solve.**

The distance *(d)* in feet that an object will fall in *t* seconds is given by the following formula:

$$d = 16t^2 \quad\longleftarrow \begin{array}{l}\textit{First square } t. \\ \textit{Then multiply by 16.}\end{array}$$

37. Use the formula to complete the chart.

38. Look at your completed chart. How far does an object fall *during* the first second? During the second second? The third? Why are the distances different?

5-2 Laws of Exponents

OBJECTIVE: To use the laws of exponents to simplify algebraic expressions.

1. Look at the Powers of 10 chart.
   a. Does $10^2 \cdot 10^3 = 10^5$?
   b. Does $10^5 \div 10^2 = 10^3$?

Powers of 10
$10^5 = 100,000$
$10^4 = 10,000$
$10^3 = 1,000$
$10^2 = 100$
$10^1 = 10$

E X A M P L E S | **Here's how to multiply powers with like bases.**

In arithmetic

$\overbrace{\text{2 factors}}$ $\overbrace{\text{3 factors}}$

A $10^2 \cdot 10^3 = (10 \cdot 10) \cdot (10 \cdot 10 \cdot 10)$

$\overbrace{(2 + 3) \text{ factors}}$

$= 10 \cdot 10 \cdot 10 \cdot 10 \cdot 10$

$= 10^5$

In algebra

$\overbrace{\text{3 factors}}$ $\overbrace{\text{4 factors}}$

B $n^3 \cdot n^4 = (n \cdot n \cdot n) \cdot (n \cdot n \cdot n \cdot n)$

$\overbrace{(3 + 4) \text{ factors}}$

$= n \cdot n \cdot n \cdot n \cdot n \cdot n \cdot n$

$= n^7$

CHECK for Understanding

2. Look at Examples A and B. Do you see a shortcut for finding the product?
   Hint: Does $10^2 \cdot 10^3 = 10^{(2+3)}$?

E X A M P L E S | **Here's how to divide powers with like bases.**

In arithmetic

$\overbrace{\text{5 factors}}$

C $\dfrac{10^5}{10^2} = \dfrac{10 \cdot 10 \cdot 10 \cdot 10 \cdot 10}{\underbrace{10 \cdot 10}_{\text{2 factors}}}$

$= \dfrac{10}{10} \cdot \dfrac{10}{10} \cdot 10 \cdot 10 \cdot 10$

$\overbrace{(5 - 2) \text{ factors}}$

$= 1 \cdot 1 \cdot 10 \cdot 10 \cdot 10$

$= 10^3$

In algebra

$\overbrace{\text{6 factors}}$

D $\dfrac{n^6}{n^4} = \dfrac{n \cdot n \cdot n \cdot n \cdot n \cdot n}{\underbrace{n \cdot n \cdot n \cdot n}_{\text{4 factors}}}$

$= \dfrac{n}{n} \cdot \dfrac{n}{n} \cdot \dfrac{n}{n} \cdot \dfrac{n}{n} \cdot n \cdot n$

$\overbrace{(6 - 4) \text{ factors}}$

$= 1 \cdot 1 \cdot 1 \cdot 1 \cdot n \cdot n$

$= n^2$

CHECK for Understanding

3. Look at Examples C and D. Do you see a shortcut for finding the quotient?
   Hint: Does $10^5 \div 10^2 = 10^{(5-2)}$?

You can use the following laws of exponents:

	In arithmetic	In algebra
To multiply expressions with like bases, **add** the **exponents**.	$10^2 \cdot 10^3 = 10^{2+3} = 10^5$	$a^m \cdot a^n = a^{m+n}$
To divide expressions with like bases, **subtract** the **exponents**.	$\dfrac{10^5}{10^2} = 10^{5-2} = 10^3$	$\dfrac{a^m}{a^n} = a^{m-n}$

EXERCISES

▶ **Simplify using laws of exponents.**

4. $10^2 \cdot 10^4 = 10^{2+4} = 10^{\underline{?}}$

5. $4^3 \cdot 4^2 = 4^{3+2} = 4^{\underline{?}}$

6. $10^4 \cdot 10^3$

7. $6^2 \cdot 6^5$

8. $7^3 \cdot 7^3$

9. $8^1 \cdot 8^7$

10. $5^6 \cdot 5^1$

11. $2^2 \cdot 2^3$

12. $10^7 \cdot 10^2$

13. $d^1 \cdot d^1$

14. $e^7 \cdot e^5$

15. $f^5 \cdot f^4$

16. $n^4 \cdot n^3$

17. $m^5 \cdot m^2$

18. $r^1 \cdot r^4$

19. $t^3 \cdot t^5$

20. $y^2 \cdot y^2$

▶ **Simplify using laws of exponents.**

21. $\dfrac{10^5}{10^3} = 10^{5-3} = 10^{\underline{?}}$

22. $\dfrac{n^6}{n^2} = n^{6-2} = n^{\underline{?}}$

23. $\dfrac{10^6}{10^3}$

24. $\dfrac{3^5}{3^2}$

25. $\dfrac{4^5}{4^1}$

26. $\dfrac{6^6}{6^2}$

27. $\dfrac{8^7}{8^1}$

28. $\dfrac{7^8}{7^6}$

29. $\dfrac{5^5}{5^2}$

30. $\dfrac{d^7}{d^2}$

31. $\dfrac{e^6}{e^5}$

32. $\dfrac{f^8}{f^1}$

33. $\dfrac{n^6}{n^4}$

34. $\dfrac{m^5}{m^3}$

35. $\dfrac{r^6}{r^2}$

36. $\dfrac{t^8}{t^5}$

Challenge! NUMBER SENSE

1000	100	10	1	0.1	0.01	0.001
10^3	10^2	10^1	10^0	10^{-1}	10^{-2}	10^{-3}

▶ **Give the missing exponent.**

Hint: Use the information in the chart above.

37. $10^{-1} \cdot 10^{-2} = 0.1 \cdot 0.01 = 0.001 = 10^{\underline{?}}$

38. $10^3 \cdot 10^{-2} = 10^{\underline{?}}$

39. $10^2 \cdot 10^{-2} = 10^{\underline{?}}$

40. $10^2 \cdot 10^{-3} = 10^{\underline{?}}$

41. $\dfrac{10^2}{10^{-1}} = \dfrac{100}{0.1} = 1000 = 10^{\underline{?}}$

42. $\dfrac{10^3}{10^{-1}} = 10^{\underline{?}}$

43. $\dfrac{10^3}{10^3} = 10^{\underline{?}}$

44. $\dfrac{10^2}{10^3} = 10^{\underline{?}}$

OBJECTIVE: To write numbers in scientific notation.

About 3,500,000 gallons of water pour over Niagara Falls every second.

The water levels of the oceans rise about 0.0029 of a foot every year.

Very large and very small numbers can be written in scientific notation.

EXAMPLES | **Here's how to write numbers in scientific notation.**

To write a number in **scientific notation,** write it as a product so that the first factor is between 1 and 10 and the second factor is a power of 10.

Positive powers of 10	$10^1 = 10$	$10^2 = 100$	$10^3 = 1000$	$10^4 = 10,000$	etc.
Negative powers of 10	$10^{-1} = 0.1$	$10^{-2} = 0.01$	$10^{-3} = 0.001$	$10^{-4} = 0.0001$	etc.

	A	B
	3,500,000 (Large number)	**0.0029** (Small number)

Step 1. Position the decimal point to get a number between 1 and 10.

3.500000   0.002.9

Step 2. Count the number of places the decimal point was moved to get the number between 1 and 10.

3.500000   0.002.9
6 places   *3 places*
to the left   *to the right*

Step 3. Write the number as a product of a number between 1 and 10 and a power of 10.

$3.5 \cdot 10^6$ $2.9 \cdot 10^{-3}$

$$3,500,000 = 3.5 \cdot 10^6 \qquad 0.0029 = 2.9 \cdot 10^{-3}$$

| *A number between 1 and 10* | *A power of 10* | *Read as "2.9 times 10 to the negative 3rd power."* |

C $610,000,000 = 6.1 \times 10^8$ **D** $0.0000875 = 8.75 \times 10^{-5}$

CHECK for Understanding

1. Look at Example A. When 3,500,000 is written in scientific notation, the factor between 1 and 10 is __?__.

2. Look at Example B. When 0.0029 is expressed in scientific notation, the factor that is a power of 10 is __?__.

EXERCISES

▶ **Give each missing exponent.**

3. $40{,}000{,}000 = 4 \cdot 10^{\underline{?}}$

4. $83{,}700 = 8.37 \cdot 10^{\underline{?}}$

5. $7{,}400{,}000 = 7.4 \cdot 10^{\underline{?}}$

6. $326{,}000 = 3.26 \cdot 10^{\underline{?}}$

7. $32{,}040{,}000 = 3.204 \cdot 10^{\underline{?}}$

8. $1420 = 1.42 \cdot 10^{\underline{?}}$

9. $0.00039 = 3.9 \cdot 10^{\underline{?}}$

10. $0.032 = 3.2 \cdot 10^{\underline{?}}$

11. $0.00239 = 2.39 \cdot 10^{\underline{?}}$

12. $0.0000083 = 8.3 \cdot 10^{\underline{?}}$

13. $0.00008 = 8 \cdot 10^{\underline{?}}$

14. $0.000302 = 3.02 \cdot 10^{\underline{?}}$

▶ **Write in scientific notation.**

15. 31,000

16. 400,000

17. 5,460,000

18. 8800

19. 236,000

20. 85,000,000

21. 51,000

22. 9,634,000

23. 700,000,000

24. 302,000

25. 1,110,000

26. 70,000,000

27. 0.00075

28. 0.00381

29. 0.00006

30. 0.00075

31. 0.0317

32. 0.0000043

33. 0.00649

34. 0.000045

▶ **Write each number in scientific notation.**

35. An average-size thundercloud holds about 6,000,000,000,000 raindrops.

36. It takes 0.00025 of a second for a stick of dynamite to detonate.

37. Lightning strikes the earth about 8,640,000 times every day.

38. The thickness of a human hair is about 0.00366 of a foot.

Problem Solving USING SCIENTIFIC NOTATION

▶ **Solve. Use the large number fact at the top of page 126.**

39. About how many gallons of water pour over Niagara Falls in a year ($3.1536 \cdot 10^{7}$ seconds)?

40. About how many hours does it take for 7 billion gallons of water to go over Niagara Falls?

Challenge! CALCULATOR APPLICATION

Anita's scientific calculator uses an abbreviated scientific notation to express very large and very small numbers. The answers she got when she solved two problems are shown below.

▶ **Write the answers to the two problems in standard scientific notation.**

41. $500{,}000 \boxed{\times} 10{,}000 \boxed{=}$ (5.   9)

42. $0.000005 \boxed{\div} 100{,}000 \boxed{=}$ (5.   −11)

OBJECTIVE: To test for divisibility by 2, 3, 4, 5, 6, 8, 9, and 10.

If you divide 144 by 8, you get a quotient of 18 and a remainder of 0. Since the remainder is 0, we say that 144 is **divisible** by 8.

You can tell whether a whole number is divisible by certain other whole numbers without actually dividing.

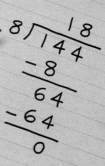

EXAMPLES | Here are some rules for divisibility.

A whole number is divisible by

A 2 if its last digit is divisible by 2.

9436 is divisible by 2.

A number that is divisible by 2 is called an **even number.** A number that is not divisible by 2 is called an **odd number.**

B 3 if the sum of the digits is divisible by 3.

1545 is divisible by 3.

$$1 + 5 + 4 + 5 = 15$$

C 4 if its last two digits are divisible by 4.

7316 is divisible by 4.

D 5 if the last digit is divisible by 5, that is, if the last digit is 0 or 5.

8490 is divisible by 5.

E 6 if it is divisible by both 2 and 3.

5202 is divisible by 6.

F 8 if its last three digits are divisible by 8.

6104 is divisible by 8.

G 9 if the sum of its digits is divisible by 9.

2835 is divisible by 9.

$$2 + 8 + 3 + 5 = 18$$

H 10 if its last digit is divisible by 10, that is, if the last digit is 0.

5980 is divisible by 10.

EXERCISES

▶ **Is the number divisible by 2?**
   1. 2344   **2.** 2463   **3.** 1670   **4.** 38,166   **5.** 84,998   **6.** 78,367

▶ **Is the number divisible by 3?**
   7. 2016   **8.** 3805   **9.** 6822   **10.** 27,492   **11.** 50,646   **12.** 63,275

▶ **Is the number divisible by 4?**
   13. 3616   **14.** 7532   **15.** 5314   **16.** 41,342   **17.** 83,560   **18.** 90,388

▶ **Is the number divisible by 5?**
   19. 7352   **20.** 6895   **21.** 9070   **22.** 23,842   **23.** 53,295   **24.** 47,000

▶ **Is the number divisible by 6?**
   25. 9142   **26.** 4308   **27.** 6183   **28.** 60,210   **29.** 75,444   **30.** 52,890

▶ **Is the number divisible by 8?**
   31. 6320   **32.** 5136   **33.** 9208   **34.** 19,176   **35.** 32,048   **36.** 88,436

▶ **Is the number divisible by 9?**
   37. 1926   **38.** 5699   **39.** 6237   **40.** 24,894   **41.** 79,254   **42.** 42,783

▶ **Is the number divisible by 10?**
   43. 7830   **44.** 5628   **45.** 4864   **46.** 37,480   **47.** 65,370   **48.** 99,880

▶ **True or false?**

49. If a whole number is divisible by 2, then its last digit is even.

50. If a whole number is divisible by 3, then its last digit is odd.

51. If a number is divisible by 10, then it is divisible by 5.

52. If a number is divisible by 5, then it is divisible by 10.

53. All numbers divisible by 4 are even numbers.

54. Some numbers divisible by 5 are odd numbers.

Challenge!   LOGICAL REASONING

▶ **Find the whole number.**

55. *Clues:*
It is a 2-digit number.
It is greater than 90.
It is divisible by 2.
It is divisible by 3.

56. *Clues:*
It is a 4-digit number.
It is less than 1060.
It is divisible by 5.
It is divisible by 9.

5-5 Prime and Algebraic Factorization

OBJECTIVE: To write the prime factorization of a whole number. To write the algebraic factorization of an expression.

A whole number that has exactly two whole-number factors is called a **prime number.**

Here are the first few prime numbers:

$$2 \quad 3 \quad 5 \quad 7 \quad 11 \quad 13$$

A whole number (other than 0) that has more than two whole-number factors is called a **composite number.**

Here are the first few composite numbers:

$$4 \quad 6 \quad 8 \quad 9 \quad 10 \quad 12$$

0 and 1 are neither prime nor composite. Every number is a factor of 0 ($0 \cdot 1 = 0$, $0 \cdot 2 = 0$, $0 \cdot 3 = 0$, etc.), and the only factor of 1 is 1 itself.

Every composite number can be factored into a product of prime numbers. To express a composite number as a product of prime numbers is to give the **prime factorization** of the number.

EXAMPLES | **Here's how to give the prime factorization of a composite number.**

$\boxed{A}$ $18 = 2 \cdot 9$ Factor. Since 9 is not prime, factor again.

     $= 2 \cdot 3 \cdot 3$ All factors are prime!

$\boxed{B}$ $60 = 6 \cdot 10$

     $= 2 \cdot 3 \cdot 2 \cdot 5$, or $2^2 \cdot 3 \cdot 5$

CHECK for Understanding

1. Look at Example A. The prime factorization of 18 is $2 \cdot 3 \cdot$ __?__ .

2. Complete these examples.

   a. $52 = 2 \cdot 26$ **b.** $72 = 8 \cdot 9$ **c.** $50 = 2 \cdot 25$

      $= 2 \cdot 2 \cdot$ __?__ $= 2 \cdot 2 \cdot$ __?__ $\cdot 3 \cdot 3$ $= 2 \cdot$ __?__ $\cdot 5$

EXAMPLES | **Here's how to give the algebraic factorization of an expression.**

First write the prime factorization of the whole number, and then write the variables in alphabetical order.

$\boxed{C}$ $6a^2 = 2 \cdot 3 \cdot a \cdot a$ $\boxed{D}$ $10b^3 = 2 \cdot 5 \cdot b \cdot b \cdot b$

$\boxed{E}$ $20c^2d = 2 \cdot 2 \cdot 5 \cdot c \cdot c \cdot d$ $\boxed{F}$ $5e^3f^2 = 5 \cdot e \cdot e \cdot e \cdot f \cdot f$

CHECK for Understanding

3. Look at Example C. The algebraic factorization of $6a^2$ is $2 \cdot$ __?__ $\cdot a \cdot a$.

4. Complete these examples.

   a. $20c^2 = 2 \cdot 2 \cdot 5 \cdot c \cdot$ __?__   **b.** $9f^3 = 3 \cdot$ __?__ $\cdot f \cdot f \cdot f$

   c. $7r^2s^3 = 7 \cdot r \cdot$ __?__ $\cdot s \cdot s \cdot s$

EXERCISES

▶ **Give the prime factorization of each number.** *Example:* 4   *2 · 2*

5. 35	6. 9	7. 10	8. 6	9. 15	10. 14
11. 24	**12.** 18	**13.** 22	**14.** 20	**15.** 8	**16.** 25
17. 21	**18.** 30	**19.** 12	**20.** 27	**21.** 50	**22.** 28
23. 44	**24.** 26	**25.** 38	**26.** 32	**27.** 16	**28.** 42
29. 48	**30.** 39	**31.** 33	**32.** 36	**33.** 49	**34.** 57
35. 45	**36.** 55	**37.** 58	**38.** 52	**39.** 34	**40.** 46

▶ **Give the prime factorization using exponents.** *Example:* 60 $2^2 \cdot 3 \cdot 5$

41. 25	**42.** 50	**43.** 40	**44.** 35	**45.** 64	**46.** 24
47. 66	**48.** 63	**49.** 54	**50.** 56	**51.** 62	**52.** 74
53. 65	**54.** 75	**55.** 69	**56.** 76	**57.** 70	**58.** 68

▶ **Give the algebraic factorization.** *Example:* $9b^2$ $3 \cdot 3 \cdot b \cdot b$

59. $4x^3$	**60.** $9m^3$	**61.** $10c^2$	**62.** $5a^3$	**63.** $12d$
64. $7f^3$	**65.** $15g^2$	**66.** $42w^2$	**67.** $36y$	**68.** $40x^3$
69. $35t^3$	**70.** $25r^4$	**71.** $32u$	**72.** $27q$	**73.** $45d^3$
74. $14f^2$	**75.** $11g$	**76.** $30h^4$	**77.** $24j^2$	**78.** $11ab^2$
79. $32a^2b$	**80.** $56ab$	**81.** $7ab^3$	**82.** $38a^2b$	**83.** $54a^2b^2$

Problem Solving USING A FORMULA

$$h = \frac{d^2}{8000}$$ ◀— *First square d;
then divide
by 8000.*

▶ **Use the formula to complete the chart.**
The height (*h*) in miles at which a pilot must fly in order to see a distance of *d* miles is given by the formula on the left.

	Distance (*d*) seen from airplane in miles	Height (*h*) of the airplane in miles
84. a.	20	?
b.	40	?
c.	80	?
d.	200	?

Algebra Using Exponents and Fractions | **131**

OBJECTIVE: To find the greatest common factor of two algebraic expressions. To find the least common multiple of two algebraic expressions.

FACTORS OF 18: 1, 2, 3, 6, 9, 18
FACTORS OF 24: 1, 2, 3, 4, 6, 8, 12, 24
1, 2, 3, and 6 are factors of both 18 and 24. They are called **common factors** of 18 and 24.

1. The **greatest common factor (GCF)** of two numbers or expressions is the largest of their common factors. What is the GCF of 18 and 24?

EXAMPLES	Here's how to find the GCF by factoring.

In arithmetic

A Find the GCF of 18 and 24.

First write the prime factorization of each number.

$$18 = 2 \cdot 3 \cdot 3$$

$$24 = 2 \cdot 2 \cdot 2 \cdot 3$$

Then multiply the prime factors that are common to both 18 and 24.

$$GCF = 2 \cdot 3 = 6$$

In algebra

B Find the GCF of $10r^3s$ and $15r^2s^2$.

First write the algebraic factorization of each expression.

$$10r^3s = 2 \cdot 5 \cdot r \cdot r \cdot r \cdot s$$

$$15r^2s^2 = 3 \cdot 5 \cdot r \cdot r \cdot s \cdot s$$

Then multiply the prime factors that are common to both $10r^3s$ and $15r^2s^2$.

$$GCF = 5 \cdot r \cdot r \cdot s = 5r^2s$$

MULTIPLES OF 18: 18, 36, 54, 72, 90, 108, 126, 144, . . .
MULTIPLES OF 24: 24, 48, 72, 96, 120, 144, . . .
72 and 144 are multiples of both 18 and 24. They are called **common multiples** of 18 and 24.

2. The **least common multiple (LCM)** of two numbers or expressions is the smallest of their common multiples. What is the LCM of 18 and 24?

EXAMPLES	Here's how to find the LCM by factoring.

In arithmetic

C Find the LCM of 18 and 24.

$$18 = 2 \cdot 3 \cdot 3$$

$$24 = 2 \cdot 2 \cdot 2 \cdot 3$$

Multiply the factors that are common to both numbers by the factors that are not common.

$$LCM = 2 \cdot 3 \cdot 3 \cdot 2 \cdot 2 = 72$$

In algebra

D Find the LCM of $10r^3s$ and $15r^2s^2$.

$$10r^3s = 2 \cdot 5 \cdot r \cdot r \cdot r \cdot s$$

$$15r^2s^2 = 3 \cdot 5 \cdot r \cdot r \cdot s \cdot s$$

Multiply the factors that are common to both expressions by the factors that are not common.

$$LCM = 2 \cdot 3 \cdot 5 \cdot r \cdot r \cdot r \cdot s \cdot s = 30r^3s^2$$

EXERCISES

▶ **Give the greatest common factor.**

Here are scrambled answers for the next row of exercises: *8   3   5   4   6*

3. 4, 8	**4.** 6, 9	**5.** 12, 18	**6.** 20, 15	**7.** 8, 24
8. 3, 10	**9.** 12, 24	**10.** 25, 20	**11.** 15, 24	**12.** 6, 14
13. 18, 20	**14.** 32, 6	**15.** 36, 24	**16.** 28, 32	**17.** 40, 15

★**18.** 12, 20, 28   ★**19.** 12, 21, 27   ★**20.** 18, 30, 42   ★**21.** 20, 24, 27   ★**22.** 24, 56, 72

Here are scrambled answers for the next row of exercises: *3d   2   d   3*

23. 2, $4a$	**24.** 6, $9c$	**25.** $3d$, $12d$	**26.** d, $15d$
27. $6x$, $8x$	**28.** $9y$, $5y$	**29.** $4xy$, x	**30.** $10e$, $14e$
31. $10u^2$, $5u$	**32.** $16a^2b$, $20ab$	**33.** $6c^2d$, $15c^2$	**34.** $18y^2$, $6y$
35. $11y^2z$, $25yz$	**36.** $35m^3n$, $25m^2n^2$	**37.** $20ab^2$, $24a^2$	**38.** $27rs$, $36r^2s$

▶ **Give the least common multiple.**

Here are scrambled answers for the next row of exercises: *24   28   6   12   18*

39. 2, 3	**40.** 3, 8	**41.** 4, 6	**42.** 6, 9	**43.** 4, 7
44. 4, 5	**45.** 6, 8	**46.** 5, 6	**47.** 7, 5	**48.** 8, 12
49. 6, 12	**50.** 5, 10	**51.** 3, 11	**52.** 7, 6	**53.** 12, 4

★**54.** 5, 8, 10   ★**55.** 3, 4, 10   ★**56.** 6, 8, 9   ★**57.** 4, 5, 6   ★**58.** 6, 10, 15

Here are scrambled answers for the next row of exercises: *18x   10w   16x*

59. 5, $10w$	**60.** $9x$, 6	**61.** $4x$, $16x$
62. $6z$, $7z$	**63.** $9b$, $5b^2$	**64.** $12c$, c^2
65. $24d^2e$, $6d$	**66.** $10ab$, $15a^2$	**67.** $12x^2y$, $16xy$
68. $25uv$, $10u^2v^2$	**69.** $11x^2y$, $5x^2y$	**70.** $6w^3x$, w^2x^2

Challenge!   LOGICAL REASONING

▶ **Solve.**

71. Use the clues to find Terry's age.

Clues:
Sara is 25.
The LCM of their ages is 150.
The GCF of their ages is 5.

You will use a variety of problem-solving strategies on these two pages. For some problems, you may wish to use a calculator.

1 SOLVE A SIMPLER PROBLEM

Eight people have entered a chess tournament. Each person is scheduled to play every other person once. How many games are scheduled for the tournament?

✓ Problem Solving Tips

Solve a Simpler Problem.* Solve the same problem but with smaller numbers. Then look for a pattern.

Number of players	Number of games needed
2	1
3	3
4	6
•	•
•	•
•	•
8	?

*At least one more problem on these two pages can be solved by first solving a simpler problem.

2 INTEGER PUZZLES

Use the number cards to get the answers. Use each card only once in each puzzle.

a. $(\boxed{?} + \boxed{?}) \times (\boxed{?} + \boxed{?}) = {}^{-}12$
b. $(\boxed{?} + \boxed{?}) \times (\boxed{?} - \boxed{?}) = {}^{-}40$
c. $(\boxed{?} + \boxed{?}) \div (\boxed{?} + \boxed{?}) = {}^{-}3$
d. $(\boxed{?} - \boxed{?}) \div (\boxed{?} + \boxed{?}) = {}^{-}2$

3 DOMINO DILEMMA

Carlos and Bill were playing dominoes using a double-twelve set. Carlos asked Bill, "How many dominoes are in this set?" Study this drawing of a double-three set, then answer Carlos's question.
You may wish to draw more dominoes.

4 GEOMETRY—VISUAL THINKING

In each picture, a clock's face is shown reflected in a mirror. What is the true time for each picture?

b.

c.

d.

a.

5 CHAIN LINKS

It costs $2 to open a link and $3 to weld it closed. You want to join these five pieces of chain into one long chain. If you have it done the cheapest way possible, how much will it cost?

6 BACK-TO-THE-FUTURE MULTIPLES

On my 14th birthday, my father was 3 times as old as I was then. How old will I be when my father is twice as old as I am? *Hint: Use a table.*

7 ALPHAMETICS

In this product, each different letter stands for a different digit. What digit does each letter stand for?

$$\begin{array}{r} ABCD \\ \times 4 \\ \hline DCBA \end{array}$$

APPLICATIONS

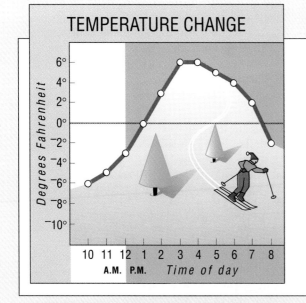

TEMPERATURE CHANGE

8 At what time was the temperature ⁻4°F?

9 What was the temperature at 6:30 P.M.?

10 What integer describes the temperature change from 2 P.M. to 8 P.M.?

Rate Your Problem-Solving Power: *6–7 correct = Good;   8–9 correct = Excellent;   10 correct = Exceptional*

OBJECTIVE: To use a table of data to solve problems.

		Market Value		
Indian Head Pennies				
Year	**Good condition**	**Fine condition**	**Quantity minted**	
1865	$2.50	$6.50	35,429,000	
1866	$7.50	$22.50	9,826,500	
1867	$7.75	$22.75	9,821,000	
1868	$7.25	$22.00	10,266,500	
1869	$13.50	$53.50	6,420,000	
1870	$13.00	$40.00	5,275,000	

▶ **Use the chart to answer the questions.**

1. In which year were the fewest Indian Head pennies minted?

2. What is the market value of an 1867 Indian Head penny that is in good condition?

3. How much more is the market value of an 1868 penny that is in fine condition than the market value of one that is in good condition?

▶ **Read the facts. Then complete the steps to answer the question.**

4. *Facts:* The market value of a 1900 nickel is $2.90. The total market value of three 1930 nickels and the one 1900 nickel is $5.15.
   Question: What is the market value of a 1930 nickel?
   a. Let m equal the market value of a 1930 nickel. Then __?__ $m + 2.90$ equals the market value of all four coins.
   b. Two equal expressions for the market value of all four coins are __?__ $m + 2.90$ and 5.15.
   c. To find the market value of a 1930 nickel, you can solve the equation $3m + 2.90 =$ __?__ .
   d. Solve the equation in Part c. What is the market value of a 1930 nickel?

5. *Facts:* The market value of an 1890 nickel is $.80 more than 4 times the market value of a 1910 nickel. The market value of the 1890 nickel is $8.80.
   Question: What is the market value of a 1910 nickel?
   a. Let n equal the market value of a 1910 nickel. Then $4n +$ __?__ equals the market value of an 1890 nickel.
   b. Two equal expressions for the market value of an 1890 nickel are $4n +$ __?__ and 8.80.
   c. To find the market value of a 1910 nickel, you can solve the equation $4n + .80 =$ __?__ .
   d. Solve the equation in Part c. What is the market value of a 1910 nickel?

▶ **Write an equation and solve the problem.**

6. The market value of an 1867 Indian Head penny that is in good condition is $.25 more than 6 times the market value of a 1920 nickel. What is the market value of a 1920 nickel? *Hint: Use the chart to find the market value of the 1867 penny.*

7. You have an 1865 Indian Head penny that is in fine condition. You also have five 1950 nickels in fine condition. Together, your six coins have a market value of $15.25. What is the market value of each 1950 nickel? *Hint: Use the chart to find the market value of the 1865 penny.*

▶ **Solve and check.** *(page 76)*

1. $3y - 6 = 15$ **2.** $\dfrac{k}{8} + 1 = 3$ **3.** $5x + 6 = 41$ **4.** $\dfrac{n}{4} - 3 = 0$

5. $44 = 6m + 2$ **6.** $2 = \dfrac{j}{3} - 2$ **7.** $10 = \dfrac{r}{12} + 6$ **8.** $56 = 12y - 4$

9. $\dfrac{d}{3} + 15 = 21$ **10.** $1 = \dfrac{k}{4} - 14$ **11.** $11p + 12 = 111$ **12.** $152 = 16r - 8$

13. $5t + 4.3 = 11.7$ **14.** $\dfrac{n}{6} - 1.5 = 2.4$ **15.** $18.4 = 4w + 5.6$ **16.** $0.6 = \dfrac{z}{8} - 2.7$

▶ **Simplify.** *(page 80)*

17. $7k + 5k$ **18.** $11j + j$ **19.** $r + r$

20. $3s + 2s + 2s$ **21.** $5x + 3x + x$ **22.** $t + t + t$

23. $5y + 4y + 3$ **24.** $5y + 4 + 3$ **25.** $5y + 4y + 3y$

26. $b + 4b + 3b + 2b$ **27.** $n + n + 4n + 6n$ **28.** $3j + 2j + 9 + 6$

▶ **Solve and check.** *(page 82)*

29. $3x + 2x = 105$ **30.** $2w + 4w = 102$ **31.** $7z + z = 144$ **32.** $y + y = 86$

33. $105 = 3b + 4b$ **34.** $220 = 9f + f$ **35.** $162 = c + 8c$ **36.** $104 = c + c$

37. $g + 3g = 6.4$ **38.** $6h + 4h = 12.5$ **39.** $8.4 = j + 7j$ **40.** $16.3 = k + k$

41. $4d - 3.8 = 6.2$ **42.** $10 = 5f + 7.4$ **43.** $\dfrac{n}{2} + 6.1 = 20.7$ **44.** $14.5 = \dfrac{q}{3} - 2.5$

▶ **< or >?** *(page 92)*

45. $^+4 \diamond ^+5$ **46.** $^+3 \diamond ^-6$ **47.** $^-6 \diamond ^+2$ **48.** $^-8 \diamond ^-1$ **49.** $^-6 \diamond ^+3$

50. $^+20 \diamond ^-23$ **51.** $^-25 \diamond ^-22$ **52.** $^-26 \diamond ^+21$ **53.** $^+24 \diamond ^-24$ **54.** $^-2 \diamond 0$

▶ **Give each absolute value.** *(page 92)*

55. $|^+6|$ **56.** $|^-3|$ **57.** $|0|$ **58.** $|^+8|$ **59.** $|^-11|$

▶ **Give each sum.** *(page 96)*

60. $^+6 + ^+2$ **61.** $^+6 + ^-2$ **62.** $0 + ^-2$ **63.** $^-4 + ^-8$ **64.** $^-4 + ^+8$

65. $^-7 + ^-10$ **66.** $^+12 + ^-12$ **67.** $^+15 + ^+11$ **68.** $^-18 + ^+16$ **69.** $^-18 + ^-16$

OBJECTIVE: To change an algebraic fraction to an equivalent one by multiplying the numerator and denominator by the same number (not 0) or expression.

Famous face	Fraction who identified face
News anchor	$\dfrac{1}{4}$
Tennis player	$\dfrac{2}{3}$
U.S. president	$\dfrac{5}{6}$
Singer	$\dfrac{5}{8}$

FAMOUS-FACES QUIZ

U.S. President   News anchor

Singer   Tennis player

Whose famous face is in the picture?

Twenty-four students took the Famous - Faces Quiz. The results are shown in the chart.

1. What fraction of those surveyed knew that the news anchor was Connie Chung?

2. What fraction of those surveyed knew that the tennis player was Boris Becker?

3. Fifteen students identified Paula Abdul as the singer. Do you agree or disagree that $\frac{5}{8}$, or $\frac{15}{24}$, of those surveyed identified the singer?

EXAMPLES | **Here's how to change a fraction to an equivalent fraction.**

In arithmetic

To change a fraction to an equivalent fraction, you can multiply both numerator and denominator by the same number (not 0).

$\boxed{A}$ numerator $\longrightarrow$ $\dfrac{5}{8} = \dfrac{5 \cdot 3}{8 \cdot 3}$ $\longleftarrow$ denominator

$= \dfrac{15}{24}$

In algebra

To change an algebraic fraction to an equivalent fraction, you can multiply both numerator and denominator by the same number (not 0) or expression.

$\boxed{B}$ $\dfrac{a}{*3b} = \dfrac{a \cdot 2b}{3b \cdot 2b}$

$= \dfrac{2ab}{6b^2}$

*You can assume that the expression in the denominator is not equal to 0.

CHECK for Understanding

4. Example A shows that $\frac{5}{8}$ and __?__ are equivalent fractions.

5. To change $\dfrac{a}{3b}$ to the equivalent fraction $\dfrac{2ab}{6b^2}$ in Example B, both numerator and denominator were multipled by __?__.

6. Complete these examples.

a. $\dfrac{5}{3} = \dfrac{5 \cdot 6}{3 \cdot ?}$
$= \dfrac{30}{18}$

b. $\dfrac{2}{5} = \dfrac{2 \cdot ?}{5 \cdot ?}$
$= \dfrac{6}{15}$

c. $\dfrac{b}{c} = \dfrac{b \cdot ?}{c \cdot 2c}$
$= \dfrac{2bc}{2c^2}$

d. $\dfrac{2n}{3m} = \dfrac{2n \cdot ?}{3m \cdot ?}$
$= \dfrac{4nt}{6mt}$

EXERCISES

▶ **Complete to get an equivalent fraction.**

7. $\dfrac{4}{3} = \dfrac{?}{6}$ **8.** $\dfrac{5}{4} = \dfrac{?}{12}$ **9.** $\dfrac{7}{8} = \dfrac{?}{24}$ **10.** $\dfrac{3}{4} = \dfrac{?}{16}$ **11.** $\dfrac{1}{8} = \dfrac{?}{48}$

12. $\dfrac{1}{3} = \dfrac{?}{18}$ **13.** $\dfrac{5}{8} = \dfrac{?}{16}$ **14.** $\dfrac{7}{8} = \dfrac{?}{16}$ **15.** $\dfrac{2}{3} = \dfrac{?}{18}$ **16.** $\dfrac{3}{5} = \dfrac{?}{15}$

17. $\dfrac{3}{8} = \dfrac{?}{24}$ **18.** $\dfrac{7}{2} = \dfrac{?}{16}$ **19.** $\dfrac{7}{9} = \dfrac{?}{54}$ **20.** $\dfrac{4}{3} = \dfrac{?}{6}$ **21.** $\dfrac{5}{6} = \dfrac{?}{36}$

22. $\dfrac{16}{5} = \dfrac{?}{10}$ **23.** $\dfrac{5}{6} = \dfrac{?}{24}$ **24.** $\dfrac{3}{2} = \dfrac{?}{8}$ **25.** $\dfrac{1}{7} = \dfrac{?}{21}$ **26.** $\dfrac{5}{6} = \dfrac{?}{18}$

27. $\dfrac{10}{9} = \dfrac{?}{45}$ **28.** $\dfrac{15}{12} = \dfrac{?}{24}$ **29.** $\dfrac{1}{3} = \dfrac{?}{30}$ **30.** $\dfrac{7}{9} = \dfrac{?}{45}$ **31.** $\dfrac{4}{5} = \dfrac{?}{50}$

▶ **Complete to get an equivalent fraction.**

32. $\dfrac{2}{a} = \dfrac{?}{5a}$ **33.** $\dfrac{5}{d} = \dfrac{?}{3d}$ **34.** $\dfrac{2}{b} = \dfrac{?}{6b}$ **35.** $\dfrac{10}{d} = \dfrac{?}{4d}$ **36.** $\dfrac{4}{g} = \dfrac{?}{4g}$

37. $\dfrac{x}{y} = \dfrac{?}{y^2}$ **38.** $\dfrac{r^2}{s} = \dfrac{?}{s^2}$ **39.** $\dfrac{t}{u^2} = \dfrac{?}{9u^2}$ **40.** $\dfrac{3r}{s^2} = \dfrac{?}{8s^2}$ **41.** $\dfrac{5m}{3n^2} = \dfrac{?}{15n^2}$

42. $\dfrac{4}{x^2y} = \dfrac{?}{x^2y^2}$ **43.** $\dfrac{2}{3ac} = \dfrac{?}{6a^2c}$ **44.** $\dfrac{7rt}{4} = \dfrac{?}{4rt}$ **45.** $\dfrac{3}{5y} = \dfrac{?}{5yz^2}$ **46.** $\dfrac{7f}{4g} = \dfrac{?}{8g^2}$

47. $\dfrac{5}{7a} = \dfrac{?}{14ab}$ **48.** $\dfrac{8m}{3n} = \dfrac{?}{12mn}$ **49.** $\dfrac{4}{11c} = \dfrac{?}{11c^2d}$ **50.** $\dfrac{9s^2}{8} = \dfrac{?}{24t}$ **51.** $\dfrac{5w}{6w} = \dfrac{?}{24wx}$

Problem Solving USING MENTAL MATH

▶ **Solve. Use the chart on page 138.**

52. How many of the 24 students identified the news anchor?

53. How many of the 24 students identified the tennis player?

54. How many more students identified the president than the singer?

55. Which face was identified by the fewest students? The most students?

5-9 Writing Fractions in Lowest Terms

OBJECTIVE: To write algebraic fractions in lowest terms.

1. What is the greatest common factor (GCF) of 15 and 24?

2. What is the GCF of $8x^2$ and $6xy$?

EXAMPLES | **Here's how to write a fraction in lowest terms.**

To write a fraction in lowest terms, divide both terms (the numerator and denominator) by their greatest common factor.

In arithmetic

$\boxed{A}$ $\boxed{GCF = 3}$ $\rightarrow \dfrac{15}{24} = \dfrac{15 \div 3}{24 \div 3}$

$= \dfrac{5}{8}$

In algebra

$\boxed{B}$ $\boxed{GCF = 2x}$ $\rightarrow \dfrac{8x^2}{6xy} = \dfrac{8x^2 \div 2x}{6xy \div 2x}$

$= \dfrac{4x}{3y}$

Notice that a fraction is in lowest terms when the GCF of the numerator and denominator is 1.

CHECK for Understanding

3. Look at the example above. To write $\dfrac{15}{24}$ in lowest terms, the numerator and denominator were __?__ by 3.

4. To write $\dfrac{8x^2}{6xy}$ in lowest terms, the numerator and denominator were divided by __?__.

You can also use the following method to write a fraction in lowest terms.

EXAMPLES | **Here's how to write a fraction in lowest terms by factoring and canceling.**

In arithmetic

$\boxed{C}$ $\dfrac{15}{24} = \dfrac{\overset{1}{3} \cdot 5}{2 \cdot 2 \cdot 2 \cdot \underset{1}{3}}$

$= \dfrac{5}{8}$

In algebra

$\boxed{D}$ $\dfrac{8x^2}{6xy} = \dfrac{\overset{1}{2} \cdot 2 \cdot \overset{1}{2} \cdot \overset{1}{x} \cdot x}{\underset{1}{2} \cdot 3 \cdot \underset{1}{x} \cdot y}$

$= \dfrac{4x}{3y}$

CHECK for Understanding

5. Look at Example C. Canceling a 3 in the numerator and denominator gives the same result as dividing both numerator and denominator by __?__.

6. Look at Example D. Canceling a 2 and an x gives the same result as dividing both terms by __?__.

EXERCISES

▶ **Write each fraction in lowest terms.**

Here are scrambled answers for the next row of exercises: $\dfrac{2}{5}$ $\dfrac{5}{12}$ $\dfrac{9}{5}$ $\dfrac{1}{2}$ $\dfrac{2}{3}$ $\dfrac{1}{3}$ $\dfrac{6}{5}$

7. $\dfrac{8}{12}$ **8.** $\dfrac{36}{20}$ **9.** $\dfrac{16}{40}$ **10.** $\dfrac{12}{10}$ **11.** $\dfrac{15}{36}$ **12.** $\dfrac{20}{40}$ **13.** $\dfrac{14}{42}$

14. $\dfrac{20}{36}$ **15.** $\dfrac{10}{15}$ **16.** $\dfrac{6}{8}$ **17.** $\dfrac{4}{6}$ **18.** $\dfrac{3}{9}$ **19.** $\dfrac{22}{33}$ **20.** $\dfrac{20}{24}$

21. $\dfrac{22}{10}$ **22.** $\dfrac{40}{30}$ **23.** $\dfrac{10}{6}$ **24.** $\dfrac{18}{14}$ **25.** $\dfrac{18}{21}$ **26.** $\dfrac{12}{42}$ **27.** $\dfrac{25}{20}$

▶ **Write each algebraic fraction in lowest terms.**

Here are scrambled answers for the next row of exercises: $\dfrac{2}{3a}$ $\dfrac{3}{2a}$ $\dfrac{1}{2a}$ $\dfrac{3a}{2}$ $\dfrac{4a}{3}$ $\dfrac{1}{b}$ $\dfrac{2a}{3}$

28. $\dfrac{2}{4a}$ **29.** $\dfrac{9a}{6}$ **30.** $\dfrac{8}{12a}$ **31.** $\dfrac{10a}{15}$ **32.** $\dfrac{18}{12a}$ **33.** $\dfrac{20a}{15}$ **34.** $\dfrac{a}{ab}$

35. $\dfrac{d}{2d}$ **36.** $\dfrac{r}{rs}$ **37.** $\dfrac{x}{4x}$ **38.** $\dfrac{z}{wz}$ **39.** $\dfrac{mn}{5n}$ **40.** $\dfrac{9rs}{21r^2}$ **41.** $\dfrac{4jk^2}{7jk}$

42. $\dfrac{13p^2}{10p^2}$ **43.** $\dfrac{9q}{7q^2}$ **44.** $\dfrac{16r^2s}{20s}$ **45.** $\dfrac{11mn}{24mn}$ **46.** $\dfrac{20m}{24mn}$ **47.** $\dfrac{15n}{10m^2n}$ **48.** $\dfrac{16r^2s}{9rs}$

49. $\dfrac{21m^2}{14n}$ **50.** $\dfrac{10m}{8m^2}$ **51.** $\dfrac{9t^2}{24st}$

52. $\dfrac{16yz^2}{18z}$ **53.** $\dfrac{y^2z^2}{3z^2}$ **54.** $\dfrac{9xy^2}{27y}$

55. $\dfrac{7xy}{22xy}$ **56.** $\dfrac{18jk^2}{12j^2}$ **57.** $\dfrac{20j^2k}{23j^2k^2}$

Problem Solving USING A GRAPH

▶ **Solve.**

Ninety people were asked to name a state with great skiing. The circle graph shows the result of the survey.

58. Which state was named by $\dfrac{1}{6}$ of the people?

59. Which state was named by $\dfrac{2}{9}$ of the people?

60. Which state was named by $\dfrac{3}{10}$ of the people?

California: 15 people
Utah: 20 people
Other: 18 people
Vermont: 10 people
Colorado: 27 people

OBJECTIVE: To identify the least common denominator of a pair of algebraic fractions.

In order to add, subtract, or compare fractions, you will need to know how to find the least common denominator.

EXAMPLES | **Here's how to find the least common denominator of two fractions.**

To find the **least common denominator (LCD)** of two fractions, find the least common multiple of the denominators.

In arithmetic

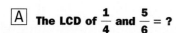 **The LCD of** $\frac{1}{4}$ **and** $\frac{5}{6}$ **= ?**

Find the least common multiple of 4 and 6.

$$4 = 2 \cdot 2$$
$$6 = 2 \cdot 3$$

To find the LCM of the denominators, multiply the factors that are common to both 4 and 6 by the factors that are not common to both.

$$LCM = 2 \cdot 2 \cdot 3 = 12$$

So, the LCD of $\frac{1}{4}$ and $\frac{5}{6}$ is 12.

In algebra

B **The LCD of** $\frac{2}{b}$ **and** $\frac{1}{3b}$ **= ?**

Find the least common multiple of b and $3b$.

$$b = b$$
$$3b = 3 \cdot b$$

To find the LCM of the denominators, multiply the factors that are common to both b and $3b$ by the factors that are not common to both.

$$LCM = 3 \cdot b = 3b$$

So, the LCD of $\frac{2}{b}$ and $\frac{1}{3b}$ is $3b$.

CHECK for Understanding

1. Look at Example A. To find the LCD of $\frac{1}{4}$ and $\frac{5}{6}$, find the least common multiple of the denominators 4 and __?__. The LCD of $\frac{1}{4}$ and $\frac{5}{6}$ is __?__.

2. Look at Example B. To find the LCD of $\frac{2}{b}$ and $\frac{1}{3b}$, find the least common multiple of the denominators __?__ and __?__. The LCD of $\frac{2}{b}$ and $\frac{1}{3b}$ is __?__.

3. Complete these examples.

a. Find the LCD: $\frac{3}{8}, \frac{5}{12}$.

$8 = 2 \cdot 2 \cdot 2$
$12 = 2 \cdot 2 \cdot 3$

$LCM = 2 \cdot 2 \cdot 2 \cdot 3 = $ __?__

LCD of $\frac{3}{8}, \frac{5}{12}$ is __?__

b. Find the LCD: $\frac{2}{9n}, \frac{1}{6}$.

$9n = 3 \cdot 3 \cdot n$
$6 = 3 \cdot 2$

$LCM = 3 \cdot 3 \cdot 2 \cdot n = $ __?__

LCD of $\frac{2}{9n}, \frac{1}{6}$ is __?__

c. Find the LCD: $\frac{c}{a^2b}, \frac{c}{ab^2}$.

$a^2b = a \cdot a \cdot b$
$ab^2 = a \cdot b \cdot b$

$LCM = a \cdot a \cdot b \cdot b = a^2b^2$

LCD of $\frac{c}{a^2b}, \frac{c}{ab^2}$ is __?__

EXERCISES

▶ **Find the least common denominator.**

Here are scrambled answers for the next row of exercises: 20 18 30 12 8

4. $\dfrac{2}{3}, \dfrac{1}{4}$ **5.** $\dfrac{5}{9}, \dfrac{1}{2}$ **6.** $\dfrac{1}{10}, \dfrac{3}{4}$ **7.** $\dfrac{1}{2}, \dfrac{5}{8}$ **8.** $\dfrac{1}{6}, \dfrac{1}{5}$

9. $\dfrac{1}{6}, \dfrac{3}{8}$ **10.** $\dfrac{2}{5}, \dfrac{3}{10}$ **11.** $\dfrac{4}{3}, \dfrac{1}{8}$ **12.** $\dfrac{2}{3}, \dfrac{4}{9}$ **13.** $\dfrac{4}{3}, \dfrac{5}{6}$

14. $\dfrac{1}{2}, \dfrac{3}{20}$ **15.** $\dfrac{1}{12}, \dfrac{1}{8}$ **16.** $\dfrac{3}{4}, \dfrac{7}{6}$ **17.** $\dfrac{5}{9}, \dfrac{7}{6}$ **18.** $\dfrac{1}{20}, \dfrac{2}{15}$

19. $\dfrac{2}{15}, \dfrac{1}{10}$ **20.** $\dfrac{1}{15}, \dfrac{5}{6}$ **21.** $\dfrac{3}{10}, \dfrac{1}{6}$ **22.** $\dfrac{1}{20}, \dfrac{3}{25}$ **23.** $\dfrac{2}{25}, \dfrac{7}{30}$

▶ **Find the least common denominator.**

Here are scrambled answers for the next row of exercises: $8n^2$ $6n$ $4n$ n^2 $5n$

24. $\dfrac{3}{n}, \dfrac{1}{4n}$ **25.** $\dfrac{3}{5n}, \dfrac{m}{n}$ **26.** $\dfrac{5}{n}, \dfrac{3}{n^2}$ **27.** $\dfrac{t}{2n}, \dfrac{3t}{3n}$ **28.** $\dfrac{5}{4n^2}, \dfrac{3}{8n}$

29. $\dfrac{1}{3d}, \dfrac{4a}{d}$ **30.** $\dfrac{7}{8t}, \dfrac{3}{2t}$ **31.** $\dfrac{t}{r^2}, \dfrac{4}{4}$ **32.** $\dfrac{7}{2u}, \dfrac{1}{4}$ **33.** $\dfrac{3}{a}, \dfrac{4c}{a^3}$

34. $\dfrac{5}{a^2}, \dfrac{6b}{4a}$ **35.** $\dfrac{f}{3b}, \dfrac{2}{6b}$ **36.** $\dfrac{4}{e^2}, \dfrac{1}{5e^2}$ **37.** $\dfrac{b}{6a}, \dfrac{b}{3}$ **38.** $\dfrac{7}{c^2}, \dfrac{1}{c^3}$

39. $\dfrac{1}{2a}, \dfrac{1}{3a^2}$ **40.** $\dfrac{3}{5n^2}, \dfrac{1}{10n}$ **41.** $\dfrac{r}{6t^2}, \dfrac{s}{4t}$ **42.** $\dfrac{3}{3t^3}, \dfrac{1}{4t}$ **43.** $\dfrac{a}{9b}, \dfrac{a^2}{6b^2}$

 **ANALYZING DATA** *How Do You Measure Up?*

▶ **Work in a small group.**

44. Measure and record your height to the nearest inch.

45. a. Record the sex and height of each class member. For example, B58 would represent a boy having a height of 58 inches.
  b. Order the data from least to greatest for both boys and girls.

46. a. Compute the average height for both boys and girls.
  b. How do the averages compare?
  c. Was your height above or below the average for your sex?

47. Write a paragraph that summarizes the data.

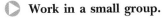

OBJECTIVE: To compare two fractions.

Some high school students were asked to name one person whose face appears on United States currency. The circle graph shows the results of the poll.

1. What fraction of the students named George Washington?

2. Whose portrait appears on the bill that was named by $\frac{2}{9}$ of the students?

3. What two fractions would you compare to decide whether more students named Washington or Hamilton?

EXAMPLES | Here's how to compare two fractions.

To compare fractions with a common denominator, compare the numerators.

$$\boxed{A} \qquad \frac{4}{9} > \frac{2}{9}$$

4 is greater than 2. So, $\frac{4}{9}$ is greater than $\frac{2}{9}$.

To compare fractions with different denominators, compare fractions with the same denominator that are equivalent to them.

$$\boxed{B} \qquad \frac{2}{9} \, \boxed{?} \, \frac{1}{3} \qquad\qquad \frac{2}{9} \, \boxed{?} \, \frac{3}{9} \qquad\qquad \frac{2}{9} < \frac{3}{9} \qquad\qquad \frac{2}{9} < \frac{1}{3}$$

9 is the LCD.

Write two fractions with the same denominator. Use the LCD.

Compare the numerators.

So, $\frac{2}{9}$ is less than $\frac{1}{3}$.

CHECK for Understanding

4. Look at Example B. Did more students name Lincoln or Hamilton?

EXERCISES

▶ < or >?

5. $\dfrac{2}{3} \diamond \dfrac{1}{3}$ **6.** $\dfrac{5}{9} \diamond \dfrac{7}{9}$ **7.** $\dfrac{5}{7} \diamond \dfrac{3}{7}$ **8.** $\dfrac{5}{4} \diamond \dfrac{7}{4}$ **9.** $\dfrac{2}{5} \diamond \dfrac{3}{5}$

10. $\dfrac{8}{8} \diamond \dfrac{9}{8}$ **11.** $\dfrac{7}{4} \diamond \dfrac{8}{4}$ **12.** $\dfrac{0}{9} \diamond \dfrac{1}{9}$ **13.** $\dfrac{7}{8} \diamond \dfrac{5}{8}$ **14.** $\dfrac{6}{5} \diamond \dfrac{7}{5}$

▶ <, >, or =? *Hint: First, write equivalent fractions with the same denominator.*

15. $\dfrac{1}{5} \diamond \dfrac{3}{10}$ **16.** $\dfrac{5}{8} \diamond \dfrac{3}{4}$ **17.** $\dfrac{1}{3} \diamond \dfrac{2}{7}$ **18.** $\dfrac{1}{4} \diamond \dfrac{1}{3}$ **19.** $\dfrac{5}{4} \diamond \dfrac{8}{7}$

$\boxed{\dfrac{2}{10}}\,\boxed{\dfrac{3}{10}}$ $\boxed{\dfrac{5}{8}}\,\boxed{\dfrac{6}{8}}$ $\boxed{\dfrac{7}{21}}\,\boxed{\dfrac{6}{21}}$ $\boxed{\dfrac{?}{12}}\,\boxed{\dfrac{?}{12}}$ $\boxed{\dfrac{?}{28}}\,\boxed{\dfrac{?}{28}}$

20. $\dfrac{2}{5} \diamond \dfrac{1}{4}$ **21.** $\dfrac{2}{9} \diamond \dfrac{6}{27}$ **22.** $\dfrac{1}{7} \diamond \dfrac{1}{8}$ **23.** $\dfrac{3}{5} \diamond \dfrac{3}{10}$ **24.** $\dfrac{3}{10} \diamond \dfrac{1}{3}$

25. $\dfrac{0}{3} \diamond \dfrac{0}{7}$ **26.** $\dfrac{2}{3} \diamond \dfrac{5}{9}$ **27.** $\dfrac{3}{4} \diamond \dfrac{3}{5}$ **28.** $\dfrac{5}{6} \diamond \dfrac{3}{4}$ **29.** $\dfrac{3}{4} \diamond \dfrac{2}{3}$

30. $\dfrac{15}{12} \diamond \dfrac{5}{4}$ **31.** $\dfrac{7}{8} \diamond \dfrac{8}{9}$ **32.** $\dfrac{4}{7} \diamond \dfrac{5}{8}$ **33.** $\dfrac{9}{16} \diamond \dfrac{5}{8}$ **34.** $\dfrac{3}{7} \diamond \dfrac{9}{21}$

▶ <, >, or =? Let $a = 4$, $b = 5$, $c = 3$, $d = 9$.

35. $\dfrac{1}{a} \diamond \dfrac{3}{a}$ **36.** $\dfrac{8}{b} \diamond \dfrac{7}{b}$ **37.** $\dfrac{3}{c} \diamond \dfrac{4}{c}$ **38.** $\dfrac{c}{4} \diamond \dfrac{c}{5}$ **39.** $\dfrac{d}{6} \diamond \dfrac{d}{5}$

40. $\dfrac{b+1}{c} \diamond \dfrac{b}{c}$ **41.** $\dfrac{a}{b} \diamond \dfrac{a+1}{b}$ **42.** $\dfrac{c+1}{d} \diamond \dfrac{c+2}{d}$ **43.** $\dfrac{c-1}{d} \diamond \dfrac{c}{d}$ **44.** $\dfrac{b}{c} \diamond \dfrac{b-1}{c}$

45. $\dfrac{c}{b} \diamond \dfrac{a}{d}$ **46.** $\dfrac{c}{b} \diamond \dfrac{b}{d}$ **47.** $\dfrac{c-1}{c} \diamond \dfrac{d-3}{d}$ **48.** $\dfrac{c}{a} \diamond \dfrac{b}{d-1}$ **49.** $\dfrac{b}{b+1} \diamond \dfrac{d-2}{d}$

Challenge! CALCULATOR

50. Match the fractions to the tags.
Hint: Divide to find the decimal name for each fraction.

$5 \div 14 = \boxed{0.3571428}$

$\dfrac{5}{14}$	$\dfrac{35}{26}$	$\dfrac{22}{13}$
$\dfrac{103}{38}$	$\dfrac{55}{48}$	$\dfrac{37}{18}$
$\dfrac{15}{22}$	$\dfrac{65}{27}$	$\dfrac{9}{17}$

OBJECTIVE: To use equations with like terms to solve problems.

In this lesson, you will solve problems by writing equations to represent the facts in given situations.

| EXAMPLES | Here's how to use an equation to solve a problem. |

Problem: Cora has 40 more stamps in her collection than Fran has in hers. Together they have 400 stamps. How many stamps does Fran have?

Step 1. Choose a variable. Use it and the facts to represent the numbers in the problems.

Let n = the number of stamps Fran has. Then $n + 40$ = the number of stamps Cora has.

| Number of Fran's stamps | Number of Cora's stamps |

So, $n + \overline{n + 40}$ = the number of stamps they have together.

Step 2. Write an equation based on the facts.

$$n + n + 40 = 400$$

These are two equal expressions for the total number of stamps.

Step 3. Solve the equation.

Fran has 180 stamps.

$$n + n + 40 = 400$$
$$2n + 40 = 400$$
$$2n + 40 - 40 = 400 - 40$$
$$2n = 360$$
$$\frac{2n}{2} = \frac{360}{2}$$
$$n = 180$$

CHECK for Understanding

1. Look at the example above.
   a. In Step 1, we let n equal the number of stamps Fran has. Then the expression $n + \underline{?}$ equals the number of stamps Cora has and $\underline{?} + n + 40$ equals the number of stamps Fran and Cora have together.
   b. In Step 2, the two equal expressions for the total number of stamps are $\underline{?}$ and 400.
   c. An equation for the problem is $n + n + 40 = 400$, or $\underline{?}\, n + 40 = \underline{?}$.
   d. Is *180 stamps* a reasonable answer? Explain how to check it.

EXERCISES

 Read the facts. Then complete the steps to answer the question.

2. *Facts:* Ken has 20 more dimes in his collection than Emily has in hers. Together they have 150 dimes.

Question: How many dimes does Emily have?

a. If you let *d* equal the number of dimes that Emily has, then $d +$ _?_ equals the number of dimes that Ken has and $d + d +$ _?_ equals the total number of dimes Emily and Ken have.

b. To find the number of dimes Emily has, you can solve the equation $d + d + 20 =$ _?_ .

c. Solve the equation in Part b. Emily has _?_ dimes.

3. *Facts:* Lou has 30 fewer nickels in his collection than Nara has in hers. Together they have 124 nickels.

Question: How many nickels does Nara have?

a. If you let *n* equal the number of nickels that Nara has, then $n -$ _?_ equals the number of nickels that Lou has and $n + n -$ _?_ equals the total number of nickels Nara and Lou have.

b. To find the number of nickels Nara has, you can solve the equation $n + n - 30 =$ _?_ .

c. Solve the equation in Part b. Nara has _?_ nickels.

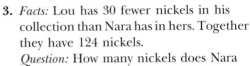

 Write an equation and solve the problem.

4. Manuel scored 8 fewer points than Will scored. Together they scored 44 points. How many points did Will score?
(Let $w =$ the number of points that Will scored.)

5. Mary Kae scored 9 more points than Emma scored. Together they scored 77 points. How many points did Emma score?
(Let $e =$ the number of points that Emma scored.)

6. A bottle cost 70¢ more than a cork. Together they cost 90¢. How much did the cork cost?
(Let $c =$ the cost of the cork.)

7. An eraser cost 10¢ less than a pencil. Together they cost 80¢. How much did the pencil cost?
(Let $p =$ the cost of the pencil.)

8. Together Tom and Melda weigh 262 pounds. Tom weighs 20 pounds more than Melda. How much does Melda weigh?
(Let $m =$ Melda's weight in pounds.)

9. Together Virginia and Tim weigh 267 pounds. Virginia weighs 45 pounds less than Tim. How much does Tim weigh?
(Let $t =$ Tim's weight in pounds.)

10. There are 52 students on the bus. There are 4 more boys than girls on the bus. How many girls are on the bus?
(Let $g =$ the number of girls on the bus.)

11. There are 128 pages in the book. There are 6 fewer pages without drawings than pages with drawings. How many pages with drawings are there?
(Let $d =$ the number of pages with drawings.)

OBJECTIVE: To solve problems using a calculator when appropriate.

Suppose that 100 years ago your great-great-grandfather deposited $100 in a savings account and never made any withdrawals. At the end of each year the bank multiplied the amount in the account by 1.05. (That is, the bank paid 5% interest compounded annually.)

You can use this formula to find how much money would be in the account at the end of each year.

Formula:

$$\underset{\substack{\nearrow \\ \text{original} \\ \$100}}{100} \cdot \underset{\substack{\uparrow \\ \text{growth} \\ \text{factor}}}{1.05^{\overset{\text{number of years}}{\underset{\downarrow}{n}}}} = \underset{\substack{\nwarrow \\ \text{Money in account} \\ \text{at end of n years}}}{M}$$

To find $100 \cdot 1.05^4$ (the amount of money in the account at the end of the 4th year) you can use two different calculator key sequences.

Using repeated multiplication:

100 ⊠ 1.05 ⊠ 1.05 ⊠ 1.05 ⊠ 1.05 ⊟ (*121.55063*)

Using the ⟦y^x⟧ key:

100 ⊠ 1.05 ⟦y^x⟧ 4 ⊟ (*121.55063*)

▶ **Solve. Use the formula.**

1. Look at the example above. To the nearest cent, how much money was in the account at the end of the 4th year?

2. How much money was in the account at the end of the 20th year?

3. How much more money was in the account at the end of the 30th year than at the end of the 20th year?

4. How much more money was in the account at the end of the 40th year than at the end of the 30th year?

5. At the end of which year did the amount in the account exceed $10,000?

6. **Write a problem** about great-great-grandfather's savings account that can be solved using the ⟦y^x⟧ key.

▶ **Give each difference.** *(page 98)*

1. $^+4 - {}^+2$ **2.** $^+4 - {}^-2$ **3.** $^-5 - {}^+8$ **4.** $^+5 - {}^+8$ **5.** $^-8 - 0$

6. $^-12 - {}^+10$ **7.** $^-13 - {}^-13$ **8.** $^+18 - {}^-11$ **9.** $^+15 - {}^+19$ **10.** $^+21 - {}^-3$

▶ **Give each product.** *(page 100)*

11. $^+3 \times {}^+5$ **12.** $^+3 \times {}^-5$ **13.** $^-7 \times {}^-4$ **14.** $^+7 \times {}^+4$ **15.** $^+7 \times {}^-1$

16. $^-11 \times {}^+8$ **17.** $^-12 \times {}^-12$ **18.** $^+15 \times {}^-6$ **19.** $^+16 \times {}^+10$ **20.** $^-21 \times {}^+10$

▶ **Give each quotient.** *(page 102)*

21. $^+12 \div {}^+3$ **22.** $^+12 \div {}^-3$ **23.** $^-12 \div {}^+3$ **24.** $^+8 \div {}^+1$ **25.** $^+16 \div {}^+2$

26. $^-32 \div {}^-4$ **27.** $^-36 \div {}^+9$ **28.** $^+45 \div {}^+5$ **29.** $^+54 \div {}^-9$ **30.** $^-63 \div {}^-9$

▶ **Complete.** *(pages 96, 98, 100, 102)*

31. $^+9 + {}^-3$ **32.** $^+8 - {}^+3$ **33.** $^+9 \times 0$ **34.** $^+24 \div {}^+3$ **35.** $^-11 + {}^-4$

36. $^-6 - {}^+10$ **37.** $^-4 \times {}^-11$ **38.** $^-48 \div {}^-12$ **39.** $^+17 + {}^+17$ **40.** $^-5 - {}^-9$

41. $^+10 \times {}^-11$ **42.** $^-60 \div {}^+5$ **43.** $^-20 + {}^+12$ **44.** $^-8 - {}^+15$ **45.** $^+10 \times {}^-15$

▶ **Solve and check.** *(page 108)*

46. $n + 7 = 4$ **47.** $x + 9 = {}^-3$ **48.** $a + 6 = 6$ **49.** $y + 8 = 8$

50. $r + 9 = 5$ **51.** $l - 12 = 7$ **52.** $12 + v = {}^-1$ **53.** $m - 9 = {}^-8$

▶ **Solve and check.** *(page 110)*

54. $6n = 42$ **55.** $5c = {}^-35$ **56.** $8d = 24$ **57.** $^-9x = 81$

58. $\dfrac{s}{3} = 12$ **59.** $\dfrac{v}{^-6} = 8$ **60.** $\dfrac{y}{^-10} = {}^-20$ **61.** $\dfrac{x}{^-30} = {}^-90$

▶ **Solve and check.** *(page 112)*

62. $\dfrac{c}{^-3} + 7 = {}^-8$ **63.** $^-7x + 4 = 32$ **64.** $\dfrac{y}{^-4} - {}^-2 = 10$ **65.** $9a - 4 = 23$

66. $\dfrac{d}{9} + 9 = 0$ **67.** $8r + {}^-5 = 11$ **68.** $^-10y + 6 = {}^-14$ **69.** $\dfrac{s}{^-7} - 3 = 8$

70. $^-4x + 8 = 0$ **71.** $\dfrac{n}{^-4} + {}^-6 = 1$ **72.** $12s - {}^-3 = {}^-9$ **73.** $\dfrac{n}{7} + {}^-5 = 0$

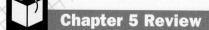

Here are scrambled answers for the review exercises:

$^-3$	6	$2c^2$	common	equivalent	multiply
1	10	$2y$	composite	exponent	power
3	36	add	denominator	factorization	prime
5	$2c$	algebraic	digits	multiple	subtract

1. In 4^3, 4 is called the base and 3 is called the __?__. The exponent tells you how many times the base is used as a factor. You can read n^3 as "n to the third __?__." *(page 122)*

2. To multiply expressions with like bases, __?__ the exponents. To divide expressions with like bases, __?__ the exponents. *(page 124)*

3. To express a number in scientific notation, you write it as a product so that the first factor is between __?__ and 10 and the second factor is a power of __?__. **390,000 = 3.9 × 10$^{\underline{?}}$**   **0.00465 = 4.65 × 10$^{\underline{?}}$** *(page 126)*

4. A whole number is divisible by 6 if it is divisible by both 2 and __?__. A whole number is divisible by 9 if the sum of its __?__ is divisible by 9. *(page 128)*

5. A whole number that has exactly two factors is called a __?__ number. A whole number (other than 0) that has more than two factors is called a __?__ number. The prime __?__ of 24 is $2 \cdot 2 \cdot 2 \cdot 3$. The __?__ factorization of $6a^2b$ is $2 \cdot 3 \cdot a \cdot a \cdot b$. *(page 130)*

6. Look at the prime factorization of these two numbers:
   $12 = 2 \cdot 2 \cdot 3$ $18 = 2 \cdot 3 \cdot 3$
   To find the greatest common factor, __?__ the factors that are common to both 12 and 18. The GCF is __?__. To find the least common multiple, multiply the factors that are common to both 12 and 18 by the factors that are not __?__ to both. The LCM is __?__. *(page 132)*

7. To change a fraction to an equivalent fraction, you can multiply both numerator and __?__ by the same number (not 0). If you multiply both the numerator and denominator of $\dfrac{x}{4y}$ by __?__ you get the equivalent fraction $\dfrac{2xy}{8y^2}$. *(page 138)*

8. To write $\dfrac{4c^2}{6cd}$ in lowest terms, you would divide both terms by __?__. *(page 140)*

9. To find the least common denominator of two fractions, find the least common __?__ of the denominators. The least common denominator of $\dfrac{b}{2}$ and $\dfrac{3b}{c^2}$ is __?__. *(page 142)*

10. To compare fractions with different denominators, compare __?__ fractions with the same denominator. *(page 144)*

▶ **Write using exponents.** *(page 122)*

1. $7 \cdot 7 \cdot 7$

2. $8 \cdot 8 \cdot 10 \cdot 10 \cdot 10$

3. $x \cdot x \cdot x \cdot y$

4. $y \cdot 3 \cdot x \cdot y \cdot x \cdot y$

▶ **Simplify using laws of exponents.** *(page 124)*

5. $3^4 \cdot 3^5$

6. $r^2 \cdot r^3$

7. $\dfrac{5^4}{5^1}$

8. $\dfrac{s^7}{s^3}$

▶ **Write in scientific notation.** *(page 126)*

9. 500,000

10. 683,000,000

11. 0.00067

12. 0.0000097

▶ **True or false?** *(page 128)*

13. 635 is divisible by 5.

14. 2842 is divisible by 8.

15. 1305 is divisible by 9.

▶ **Give the prime factorization or algebraic factorization.** *(page 130)*

16. 18

17. 24

18. $12a^3$

19. $30ab^2$

▶ **Give the greatest common factor and the least common multiple.** *(page 132)*

20. 8, 24

21. 15, 20

22. $9x$, 12

23. $7x^2$, $5xy^2$

▶ **Complete to get an equivalent fraction.** *(page 138)*

24. $\dfrac{5}{3} = \dfrac{?}{15}$

25. $\dfrac{7}{9} = \dfrac{?}{36}$

26. $\dfrac{10}{t} = \dfrac{?}{3t^2}$

27. $\dfrac{9s}{2t} = \dfrac{?}{4st^2}$

▶ **Write in lowest terms.** *(page 140)*

28. $\dfrac{9}{12}$

29. $\dfrac{15}{10}$

30. $\dfrac{27}{45}$

31. $\dfrac{8}{4y}$

32. $\dfrac{3xy^2}{15y}$

33. $\dfrac{12x^2y}{20xy^2}$

▶ **Give the least common denominator.** *(page 142)*

34. $\dfrac{5}{12}, \dfrac{3}{8}$

35. $\dfrac{4}{9}, \dfrac{7}{6}$

36. $\dfrac{1}{3w}, \dfrac{1}{4w^2}$

37. $\dfrac{3}{5wx}, \dfrac{1}{4w^2}$

▶ **< or >?** *(page 144)*

38. $\dfrac{3}{8} \diamondsuit \dfrac{5}{8}$

39. $\dfrac{1}{2} \diamondsuit \dfrac{1}{3}$

40. $\dfrac{2}{3} \diamondsuit \dfrac{3}{4}$

41. $\dfrac{6}{5} \diamondsuit \dfrac{9}{8}$

▶ **Write an equation and solve the problem.** *(page 146)*

42. Sid has 3 fewer dollars than Anne. Together they have 33 dollars. How many dollars does Anne have?
(Let n = the number of dollars Anne has.)

43. Caitlin has 3 more dollars than Elaine. Together they have 33 dollars. How many dollars does Elaine have?
(Let n = the number of dollars Elaine has.)

▶ **Choose the correct letter.**

1. Solve.

$$10 = \frac{y}{3} - 5$$

A. 5

B. 15

C. 45

D. none of these

2. $5x + 2 + 1 + 2x = \underline{\ ?\ }$

A. $3x + 5$

B. $5x + 3$

C. $2x + 3$

D. none of these

3. Solve.

$$16.4 = n + n$$

A. 8.2

B. 32.8

C. 16.4

D. none of these

4. Give the sum.

$$^-6 + {}^+2$$

A. $^+8$

B. $^-8$

C. $^-4$

D. none of these

5. Give the difference.

$$^+3 - {}^-5$$

A. $^+8$

B. $^-8$

C. $^-2$

D. none of these

6. Give the product.

$$^-6 \times {}^-1$$

A. $^+6$

B. $^-6$

C. $^+7$

D. none of these

7. Give the quotient.

$$^-10 \div {}^+2$$

A. $^+8$

B. $^-8$

C. $^+5$

D. none of these

8. Solve.

$$n - 7 = {}^-3$$

A. 10

B. $^-10$

C. 4

D. none of these

9. Solve.

$$n + 6 = 5$$

A. 11

B. $^-11$

C. 1

D. none of these

10. Solve.

$$^-4n = {}^-20$$

A. 80

B. $^-80$

C. 5

D. none of these

11. Solve.

$$\frac{n}{3} + 6 = {}^-15$$

A. $^-63$

B. $^-27$

C. $^-3$

D. none of these

12. Choose the equation.

You had $18. After being paid for working 6 hours, you then had a total of $39. How much were you paid per hour?

A. $\dfrac{n}{6} + 18 = 39$

B. $\dfrac{n}{6} - 18 = 39$

C. $6n + 18 = 39$

D. $6n - 18 = 39$

$$w = \frac{s}{8}$$

The amount of water (in inches) obtained from melting snow is equal to the amount of snow (in inches) divided by 8. (page 157)

S ome winter storms can produce 3 feet of snow in 24 hours. Winds can build snow-drifts that are over 20 feet deep.

OBJECTIVE: To write whole numbers and mixed numbers as fractions.

1. Look at a whole page of ads.

Does $1 = \dfrac{4}{4}$?

2. Look at 2 whole pages.

Does $2 = \dfrac{8}{4}$?

3. In all, there are ads on 2 whole pages plus __?__ fourths of a page. The mixed number $2\dfrac{3}{4}$ can be used to tell how many pages of ads. Read "$2\dfrac{3}{4}$" as "2 and $\dfrac{3}{4}$."

E X A M P L E S | **Here's how to change a whole number or a mixed number to an equivalent fraction.**

A **Change 2 to fourths.** $2 = \dfrac{?}{4}$

Write the whole number over the denominator 1.

$$\dfrac{2}{1}$$

Multiply both the numerator and denominator by 4.

$$\dfrac{2}{1} = \dfrac{2 \cdot 4}{1 \cdot 4}$$
$$= \dfrac{8}{4}$$

B **Change $2\dfrac{3}{4}$ to a fraction.** $2\dfrac{3}{4} = ?$

Multiply the denominator by the whole number. This gives the number of fourths in 2.

$$2\dfrac{3}{4}$$

Add the numerator. This gives the number of fourths in $2\dfrac{3}{4}$.

$$2\dfrac{3}{4} = \dfrac{11}{4}$$

There are 11 fourths in $2\dfrac{3}{4}$.

CHECK for Understanding

4. Look at Example B. To change $2\dfrac{3}{4}$ to a fraction, you would first __?__ 2 by 4 and then __?__ 3.

EXERCISES

▶ **Change to halves.**
Here are scrambled answers for the next row of exercises: $\frac{18}{2}$ $\frac{6}{2}$ $\frac{2}{2}$ $\frac{10}{2}$ $\frac{8}{2}$ $\frac{12}{2}$

5. 4 **6.** 1 **7.** 6 **8.** 9 **9.** 5 **10.** 3

11. 7 **12.** 12 **13.** 16 **14.** 2 **15.** 20 **16.** 10

▶ **Change to fifths.**

17. 3 **18.** 5 **19.** 1 **20.** 2 **21.** 7 **22.** 4

23. 12 **24.** 8 **25.** 11 **26.** 20 **27.** 9 **28.** 25

▶ **Change each mixed number to a fraction.**

29. $1\frac{1}{2}$ **30.** $2\frac{1}{3}$ **31.** $1\frac{1}{3}$ **32.** $2\frac{1}{4}$ **33.** $2\frac{1}{2}$ **34.** $1\frac{1}{4}$

35. $1\frac{3}{4}$ **36.** $1\frac{2}{5}$ **37.** $2\frac{4}{5}$ **38.** $1\frac{2}{3}$ **39.** $3\frac{1}{3}$ **40.** $8\frac{1}{6}$

41. $6\frac{3}{4}$ **42.** $6\frac{5}{6}$ **43.** $5\frac{3}{4}$ **44.** $10\frac{1}{8}$ **45.** $5\frac{1}{6}$ **46.** $5\frac{2}{5}$

47. $13\frac{1}{2}$ **48.** $20\frac{3}{8}$ **49.** $16\frac{1}{2}$ **50.** $25\frac{1}{4}$ **51.** $11\frac{3}{5}$ **52.** $12\frac{2}{3}$

Problem Solving USING DATA

▶ **Solve. Use the information in the chart.**

53. Who sold $1\frac{1}{4}$ pages of ads?

54. Who sold 2 pages of ads?

55. Which two people sold a total of 4 pages of ads?

56. Which two people sold a total of $3\frac{1}{4}$ pages of ads?

Salesperson	Number of $\frac{1}{4}$-page ads sold
Nevaro Clark	7
Stuart Long	5
Jane Felton	8
Brian Tully	1
Edita Sosa	9

Challenge! LOGICAL REASONING

▶ **Solve.**

57. Stuart sold his $\frac{1}{4}$-page ads two days before yesterday. Jane sold her ads the day after Stuart did. The day after tomorrow is Friday. On what day did Jane sell her ads?

OBJECTIVE: To write fractions as whole numbers or as mixed numbers.

A fraction can be changed to a whole number or mixed number if the denominator is less than or equal to the numerator.

Notice that 4 pieces make $\frac{4}{4}$, or 1, pizza and 7 pieces make $\frac{7}{4}$, or $1\frac{3}{4}$, pizzas.

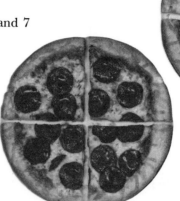

E X A M P L E S	**Here's how to change a fraction to an equivalent whole number or mixed number.**

To change a fraction to a whole number or mixed number, divide the numerator by the denominator.

A $\frac{15}{3} = ?$

Number of thirds in one → $3\overline{)15}$ ← *Number of thirds in all*

$$\frac{5}{3\overline{)15}}$$

$$\frac{15}{3} = 5$$

B $\frac{17}{5} = ?$

$$\begin{array}{r} 3 \\ 5\overline{)17} \\ -15 \\ \hline 2 \end{array}$$ ← *Number of fifths left over*

$$\frac{17}{5} = 3\frac{2}{5}$$

EXERCISES

▶ **Change each fraction to a whole number.**

Here are scrambled answers for the next row of exercises: 6   4   8   1   3   5

1. $\frac{6}{6}$ 2. $\frac{8}{2}$ 3. $\frac{5}{1}$ 4. $\frac{9}{3}$ 5. $\frac{24}{4}$ 6. $\frac{24}{3}$

7. $\frac{12}{3}$ 8. $\frac{50}{10}$ 9. $\frac{32}{4}$ 10. $\frac{36}{6}$ 11. $\frac{28}{2}$ 12. $\frac{40}{8}$

13. $\frac{75}{5}$ 14. $\frac{48}{4}$ 15. $\frac{42}{2}$ 16. $\frac{64}{8}$ 17. $\frac{30}{3}$ 18. $\frac{40}{10}$

▶ **Change each fraction to a whole number or mixed number.**

Here are scrambled answers for the next row of exercises: $1\frac{1}{4}$ $1\frac{5}{6}$ 2 9 3 $1\frac{1}{2}$

19. $\dfrac{9}{3}$ **20.** $\dfrac{16}{8}$ **21.** $\dfrac{5}{4}$ **22.** $\dfrac{3}{2}$ **23.** $\dfrac{11}{6}$ **24.** $\dfrac{81}{9}$

25. $\dfrac{13}{5}$ **26.** $\dfrac{7}{4}$ **27.** $\dfrac{11}{9}$ **28.** $\dfrac{13}{10}$ **29.** $\dfrac{5}{2}$ **30.** $\dfrac{10}{3}$

31. $\dfrac{16}{4}$ **32.** $\dfrac{17}{8}$ **33.** $\dfrac{11}{2}$ **34.** $\dfrac{14}{3}$ **35.** $\dfrac{27}{10}$ **36.** $\dfrac{19}{5}$

37. $\dfrac{27}{7}$ **38.** $\dfrac{15}{4}$ **39.** $\dfrac{18}{3}$ **40.** $\dfrac{30}{6}$ **41.** $\dfrac{35}{5}$ **42.** $\dfrac{36}{6}$

43. $\dfrac{25}{3}$ **44.** $\dfrac{36}{4}$ **45.** $\dfrac{35}{2}$ **46.** $\dfrac{29}{6}$ **47.** $\dfrac{37}{10}$ **48.** $\dfrac{19}{6}$

49. $\dfrac{144}{12}$ **50.** $\dfrac{182}{25}$ **51.** $\dfrac{137}{20}$ **52.** $\dfrac{195}{16}$ **53.** $\dfrac{225}{15}$ **54.** $\dfrac{253}{35}$

Problem Solving USING A FORMULA

▶ **Solve.**

55. If you melt 1 inch of snow, do you get more or less than 1 inch of water? *Hint: The formula below shows that 8 inches of snow contains 1 inch of water.*

56. Substitute in the formula at the left to complete this chart. Write the number of inches as a whole number or mixed number.

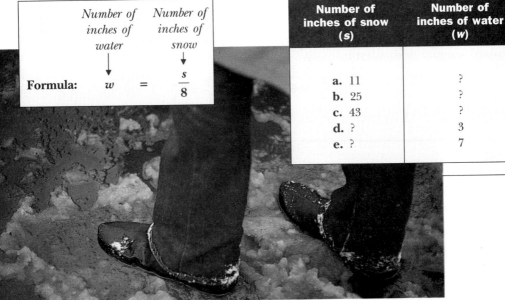

Number of inches of water	Number of inches of snow
↓	↓

Formula: $w = \dfrac{s}{8}$

Number of inches of snow (s)	Number of inches of water (w)
a. 11	?
b. 25	?
c. 43	?
d. ?	3
e. ?	7

Algebra Using Sums and Differences of Fractions | **157**

OBJECTIVE: To review writing fractions and mixed numbers in simplest form and extend that skill to algebraic fractions.

If the numerator of a fraction is greater than the denominator, then the fraction is greater than 1. These fractions are greater than 1.

1. Think about a fraction that has a numerator equal to its denominator. Is the fraction equal to 1?

2. Think about a fraction that has a numerator less than its denominator. Is the fraction less than 1?

EXAMPLES | **Here's how to write fractions and mixed numbers in simplest form.**

In arithmetic

Write fractions less than 1 in lowest terms.

$$\boxed{A} \quad \frac{10}{15} = \frac{2}{3}$$

In algebra

Write algebraic fractions in lowest terms.

$$\boxed{B} \quad \frac{10a^2}{6ab} = \frac{5a}{3b}$$

Write mixed numbers with the fraction part less than 1 and in lowest terms.

$$\boxed{C} \quad 4\frac{6}{8} = 4\frac{3}{4}$$

Write fractions that are greater than or equal to 1 as a whole number or as a mixed number in simplest form.

$$\boxed{D} \quad \frac{24}{4} = 6 \qquad \boxed{E} \quad \frac{17}{5} = 3\frac{2}{5}$$

EXERCISES

Write in simplest form.

Here are scrambled answers for the next row of exercises: $\frac{3}{5}$ $\frac{5}{6}$ $\frac{1}{6}$ $\frac{1}{2}$ $\frac{3}{4}$ $\frac{4}{5}$

3. $\frac{6}{8}$ 4. $\frac{15}{18}$ 5. $\frac{6}{10}$ 6. $\frac{5}{10}$ 7. $\frac{20}{25}$ 8. $\frac{2}{12}$

9. $\frac{16}{24}$ 10. $\frac{10}{12}$ 11. $\frac{6}{18}$ 12. $\frac{6}{9}$ 13. $\frac{8}{14}$ 14. $\frac{5}{20}$

▶ **Write in simplest form.**

15. $2\frac{4}{6}$　　**16.** $3\frac{6}{8}$　　**17.** $5\frac{10}{12}$　　**18.** $2\frac{2}{4}$　　**19.** $4\frac{2}{6}$　　**20.** $4\frac{2}{8}$

21. $6\frac{5}{10}$　　**22.** $5\frac{3}{12}$　　**23.** $7\frac{10}{15}$　　**24.** $2\frac{3}{9}$　　**25.** $8\frac{8}{10}$　　**26.** $5\frac{9}{12}$

▶ **Write in simplest form.**

27. $\dfrac{6}{3}$　　**28.** $\dfrac{9}{2}$　　**29.** $\dfrac{17}{3}$　　**30.** $\dfrac{8}{10}$　　**31.** $\dfrac{33}{6}$　　**32.** $\dfrac{8}{12}$

33. $\dfrac{10}{2}$　　**34.** $\dfrac{8}{3}$　　**35.** $\dfrac{15}{4}$　　**36.** $\dfrac{10}{8}$　　**37.** $\dfrac{36}{3}$　　**38.** $\dfrac{6}{24}$

39. $\dfrac{12}{4}$　　**40.** $\dfrac{12}{8}$　　**41.** $\dfrac{7}{4}$　　**42.** $\dfrac{16}{3}$　　**43.** $\dfrac{24}{36}$　　**44.** $\dfrac{8}{1}$

45. $\dfrac{36a}{3b}$　　**46.** $\dfrac{10x^2}{3xy}$　　**47.** $\dfrac{16x}{12xy^2}$　　**48.** $\dfrac{9b^2}{8b}$　　**49.** $\dfrac{22rs}{16s^2}$　　**50.** $\dfrac{25cd}{10cd}$

51. $\dfrac{24ab}{32b^2}$　　**52.** $\dfrac{24m^2}{8n^2}$　　**53.** $\dfrac{32xy}{24y^2}$　　**54.** $\dfrac{18c}{12d^2}$　　**55.** $\dfrac{14a}{7a}$　　**56.** $\dfrac{6r}{rs}$

57. $\dfrac{8m^2}{10m}$　　**58.** $\dfrac{12c^2}{9c}$　　**59.** $\dfrac{35m^2}{7m^2n^2}$　　**60.** $\dfrac{35st^2}{15st}$　　**61.** $\dfrac{24y^2z}{8yz}$　　**62.** $\dfrac{21a^2b}{14b^2}$

Group Project　　**ANALYZING DATA**　*Are You An Average Sleeper?*

▶ **Work in a small group.**

63. Estimate whether you sleep more or less than most members of your class.

64. a. List how many hours you slept each night last week. Round each time to the nearest half hour.
b. Compute your average sleep time. Round your average sleep time to the nearest half hour.

65. a. List the average sleep time of each class member.
b. Order the data from least to greatest.

66. Did you sleep more or less than most members of your class?

67. Write a paragraph that summarizes the data.

Algebra Using Sums and Differences of Fractions | **159**

OBJECTIVE: To change fractions to decimals. To change decimals to fractions.

Name	Number of hits	Number of at bats
Carlo	6	20
Brenner	7	21
McGrath	6	21
Rivera	6	16
Shepard	5	16

1. Rivera got a hit $\frac{6}{16}$ of his times at bat. What fraction of his times at bat did Brenner get a hit?

The batting average of a baseball player is the quotient of the number of "hits" divided by the number of "at bats." It is generally given as a decimal rounded to the nearest thousandth.

EXAMPLES | Here's how to change a fraction to a decimal.

To change a fraction to a decimal, divide the numerator by the denominator.

A **Rivera's average**
Pencil and paper:

$$\frac{6}{16} = ?$$

$$
\begin{array}{r}
0.375 \\
16\overline{)6.000} \\
-48 \\
\hline
120 \\
-112 \\
\hline
80 \\
-80 \\
\hline
0
\end{array}
$$

B **Brenner's average**
Pencil and paper:

$$\frac{7}{21} = ?$$

$$
\begin{array}{r}
0.333\frac{1}{3} \\
21\overline{)7.000} \\
-63 \\
\hline
70 \\
-63 \\
\hline
70 \\
-63 \\
\hline
7
\end{array}
$$

Write the remainder over the divisor in simplest form.

Calculator:

6 ÷ 16 = (0.375)

Calculator:

7 ÷ 21 = (0.3333333)

CHECK for Understanding

2. What is Rivera's batting average? Give your answer as a decimal.

3. What is Brenner's average rounded to the nearest thousandth?

EXAMPLES | Here's how to change a decimal to a fraction or mixed number in simplest form.

Read the decimal.   Write as a fraction or mixed number.   Write in simplest form.

C 0.4 ← *4 tenths* $0.4 = \frac{4}{10}$ $0.4 = \frac{4}{10} = \frac{2}{5}$

D 2.75 ← *2 and 75 hundredths* $2.75 = 2\frac{75}{100}$ $2.75 = 2\frac{75}{100} = 2\frac{3}{4}$

EXERCISES

▷ Change each fraction to a decimal. If the division does not come out evenly, express your answer in thousandths and the remainder as a fraction as in Example B.

Here are scrambled answers for the next row of exercises:

$0.375 \quad 0.444\frac{4}{9} \quad 0.4 \quad 0.25 \quad 0.333\frac{1}{3} \quad 0.833\frac{1}{3}$

4. $\dfrac{1}{4}$ **5.** $\dfrac{2}{5}$ **6.** $\dfrac{1}{3}$ **7.** $\dfrac{3}{8}$ **8.** $\dfrac{5}{6}$ **9.** $\dfrac{4}{9}$

10. $\dfrac{1}{8}$ **11.** $\dfrac{2}{9}$ **12.** $\dfrac{3}{4}$ **13.** $\dfrac{1}{5}$ **14.** $\dfrac{1}{2}$ **15.** $\dfrac{5}{3}$

16. $\dfrac{5}{12}$ **17.** $\dfrac{2}{3}$ **18.** $\dfrac{1}{6}$ **19.** $\dfrac{5}{8}$ **20.** $\dfrac{5}{4}$ **21.** $\dfrac{7}{6}$

▷ Change each decimal to a fraction or mixed number in simplest form.

Here are scrambled answers for the next row of exercises: $\dfrac{1}{5} \quad 1\dfrac{1}{4} \quad \dfrac{5}{8} \quad \dfrac{6}{25}$

Example

$0.33\dfrac{1}{3} = \dfrac{33\dfrac{1}{3}}{100}$

$= 33\dfrac{1}{3} \div 100$

$= \dfrac{1}{\cancel{100}} \cdot \dfrac{1}{\cancel{100}}$
$\qquad 3$

$= \dfrac{1}{3}$

22. 0.2   **23.** 0.24   **24.** 1.25   **25.** 0.625

26. 2.5   **27.** 0.5   **28.** 0.05   **29.** 0.16

30. 3.4   **31.** 1.75   **32.** 0.008   **33.** 0.875

34. 0.375   **35.** 0.75   **36.** 0.35   **37.** 2.25

38. $0.33\dfrac{1}{3}$ **39.** $0.66\dfrac{2}{3}$ **40.** $1.37\dfrac{1}{2}$ **41.** $0.08\dfrac{1}{3}$

Problem Solving USING DATA

▷ Solve. Use the table on page 160. Give batting averages as decimals rounded to the nearest thousandth. For some problems, you may wish to use a calculator.

42. What is Carlo's batting average?

43. What is Shepard's batting average?

44. Without computing, decide which player who got 6 hits has the highest batting average.

45. Rank the players in order from the highest batting average to the lowest batting average.

You will use a variety of problem-solving strategies on these two pages. For some problems, you may wish to use a calculator.

1 MAKE AN ORGANIZED LIST

The kayakers are ready to go downstream. In how many ways can they line up single file?

✓ *Problem Solving Tips*

Make an organized list.* List the ways that are possible with the white kayak first in line. Then list the ways that are possible with the blue, green, and red kayaks first.

WBGR	BWGR	GWBR	RWBG
WBRG	BWRG	•	•
WGBR	•	•	•
WGRB	•	•	•
WRBG	•		
WRGB			

*At least one more problem on these two pages can be solved by making an organized list.

2 DIGIT DILEMMA

A number consists of three digits: 9, 5, and another. If these digits are reversed and then subtracted from the original, the result will have the same digits, arranged in yet another order.

What is the missing digit?

3 GOT CHANGE FOR A QUARTER?

How many different ways can you make change for a quarter using pennies, nickels, and dimes?
Hint: Make an organized list.

4 AN ODD AVERAGE

Find the average of the first 50 odd numbers.

Hint: First solve a simpler problem. Find the average of the first 5 odd numbers: 1, 3, 5, 7, and 9.

5 YOUR ORDER, PLEASE

Put in grouping symbols to make the equation true.

a. $14 + {}^-5 \times {}^-3 - 6 = {}^-81$
b. $14 + {}^-5 \times {}^-3 - 6 = 23$
c. $14 + {}^-5 \times {}^-3 - 6 = 59$
d. $14 + {}^-5 \times {}^-3 - 6 = {}^-33$

6 GEOMETRY—VISUAL THINKING

How many triangles are in this figure?

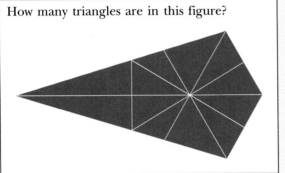

7 RELAY ROUNDUP

The track coach uses these guidelines to form a four-woman relay team.

- Choose from three freshmen: Ann, Barbara, and Carol, and three sophomores: Darlene, Edith, and Francine.
- Use two freshmen and two sophomores.

How many different four-woman relay teams are possible?

APPLICATIONS

A round-robin schedule has each team play each of the other teams once. The total number of games required can be determined by using this formula:

Number of games $\searrow$ Number of teams $\downarrow$
$$g = \frac{t(t-1)}{2}$$

8 Use the formula to complete the chart.

Sports league	Number of teams	Number of games in a round-robin schedule	Number of games in the official schedule
National Football League	28	**a.**	224
Major League Baseball	26	**b.**	2106
National Basketball Association	**c.***	351	1107
National Hockey League	**d.***	210	840

*Use the formula and the guess-and-check strategy.

9
The professional sports league with the greatest number of teams is the __?__ .

10
Would it be reasonable for any of the professional sports leagues to play a round-robin schedule? Why or why not?

Rate Your Problem-Solving Power: *6–7 correct = Good;   8–9 correct = Excellent;   10 correct = Exceptional*

OBJECTIVE: To read data from a bar graph to solve problems.

▶ **Use the bar graph to solve the problems.**

1. Which item cost $2 more in 1990 than it did in 1985?

2. Which item cost $2 less in 1990 than it did in 1985?

3. Which had the greater increase in price from 1985 to 1990, swim goggles or a bike lock?

4. In which year could you have paid for 3 flashlights with a $10 bill and received $1.75 in change?

5. In 1985, would $20 have been enough money to buy 2 flashlights, a calculator, and a pair of swim goggles?

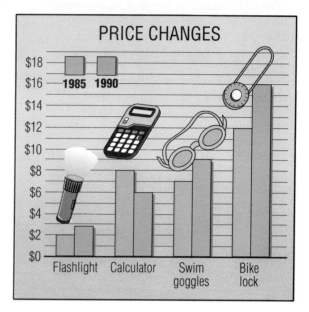

PRICE CHANGES

1985 1990

Flashlight   Calculator   Swim goggles   Bike lock

▶ **Read the facts. Then complete the steps to answer the question.**

6. *Facts:* The cost of a movie ticket in 1990 was $6.50. That is $.70 less than 2 times the cost of a movie ticket in 1980.
   Question: What was the cost of a movie ticket in 1980?
   a. Let t equal the cost in dollars of a movie ticket in 1980. Then __?__ $t - .70$ equals the cost of a movie ticket in 1990.
   b. Two equal expressions for the cost of a 1990 ticket are __?__ $t - .70$ and 6.50.
   c. To find the cost of a 1980 ticket, you can solve the equation $2t - .70 = $ __?__ .
   d. Solve the equation in Part c. What was the cost of a ticket in 1980?

7. *Facts:* The cost of a burger in 1990 was $1.20. That is $.30 less than 3 times the cost of a burger in 1980.
   Question: What was the cost of a burger in 1980?
   a. Let b equal the cost in dollars of a burger in 1980. Then $3b - $ __?__ equals the cost of a burger in 1990.
   b. Two equal expressions for the cost of a burger in 1990 are $3b - $ __?__ and 1.20.
   c. To find the cost of a burger in 1980, you can solve the equation $3b - .30 = $ __?__ .
   d. Solve the equation in Part c. What was the cost of a burger in 1980?

▶ **Write an equation and solve the problem.**

8. The cost of swim goggles in 1990 was $.50 less than 2 times the cost of swim goggles in 1980. What was the cost of swim goggles in 1980? *Hint: Use the bar graph to find the cost of swim goggles in 1990.*
   (Let s = the cost in dollars of swim goggles in 1980.)

9. If you multiply the cost of a bike lock in 1980 by 1.7 and then add $.87, you get the cost of a bike lock in 1990. What was the cost of a bike lock in 1980?
   Hint: Use the bar graph to find the cost of a bike lock in 1990.
   (Let b = the cost in dollars of a bike lock in 1980.)

▶ **Solve and check.** *(page 52)*

1. $51 = t - 16$ **2.** $f + 38 = 38$ **3.** $17 = x + 6.8$ **4.** $\dfrac{m}{5} = 42$

5. $22 = 8e$ **6.** $3r = 4.8$ **7.** $c - 8.2 = 10$ **8.** $25 = \dfrac{d}{4}$

9. $6.4 = y - 6.4$ **10.** $\dfrac{n}{8} = 41$ **11.** $1.2 = 12g$ **12.** $12 = y + 4.6$

▶ **Solve and check.** *(page 76)*

13. $3y + 6 = 15$ **14.** $\dfrac{k}{8} - 1 = 3$ **15.** $4x - 7 = 41$ **16.** $\dfrac{n}{4} - 20 = 0$

17. $40 = 6m - 2$ **18.** $2 = \dfrac{j}{3} - 1$ **19.** $8 = \dfrac{r}{9} - 6$ **20.** $64 = 12y + 4$

21. $\dfrac{d}{3} - 15 = 21$ **22.** $14 = \dfrac{k}{4} + 1$ **23.** $11p - 12 = 54$ **24.** $152 = 16r + 8$

25. $5t - 4.3 = 11.7$ **26.** $\dfrac{n}{6} - 1.5 = 2.4$ **27.** $18.4 = 4w - 5.6$ **28.** $3.6 = \dfrac{a}{8} + 2.7$

▶ **Solve and check.** *(page 82)*

29. $3x + 2x = 85$ **30.** $2w + 4w = 108$ **31.** $7z + z = 152$ **32.** $y + y = 76$

33. $98 = 3b + 4b$ **34.** $210 = 9f + f$ **35.** $153 = c + 8c$ **36.** $106 = c + c$

37. $g + 3g = 6.8$ **38.** $6h + 4h = 13.5$ **39.** $6.3 = j + 5j$ **40.** $14.3 = k + k$

41. $4d - 5.8 = 8.2$ **42.** $10 = 5f + 6.4$ **43.** $\dfrac{n}{2} + 4.1 = 20.7$ **44.** $14.5 = \dfrac{q}{3} - 4.5$

▶ **Give each sum.** *(page 96)*

45. $^{+}8 + {}^{+}6$ **46.** $^{-}6 + {}^{+}9$ **47.** $^{-}7 + {}^{-}5$ **48.** $^{+}8 + 0$

49. $^{+}11 + {}^{-}3$ **50.** $0 + {}^{-}15$ **51.** $^{+}2 + {}^{+}9$ **52.** $^{-}16 + {}^{+}16$

53. $^{-}30 + {}^{+}24$ **54.** $^{-}18 + {}^{-}12$ **55.** $^{+}29 + {}^{-}11$ **56.** $^{+}15 + {}^{+}18$

▶ **Give each product.** *(page 100)*

57. $^{+}10 \cdot {}^{+}6$ **58.** $^{-}9 \cdot {}^{+}7$ **59.** $^{+}8 \cdot {}^{-}6$ **60.** $^{-}9 \cdot {}^{-}4$

61. $^{-}12 \cdot 0$ **62.** $^{+}11 \cdot {}^{+}6$ **63.** $^{+}16 \cdot {}^{-}2$ **64.** $^{-}13 \cdot {}^{-}3$

65. $^{-}15 \cdot {}^{+}3$ **66.** $0 \cdot {}^{-}18$ **67.** $^{-}16 \cdot {}^{-}10$ **68.** $^{+}25 \cdot {}^{-}4$

OBJECTIVE: To review adding and subtracting fractions with common denominators and extend those skills to algebraic fractions.

This sign at the trail entrance shows the distance between the trail entrance and four scenic points.

1. What is the hiking distance to Shady Grove?
2. What two fractions would you add to compute the distance from Eagle Point to Beaver Dam?

| EXAMPLES | Here's how to add fractions with common denominators. |

To add fractions with common denominators, write the sum of the numerators over the common denominator.

In arithmetic

$\boxed{A}$ $\dfrac{5}{8} + \dfrac{7}{8} = \dfrac{12}{8} = 1\dfrac{1}{2}$ ← *Simplest form*

In algebra

$\boxed{B}$ $\dfrac{11}{c} + \dfrac{5}{c} = \dfrac{11+5}{c} = \dfrac{16}{c}$ ← *Simplest form*

CHECK for Understanding

3. Look at Example A. How far is it from Eagle Point to Beaver Dam?

4. Complete these examples.

   a. $\dfrac{3}{5} + \dfrac{4}{5} = \dfrac{?}{5} = 1\dfrac{?}{5}$

   b. $\dfrac{r}{t} + \dfrac{s}{t} = \dfrac{r+s}{?}$

   c. $\dfrac{7}{z} + \dfrac{8}{z} = \dfrac{7+8}{?} = \dfrac{?}{z}$

5. Look at the sign. What two fractions would you subtract to compute the distance from Shady Grove to Eagle Point?

| EXAMPLES | Here's how to subtract fractions with common denominators. |

To subtract fractions with common denominators, write the difference of the numerators over the common denominator.

In arithmetic

$\boxed{C}$ $\dfrac{5}{8} - \dfrac{3}{8} = \dfrac{2}{8} = \dfrac{1}{4}$ ← *Simplest form*

In algebra

$\boxed{D}$ $\dfrac{12}{c} - \dfrac{4}{c} = \dfrac{12-4}{c} = \dfrac{8}{c}$ ← *Simplest form*

CHECK for Understanding

6. Look at Example C. How far is it from Shady Grove to Eagle Point?

7. Explain how to get the number and letter printed in red in Examples C and D.

EXERCISES

▶ **Give each sum in simplest form.**

Here are scrambled answers for the next row of exercises: $\frac{2}{3}$ $1\frac{1}{4}$ $1\frac{2}{7}$ 1 $1\frac{1}{7}$

8. $\frac{1}{3} + \frac{1}{3}$ **9.** $\frac{4}{7} + \frac{5}{7}$ **10.** $\frac{5}{6} + \frac{1}{6}$ **11.** $\frac{6}{7} + \frac{2}{7}$ **12.** $\frac{3}{4} + \frac{2}{4}$

13. $\frac{3}{8} + \frac{7}{8}$ **14.** $\frac{1}{4} + \frac{3}{4}$ **15.** $\frac{3}{10} + \frac{1}{10}$ **16.** $\frac{7}{16} + \frac{5}{16}$ **17.** $\frac{1}{8} + \frac{1}{8}$

18. $\frac{b}{d} + \frac{c}{d}$ **19.** $\frac{r}{t} + \frac{s}{t}$ **20.** $\frac{2a}{c} + \frac{3}{c}$ **21.** $\frac{3x}{y} + \frac{w}{y}$ **22.** $\frac{3b}{d} + \frac{7}{d}$

23. $\frac{5}{a^2} + \frac{7}{a^2}$ **24.** $\frac{10}{xy} + \frac{6}{xy}$ **25.** $\frac{12}{pq} + \frac{3}{pq}$ **26.** $\frac{18}{k^2} + \frac{6}{k^2}$ **27.** $\frac{20}{a^2b} + \frac{12}{a^2b}$

▶ **Give each difference in simplest form.**

Here are scrambled answers for the next row of exercises: $\frac{1}{3}$ $\frac{2}{3}$ 0 2 $\frac{1}{5}$

28. $\frac{4}{15} - \frac{1}{15}$ **29.** $\frac{5}{9} - \frac{2}{9}$ **30.** $\frac{7}{12} - \frac{7}{12}$ **31.** $\frac{10}{9} - \frac{4}{9}$ **32.** $\frac{11}{4} - \frac{3}{4}$

33. $\frac{9}{8} - \frac{3}{8}$ **34.** $\frac{6}{4} - \frac{2}{4}$ **35.** $\frac{12}{10} - \frac{7}{10}$ **36.** $\frac{11}{12} - \frac{7}{12}$ **37.** $\frac{8}{3} - \frac{2}{3}$

38. $\frac{r}{t} - \frac{s}{t}$ **39.** $\frac{a}{d} - \frac{c}{d}$ **40.** $\frac{3x}{y} - \frac{w}{y}$ **41.** $\frac{a}{g} - \frac{2b}{g}$ **42.** $\frac{4d}{e} - \frac{5}{e}$

43. $\frac{11}{w^2} - \frac{4}{w^2}$ **44.** $\frac{9}{mn} - \frac{3}{mn}$ **45.** $\frac{20}{rs} - \frac{14}{rs}$ **46.** $\frac{16}{j^2} - \frac{2}{j^2}$ **47.** $\frac{25}{c^2d} - \frac{8}{c^2d}$

Problem Solving USING DATA

▶ **Solve. Use the trail sign on page 166.**

48. What is the hiking distance from Shady Grove to Beaver Dam?

49. What is the hiking distance from Crystal Cave to Beaver Dam?

50. How far is a round-trip from the trail entrance to Beaver Dam and back again?

★**51.** If you hiked at 1 mile per hour, could you hike from Crystal Cave to Eagle Point in less than $\frac{1}{2}$ hour?

Challenge! NUMBER SENSE

▶ **Tell whether the value of the expression increases or decreases when the value of *x* increases.**

52. $\frac{x}{2}$ **53.** $\frac{x}{-3}$ **54.** $10 - x$ **55.** x^3 **56.** $\frac{x}{3} + \frac{2x}{3}$

OBJECTIVE: To review adding and subtracting fractions with different denominators and extend those skills to algebraic fractions.

As part of their driver's test, some students were asked to identify 20 road signs. The table shows the results.

Signs identified	Number of students	Fraction of students
20	ЖΙ	$\frac{1}{8}$
19	ЖЖЖЖ	$\frac{5}{12}$
18	ЖЖ ΙΙ	$\frac{1}{4}$
17	Ж ΙΙΙ	$\frac{1}{6}$
16 or fewer	ΙΙ	$\frac{1}{24}$

1. Which two fractions would you add to find what fraction of the students identified 19 or more road signs?

EXAMPLES | **Here's how to add fractions with different denominators.**

To add fractions with different denominators, find the least common denominator (LCD), change to equivalent fractions, and add.

In arithmetic

$\boxed{A}$ $\frac{5}{12} + \frac{1}{8} = \frac{10}{24} + \frac{3}{24} = \frac{13}{24}$

The LCD is 24.

In algebra

$\boxed{B}$ $\frac{a}{6} + \frac{b}{4} = \frac{2a}{12} + \frac{3b}{12} = \frac{2a + 3b}{12}$

The LCD is 12.

CHECK for Understanding

2. Look at Example A. What fraction of the students identified 19 or more signs?

3. What fraction would you subtract from 1 to find the fraction of students who did not identify all 20 signs?

EXAMPLES | **Here's how to subtract fractions with different denominators.**

To subtract fractions with different denominators, find the least common denominator (LCD), change to equivalent fractions, and subtract.

In arithmetic

$\boxed{C}$ $1 - \frac{1}{8} = \frac{8}{8} - \frac{1}{8} = \frac{7}{8}$

The LCD is 8.

In algebra

$\boxed{D}$ $\frac{x}{6} - \frac{y}{9} = \frac{3x}{18} - \frac{2y}{18} = \frac{3x - 2y}{18}$

The LCD is 18.

CHECK for Understanding

4. Look at Example C. What fraction of the students did not identify all 20 signs?

EXERCISES

▶ **Give each sum in simplest form.**

Here are scrambled answers for the next row of exercises: $1\frac{1}{15}$ $1\frac{5}{24}$ $\frac{5}{8}$ $1\frac{1}{8}$ $\frac{5}{6}$

5. $\frac{1}{2} + \frac{1}{3}$ **6.** $\frac{3}{8} + \frac{1}{4}$ **7.** $\frac{2}{5} + \frac{2}{3}$ **8.** $\frac{5}{8} + \frac{1}{2}$ **9.** $\frac{5}{6} + \frac{3}{8}$

10. $\frac{5}{9} + \frac{1}{6}$ **11.** $\frac{3}{5} + \frac{1}{5}$ **12.** $\frac{7}{16} + \frac{1}{2}$ **13.** $\frac{5}{8} + \frac{1}{6}$ **14.** $\frac{2}{5} + \frac{1}{4}$

15. $\frac{a}{2} + \frac{b}{3}$ **16.** $\frac{n}{3} + \frac{3}{4}$ **17.** $\frac{c}{4} + \frac{d}{2}$ **18.** $\frac{5}{3} + \frac{k}{5}$ **19.** $\frac{r}{4} + \frac{s}{5}$

20. $\frac{2f}{5} + \frac{g}{6}$ **21.** $\frac{4}{5} + \frac{3q}{2}$ **22.** $\frac{3s}{4} + \frac{5}{6}$ **23.** $\frac{u}{5} + \frac{5v}{8}$ **24.** $\frac{5n}{2} + \frac{1}{8}$

25. $\frac{5b}{6} + \frac{7c}{9}$ **26.** $\frac{3r}{8} + \frac{5s}{4}$ **27.** $\frac{3g}{5} + \frac{7a}{10}$ **28.** $\frac{11m}{6} + \frac{7n}{3}$ **29.** $\frac{2x}{9} + \frac{7y}{12}$

▶ **Give each difference in simplest form.**

Here are scrambled answers for the next row of exercises: $\frac{3}{10}$ $\frac{7}{24}$ $\frac{1}{4}$ $\frac{1}{9}$ $\frac{5}{12}$

30. $\frac{3}{4} - \frac{1}{3}$ **31.** $\frac{7}{12} - \frac{1}{3}$ **32.** $\frac{5}{8} - \frac{1}{3}$ **33.** $\frac{9}{10} - \frac{3}{5}$ **34.** $\frac{2}{3} - \frac{5}{9}$

35. $\frac{2}{3} - \frac{7}{12}$ **36.** $\frac{5}{6} - \frac{2}{3}$ **37.** $\frac{7}{10} - \frac{1}{5}$ **38.** $\frac{1}{2} - \frac{1}{3}$ **39.** $\frac{7}{8} - \frac{0}{6}$

40. $\frac{c}{4} - \frac{d}{2}$ **41.** $\frac{m}{5} - \frac{1}{6}$ **42.** $\frac{k}{2} - \frac{5}{3}$ **43.** $\frac{x}{6} - \frac{y}{4}$ **44.** $\frac{u}{2} - \frac{v}{8}$

45. $\frac{2a}{3} - \frac{b}{4}$ **46.** $\frac{7c}{8} - \frac{d}{4}$ **47.** $\frac{3k}{2} - \frac{2}{5}$ **48.** $\frac{y}{3} - \frac{4z}{5}$ **49.** $\frac{5n}{6} - \frac{1}{3}$

50. $\frac{5c}{6} - \frac{7d}{9}$ **51.** $\frac{3x}{4} - \frac{2y}{5}$ **52.** $\frac{5r}{12} - \frac{2t}{9}$ **53.** $\frac{2m}{5} - \frac{7n}{8}$ **54.** $\frac{4j}{5} - \frac{9k}{10}$

Problem Solving USING DATA

▶ **Solve. Use the table on page 168.**

55. What fraction of the students identified 17 or fewer road signs?

56. What fraction of the students missed 2 or fewer of the road signs?

57. What fraction of the students identified 17 or more of the road signs?

58. What fraction of the students missed at least 1 of the road signs?

OBJECTIVE: To add and subtract fractions with different algebraic denominators.

Katy is trying to use what she knows about adding fractions in order to combine fractions with different algebraic denominators.

$$\frac{x}{2} + \frac{5}{y} = \frac{}{2y} + \frac{}{2y}$$

1. Look at Katy's work. The least common denominator of $\frac{x}{2}$ and $\frac{5}{y}$ is __?__.

E X A M P L E S	Here's how to add fractions with different algebraic denominators.

To add fractions with different algebraic denominators, find the least common denominator (LCD), change to equivalent fractions, and add.

$\boxed{A}$ $\dfrac{x}{2} + \dfrac{5}{y} = \dfrac{x \cdot y}{2 \cdot y} + \dfrac{2 \cdot 5}{2 \cdot y} = \dfrac{xy + 10}{2y}$

The LCD is 2y.

$\boxed{B}$ $\dfrac{2}{a} + \dfrac{3a}{b} = \dfrac{2 \cdot b}{a \cdot b} + \dfrac{3a \cdot a}{a \cdot b} = \dfrac{2b + 3a^2}{ab}$

The LCD is ab.

CHECK for Understanding

2. Look at Example A. To change $\frac{x}{2}$ to the equivalent fraction $\frac{xy}{2y}$, both numerator and denominator were multiplied by __?__.

3. Complete these examples.

   a. $\dfrac{5}{4x} + \dfrac{y}{6x} = \dfrac{15}{12x} + \dfrac{?}{12x} = \dfrac{15 + ?}{12x}$

   b. $\dfrac{3}{s} + \dfrac{2}{r} = \dfrac{?}{rs} + \dfrac{2s}{rs} = \dfrac{? + 2s}{rs}$

E X A M P L E S	Here's how to subtract fractions with different algebraic denominators.

To subtract fractions with different algebraic denominators, find the least common denominator, change to equivalent fractions, and subtract.

$\boxed{C}$ $\dfrac{7}{x^2} - \dfrac{5}{x} = \dfrac{7}{x^2} - \dfrac{5 \cdot x}{x \cdot x} = \dfrac{7 - 5x}{x^2}$

The LCD is x^2.

$\boxed{D}$ $\dfrac{a}{3} - \dfrac{2}{b} = \dfrac{a \cdot b}{3 \cdot b} - \dfrac{2 \cdot 3}{3 \cdot b} = \dfrac{ab - 6}{3b}$

The LCD is 3b.

CHECK for Understanding

4. Look at Example D. To change $\frac{a}{3}$ to the equivalent fraction $\frac{ab}{3b}$, both numerator and denominator were multiplied by __?__.

5. Complete these examples.

   a. $\dfrac{3}{2x} - \dfrac{y}{3x} = \dfrac{9}{6x} - \dfrac{?}{6x} = \dfrac{9 - ?}{6x}$

   b. $\dfrac{5}{m} - \dfrac{3}{n} = \dfrac{?}{mn} - \dfrac{3m}{mn} = \dfrac{? - 3m}{mn}$

EXERCISES

Add.
Here are scrambled answers for the next row of exercises. $\dfrac{2+ab}{2a}$ $\dfrac{ab+2}{b}$ $\dfrac{2+b}{ab}$ $\dfrac{a+2b}{ab}$ $\dfrac{ab+2}{2b}$

6. $\dfrac{1}{a}+\dfrac{b}{2}$ **7.** $\dfrac{1}{b}+\dfrac{2}{a}$ **8.** $\dfrac{a}{1}+\dfrac{2}{b}$ **9.** $\dfrac{2}{ab}+\dfrac{1}{a}$ **10.** $\dfrac{a}{2}+\dfrac{1}{b}$

11. $\dfrac{2}{x}+\dfrac{y}{3}$ **12.** $\dfrac{3}{x}+\dfrac{1}{y}$ **13.** $\dfrac{x}{1}+\dfrac{3}{y}$ **14.** $\dfrac{1}{x}+\dfrac{2}{xy}$ **15.** $\dfrac{1}{xy}+\dfrac{2}{y}$

16. $\dfrac{a}{3}+\dfrac{2}{b}$ **17.** $\dfrac{3}{a}+\dfrac{2a}{b}$ **18.** $\dfrac{5}{2a}+\dfrac{b}{3a}$ **19.** $\dfrac{3}{ab}+\dfrac{2}{b}$ **20.** $\dfrac{a}{3b}+\dfrac{2}{b}$

21. $\dfrac{1}{r}+\dfrac{2}{s}$ **22.** $\dfrac{2}{rs}+\dfrac{3}{r}$ **23.** $\dfrac{5}{s}+\dfrac{r}{2s}$ **24.** $\dfrac{s}{1}+\dfrac{s}{r}$ **25.** $\dfrac{2}{r^2}+\dfrac{5}{r}$

Subtract.
Here are scrambled answers for the next row of exercises. $\dfrac{a^2-b^2}{ab}$ $\dfrac{ab-2}{2a}$ $\dfrac{ab-1}{b}$ $\dfrac{1-2b}{ab}$ $\dfrac{2b-3a}{ab}$

26. $\dfrac{b}{2}-\dfrac{1}{a}$ **27.** $\dfrac{a}{1}-\dfrac{1}{b}$ **28.** $\dfrac{2}{a}-\dfrac{3}{b}$ **29.** $\dfrac{a}{b}-\dfrac{b}{a}$ **30.** $\dfrac{1}{ab}-\dfrac{2}{a}$

31. $\dfrac{2}{x}-\dfrac{y}{5}$ **32.** $\dfrac{3}{x}-\dfrac{1}{y}$ **33.** $\dfrac{x}{1}-\dfrac{3}{y}$ **34.** $\dfrac{1}{xy}-\dfrac{2}{y}$ **35.** $\dfrac{1}{x}-\dfrac{2}{xy}$

36. $\dfrac{a}{2}-\dfrac{3}{b}$ **37.** $\dfrac{2}{a}-\dfrac{3a}{b}$ **38.** $\dfrac{2}{3a}-\dfrac{b}{5a}$ **39.** $\dfrac{3}{ab}-\dfrac{2}{b}$ **40.** $\dfrac{2}{b}-\dfrac{1}{2b}$

41. $\dfrac{1}{s}-\dfrac{3}{r}$ **42.** $\dfrac{3}{r}-\dfrac{2}{rs}$ **43.** $\dfrac{s}{1}-\dfrac{2}{r}$ **44.** $\dfrac{3}{rs}-\dfrac{1}{s}$ **45.** $\dfrac{3}{r^2}-\dfrac{2}{r}$

Problem Solving FINDING ERRORS

Write a sentence telling what is incorrect in each exercise. Then show the correct answer.

46. $\dfrac{x}{2}+\dfrac{5}{y}=\dfrac{x}{2y}+\dfrac{5}{2y}=\dfrac{x+5}{2y}$

47. $\dfrac{7}{r}-\dfrac{6}{s}=\dfrac{7}{rs}-\dfrac{6s}{rs}=\dfrac{7r-6s}{rs}$

48. $\dfrac{6}{z}-5=\dfrac{6-5}{z}=\dfrac{1}{z}$

49. $\dfrac{p}{a}+\dfrac{h^2}{a^2}=\dfrac{p^2}{a^2}+\dfrac{h^2}{a^2}=\dfrac{p^2+h^2}{a^2}$

OBJECTIVE: To add and subtract mixed numbers without regrouping.

1. A balloonist who weighs $145\frac{1}{2}$ pounds plans to take $23\frac{1}{4}$ pounds of equipment on a balloon ride. What would you do to compute the total weight of the balloonist and the equipment?

2. Suppose that you are in a balloon that is 1000 feet above the ground. Suppose also that your friend is in a balloon that is 200 feet above the ground. What would you do to compute how much farther you can see than your friend? Look at the chart.

Height of balloon in feet	10	50	100	200	300	1000	3000
Distance seen from balloon in miles	$3\frac{7}{8}$	$8\frac{1}{2}$	$12\frac{3}{8}$	$17\frac{3}{8}$	22	$38\frac{3}{4}$	$67\frac{1}{8}$

| EXAMPLES | Here's how to add and subtract mixed numbers. |

To add (or subtract) mixed numbers with different denominators, first write fractions with a common denominator. Next add (or subtract) the equivalent fractions. Then add (or subtract) the whole numbers.

$\boxed{A}$ $145\frac{1}{2} + 23\frac{1}{4} = ?$ $\boxed{B}$ $38\frac{3}{4} - 17\frac{3}{8} = ?$

$$145\frac{1}{2} = 145\frac{2}{4}$$
$$+ 23\frac{1}{4} = + 23\frac{1}{4}$$
$$\overline{\qquad\qquad 168\frac{3}{4}}$$

Change to a common denominator. Add the fractions. Add the whole numbers.

$$38\frac{3}{4} = 38\frac{6}{8}$$
$$-17\frac{3}{8} = -17\frac{3}{8}$$
$$\overline{\qquad\qquad 21\frac{3}{8}}$$

Change to a common denominator. Subtract the fractions. Subtract the whole numbers.

CHECK for Understanding

3. Look at the examples above. What is the total weight? How much farther can you see?

4. Complete these examples.

a. $4\frac{2}{5}$
$+2\frac{1}{5}$
$\overline{\quad 6\frac{?}{}}$

b. $4\frac{1}{6} = 4\frac{4}{24}$
$+2\frac{3}{8} = +2\frac{9}{24}$
$\overline{\quad ?\frac{13}{24}}$

c. $8\frac{7}{9}$
$-3\frac{1}{9}$
$\overline{\quad 5\frac{6}{9} = 5\frac{?}{}}$

d. $10\frac{1}{3} = 10\frac{4}{12}$
$-2\frac{1}{4} = -2\frac{?}{}$
$\overline{\quad 8\frac{1}{12}}$

EXERCISES

▶ **Add. Write the sum in simplest form.**
Here are scrambled answers for the next row of exercises: $5\frac{7}{8}$ $6\frac{7}{8}$ $7\frac{19}{20}$ $4\frac{5}{6}$

5. $3\frac{1}{3} + 1\frac{1}{2}$ **6.** $4\frac{1}{2} + 2\frac{3}{8}$ **7.** $1\frac{3}{4} + 4\frac{1}{8}$ **8.** $5\frac{3}{4} + 2\frac{1}{5}$

9. $4\frac{1}{2} + 3\frac{1}{5}$ **10.** $4\frac{1}{3} + 8\frac{1}{5}$ **11.** $8\frac{5}{12} + 1\frac{1}{8}$ **12.** $6 + 3\frac{3}{8}$

13. $7\frac{3}{8} + 9\frac{1}{6}$ **14.** $9\frac{1}{5} + 5\frac{3}{10}$ **15.** $2\frac{1}{8} + 2\frac{3}{8}$ **16.** $2\frac{5}{8} + 1\frac{1}{5}$

17. $1\frac{1}{4} + 3\frac{2}{5}$ **18.** $6\frac{1}{2} + 3\frac{1}{9}$ **19.** $3\frac{5}{12} + 4\frac{1}{2}$ **20.** $8\frac{1}{6} + 2\frac{5}{9}$

▶ **Subtract. Write the difference in simplest form.**
Here are scrambled answers for the next row of exercises: $3\frac{1}{8}$ $1\frac{1}{4}$ $2\frac{3}{8}$ $4\frac{1}{4}$

21. $4\frac{5}{8} - 2\frac{1}{4}$ **22.** $2\frac{1}{2} - 1\frac{1}{4}$ **23.** $5\frac{7}{8} - 2\frac{3}{4}$ **24.** $6\frac{3}{4} - 2\frac{1}{2}$

25. $7\frac{1}{2} - 4\frac{1}{8}$ **26.** $5\frac{1}{3} - 2\frac{1}{4}$ **27.** $15\frac{1}{2} - 8\frac{1}{4}$ **28.** $83\frac{1}{2} - 12\frac{3}{10}$

29. $15\frac{3}{4} - 6\frac{5}{8}$ **30.** $8\frac{4}{5} - 1\frac{1}{4}$ **31.** $2\frac{7}{10} - 1\frac{3}{5}$ **32.** $18\frac{1}{4} - 8\frac{1}{6}$

33. $5\frac{7}{8} - 4\frac{1}{2}$ **34.** $6\frac{4}{9} - 4\frac{1}{3}$ **35.** $9\frac{5}{8} - 2\frac{1}{6}$ **36.** $9\frac{7}{10} - 3\frac{1}{4}$

Problem Solving USING EQUATIONS

▶ **First choose the equation and solve the problem. Then reread the problem to see if your answer makes sense.**

| Equation A: | $n + 1\frac{1}{4} = 2\frac{1}{2}$ |
| Equation B: | $n - 1\frac{1}{4} = 2\frac{1}{2}$ |

37. Your hot-air balloon is $1\frac{1}{4}$ miles above another balloon. If you are at a height of $2\frac{1}{2}$ miles, what is the height of the other balloonist?

38. The difference in height of two balloons is $2\frac{1}{2}$ miles. If the lower balloon is at a height of $1\frac{1}{4}$ miles, what is the height of the other balloon?

39. Your balloon landed $1\frac{1}{4}$ miles closer to the target than another balloon did. If you were $2\frac{1}{2}$ miles from the target, how far was the other balloon from the target?

40. Your balloon ride took $1\frac{1}{4}$ hours less than your friend's ride did. If your ride took $2\frac{1}{2}$ hours, how many hours did your friend's ride take?

Algebra Using Sums and Differences of Fractions | **173**

OBJECTIVE: To add and subtract mixed numbers with regrouping.

Recommended Daily Amounts of Food			
Dog Size	Breed	Age 5–9 weeks	Age 10–15 weeks
Small	Pug Dachshund	$1\frac{2}{3}$ cups	$2\frac{1}{4}$ cups
Medium	Cocker Spaniel Border Collie	$2\frac{3}{4}$ cups	$4\frac{1}{2}$ cups
Large	Old English Sheepdog Retriever	$3\frac{1}{2}$ cups	$6\frac{1}{3}$ cups

1. You have a 6-week-old dachshund and a 7-week-old retriever. What two mixed numbers would you add to find the total number of cups of food you should feed them each day?

2. What two numbers would you subtract to find how much more food each day you should feed a 14-week-old retriever than a 12-week-old cocker spaniel?

EXAMPLES | Here's how to add and subtract mixed numbers with regrouping.

$\boxed{A}$ Find the sum. $\quad 1\frac{2}{3} + 3\frac{1}{2} = ?$

$$1\frac{2}{3} = 1\frac{4}{6}$$

$$+3\frac{1}{2} = +3\frac{3}{6}$$

$$\overline{4\frac{7}{6}} = 5\frac{1}{6}$$

Change to a common denominator.

Since $\frac{7}{6} = 1\frac{1}{6}$, regroup.

$\boxed{B}$ Find the difference. $\quad 6\frac{1}{3} - 4\frac{1}{2} = ?$

$$6\frac{1}{3} = 6\overset{5\ \ 8}{\frac{2}{6}}$$

$$-4\frac{1}{2} = -4\frac{3}{6}$$

$$\overline{1\frac{5}{6}}$$

Not enough sixths! So regroup 1 to get $\frac{6}{6}$.

CHECK for Understanding

3. Look at Examples A and B.
   a. How many cups of food should you feed a 6-week-old dachshund and a 7-week-old retriever each day?
   b. How much more food each day should you feed a 14-week-old retriever than a 12-week-old cocker spaniel?

4. Complete these examples.

a.
$$4\frac{5}{6} = 4\frac{20}{24}$$
$$+4\frac{3}{8} = +4\frac{9}{24}$$
$$\overline{8\frac{29}{24}} = 9\frac{?}{?}$$

b.
$$14 = 13\frac{2}{2}$$
$$-3\frac{1}{2} = -3\frac{1}{2}$$
$$\overline{?\frac{1}{2}}$$

Regroup 1 to get $\frac{2}{2}$.

c.
$$6\frac{1}{2} = 6\overset{5\ \ ?}{\frac{4}{8}}$$
$$-2\frac{7}{8} = -2\frac{7}{8}$$
$$\overline{3\frac{5}{8}}$$

EXERCISES

Add. Write the sum in simplest form.

Here are scrambled answers for the next row of exercises: $9\frac{3}{8}$ $6\frac{7}{8}$ $6\frac{1}{6}$ $5\frac{1}{2}$ $9\frac{1}{8}$

5. $4\frac{2}{3} + 1\frac{1}{2}$ **6.** $2\frac{1}{2} + 6\frac{5}{8}$ **7.** $3\frac{3}{4} + 5\frac{5}{8}$ **8.** $1\frac{1}{8} + 5\frac{3}{4}$ **9.** $2\frac{5}{6} + 2\frac{2}{3}$

10. $6\frac{3}{5} + 2\frac{3}{10}$ **11.** $3\frac{1}{5} + 6\frac{7}{10}$ **12.** $2\frac{9}{10} + 4\frac{1}{2}$ **13.** $2\frac{1}{2} + 5\frac{2}{5}$ **14.** $7\frac{11}{12} + 2\frac{1}{4}$

15. $2\frac{1}{6} + 2\frac{7}{8}$ **16.** $3\frac{5}{8} + 2\frac{1}{2}$ **17.** $1\frac{1}{4} + 3\frac{4}{5}$ **18.** $6\frac{1}{2} + 3\frac{2}{3}$ **19.** $3\frac{7}{12} + 4\frac{1}{2}$

Subtract. Write the difference in simplest form.

Here are scrambled answers for the next row of exercises: $4\frac{3}{8}$ $2\frac{3}{4}$ $2\frac{1}{4}$ $4\frac{3}{4}$ $4\frac{7}{8}$

20. $13\frac{1}{4} - 8\frac{1}{2}$ **21.** $7\frac{1}{8} - 2\frac{3}{4}$ **22.** $10 - 5\frac{1}{8}$ **23.** $6\frac{5}{8} - 3\frac{7}{8}$ **24.** $9\frac{1}{8} - 6\frac{7}{8}$

25. $8\frac{1}{3} - 2\frac{1}{2}$ **26.** $10\frac{1}{4} - 1\frac{2}{3}$ **27.** $7\frac{7}{8} - 1\frac{3}{4}$ **28.** $15\frac{1}{6} - 3\frac{2}{3}$ **29.** $9 - 7\frac{1}{2}$

30. $20 - 8\frac{3}{4}$ **31.** $5\frac{5}{8} - 2\frac{3}{4}$ **32.** $4\frac{2}{3} - 1\frac{1}{2}$ **33.** $4\frac{1}{10} - 3\frac{1}{2}$ **34.** $9\frac{2}{3} - 8\frac{1}{2}$

35. $2\frac{2}{9} - 1\frac{1}{3}$ **36.** $18 - 8\frac{5}{8}$ **37.** $6\frac{1}{2} - 2\frac{1}{8}$ **38.** $9\frac{1}{10} - 5\frac{3}{5}$ **39.** $10\frac{1}{4} - 5\frac{2}{3}$

Problem Solving USING DATA

Solve. Use the chart on page 174.

40. Would $9\frac{1}{2}$ cups of food each day be enough food for an 8-week-old Old English sheepdog and a 12-week-old retriever?

41. If dog food costs 18¢ per cup, how much does it cost each day to feed a 15-week-old pug and a 5-week-old border collie?

Challenge! MAKE AN ORGANIZED LIST

42. A dog-food manufacturer is testing which of four new formulations dogs prefer. The dishes are put in a different order for each test. How many different orders are possible?

Algebra Using Sums and Differences of Fractions | **175**

OBJECTIVE: To use equations with mixed numbers to solve problems.

In earlier chapters, you wrote addition and subtraction equations with whole numbers to solve problems. You can use the same method to solve problems with mixed numbers.

EXAMPLE	Here's how to use an equation to solve a problem.

Problem: Eve is $4\frac{1}{2}$ inches taller than Theresa. Eve is $67\frac{3}{4}$ inches tall. How tall is Theresa?

Step 1. Choose a variable. Use it and the facts to represent the numbers in the problem.

Let t = Theresa's height in inches. Then $t + 4\frac{1}{2}$ = Eve's height in inches.

Step 2. Write an equation based on the facts.

$$t + 4\frac{1}{2} = 67\frac{3}{4}$$

These are two equal expressions for Eve's height.

Step 3. Solve the equation.

$$t + 4\frac{1}{2} = 67\frac{3}{4}$$
$$t + 4\frac{1}{2} - 4\frac{1}{2} = 67\frac{3}{4} - 4\frac{1}{2}$$
$$t = 63\frac{1}{4}$$

Theresa is $63\frac{1}{4}$ inches tall.

CHECK for Understanding

1. Look at the example above.
   a. Read the problem. What is the relevant information?
   b. In Step 1, we let t equal __?__ height in inches. Then the expression $t +$ __?__ equals Eve's height in inches.
   c. In Step 2, the two equal expressions for Eve's height in inches are __?__ and __?__.
   d. An equation for the problem is __?__

2. To check the solution, ask yourself if the answer fits the facts. Is Theresa $63\frac{1}{4}$ inches tall?

EXERCISES

▶ First choose the equation and solve the problem. Then reread the problem to see if your answer makes sense.

> **Equation A:** $n + 5\frac{3}{4} = 68\frac{1}{2}$
>
> **Equation B:** $n - 5\frac{3}{4} = 68\frac{1}{2}$

3. If Berni grows $5\frac{3}{4}$ inches, he will be as tall as Amber. If Amber is $68\frac{1}{2}$ inches tall, how tall is Berni?

4. Beryl is $5\frac{3}{4}$ inches shorter than Jack. If Beryl is $68\frac{1}{2}$ inches tall, how tall is Jack?

5. The sum of a number n and $5\frac{3}{4}$ is $68\frac{1}{2}$. What is the number?

6. A number n decreased by $5\frac{3}{4}$ is $68\frac{1}{2}$. What is the number?

7. A number n increased by $5\frac{3}{4}$ is $68\frac{1}{2}$. What is the number?

8. A number n minus $5\frac{3}{4}$ equals $68\frac{1}{2}$. What is the number?

9. The difference between two numbers is $68\frac{1}{2}$. If the smaller number is $5\frac{3}{4}$, what is the larger number?

10. $5\frac{3}{4}$ more than a number n is $68\frac{1}{2}$. What is the number?

11. Write a problem that can be solved using Equation A. Then solve the problem.

12. Write a problem that can be solved using Equation B. Then solve the problem.

▶ Write an equation and solve the problem.

13. A number decreased by $3\frac{1}{8}$ is $14\frac{1}{2}$. What is the number?
(Let n = the number.)

14. A number increased by $12\frac{3}{10}$ is $20\frac{2}{5}$. What is the number?
(Let m = the number.)

15. The sum of two numbers is $20\frac{2}{3}$. If one of the numbers is $6\frac{1}{2}$, what is the other number?
(Let n = the other number.)

16. The difference between two numbers is $2\frac{3}{5}$. If the smaller number is $6\frac{7}{10}$, what is the larger number?
(Let n = the larger number.)

17. Dana worked $3\frac{1}{2}$ more hours than Celia did. If Dana worked $26\frac{1}{4}$ hours, how many hours did Celia work?
(Let c = the number of hours Celia worked.)

18. Mario worked $10\frac{3}{4}$ hours less than Jean did. If Mario worked $28\frac{1}{2}$ hours, how many hours did Jean work?
(Let j = the number of hours Jean worked.)

19. Dennis worked 5 more than twice as many hours as Helen. If he worked 43 hours, how many hours did Helen work?
(Let h = the number of hours Helen worked.)

20. Kari worked 3 times as many hours as Neil. Together they worked 64 hours. How many hours did Neil work?
(Let n = the number of hours Neil worked.)

Algebra Using Sums and Differences of Fractions | **177**

OBJECTIVE: To solve problems using a calculator when appropriate.

The amount you pay for a TV is the **retail price.** What the appliance store paid was the **wholesale price.** The difference between the retail price and the wholesale price is the **markup.**

Look at the price tags. The store used this letter code to keep the wholesale prices a secret.

Wholesale-Price Code										
A	B	C	D	E	F	G	H	I	J	K
1	2	3	4	5	6	7	8	9	0	.

EFIKGI $389.99

BACKAI $319.99

▶ **Use the price-tag information and the wholesale-price code to solve these problems.**

1. What is the wholesale price of the TV that costs $319.99?

2. What is the markup on the $389.99 TV?

3. If the markup is $287.75 for a refrigerator that is coded DFCKDI, what is the retail price?

4. What wholesale-price code would appear on a $729.95 refrigerator that has been marked up $279.89?

▶ **Use the formulas. Fill in the facts that fit the calculator sequence.**

5. 39.95 ⊟ 7.79 ⊟ (32.16)

   With a $7.79 coupon, you can buy a $___**a.**___
   (number)
   radio at the sale price of $___**b.**___.
   (number)

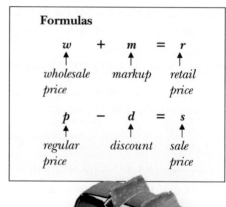

Formulas

$$w + m = r$$
wholesale price · markup · retail price

$$p - d = s$$
regular price · discount · sale price

▶ **Use the formulas. Decide when a calculator would be helpful.**

6. What is the retail price of a microwave oven that has a $99 markup on a $261.14 wholesale price?

7. A toaster that is on sale for $37.99 has been discounted $9.99. What is the regular price?

8. With a $7.50-off coupon, you can buy a coffee maker for $39.59. What is the regular price?

9. **Write a problem** that can be solved using the formulas. Then solve the problem.

▶ **Solve.** *(page 110)*

1. $5n - 10 = {}^-30$

2. ${}^-6y + 3 = {}^-27$

3. $2d - 16 = 20$

4. $4k + 8 = {}^-24$

5. $\dfrac{c}{-2} + 4 = {}^-3$

6. $\dfrac{x}{5} - 6 = 5$

7. $\dfrac{d}{-3} - 4 = 0$

8. $\dfrac{f}{6} + 8 = 5$

9. ${}^-2g - 4 = 16$

10. $\dfrac{r}{8} - {}^-1 = 9$

11. $7a + 3 = {}^-11$

12. $\dfrac{s}{-6} + {}^-2 = 0$

13. $\dfrac{d}{6} + 8 = 2$

14. ${}^-4y + 2 = {}^-30$

15. $\dfrac{w}{-3} - 6 = {}^-1$

16. $5y - {}^-3 = 18$

▶ **Write using exponents.** *(page 122)*

17. $4 \cdot 4 \cdot 4$

18. $10 \cdot 10$

19. $6 \cdot 6 \cdot 8 \cdot 8$

20. $a \cdot a \cdot b$

21. $x \cdot x \cdot x \cdot y \cdot y$

22. $3 \cdot x \cdot x \cdot y$

23. $2 \cdot y \cdot z \cdot y \cdot z$

24. $6 \cdot r \cdot s \cdot s \cdot s$

25. $z \cdot y \cdot 6 \cdot y \cdot z \cdot y$

▶ **Give the prime or algebraic factorization.** *(page 130)*

26. 6

27. 14

28. 18

29. 16

30. 24

31. 42

32. 36

33. 45

34. 48

35. $9x^2$

36. $10g^3$

37. $15d$

38. $11c^3$

39. $8n^3$

40. $20xy^2$

41. $25r^2s^2$

42. $18cd^3$

43. $27y^3z^2$

▶ **Complete to get an equivalent fraction.** *(page 138)*

44. $\dfrac{5}{8} = \dfrac{?}{32}$

45. $\dfrac{3}{4} = \dfrac{?}{32}$

46. $\dfrac{7}{8} = \dfrac{?}{24}$

47. $\dfrac{5}{9} = \dfrac{?}{36}$

48. $\dfrac{7}{16} = \dfrac{?}{48}$

49. $\dfrac{5}{c} = \dfrac{?}{3c}$

50. $\dfrac{b}{4} = \dfrac{?}{4b}$

51. $\dfrac{3r}{2s} = \dfrac{?}{8s^2}$

52. $\dfrac{5y}{2x} = \dfrac{?}{6xy}$

53. $\dfrac{3a}{4b} = \dfrac{?}{12b^2}$

54. $\dfrac{3mn}{m^2n} = \dfrac{?}{5m^2n^2}$

55. $\dfrac{8y}{3yz} = \dfrac{?}{6y^2z}$

56. $\dfrac{3a^2b}{7} = \dfrac{?}{21b}$

57. $\dfrac{8fg}{3} = \dfrac{?}{12fg}$

58. $\dfrac{6y^2}{5x} = \dfrac{?}{20xy}$

▶ **<, >, or =?** Let $a = 3$, $b = 4$, $c = 2$, $d = 7$. *(page 144)*

59. $\dfrac{3}{d} \diamond \dfrac{1}{d}$

60. $\dfrac{2}{b} \diamond \dfrac{3}{b}$

61. $\dfrac{b}{9} \diamond \dfrac{c}{9}$

62. $\dfrac{b}{2} \diamond \dfrac{b}{4}$

63. $\dfrac{c}{3} \diamond \dfrac{d}{7}$

64. $\dfrac{3}{b} \diamond \dfrac{3}{d}$

65. $\dfrac{1}{a} \diamond \dfrac{1}{c}$

66. $\dfrac{1}{b} \diamond \dfrac{c}{8}$

67. $\dfrac{2}{5} \diamond \dfrac{b}{10}$

68. $\dfrac{2}{a} \diamond \dfrac{b}{4}$

69. $\dfrac{c}{b} \diamond \dfrac{a}{d}$

70. $\dfrac{c}{d} \diamond \dfrac{c}{a}$

71. $\dfrac{b}{d} \diamond \dfrac{c}{a}$

72. $\dfrac{b-1}{d} \diamond \dfrac{c+1}{d}$

73. $\dfrac{d+1}{b} \diamond \dfrac{d-1}{b}$

Here are scrambled answers for the review exercises:

1	5	denominator	fraction	multiply	subtract
2	4t	difference	form	numerators	sum
3	add	divide	least	regroup	terms
4	common	equivalent	lowest	simplest	whole

1. To change a whole number to a fraction, write the whole number over the denominator 1 and __?__ both numerator and denominator by the same number. To change $3\frac{5}{6}$ to a fraction, multiply 6 by __?__ and add __?__ to the product. *(page 154)*

2. To change $\frac{14}{3}$ to a mixed number, first __?__ 14 by 3.
The remainder __?__ tells the number of thirds left over. *(page 156)*

$$\begin{array}{r} 4 \\ 3\overline{)14} \\ -12 \\ \hline 2 \end{array}$$

3. To write fractions and mixed numbers in simplest form: Write fractions less than 1 in __?__ terms. Write algebraic fractions in lowest __?__. Write mixed numbers with the __?__ part less than 1 and in lowest terms. *(page 158)*

4. To change a fraction to a decimal, divide the numerator by the __?__. *(page 160)*

5. To add fractions with common denominators, write the __?__ of the numerators over the __?__ denominator. To subtract fractions with common denominators, write the __?__ of the __?__ over the common denominator. *(page 166)*

6. To add fractions with different denominators, find the __?__ common denominator, change to equivalent fractions, and __?__. To subtract fractions with different denominators, find the least common denominator, change to __?__ fractions, and __?__. *(page 168)*

7. To find the sum $\frac{2}{t} + \frac{r}{4t}$, multiply the numerator and denominator of the first fraction by __?__, because the least common denominator of the fractions is __?__. *(page 170)*

8. To find this sum, first add the fractions and then add the whole numbers. The last step would be to write the sum in __?__ form. *(page 172)*

$$\begin{array}{r} 4\frac{1}{2} = 4\frac{3}{6} \\ +3\frac{1}{6} = +3\frac{1}{6} \\ \hline \end{array}$$

9. To find this difference, first subtract the fractions and then subtract the __?__ numbers. The last step would be to write the difference in simplest __?__. *(page 172)*

$$\begin{array}{r} 5\frac{2}{3} = 5\frac{4}{6} \\ -2\frac{1}{6} = -2\frac{1}{6} \\ \hline \end{array}$$

10. To find this sum, you would next __?__ $\frac{5}{4}$ to get $1\frac{1}{4}$. *(page 174)*

$$\begin{array}{r} 5\frac{1}{2} = 5\frac{2}{4} \\ +2\frac{3}{4} = +2\frac{3}{4} \\ \hline 7\frac{5}{4} \end{array}$$

11. To find this difference, you would next regroup __?__ to get $\frac{8}{8}$ and then subtract. *(page 174)*

$$\begin{array}{r} 6\frac{1}{8} = 6\frac{1}{8} \\ -3\frac{1}{4} = -3\frac{2}{8} \\ \hline \end{array}$$

▶ **Change each mixed number to a fraction.** *(page 154)*

1. $1\frac{1}{3}$ **2.** $1\frac{3}{4}$ **3.** $2\frac{1}{5}$ **4.** $3\frac{2}{3}$ **5.** $2\frac{5}{6}$ **6.** $4\frac{5}{8}$

▶ **Change each fraction to a whole number or mixed number.** *(page 156)*

7. $\frac{4}{2}$ **8.** $\frac{5}{3}$ **9.** $\frac{12}{4}$ **10.** $\frac{15}{4}$ **11.** $\frac{17}{6}$ **12.** $\frac{29}{8}$

▶ **Write in simplest form.** *(page 158)*

13. $\frac{9}{3}$ **14.** $\frac{12}{16}$ **15.** $\frac{22}{8}$ **16.** $\frac{12b}{2b}$ **17.** $\frac{5e}{ef}$ **18.** $\frac{8c^2}{10cd}$

▶ **Change each fraction to a decimal.** *(page 160)*

19. $\frac{1}{5}$ **20.** $\frac{3}{4}$ **21.** $\frac{5}{8}$ **22.** $\frac{2}{3}$ **23.** $\frac{3}{2}$ **24.** $\frac{11}{8}$

▶ **Change each decimal to a fraction or mixed number in simplest form.**
(page 160)

25. 0.08 **26.** 0.25 **27.** $0.33\frac{1}{3}$ **28.** 0.5 **29.** $0.16\frac{2}{3}$ **30.** 1.375

▶ **Give each sum or difference in simplest form.** *(pages 166, 168, 170, 172, 174)*

31. $\frac{1}{8} + \frac{5}{8}$ **32.** $\frac{5}{6} + \frac{5}{6}$ **33.** $\frac{r}{s} - \frac{t}{s}$ **34.** $\frac{8}{ab} + \frac{4}{ab}$ **35.** $\frac{2c}{d^2} - \frac{3}{d^2}$

36. $\frac{1}{3} - \frac{1}{4}$ **37.** $\frac{3}{8} + \frac{3}{4}$ **38.** $\frac{c}{6} + \frac{d}{3}$ **39.** $\frac{3x}{4} - \frac{y}{6}$ **40.** $\frac{4e}{9} + \frac{5f}{6}$

41. $\frac{1}{ab} + \frac{3}{a}$ **42.** $\frac{d}{c} - \frac{2}{5c}$ **43.** $\frac{e}{4} - \frac{f}{e}$ **44.** $\frac{5}{2g} + \frac{h}{3g}$ **45.** $\frac{5}{n^2} - \frac{1}{n}$

46. $15\frac{5}{6} + 11\frac{1}{8}$ **47.** $18\frac{7}{8} - 12\frac{1}{3}$ **48.** $5\frac{3}{4} + 2\frac{1}{3}$ **49.** $6 - 4\frac{3}{5}$ **50.** $12\frac{1}{5} - 8\frac{7}{10}$

▶ **Write an equation and solve the problem.** *(page 176)*

51. A number decreased by $3\frac{1}{2}$ is $6\frac{3}{4}$. What is the number?

52. A number increased by $3\frac{1}{2}$ is $6\frac{3}{4}$. What is the number?

53. The difference between two numbers is $3\frac{1}{2}$. If the smaller number is $6\frac{3}{4}$, what is the larger number?

54. The sum of two numbers is $6\frac{3}{4}$. If one of the numbers is $3\frac{1}{2}$, what is the other number?

▶ **Choose the correct letter.**

---

1. Solve.

$$58 = 26 + j$$

A. 32

B. 84

C. 58

D. none of these

---

2. Solve.

$$12 = \frac{n}{6} - 4$$

A. 72

B. 48

C. 96

D. none of these

---

3. Solve.

$$60 = 3c + c$$

A. 15

B. 240

C. 20

D. none of these

---

4. Give the sum.

$$^-8 + {}^+5$$

A. $^+13$

B. $^-13$

C. $^+3$

D. $^-3$

---

5. Give the product.

$$^-9 \cdot {}^-3$$

A. $^+12$

B. $^-12$

C. $^+27$

D. $^-27$

---

6. Solve.

$$\frac{t}{^-3} - 2 = 1$$

A. $^-1$

B. $^-9$

C. 9

D. none of these

---

7. Solve.

$$3n + 6 = {}^-18$$

A. 8

B. 4

C. $^-4$

D. none of these

---

8. $r \cdot r \cdot r \cdot s \cdot s = \underline{\ ?\ }$

A. $r^2 s^3$

B. $r^3 s$

C. $r^3 s^2$

D. none of these

---

9. The algebraic factorization of $15x^2y$ is

A. $15 \cdot x \cdot x \cdot y$

B. $3 \cdot 5 \cdot x^2 \cdot y$

C. $3 \cdot 5 \cdot x \cdot x \cdot y$

D. none of these

---

10. Complete.

$$\frac{3c}{2d} = \frac{?}{4cd}$$

A. $2c$

B. $4c$

C. $2cd$

D. none of these

---

11. $\dfrac{3}{4} > \underline{\ ?\ }$

A. $\dfrac{4}{5}$

B. $\dfrac{5}{7}$

C. $\dfrac{7}{9}$

D. $\dfrac{9}{10}$

---

12. Choose the equation.

Twelve more than the quotient of a number divided by 6 is 18. What is the number?

A. $6n + 12 = 18$

B. $6n - 12 = 18$

C. $\dfrac{n}{6} - 12 = 18$

D. $\dfrac{n}{6} + 12 = 18$

Algebra Using Products and Quotients of Fractions

7

$b = c \cdot h \cdot w$

The number of calories a person burns during an activity equals the number of calories burned per pound in one hour (for that activity) times the number of hours spent in the activity times the person's weight in pounds.

(page 204)

One hour of dancing burns 2.2 calories per pound. A 120-pound person burns 132 calories during a half hour of dancing, while a 150-pound person burns 165 calories.

183

OBJECTIVE: To review multiplying fractions and extend that skill to multiplying algebraic fractions.

1. What fraction of the green page is covered with baseball cards?

2. To find what fraction of the page is covered with baseball cards, you would multiply $\frac{3}{4}$ by what number?

EXAMPLES | Here's how to multiply fractions.

To multiply fractions, multiply the numerators to get the numerator of the product and multiply the denominators to get the denominator of the product.

In arithmetic

$$\boxed{A} \quad \frac{3}{4} \cdot \frac{2}{3} = \frac{3 \cdot 2}{4 \cdot 3}$$
$$= \frac{6}{12}$$
$$= \frac{1}{2} \leftarrow \boxed{\textit{Simplest form}}$$

In algebra

$$\boxed{B} \quad \frac{x}{y} \cdot \frac{y}{z} = \frac{x \cdot y}{y \cdot z}$$
$$= \frac{x}{z} \leftarrow \boxed{\textit{Simplest form}}$$

Here is a shortcut called canceling that can be used when multiplying fractions.

$$\boxed{C} \quad \frac{\overset{1}{\cancel{3}}}{\underset{2}{\cancel{4}}} \cdot \frac{\overset{1}{\cancel{2}}}{\underset{1}{\cancel{3}}} = \frac{1}{2}$$

Divide numerator and denominator by the common factors 3 and 2 before multiplying.

$$\boxed{D} \quad \frac{x}{\cancel{y}} \cdot \frac{\overset{1}{\cancel{y}}}{z} = \frac{x}{z}$$

Divide numerator and denominator by y.

CHECK for Understanding

3. Complete these examples.

a. $\dfrac{5}{\underset{2}{\cancel{6}}} \cdot \overset{1}{\cancel{3}} = \dfrac{5}{2}$

$= 2\dfrac{?}{2}$

b. $\dfrac{\overset{3}{\cancel{6}}}{5} \cdot \dfrac{15}{\underset{4}{\cancel{8}}} = \dfrac{9}{4}$

$= 2\dfrac{?}{4}$

c. $\overset{1}{\cancel{t}} \cdot \dfrac{t}{\underset{1}{\cancel{t}}} = \underline{\ ?\ }$

d. $\dfrac{\overset{1}{\cancel{2}}a}{b^2} \cdot \dfrac{b}{\underset{2}{\cancel{4}}} = \dfrac{?}{2b}$

EXERCISES

▷ **Give each product in simplest form.**
Here are scrambled answers for the next row of exercises: $2\frac{1}{4}$ $\frac{1}{2}$ 6 1 $\frac{3}{8}$ $\frac{1}{4}$

4. $\frac{3}{2} \cdot \frac{1}{4}$ **5.** $\frac{5}{8} \cdot \frac{4}{5}$ **6.** $\frac{1}{3} \cdot \frac{3}{4}$ **7.** $\frac{3}{4} \cdot 3$ **8.** $\frac{5}{2} \cdot \frac{2}{5}$ **9.** $\frac{9}{2} \cdot \frac{4}{3}$

10. $6 \cdot \frac{3}{10}$ **11.** $\frac{1}{2} \cdot \frac{1}{3}$ **12.** $2 \cdot \frac{1}{2}$ **13.** $5 \cdot \frac{2}{5}$ **14.** $\frac{5}{8} \cdot \frac{2}{5}$ **15.** $\frac{1}{4} \cdot \frac{1}{4}$

16. $\frac{2}{3} \cdot 2$ **17.** $\frac{5}{12} \cdot \frac{3}{2}$ **18.** $\frac{1}{2} \cdot \frac{2}{5}$ **19.** $\frac{1}{2} \cdot \frac{1}{2}$ **20.** $\frac{3}{2} \cdot \frac{0}{2}$ **21.** $\frac{3}{5} \cdot \frac{1}{3}$

22. $\frac{3}{5} \cdot \frac{5}{3}$ **23.** $\frac{7}{10} \cdot \frac{5}{4}$ **24.** $\frac{11}{12} \cdot 3$ **25.** $4 \cdot \frac{3}{4}$ **26.** $\frac{6}{5} \cdot \frac{15}{2}$ **27.** $\frac{1}{2} \cdot \frac{2}{3}$

▷ **Give each product in simplest form.**
Here are scrambled answers for the next row of exercises: $\frac{p}{q}$ $\frac{2n}{mp}$ $\frac{u^2}{xy}$ $\frac{uv}{xy}$ $\frac{2m}{n^2}$

28. $\frac{u}{x} \cdot \frac{v}{y}$ **29.** $\frac{6}{q} \cdot \frac{p}{6}$ **30.** $\frac{u}{x} \cdot \frac{u}{y}$ **31.** $\frac{m}{n} \cdot \frac{2}{n}$ **32.** $\frac{n}{m} \cdot \frac{2}{p}$

33. $\frac{x}{5} \cdot \frac{1}{x}$ **34.** $\frac{b}{a} \cdot c$ **35.** $\frac{8}{j} \cdot 3$ **36.** $\frac{m}{3} \cdot \frac{9}{n}$ **37.** $\frac{5}{j} \cdot \frac{k}{10}$

38. $\frac{2}{5} \cdot \frac{15}{w}$ **39.** $12 \cdot \frac{5x}{3}$ **40.** $\frac{6}{5} \cdot \frac{5u}{v}$ **41.** $\frac{2u}{y} \cdot \frac{v}{2y}$ **42.** $\frac{4}{9} \cdot \frac{3x}{y}$

43. $\frac{v}{2w} \cdot \frac{2}{z}$ **44.** $\frac{r}{s} \cdot \frac{s}{r}$ **45.** $\frac{x}{y} \cdot \frac{y}{x}$ **46.** $\frac{a}{b} \cdot \frac{b}{a}$ **47.** $\frac{2}{9e} \cdot \frac{3d}{4}$

48. $\frac{4v}{3} \cdot \frac{1}{8}$ **49.** $\frac{2r}{s} \cdot \frac{8}{2r}$ **50.** $\frac{a}{c} \cdot \frac{b}{a^2}$ **51.** $\frac{b^2}{15} \cdot \frac{1}{b}$ **52.** $\frac{9}{c} \cdot \frac{c^2}{4}$

53. $\frac{a}{b^2} \cdot \frac{2b}{3}$ **54.** $\frac{3c}{b} \cdot \frac{b}{c^2}$ **55.** $\frac{s^2}{r} \cdot \frac{r}{s}$ **56.** $\frac{m}{n} \cdot \frac{n^2}{3}$ **57.** $\frac{7}{d} \cdot \frac{3d}{e}$

Challenge!   LOGICAL REASONING

▷ **Study the clues to find the final score.**

58. *Clues:*
  • After 6 complete innings, the Cardinals were leading the Cubs 5 to 2.
  • Only one team scored in the last 3 innings.
  • The winning team won by 3 runs.

59. *Clues:*
  • After 8 complete innings, the Orioles were leading the Indians 3 to 1.
  • The game went extra innings.
  • Both teams scored after the 8th inning.
  • The teams scored a total of 7 runs.

OBJECTIVE: To multiply mixed numbers.

1. How much flour is needed to make pizza dough?
2. Suppose that you want to make $1\frac{1}{2}$ times the recipe. What two mixed numbers would you multiply to find how much flour is needed?

PIZZA DOUGH

2 packages dry yeast $1\frac{1}{4}$ cups warm water
$4\frac{1}{2}$ cups flour $1\frac{3}{4}$ teaspoons sugar
$1\frac{1}{2}$ teaspoons salt   2 eggs
$3\frac{1}{4}$ teaspoons oil

Sprinkle yeast over water and stir until disolved. Stir in 2 cups flour, the sugar and salt. Add egg and oil and stir until smooth and glossy. Stir in about 2 cups or enough of remaining flour to keep dough from sticking. On lightly floured surface knead dough until smooth.

EXAMPLES | Here's how to multiply mixed numbers.

A $1\frac{1}{2} \cdot 4\frac{1}{2} = ?$

$1\frac{1}{2} \cdot 4\frac{1}{2} = \dfrac{3}{2} \cdot \dfrac{9}{2}$ Change each mixed number to a fraction.

$= \dfrac{27}{4}$ Multiply.

$= 6\dfrac{3}{4}$ Write the product in simplest form.

B $3\frac{1}{3} \cdot 1\frac{1}{6} = \dfrac{\overset{5}{\cancel{10}}}{3} \cdot \dfrac{7}{\underset{3}{\cancel{6}}}$

$= \dfrac{35}{9}$

$= 3\dfrac{8}{9}$

C $5\frac{3}{4} \cdot 2\frac{2}{3} = \dfrac{23}{\underset{1}{\cancel{4}}} \cdot \dfrac{\overset{2}{\cancel{8}}}{3}$

$= \dfrac{46}{3}$

$= 15\dfrac{1}{3}$

D $2\frac{5}{6} \cdot 9 = \dfrac{17}{\cancel{6}} \cdot \dfrac{\overset{3}{\cancel{9}}}{2}$

$= \dfrac{51}{2}$

$= 25\dfrac{1}{2}$

CHECK for Understanding

3. Look at Example A. How many cups of flour will you need? Does the answer seem reasonable?

4. Explain how to get the numbers printed in red in Examples B, C, and D.

EXERCISES

▶ **Give each product in simplest form.**
Here are scrambled answers for the next row of exercises: $6\frac{1}{4}$ $4\frac{1}{6}$ 2 $7\frac{7}{8}$ $6\frac{1}{8}$

5. $1\frac{2}{3} \cdot 2\frac{1}{2}$ **6.** $3\frac{1}{2} \cdot 1\frac{3}{4}$ **7.** $2\frac{1}{4} \cdot 3\frac{1}{2}$ **8.** $1\frac{1}{2} \cdot 1\frac{1}{3}$ **9.** $2\frac{1}{2} \cdot 2\frac{1}{2}$

10. $2\frac{2}{3} \cdot 3\frac{1}{6}$ **11.** $2 \cdot 3\frac{1}{2}$ **12.** $4\frac{1}{3} \cdot 3\frac{1}{2}$ **13.** $3\frac{3}{4} \cdot 6$ **14.** $1\frac{1}{3} \cdot 6$

15. $3\frac{1}{6} \cdot 2\frac{3}{4}$ **16.** $5\frac{3}{4} \cdot 2\frac{1}{2}$ **17.** $2\frac{4}{5} \cdot 3$ **18.** $4\frac{1}{2} \cdot 3$ **19.** $3\frac{2}{3} \cdot 4\frac{1}{3}$

20. $2\frac{3}{4} \cdot 2$ **21.** $5\frac{1}{2} \cdot 4\frac{3}{4}$ **22.** $4\frac{1}{5} \cdot 5\frac{3}{8}$ **23.** $2\frac{1}{5} \cdot 3$ **24.** $1\frac{1}{2} \cdot 2\frac{1}{3}$

25. $2\frac{3}{8} \cdot 4$ **26.** $3\frac{3}{8} \cdot 6\frac{3}{4}$ **27.** $2\frac{1}{2} \cdot 3\frac{1}{2}$ **28.** $1\frac{2}{3} \cdot 6$ **29.** $1\frac{5}{8} \cdot 4\frac{1}{2}$

30. $3\frac{1}{3} \cdot 4\frac{1}{3}$ **31.** $2\frac{1}{3} \cdot 4\frac{1}{2}$ **32.** $3 \cdot 5\frac{2}{3}$ **33.** $2\frac{1}{5} \cdot 1\frac{1}{2}$ **34.** $4 \cdot 2\frac{1}{2}$

Problem Solving USING DATA

▶ **Solve. Use the recipe on page 186.**

35. How much sugar is needed to double the recipe?

36. How much oil is needed to triple the recipe?

37. Suppose that you wanted to make $2\frac{1}{2}$ times the recipe. How much flour would you need?

38. Suppose that you wanted to make $1\frac{3}{4}$ times the recipe. How much water would you need?

39. You have 12 cups of flour. Can you make $2\frac{3}{4}$ times the recipe?

40. You have 4 teaspoons of salt. How much more do you need to make $3\frac{1}{2}$ times the recipe?

Challenge! MAKE A DRAWING

▶ **Solve.**

41. One of your friends ate $\frac{1}{2}$ of your pieces of pizza. Another friend ate $\frac{1}{2}$ of the remaining pieces. A third friend ate the 3 pieces that were left. How many pieces of pizza did you have to begin with?

OBJECTIVE: To review dividing fractions and extend that skill to dividing algebraic fractions.

A bicyclist is riding some practice laps on a track. The distance around the track is $\frac{1}{5}$ of a mile. She rode a total of 2 miles while riding her laps.

1. a. Divide to find how many laps she rode.

Number of miles		Number of miles around the track		Number of laps
↓		↓		↓
2	÷	$\frac{1}{5}$	=	?

b. Multiply to find how many laps she rode.

Number of miles		Number of laps in each mile		Number of laps
↓		↓		↓
2	×	5	=	?

2. Look at exercise 1. Dividing by $\frac{1}{5}$ is the same as multiplying by what number?

Two numbers are **reciprocals** if their product is 1.

Since $5 \cdot \frac{1}{5} = 1$, **5** is the reciprocal of $\frac{1}{5}$ and $\frac{1}{5}$ is the reciprocal of 5.

Since $\frac{2}{3} \cdot \frac{3}{2} = 1$, $\frac{2}{3}$ is the reciprocal of $\frac{3}{2}$ and $\frac{3}{2}$ is the reciprocal of $\frac{2}{3}$.

> The reciprocal of $\frac{a}{b}$ is $\frac{b}{a}$.
> The reciprocal of n is $\frac{1}{n}$.

EXAMPLES | **Here's how to divide fractions.**

To divide by a fraction, multiply by its reciprocal.

In arithmetic

A $\quad \dfrac{3}{5} \div \dfrac{9}{10} = \dfrac{\overset{1}{3}}{\underset{1}{5}} \cdot \dfrac{\overset{2}{10}}{\underset{3}{9}}$

$\quad = \dfrac{2}{3}$

C $\quad \dfrac{2}{5} \div 2 = \dfrac{2}{5} \cdot \dfrac{\overset{1}{1}}{\underset{1}{2}}$

$\quad = \dfrac{1}{5}$

In algebra

B $\quad \dfrac{2a}{b} \div \dfrac{4}{b^2} = \dfrac{2a}{\underset{1}{b}} \cdot \dfrac{\overset{b}{b^2}}{\underset{2}{4}}$

$\quad = \dfrac{ab}{2}$

D $\quad \dfrac{3r}{t} \div t = \dfrac{3r}{t} \cdot \dfrac{1}{t}$

$\quad = \dfrac{3r}{t^2}$

CHECK for Understanding

3. Explain how to get the numbers and letter that are printed in red in Examples C and D.

EXERCISES

▶ **Give the reciprocal of each number.**

4. 2 **5.** $\dfrac{1}{8}$ **6.** $\dfrac{1}{6}$ **7.** $\dfrac{3}{2}$ **8.** $\dfrac{4}{7}$ **9.** $\dfrac{3}{5}$

▶ **Give each quotient in simplest form.**

Here are scrambled answers for the next row of exercises: $1 \quad \dfrac{1}{6} \quad 1\dfrac{1}{8} \quad \dfrac{9}{16} \quad \dfrac{2}{3}$

10. $\dfrac{2}{3} \div 4 \leftarrow \boxed{\dfrac{2}{3} \cdot \dfrac{1}{4}}$ **11.** $\dfrac{3}{4} \div \dfrac{2}{3}$ **12.** $\dfrac{7}{5} \div \dfrac{7}{5}$ **13.** $\dfrac{3}{8} \div \dfrac{2}{3}$ **14.** $\dfrac{2}{9} \div \dfrac{1}{3}$

15. $\dfrac{5}{9} \div \dfrac{1}{3}$ **16.** $\dfrac{5}{8} \div 2$ **17.** $\dfrac{4}{5} \div \dfrac{3}{3}$ **18.** $\dfrac{2}{3} \div \dfrac{5}{9}$ **19.** $5 \div \dfrac{2}{5}$ **20.** $\dfrac{2}{7} \div \dfrac{4}{5}$

21. $\dfrac{3}{5} \div \dfrac{2}{5}$ **22.** $\dfrac{3}{2} \div \dfrac{2}{3}$ **23.** $\dfrac{7}{9} \div \dfrac{4}{3}$ **24.** $\dfrac{2}{5} \div 5$ **25.** $\dfrac{7}{8} \div \dfrac{3}{4}$ **26.** $\dfrac{3}{7} \div \dfrac{7}{9}$

27. $4 \div \dfrac{5}{8}$ **28.** $\dfrac{9}{4} \div \dfrac{7}{8}$ **29.** $6 \div \dfrac{3}{2}$ **30.** $\dfrac{0}{2} \div \dfrac{9}{4}$ **31.** $\dfrac{7}{8} \div \dfrac{5}{16}$ **32.** $8 \div \dfrac{1}{2}$

Here are scrambled answers for the next row of exercises: $\dfrac{t}{rs} \quad \dfrac{yz}{x^2} \quad \dfrac{a}{cd} \quad \dfrac{ad}{bc}$

33. $\dfrac{a}{b} \div \dfrac{c}{d} \leftarrow \boxed{\dfrac{a}{b} \cdot \dfrac{d}{c}}$ **34.** $\dfrac{1}{r} \div \dfrac{s}{t}$ **35.** $\dfrac{a}{c} \div d$ **36.** $\dfrac{y}{x} \div \dfrac{x}{z}$

37. $\dfrac{r}{s} \div \dfrac{4}{r}$ **38.** $\dfrac{2}{a} \div \dfrac{b}{c}$ **39.** $\dfrac{3m}{n} \div \dfrac{2}{q}$ **40.** $\dfrac{r}{s} \div \dfrac{5t}{2}$ **41.** $\dfrac{6d}{e} \div f$

42. $\dfrac{3x}{2y} \div \dfrac{5}{6y}$ **43.** $\dfrac{m}{p} \div \dfrac{m}{2q}$ **44.** $\dfrac{c^2}{d} \div \dfrac{c}{d}$ **45.** $\dfrac{x}{9y} \div \dfrac{z}{3y}$ **46.** $\dfrac{4a}{3b^2} \div \dfrac{6a}{b}$

Group Project **ANALYZING DATA** *On The Road Again*

▶ **Work in a small group.**

47. a. Get a map and plan a 200- to 300-mile bicycle tour.
   b. Write a paragraph that describes your route.
   c. Compute the number of miles of the tour.

48. Suppose that your tour group starts riding at 7:00 A.M. You ride for 2 hours, rest for 30 minutes, ride for 2 hours, rest for 30 minutes, etc. At what time will your group have 8 hours of riding time?

49. Suppose that you average $8\dfrac{1}{2}$ miles per hour. How far will you travel during 8 hours of riding time?

50. How many 8-hour days will your tour take?

Algebra Using Products and Quotients of Fractions | **189**

OBJECTIVE: To divide mixed numbers.

TRAIL RULES AND INFORMATION

All children under 12 years of age must be accompanied by an adult. All litter is to be carried out by hikers. All trails are closed at 6:00 P.M.

TRAIL NAME	DISTANCE (miles)	HIKING TIME (hours)
Pine Bridge	$5\frac{1}{2}$	$2\frac{1}{2}$
Blue Basin	5	$3\frac{3}{4}$
Crawford Notch	$3\frac{3}{4}$	$1\frac{1}{4}$

1. How many miles long is Crawford Notch Trail?

2. How many hours does it take to hike Crawford Notch Trail?

3. To find how many miles per hour you would average when hiking Crawford Notch Trail, you would divide $3\frac{3}{4}$ by what number?

EXAMPLE **Here's how to divide mixed numbers.**

$3\frac{3}{4} \div 1\frac{1}{4} = ?$

$3\frac{3}{4} \div 1\frac{1}{4} = \frac{15}{4} \div \frac{5}{4}$ Change each mixed number to a fraction.

$= \frac{\overset{3}{15}}{\underset{1}{4}} \cdot \frac{\overset{1}{4}}{\underset{1}{5}}$ Divide.

$= 3$ Write the quotient in simplest form.

CHECK for Understanding

4. Look at the example above. How many miles per hour would you average while hiking Crawford Notch Trail?

5. Complete these examples.

a. $5\frac{1}{4} \div 2\frac{1}{2} = \frac{21}{4} \div \frac{5}{2}$

$= \frac{21}{\underset{2}{4}} \cdot \frac{\overset{1}{2}}{5}$

$= \frac{21}{10}$

$= 2\,\underline{?}$

b. $4\frac{2}{3} \div 2\frac{1}{4} = \frac{14}{3} \div \frac{9}{4}$

$= \frac{14}{3} \cdot \underline{?}$

$= \frac{56}{27}$

$= 2\frac{2}{27}$

c. $3\frac{1}{8} \div 1\frac{3}{4} = \frac{25}{8} \div \underline{?}$

$= \frac{25}{\underset{2}{8}} \cdot \frac{\overset{1}{4}}{7}$

$= \frac{25}{14}$

$= 1\frac{11}{14}$

EXERCISES

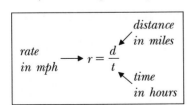

▶ **Give each quotient in simplest form.**
Here are scrambled answers for the next row of exercises: $1\frac{1}{4}$ $4\frac{1}{8}$ $1\frac{4}{5}$ $2\frac{1}{7}$ $2\frac{1}{4}$

6. $2\frac{1}{4} \div 1\frac{1}{4}$ **7.** $5\frac{1}{2} \div 1\frac{1}{3}$ **8.** $5 \div 2\frac{1}{3}$ **9.** $4\frac{1}{2} \div 2$ **10.** $2\frac{1}{2} \div 2$

11. $6\frac{1}{2} \div 2\frac{2}{3}$ **12.** $6\frac{1}{2} \div 2\frac{1}{4}$ **13.** $8 \div 2\frac{1}{4}$ **14.** $6\frac{1}{4} \div 1\frac{1}{4}$ **15.** $1\frac{1}{6} \div 1\frac{1}{2}$

16. $5\frac{7}{8} \div 1\frac{3}{4}$ **17.** $7 \div 2\frac{1}{3}$ **18.** $4\frac{1}{3} \div 2\frac{1}{2}$ **19.** $2 \div 1\frac{1}{2}$ **20.** $4\frac{1}{2} \div 4$

21. $2\frac{3}{8} \div 1\frac{1}{3}$ **22.** $6\frac{3}{4} \div 3$ **23.** $1\frac{1}{5} \div 5$ **24.** $8\frac{1}{2} \div 1\frac{3}{4}$ **25.** $6 \div 1\frac{1}{2}$

26. $3\frac{3}{5} \div 1\frac{1}{5}$ **27.** $6\frac{2}{3} \div 2$ **28.** $9\frac{1}{4} \div 2\frac{1}{4}$ **29.** $8 \div 2\frac{1}{2}$ **30.** $2 \div \frac{2}{3}$

31. $6 \div 4\frac{1}{2}$ **32.** $9 \div 2\frac{1}{4}$ **33.** $2\frac{1}{3} \div 2\frac{1}{3}$ **34.** $2\frac{1}{2} \div 1\frac{1}{4}$ **35.** $3\frac{1}{2} \div 3\frac{1}{2}$

36. $1\frac{1}{2} \div 2$ **37.** $7\frac{1}{2} \div 1\frac{1}{2}$ **38.** $3\frac{2}{3} \div 1\frac{1}{3}$ **39.** $6 \div 1\frac{1}{3}$ **40.** $1\frac{1}{2} \div \frac{1}{2}$

Problem Solving USING DATA

▶ **Solve. Use the information on page 190.**

41. How much longer is Blue Basin Trail than Crawford Notch Trail?

42. How much longer is Pine Bridge Trail than Crawford Notch Trail?

43. How many hours would you need to hike the two shorter trails?

44. It is 12 noon. Can you hike the two longer trails before the trails close?

45. How many miles per hour would you average while hiking Blue Basin Trail?

46. Which trail is the most difficult to hike (slowest average hiking rate)?

47. You can use the following formula to find your average hiking rate.

$$r = \frac{d}{t}$$

rate in mph → $r = \dfrac{d}{t}$ ← distance in miles / time in hours

Substitute in the formula above to complete this chart. Give answers in simplest form.

	distance in miles (d)	time in hours (t)	rate in mph (r)
a.	6	$2\frac{1}{2}$	?
b.	$4\frac{3}{4}$	2	?
c.	$6\frac{1}{2}$	$2\frac{1}{4}$	?
d.	?	2	$1\frac{2}{3}$
e.	?	3	$2\frac{3}{5}$

Algebra Using Products and Quotients of Fractions | **191**

PROBLEM ⚡ SOLVING ⚡ POWER

You will use a variety of problem-solving strategies on these two
pages. For some problems, you may wish to use a calculator.

1 MAKE A MODEL

The seats on a Ferris wheel are numbered
consecutively. If seat number 4 is opposite
seat number 10, how many seats are there
in all?

✓ *Problem Solving Tips*

Make a Model.* Use numbered pieces of paper
to represent seats and a circle to represent the
Ferris wheel. Arrange the numbered pieces
around the circle.

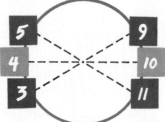

*At least one more of the problems on these two pages can
  be solved by making a model.

2 PATTERNS

Look for a pattern. Find the missing
numbers.

$$1^3 = 1 = 1^2$$
$$1^3 + 2^3 = 9 = 3^2$$
$$1^3 + 2^3 + 3^3 = \underline{\textbf{a.}} = \underline{\textbf{(b.)}}^2$$
$$1^3 + 2^3 + 3^3 + 4^3 = \underline{\textbf{c.}} = \underline{\textbf{(d.)}}^2$$
$$1^3 + 2^3 + 3^3 + 4^3 + 5^3 = \underline{\textbf{e.}} = \underline{\textbf{(f.)}}^2$$

3 COIN COMBOS

Marci said, "I have 12 coins that total 75¢.
I have only dimes and nickels." Make a
table that shows this combination and all
possible combinations of dimes and nickels
that total 75¢.

Hint:

Number of dimes	7	?	?
Number of nickels	?	?	?
Total value	75¢	75¢	75¢

4 ROW, ROW, ROW YOUR BOAT

Rob and his twin sisters, Ann and Nan, have to cross a river. They have to use a small rowboat that will carry a load of only 200 pounds. Rob weighs 180 pounds, and each of the twins weighs 90 pounds. None of the three can swim. How do they get across?

Hint: Make a model using pieces of paper.

6 SIX FLAGS OVER TEXAS

Jamie and Jill are spending the afternoon at the Six Flags Over Texas amusement park. Jamie said to Jill, "Here are 5 rides that we can ride together, but we have only enough time to go on 3 of them. I choose the Texas Chute Out, you choose the other 2 rides."

Make a list of all possible choices Jill can make.

5 GEOMETRY—VISUAL THINKING

Which of these patterns cannot be folded to make a cube?

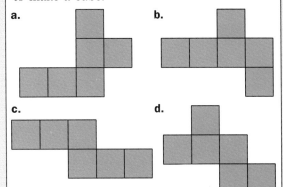

7 WEALTHY WELLINGTON

My wealthy uncle Wellington plans to visit us for 100 days. He made me this offer:

"If you walk my dog Godzilla every day, I'll pay you $1 the first day, $2 the second day, $3 the third day, and so on with a $100 payment on the final day."

How much would I earn altogether for the 100 days?

Hint: First solve a simpler problem.

APPLICATIONS—FAST-FOOD FRANCHISES

Number of Outlets (thousands)

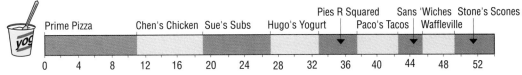

8 Together, __?__ and __?__ own about 16,000 outlets.

9 About half of the total number of outlets are owned by __?__, __?__, and __?__.

10 There are nearly twice as many __?__ outlets as there are Hugo's Yogurt outlets.

Rate Your Problem-Solving Power: *6–7 correct = Good; 8–9 correct = Excellent; 10 correct = Exceptional*

OBJECTIVE: To solve problems involving mixed numbers.

▶ **Use the information in the ad to solve the problems.**

1. How many pages of bonus stamps do you need to get a tape recorder and toaster oven?

2. You have $3\frac{1}{4}$ pages of bonus stamps. How many more pages do you need to get a designer telephone?

3. If you save $\frac{1}{4}$ pages of stamps each month, how many months will it take you to get a clock radio?

4. Randy saved $1\frac{1}{2}$ pages of stamps and his sister saved $2\frac{3}{4}$ pages. How many more pages do they need to get a designer telephone?

5. It takes 20 stamps to fill a bonus-stamp page. You have 12 stamps. What fraction of a page can you fill?

6. Rita has saved 40 stamps. How many more stamps does she need in order to get a designer telephone? (*Remember:* It takes 20 stamps to fill a page.)

SAVE ☐ BONUS ☐ STAMPS
REDEEM STAMPS FOR FREE GIFTS

☐ Designer Telephone **6 pages**

☐ Toaster Oven $9\frac{1}{2}$ **pages**

☐ Clock Radio $2\frac{3}{4}$ **pages**

☐ Tape Recorder $3\frac{1}{4}$ **pages**

B O N U S

▶ **Write an equation and solve the problem.**

7. It takes $5\frac{1}{2}$ pages of bonus stamps to get a pocket radio. That is $\frac{1}{2}$ page more than 2 times the number of pages needed to get a football. How many pages of stamps do you need to get a football?

(Let f = the number of pages needed to get the football.)

8. It takes a total of 7 pages of stamps to get a basketball and a Frisbee. A basketball takes 3 times as many pages of stamps as a Frisbee. How many pages of stamps do you need to get a Frisbee? (Let f = the number of pages needed to get the Frisbee.)

9. It takes $12\frac{1}{2}$ pages to get a telescope. That is $2\frac{1}{2}$ pages more than 5 times the number of pages needed to get a soccer ball. How many pages are needed to get a soccer ball?

(Let s = the number of pages needed to get a soccer ball.)

10. You need a total of 12 pages to get a bowling ball and bag. The ball takes 7 times as many pages as the bag. How many pages are needed to get the bag? (Let b = the number of pages needed to get the bag.)

▶ **Solve and check.** *(page 68)*

1. $19 + g = 74$ **2.** $n(9) = 108$ **3.** $135 = v - 26$ **4.** $20 = \dfrac{x}{17}$

5. $j - 83 = 26$ **6.** $16t = 176$ **7.** $m + 74 = 117$ **8.** $\dfrac{s}{18} = 31$

9. $57 = k(3)$ **10.** $19 = h - 34$ **11.** $25 = \dfrac{t}{16}$ **12.** $91 = 47 + j$

13. $15 = t + 9.4$ **14.** $42 = r - 18.6$ **15.** $2.1 = \dfrac{v}{11}$ **16.** $15.35 = q(5)$

▶ **Solve and check.** *(page 76)*

17. $\dfrac{n}{3} + 9 = 26$ **18.** $\dfrac{t}{4} - 6 = 41$ **19.** $12 = \dfrac{d}{3} + 5$ **20.** $20 + \dfrac{r}{5} = 32$

21. $4x + 10 = 58$ **22.** $72 = 3k + 12$ **23.** $60 = 11j - 17$ **24.** $15t - 30 = 0$

25. $3n + 5.8 = 9.4$ **26.** $\dfrac{n}{7} + 6.5 = 8.1$ **27.** $15.6 = \dfrac{a}{5} - 2.4$ **28.** $7.5 = 4n - 14.5$

▶ **Solve and check.** *(page 82)*

29. $2a + 3a = 35$ **30.** $48 = 2x + 2x$ **31.** $5n + n = 72$ **32.** $156 = 8m + 4m$

33. $c + c = 74$ **34.** $4b + 6b = 120$ **35.** $136 = y + 3y$ **36.** $h + 11h = 252$

37. $9k + 2k = 264$ **38.** $225 = 11m + 4m$ **39.** $200 = 19a + a$ **40.** $14j + 2j = 272$

41. $4n + n = 7.5$ **42.** $12.6 = z + 2z$ **43.** $6h + 2h = 20.8$ **44.** $32.5 = p + p$

▶ **Solve.** *(page 112)*

45. $4n - 10 = {}^-30$ **46.** ${}^-7m + 3 = 10$ **47.** $5f - 12 = {}^-22$ **48.** $8q + 7 = {}^-9$

49. $\dfrac{k}{3} + 4 = {}^-6$ **50.** $\dfrac{j}{{}^-2} + 7 = 8$ **51.** $\dfrac{x}{7} + 9 = 1$ **52.** $2n - {}^-3 = {}^-7$

53. $\dfrac{q}{{}^-8} - 2 = 5$ **54.** $\dfrac{a}{{}^-2} + 4 = 9$ **55.** ${}^-3r - 6 = 21$ **56.** $\dfrac{a}{{}^-4} - {}^-5 = 5$

▶ **Give the algebraic factorization.** *(page 130)*

57. $10a^2$ **58.** $6d^2$ **59.** $3b^3$ **60.** $8d$ **61.** $7g^3$

62. $35n^2$ **63.** $24w$ **64.** $40z^3$ **65.** $25n^3$ **66.** $10x^4$

67. $27k$ **68.** $42b^3$ **69.** $7g^4$ **70.** $52b^2$ **71.** $7r^2s$

72. $4x^2y$ **73.** $10yz^2$ **74.** $8r^2s$ **75.** $12u^2v^2$ **76.** $14a^2b^2$

OBJECTIVE: To solve an equation to find a fraction of a whole number.

1. What is the regular price of a jade plant?

2. The sale price of a jade plant is what fraction of the regular price?

3. To find the sale price of a jade plant, you would find $\frac{2}{3}$ of what price?

PLANT SALE!

Jade Plants
Regular Price $15
Pay $\frac{2}{3}$ of regular price

Spider Plants
Regular Price $12
Pay $\frac{1}{2}$ of regular price

African Violets
Regular Price $4
Pay $\frac{3}{4}$ of regular price

EXAMPLES | Here's how to find a fraction of a number.

To find a fraction of a number, you can write and solve an equation.

A Equation: $\frac{2}{3}$ of 15 = n

Cancel. $\frac{2}{3} \cdot \overset{5}{\underset{1}{\cancel{15}}} = n$

Simplify. $10 = n$

B $\frac{1}{4}$ of 15 = n

$\frac{1}{4} \cdot 15 = n$

$\frac{15}{4} = n$

$3\frac{3}{4} = n$

C $\frac{5}{8}$ of 48 = n

$\frac{5}{8} \cdot \overset{6}{\underset{1}{\cancel{48}}} = n$

$30 = n$

D $\frac{7}{10}$ of 55 = n

$\frac{7}{10} \cdot \overset{11}{\underset{2}{\cancel{55}}} = n$

$\frac{77}{2} = n$

$38\frac{1}{2} = n$

CHECK for Understanding

4. Look at Example A. What is the sale price of a jade plant?

5. Explain how to get the numbers that are printed in red in Examples B, C, and D.

EXERCISES

▶ **Solve. Give answers in simplest form.**

Here are scrambled answers for the next row of exercises: $6\frac{2}{3}$ $4\frac{4}{5}$ 4 54

6. $\frac{1}{3}$ of 12 = n

7. $\frac{2}{3}$ of 10 = n

8. $\frac{2}{5}$ of 12 = n

9. $\frac{9}{10}$ of 60 = n

10. $\frac{5}{6}$ of 30 = n

11. $\frac{1}{5}$ of 18 = n

12. $\frac{5}{8}$ of 56 = n

13. $\frac{2}{5}$ of 55 = n

14. $\frac{3}{4}$ of 36 = n

15. $\frac{3}{8}$ of 21 = n

16. $\frac{1}{3}$ of 63 = n

17. $\frac{7}{10}$ of 80 = n

18. $\frac{9}{10}$ of \$200 = n

19. $\frac{1}{2}$ of \$38 = n

20. $\frac{2}{3}$ of \$39 = n

21. $\frac{4}{9}$ of \$81 = n

22. $\frac{7}{8}$ of \$64 = n

23. $\frac{3}{8}$ of \$16 = n

24. $\frac{3}{10}$ of \$90 = n

25. $\frac{7}{8}$ of \$32 = n

▶ **Give answers to the nearest cent.**

26. $\frac{2}{5}$ of \$11 = n

27. $\frac{1}{5}$ of \$16 = n

28. $\frac{2}{3}$ of \$20 = n

29. $\frac{3}{4}$ of \$10 = n

30. $\frac{2}{3}$ of \$40 = n

31. $\frac{5}{8}$ of \$21 = n

32. $\frac{1}{6}$ of \$20 = n

33. $\frac{5}{6}$ of \$25 = n

Problem Solving USING DATA

▶ **Solve. Use the ad on page 196.**

34. What is the sale price of a spider plant?

35. What is the sale price of an African violet?

36. How much would you save by buying a jade plant on sale?

37. Which plants are on sale for $\frac{1}{4}$ off the regular price?

Challenge! CALCULATOR

38. Which calculator keys did Diane and Tony push to find $\frac{15}{22}$ of 121?

a. Diane's method:

121 _?_ 22 _?_ 15 ▤ (82.5)

b. Tony's method:

121 _?_ 15 _?_ 22 ▤ (82.5)

39. Use either Diane's or Tony's method to find n.

a. $\frac{14}{25}$ of 420 = n **b.** $\frac{27}{48}$ of 120 = n

c. $\frac{135}{182}$ of 637 = n **d.** $\frac{63}{80}$ of 992 = n

e. $\frac{144}{315}$ of 63 = n **f.** $\frac{130}{234}$ of 175.5 = n

Algebra Using Products and Quotients of Fractions | **197**

OBJECTIVE: To convert measurements within the customary system by multiplying a whole number by a mixed
number.

1. How many inches are there in 2 feet?

2. How many inches are there in $\frac{1}{2}$ of a foot?

3. If you know the number of inches in 2 feet
   and the number of inches in $\frac{1}{2}$ of a foot,
   how could you find the number of inches in
   $2\frac{1}{2}$ feet?

| 1 yard (yd) = 3 feet (ft) |
| 1 ft = 12 inches (in.) |
| 1 yd = 36 in. |

E X A M P L E S | **Here's how to find the number of inches in $2\frac{1}{2}$ feet.**

 Regular Method

Change the mixed number
to a fraction and multiply.

$2\frac{1}{2}$ feet $= 2\frac{1}{2} \times 12$ inches

$= \dfrac{5}{2} \times \overset{6}{\underset{1}{12}}$ inches

$= 30$ inches

 Shortcut Method

First find the number of inches in 2 feet
and in $\frac{1}{2}$ of a foot. Then add.

| 2×12 inches | | $\dfrac{1}{2} \times 12$ inches |

$2\frac{1}{2}$ feet $= 24$ inches $+ 6$ inches

$= 30$ inches

CHECK for Understanding

4. Look at the examples above.
   a. How many inches are in $2\frac{1}{2}$ feet?
   b. Does the answer make sense? Is it between the number of inches in 2 feet
      (24) and the number of inches in 3 feet (36)?

EXERCISES

▶ **Complete.**

Here are scrambled answers for the next row of exercises: *11   28   18   4*

5. $1\frac{1}{2}$ ft = __?__ in.

6. $1\frac{1}{3}$ yd = __?__ ft

7. $2\frac{1}{3}$ ft = __?__ in.

8. $3\frac{2}{3}$ yd = __?__ ft

9. $1\frac{1}{3}$ ft = __?__ in.

10. $2\frac{2}{3}$ yd = __?__ ft

11. $1\frac{1}{2}$ yd = __?__ in.

12. $2\frac{1}{3}$ yd = __?__ ft

13. $2\frac{1}{4}$ yd = __?__ in.

14. $1\frac{3}{4}$ ft = __?__ in.

15. $3\frac{2}{3}$ ft = __?__ in.

16. $5\frac{1}{3}$ yd = __?__ ft

> 1 day = 24 hours (h)
> 1 h = 60 minutes (min)
> 1 min = 60 seconds (s)

17. $2\frac{1}{2}$ days = __?__ h

18. $4\frac{1}{2}$ h = __?__ min

19. $1\frac{1}{3}$ h = __?__ min

20. $1\frac{1}{2}$ min = __?__ s

21. $2\frac{3}{10}$ h = __?__ min

22. $2\frac{2}{3}$ days = __?__ h

23. $1\frac{2}{3}$ min = __?__ s

24. $3\frac{1}{3}$ days = __?__ h

25. $2\frac{1}{2}$ gal = __?__ qt

26. $1\frac{1}{2}$ qt = __?__ pt

27. $3\frac{1}{2}$ pt = __?__ c

28. $3\frac{1}{2}$ qt = __?__ pt

> 1 gallon (gal) = 4 quarts (qt)
> 1 qt = 2 pints (pt)
> 1 pt = 2 cups (c)

29. $1\frac{3}{4}$ gal = __?__ qt

30. $2\frac{1}{2}$ pt = __?__ c

31. $5\frac{1}{2}$ gal = __?__ qt

32. $5\frac{1}{2}$ qt = __?__ pt

33. $7\frac{1}{2}$ pt = __?__ c

34. $3\frac{1}{2}$ gal = __?__ qt

Problem Solving   USING EXPRESSIONS

▶ **Give an expression to complete each sentence.**

35. There are __?__ hours in d days.

36. There are __?__ hours in m minutes.

37. There are __?__ minutes in s seconds.

38. There are __?__ seconds in m minutes.

39. There are __?__ inches in f feet.

40. There are __?__ yards in i inches.

41. There are __?__ inches in y yards.

42. There are __?__ yards in f feet.

43. There are __?__ pints in q quarts.

44. There are __?__ cups in p pints.

45. There are __?__ pints in c cups.

46. There are __?__ quarts in g gallons.

OBJECTIVE: To use an equation to find a number when a fraction of it is known.

1. What is the sale price of the Beginner's skateboard?

2. The sale price of the Beginner skateboard is what fraction of the regular price?

3. Is the regular price of the Beginner skateboard more or less than $27?

SKATEBOARD SALE!

PAY $\frac{3}{4}$ OF REGULAR PRICE

Beginner Sale Price $27

Intermediate Sale Price $42

Professional Sale Price $108

| EXAMPLE | Here's how to find the number when a fraction of it is known. |

To find the number when a fraction of it is known, you can write and solve an equation.

Equation: $\frac{3}{4}$ of $n = 27$

$$\frac{3}{4}n = 27$$

Divide both sides by $\frac{3}{4}$.

$$\frac{\frac{3}{4}n}{\frac{3}{4}} = \frac{27}{\frac{3}{4}}$$

Simplify.

$$n = \overset{9}{27} \cdot \frac{4}{\underset{1}{3}}$$

Simplify.

$$n = 36$$

To divide by $\frac{3}{4}$, multiply by $\frac{4}{3}$.

✔ **CHECK**

$$\frac{3}{4}n = 27$$

$$\frac{3}{4} \cdot 36 \overset{?}{=} 27$$

$$\frac{3}{4} \cdot \overset{9}{36} \overset{?}{=} 27$$
$$1$$

$$27 = 27$$

It checks!

CHECK for Understanding

4. Look at the example above. What is the regular price of the Beginner skateboard? Is the answer reasonable?

5. Complete these examples.

a. $\frac{2}{5}n = 38$

$$\frac{\frac{2}{5}n}{\frac{2}{5}} = \frac{38}{\frac{2}{5}}$$

$$n = \overset{19}{38} \cdot \frac{5}{\underset{1}{2}}$$

$$n = \underline{?}$$

b. $\frac{2}{3}n = 15$

$$\frac{\frac{2}{3}n}{\frac{2}{3}} = \frac{15}{\frac{2}{3}}$$

$$n = 15 \cdot \underline{?}$$

$$n = 22\frac{1}{2}$$

c. $\frac{4}{3}n = 10$

$$\frac{\frac{4}{3}n}{\frac{4}{3}} = \frac{10}{\frac{4}{3}}$$

$$n = 10 \cdot \underline{?}$$

$$n = \underline{?}$$

EXERCISES

▶ **Solve. Give answers in simplest form.**

Here are scrambled answers for the next row of exercises: $10\frac{2}{3}$ 24 20 15 30

6. $\frac{1}{3}n = 8$ **7.** $\frac{2}{3}n = 10$ **8.** $\frac{3}{4}n = 15$ **9.** $\frac{3}{8}n = 4$ **10.** $\frac{2}{5}n = 12$

11. $\frac{4}{3}n = 48$ **12.** $\frac{1}{4}n = 6$ **13.** $\frac{3}{5}n = 15$ **14.** $\frac{1}{7}n = 9$ **15.** $\frac{5}{2}n = 20$

16. $\frac{6}{5}n = 10$ **17.** $\frac{7}{9}n = 28$ **18.** $\frac{2}{5}n = 11$ **19.** $\frac{7}{8}n = 10$ **20.** $\frac{4}{5}n = 22$

21. $\frac{1}{2}n = 17$ **22.** $\frac{3}{2}n = 9$ **23.** $\frac{2}{9}n = 20$ **24.** $\frac{1}{5}n = 13$ **25.** $\frac{5}{6}n = 60$

26. $\frac{5}{2}n = 6$ **27.** $\frac{4}{5}n = 30$ **28.** $\frac{2}{3}n = 15$ **29.** $\frac{1}{5}n = 5$ **30.** $\frac{2}{3}n = 20$

Problem Solving USING DATA

▶ **Solve. Use the ad on page 200.**

31. What is the regular price of the Inter-mediate skateboard?

32. What is the regular price of the Profes-sional skateboard?

33. What is the difference in the regular price of the Intermediate and Beginner skateboards?

34. How much would you save by buying the Professional skateboard on sale?

▶ **First choose the equation and solve the problem. Then reread the problem to see if your answer makes sense.**

> **Equation A:** $\frac{4}{5} \cdot 24 = n$ **Equation B:** $\frac{4}{5}n = 24$
>
> **Equation C:** $\frac{4}{5} \cdot 80 = n$ **Equation D:** $\frac{4}{5}n = 80$

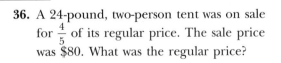

35. There were 24 tennis rackets on sale for $\frac{4}{5}$ of the regular price. What was the sale price of a racket that usually sold for $80?

36. A 24-pound, two-person tent was on sale for $\frac{4}{5}$ of its regular price. The sale price was $80. What was the regular price?

37. During the sale, $\frac{4}{5}$ of the $80 ski jackets were sold. How many ski jackets were on sale in all if 24 of them were sold?

38. There were 80 pairs of tennis shoes on sale for $\frac{4}{5}$ of the regular price. The reg-ular price was $24. What was the sale price?

Algebra Using Products and Quotients of Fractions | **201**

OBJECTIVE: To use equations with fractions to solve problems.

In this lesson you will write equations involving fractions to solve problems. *Remember:* To write an equation, you need two equal expressions.

EXAMPLE | **Here's how to use an equation to solve a problem.**

Problem: An advertising agency made a survey to determine auto trademarks that people remember. Two thirds of the adults who took part in the survey correctly identified all four auto trademarks. How many adults took part in the survey if 54 identified all the trademarks?

Step 1. Choose a variable. Use it and the facts to represent the numbers in the problem.

Let n = the number of adults who took part in the survey.

Then $\frac{2}{3}n$ = the number of adults who correctly identified all four trademarks.

Step 2. Write an equation based on the facts.

$$\frac{2}{3}n = 54$$

$\frac{2}{3}n$ *and 54 are equal expressions for the number of adults who identified all four trademarks.*

Step 3. Solve the equation.

$$\frac{2}{3}n = 54$$

$$\frac{\frac{2}{3}n}{\frac{2}{3}} = \frac{54}{\frac{2}{3}}$$

$$n = \overset{27}{\cancel{54}} \cdot \frac{3}{\underset{1}{\cancel{2}}}$$

A total of 81 adults took part in the survey.

$$n = 81$$

CHECK for Understanding

Look at the example above.

1. Read the problem. What are the relevant facts?

2. In Step 1, we let ? equal the number of adults who took part in the survey. Then the expression ? equals the number of adults who identified all four trademarks.

3. In Step 2, the two equal expressions for the number of adults who identified all four trademarks are ? and ? . Then an equation for the problem is ? .

4. Is it reasonable that the answer (81) is more than the number of adults (54) who identified all the trademarks?

EXERCISES

▶ **Decide whether Equation A, B, C, or D would be used to solve each problem. Then solve the problem.**

Equation A:	$\frac{3}{4}(60) = n$
Equation B:	$\frac{3}{4}n = 60$
Equation C:	$4n + 8 = 60$
Equation D:	$4n + 8 = 100$

5. Three fourths of the 60 people, 15-to-18-years old, identified all four companies in the trademarks survey. In the age group 14 years or under, three fourths of the group, or 60 people, identified all four automobile trademarks.

    a. How many people 15 to 18 identified the trademarks?

    b. How many people 14 or under took part in the survey?

6. In a recent survey, 60 out of 100 adults could identify the trademark of their automobile. That is 8 more than 4 times the number of people who could identify the trademark of their car-insurance company. The number of adults surveyed was 8 more than 4 times the number of people who could identify their automobile-tire trademark.

    a. How many people could identify their car-insurance trademark?

    b. How many people could identify their automobile-tire trademark?

7. Write a problem that can be solved using Equation B. Then solve the problem.

8. Write a problem that can be solved using Equation C. Then solve the problem.

▶ **Write an equation and solve the problem.**

9. One third of a number is 14. What is the number?
(Let n = the number.)

10. The product of two numbers is 40. If one number is $\frac{2}{5}$, what is the other number?
(Let m = the other number.)

11. Ten more than 3 times a number is 55. What is the number?
(Let u = the number.)

12. Divide a number by 3, then subtract 6, and you get 1. What is the number?
(Let v = the number.)

13. Ella has saved two thirds of the cost of a sweater she wants. How much does the sweater cost if she has saved $26?
(Let c = the cost of the sweater.)

14. Carol has $3 more than twice as many dollars as Bill. If Carol has $15, how many dollars does Bill have?
(Let b = the number of dollars Bill has.)

15. Three fifths of a number is 1. What is the number?
(Let a = the number.)

★**16.** Two thirds of a number is $\frac{4}{5}$. What is the number?
(Let b = the number.)

Algebra Using Products and Quotients of Fractions | **203**

OBJECTIVE: To solve problems using a calculator when appropriate.

Want to Lose Weight?
Exercise and Burn Off Calories!

When you exercise, you burn off calories. The number of calories burned depends on the type of exercise, how long you exercise, and your weight. Here is a list of activities and the number of calories you burn per pound for each hour of exercise.

Activity	Number of Calories burned per pound in one hour
BICYCLING	1.6
DANCING	2.2
FOOTBALL	3.6
SKIING	5.2
SWIMMING	4.1
TENNIS	2.6
WALKING	2.1

You can use this formula to find the calories you burn when exercising.

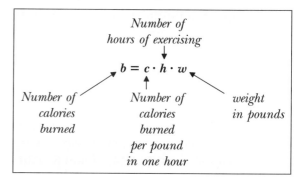

Number of hours of exercising

$$b = c \cdot h \cdot w$$

Number of calories burned

Number of calories burned per pound in one hour

weight in pounds

▶ **Use the formula and the information in the article. Fill in the facts that fit the calculator sequence.**

1. 2.1 ☒ 1.5 ☒ 110 ▣ (346.5)

Connie walks ___**a.**___ hours each day.
(number)

She weighs ___**b.**___ pounds. During
(number)

each walk she burns ___**c.**___ calories.
(number)

2. 4.1 ☒ 0.5 ☒ 140 ▣ (287)

Gary swims ___**a.**___ hours each day.
(number)

He weighs ___**b.**___ pounds. He burns
(number)

___**c.**___ calories during each
(number)

___**d.**___.
(activity)

▶ **Use the information in the article. Decide when a calculator would be helpful.**

3. Jenny likes to ski. How many calories does she burn while skiing 2 hours? Jenny weighs 120 pounds.

4. Mark plays football. How many calories does he burn while playing a 1-hour game? Mark weighs 100 pounds.

5. Julia and Eric like to dance. How many more calories will Eric burn than Julia while dancing 1.5 hours? Julia weighs 105 pounds and Eric weighs 140 pounds.

6. Don and Ruby play tennis. How many more calories will Don burn than Ruby during a 1-hour tennis match? Don weighs 150 pounds, and Ruby weighs 125 pounds.

7. To lose a pound, you have to burn 3500 calories. How many hours would a 175-pound person have to walk to lose one pound? Round your answer to the nearest tenth.

8. **Write a problem** that can be solved using the information in the article. Then solve the problem.

▶ **Give the greatest common factor.** *(page 132)*

1. 10, 15 **2.** 16, 6 **3.** 7, 27 **4.** 48, 32 **5.** 12, 18

6. $3, 6c$ **7.** $5a, a$ **8.** $10y, 22y$ **9.** $5w, 10w^2$ **10.** $3y^2, 7y$

11. $9xy, x^2$ **12.** $5u^2v^2, 15u^2$ **13.** $12c^2d, 18cd^2$ **14.** $x^3y^2, 3x^2y$ **15.** $40e^3f, 25e^2f^2$

▶ **Give the least common multiple.** *(page 132)*

16. 8, 12 **17.** 9, 12 **18.** 16, 12 **19.** 18, 12 **20.** 25, 20 **21.** 24, 16

22. $5, 10x$ **23.** $8, 24c$ **24.** $10, 5y$ **25.** $3z, 9$ **26.** $6w, 9$ **27.** $8y, 12$

28. $k, 7k$ **29.** $9j, j$ **30.** $n, 12n$ **31.** $8f, f^2$ **32.** $3m^2n, mn$ **33.** $7c^2d, 5cd^2$

▶ **Write in simplest form.** *(page 158)*

34. $\dfrac{9}{3}$ **35.** $\dfrac{6}{9}$ **36.** $\dfrac{15}{2}$ **37.** $\dfrac{15}{5}$ **38.** $\dfrac{15}{20}$ **39.** $\dfrac{23}{3}$

40. $\dfrac{30}{4}$ **41.** $\dfrac{24}{4}$ **42.** $\dfrac{15}{18}$ **43.** $\dfrac{38}{5}$ **44.** $\dfrac{50}{8}$ **45.** $\dfrac{42}{6}$

46. $\dfrac{6a}{2b}$ **47.** $\dfrac{9x^2}{12y}$ **48.** $\dfrac{10cd^2}{15c}$ **49.** $\dfrac{9x^2}{3x^2}$ **50.** $\dfrac{4rs}{12r}$ **51.** $\dfrac{8mn^2}{3m}$

▶ **Give each sum in simplest form.** *(page 168)*

52. $\dfrac{1}{4} + \dfrac{3}{8}$ **53.** $\dfrac{2}{3} + \dfrac{1}{6}$ **54.** $\dfrac{3}{10} + \dfrac{1}{5}$ **55.** $\dfrac{3}{4} + \dfrac{2}{3}$

56. $\dfrac{r}{2} + \dfrac{s}{4}$ **57.** $\dfrac{m}{5} + \dfrac{n}{2}$ **58.** $\dfrac{p}{3} + \dfrac{q}{4}$ **59.** $\dfrac{3}{4} + \dfrac{k}{5}$

60. $\dfrac{r}{6} + \dfrac{5}{9}$ **61.** $\dfrac{5m}{8} + \dfrac{n}{3}$ **62.** $\dfrac{2x}{5} + \dfrac{y}{6}$ **63.** $\dfrac{7r}{12} + \dfrac{s}{9}$

64. $\dfrac{5m}{18} + \dfrac{5n}{12}$ **65.** $\dfrac{3y}{16} + \dfrac{7z}{20}$ **66.** $\dfrac{2g}{9} + \dfrac{h}{6}$ **67.** $\dfrac{m}{5} + \dfrac{5n}{7}$

▶ **Give each difference in simplest form.** *(page 168)*

68. $\dfrac{2}{3} - \dfrac{1}{2}$ **69.** $\dfrac{5}{12} - \dfrac{1}{4}$ **70.** $\dfrac{3}{4} - \dfrac{2}{3}$ **71.** $\dfrac{11}{12} - \dfrac{5}{9}$

72. $\dfrac{r}{6} - \dfrac{s}{3}$ **73.** $\dfrac{m}{2} - \dfrac{n}{5}$ **74.** $\dfrac{x}{4} - \dfrac{y}{3}$ **75.** $\dfrac{3a}{4} - \dfrac{b}{6}$

Here are scrambled answers for the review exercises:

1	18	add	canceling	mixed	reciprocal
2	36	denominator	equation	multiply	simplest
10	a	divide	fraction	product	simplify

1. To multiply fractions, multiply the numerators to get the numerator of the product and multiply the denominators to get the _?_ of the product. To find the products at the right, a shortcut called _?_ was used before the fractions were multiplied. *(page 184)*

$$\overset{1}{\underset{5}{\cancel{2}}} \cdot \frac{3}{\underset{2}{\cancel{4}}} = \frac{3}{?} \qquad \frac{a}{\cancel{b}} \cdot \frac{\cancel{b}}{a^2} = \frac{?}{a}$$

2. To multiply mixed numbers, change each mixed number to a _?_ and multiply. The last step in finding the product at the right would be to write the product in _?_ form. *(page 186)*

$$3\frac{2}{3} \cdot 4\frac{1}{2} = \frac{11}{\underset{1}{\cancel{3}}} \cdot \frac{\overset{3}{\cancel{9}}}{2}$$
$$= \frac{33}{2}$$

3. Two numbers are reciprocals if their _?_ is 1. To divide by a fraction, _?_ by its reciprocal. To find the quotient $\frac{3}{8} \div \frac{1}{2}$, you would multiply $\frac{3}{8}$ by _?_. To find the quotient $\frac{a^2}{b} \div \frac{1}{a}$, you would multiply $\frac{a^2}{b}$ by _?_. *(page 188)*

4. To divide mixed numbers, change each _?_ number to a fraction and _?_. The next step in finding the quotient at the right would be to multiply $\frac{14}{3}$ by the _?_ of $\frac{7}{2}$. *(page 190)*

$$4\frac{2}{3} \div 3\frac{1}{2} = \frac{14}{3} \div \frac{7}{2}$$

5. To find a fraction of a number, you can write and solve an _?_. For example, to find $\frac{3}{4}$ **of 24,** you can solve the equation at the right. *(page 196)*

$$\frac{3}{4} \text{ of } 24 = n$$
$$\frac{3}{4} \cdot \overset{6}{\cancel{24}} = n$$
$$\overset{1}{} \underline{\ ?\ } = n$$

6. To find how many inches are in $3\frac{1}{4}$ feet, you could first find the number of inches in 3 feet, next find the number of inches in $\frac{1}{4}$ of a foot, and then _?_. *(page 198)*

7. To find the number when a fraction of it is known, you can write and solve an equation. For example, if you know $\frac{2}{3}$ of a number is 36, you can solve the equation at the right to find the number. To _?_ the right side of the last equation, you would multiply _?_ by $\frac{3}{2}$. *(page 200)*

$$\frac{2}{3} \text{ of } n = 36$$
$$\frac{2}{3}n = 36$$
$$\frac{\frac{2}{3}n}{\frac{2}{3}} = \frac{36}{\frac{2}{3}}$$

▷ **Give each product in simplest form.** *(pages 184, 186)*

1. $8 \cdot \dfrac{3}{4}$

2. $\dfrac{7}{16} \cdot \dfrac{12}{5}$

3. $15 \cdot \dfrac{2x}{3}$

4. $\dfrac{4m}{3n} \cdot \dfrac{p}{2m}$

5. $\dfrac{j}{2h} \cdot \dfrac{h^2}{j}$

6. $1\dfrac{1}{2} \cdot 3$

7. $2\dfrac{1}{3} \cdot 1\dfrac{1}{4}$

8. $2\dfrac{2}{3} \cdot 2\dfrac{2}{3}$

9. $2\dfrac{3}{4} \cdot 3\dfrac{1}{6}$

10. $4\dfrac{2}{3} \cdot 2\dfrac{5}{6}$

▷ **Give each quotient in simplest form.** *(pages 188, 190)*

11. $\dfrac{7}{5} \div 3$

12. $\dfrac{7}{4} \div \dfrac{3}{8}$

13. $\dfrac{a}{b} \div \dfrac{1}{b}$

14. $\dfrac{3r}{s} \div \dfrac{5}{2r}$

15. $\dfrac{2z^2}{w} \div \dfrac{8z}{3w}$

16. $7\dfrac{1}{2} \div 1\dfrac{1}{4}$

17. $4\dfrac{2}{3} \div 4$

18. $5\dfrac{1}{2} \div 1\dfrac{1}{3}$

19. $3\dfrac{5}{6} \div 1\dfrac{2}{3}$

20. $6\dfrac{7}{8} \div 2\dfrac{3}{4}$

▷ **Solve. Give answers in simplest form.** *(page 196)*

21. $\dfrac{1}{5}$ of $35 = n$

22. $\dfrac{2}{3}$ of $24 = n$

23. $\dfrac{6}{5}$ of $60 = n$

24. $\dfrac{7}{8}$ of $56 = n$

25. $\dfrac{1}{8}$ of $13 = n$

26. $\dfrac{2}{3}$ of $14 = n$

27. $\dfrac{3}{5}$ of \$30 $= n$

28. $\dfrac{1}{2}$ of \$34 $= n$

▷ **Complete.** *(page 198)*

29. $2\dfrac{1}{2}$ ft = __?__ in.

30. $1\dfrac{3}{4}$ days = __?__ h

31. $4\dfrac{3}{4}$ min = __?__ s

32. $3\dfrac{1}{2}$ gal = __?__ qt

▷ **Solve. Give answers in simplest form.** *(page 200)*

33. $\dfrac{1}{4}n = 20$

34. $\dfrac{2}{3}n = 42$

35. $\dfrac{9}{8}n = 72$

36. $\dfrac{5}{6}n = 60$

37. $\dfrac{2}{5}n = 15$

38. $\dfrac{3}{4}n = 10$

39. $\dfrac{1}{4}n = 4$

40. $\dfrac{5}{2}n = 5$

▷ **Write an equation and solve the problem.** *(page 202)*

41. The Stereo Shop put 48 pocket radios on sale for $\dfrac{3}{4}$ of the regular price. What was the sale price if the regular price for one was \$36?

42. During the sale, $\dfrac{3}{4}$ of the \$36 speakers were sold. How many of the speakers were on sale including the 48 that were sold?

43. There were 36 tape players on sale for $\dfrac{3}{4}$ of the regular price. The regular price for one was \$48. What was the sale price?

44. Forty-eight headsets were on sale for $\dfrac{3}{4}$ of the regular price. The sale price for one was \$36. What was the regular price?

Algebra Using Products and Quotients of Fractions | **207**

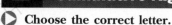

▶ **Choose the correct letter.**

---

1. Solve.

$$58 = 26 + j$$

A. 32   B. 58

C. 84   D. none of
these

---

2. Solve.

$$12 = \frac{n}{6} - 4$$

A. 72   B. 48

C. 96   D. none of
these

---

3. Solve.

$$60 = 3c + c$$

A. 15   B. 20

C. 240   D. none of
these

---

4. Solve.

$$\frac{x}{^-2} - 5 = 6$$

A. $^-22$ B. 22

C. $^-2$ D. none of
these

---

5. The algebraic
factorization of $30rs^2$ is

A. $30 \cdot r \cdot s \cdot s$

B. $3 \cdot 10 \cdot r \cdot s \cdot s$

C. $2 \cdot 3 \cdot 5 \cdot r \cdot r \cdot s$

D. none of these

---

6. The greatest common
factor of $3a^2b$ and
$15ab^2$ is

A. $3ab^2$ B. $3a^2b$

C. $3ab$ D. none of
these

---

7. The least common
multiple of $3c$ and $5c^2$ is

A. 15   B. $15c$

C. $15c^2$ D. none of
these

---

8. $\dfrac{20}{8}$ written in simplest
form is

A. $\dfrac{10}{4}$ B. $\dfrac{5}{2}$

C. $2\dfrac{4}{8}$ D. none of
these

---

9. Give the sum.

$$\frac{2}{3} + \frac{3}{4}$$

A. $1\dfrac{5}{12}$ B. $\dfrac{5}{7}$

C. $\dfrac{5}{12}$ D. none of
these

---

10. Give the sum.

$$\frac{a}{5} + \frac{b}{3}$$

A. $\dfrac{a+6}{15}$

B. $\dfrac{3a+b}{15}$

C. $\dfrac{3a+5b}{15}$

D. none of these

---

11. Give the difference.

$$\frac{3x}{4} - \frac{y}{6}$$

A. $\dfrac{9x-2y}{12}$

B. $\dfrac{3x-y}{12}$

C. $\dfrac{3x-2y}{12}$

D. none of these

---

12. Choose the equation.

Twelve more than the
quotient of a number
divided by 6 is 18. What
is the number?

A. $6n + 12 = 18$

B. $6n - 12 = 18$

C. $\dfrac{n}{6} - 12 = 18$

D. $\dfrac{n}{6} + 12 = 18$

Algebra Using Rational Numbers

8

$d = rt$

Distance traveled in miles equals the rate of traveling in miles per hour times the time spent traveling in hours.
(page 231)

In a bicycle race, cyclists can use the formula above to calculate how far they will travel if they maintain a certain rate of speed.

OBJECTIVE: To write a rational number in simplest fractional form. To compare two rational numbers.

A number that can be written as the quotient of two integers (divisor not 0) is called a **rational number.**

A rational number is used to show the net change in the closing price of a stock (the difference between yesterday's closing price and today's closing price). If a stock had a net change of $^{+}\frac{1}{2}$, the price of each share went *up* $\frac{1}{2}$ of a dollar. If a stock had a net change of $^{-}1\frac{1}{8}$, the price of each share went *down* $1\frac{1}{8}$ dollars.

Daily Stock Quotations

Stock	Close	Net Change
Disney	$128\frac{1}{2}$	$^{-}5\frac{7}{8}$
duPont	$40\frac{1}{2}$	$+1$
Exxon	$45\frac{5}{8}$	$^{-}\frac{1}{8}$
Ford	$43\frac{1}{4}$	$^{-}1$
IBM	$117\frac{5}{8}$	$^{-}3\frac{3}{8}$
Kroger	$16\frac{1}{2}$	$^{-}\frac{1}{4}$
Mobil	$65\frac{1}{4}$	$+1\frac{1}{2}$
Sears	$34\frac{1}{2}$	$+\frac{1}{4}$
TWA	$12\frac{3}{8}$	$+\frac{5}{8}$

1. What was the net change of Exxon? Of Mobil?

2. Which company had the greater gain, TWA or Sears?

EXAMPLES — **Here's how to compare two rational numbers. Notice that we will not use the raised plus sign when writing a positive rational number.**

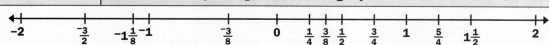

As you move to the **right** on this number line, the rational numbers get **larger.**

As you move to the **left,** the rational numbers get **smaller.**

The number line shows that

A $^{-}1\frac{1}{8} < \frac{^{-}3}{8}$ B $\frac{^{-}3}{8} < 0$ C $\frac{3}{4} > \frac{1}{4}$ D $\frac{^{-}3}{2} > ^{-}2$

$\frac{^{-}3}{2}$ *in simplest fractional form is* $^{-}1\frac{1}{2}.$

To compare rational numbers with different denominators, compare equivalent rational numbers with the same denominator.

E $\frac{^{-}1}{2} \diamond \frac{^{-}3}{8}$ F $\frac{^{-}2}{3} \diamond \frac{^{-}3}{4}$ G $2\frac{1}{3} \diamond 2\frac{3}{8}$

$\frac{^{-}4}{8} < \frac{^{-}3}{8}$ $\frac{^{-}8}{12} > \frac{^{-}9}{12}$ $2\frac{8}{24} < 2\frac{9}{24}$

CHECK for Understanding

3. Look at Example E. Is $\frac{^{-}3}{8}$ less than or greater than $\frac{^{-}1}{2}$?

4. **a.** Is 0 less than any positive rational number?
   b. Is 0 greater than any negative rational number?

EXERCISES

▶ **Write each of these rational numbers in simplest fractional form.**

Here are scrambled answers for the next row of exercises: $\dfrac{^-1}{2}$ 3 $\dfrac{5}{6}$ $^-2$ $^-3\dfrac{1}{2}$ $3\dfrac{1}{3}$

5. $\dfrac{^-8}{4}$

6. $\dfrac{10}{12}$

7. $\dfrac{9}{3}$

8. $\dfrac{^-3}{6}$

9. $\dfrac{10}{3}$

10. $\dfrac{^-14}{4}$

11. $\dfrac{^-4}{24}$

12. $\dfrac{18}{3}$

13. $\dfrac{^-19}{3}$

14. $\dfrac{6}{30}$

15. $\dfrac{24}{4}$

16. $\dfrac{^-15}{25}$

17. $\dfrac{18}{12}$

18. $\dfrac{^-22}{4}$

19. $\dfrac{^-20}{5}$

20. $\dfrac{^-8}{18}$

21. $\dfrac{9}{6}$

22. $\dfrac{^-20}{35}$

23. $\dfrac{^-11}{2}$

24. $\dfrac{17}{5}$

25. $\dfrac{^-18}{24}$

26. $\dfrac{30}{6}$

27. $\dfrac{^-42}{7}$

28. $\dfrac{^-30}{8}$

29. $\dfrac{^-9}{9}$

30. $\dfrac{15}{18}$

31. $\dfrac{0}{4}$

32. $\dfrac{^-8}{20}$

33. $\dfrac{10}{6}$

34. $\dfrac{^-25}{20}$

▶ **<, =, or >?**

35. $^-8 \diamond ^-6$

36. $8 \diamond 6$

37. $^-9 \diamond ^-10$

38. $9 \diamond 10$

39. $\dfrac{2}{3} \diamond \dfrac{3}{4}$

40. $\dfrac{^-2}{3} \diamond \dfrac{^-3}{4}$

41. $\dfrac{7}{8} \diamond \dfrac{5}{6}$

42. $\dfrac{^-7}{8} \diamond \dfrac{^-5}{6}$

43. $0 \diamond \dfrac{2}{5}$

44. $0 \diamond \dfrac{^-2}{5}$

45. $^-4 \diamond ^-4\dfrac{1}{3}$

46. $4 \diamond 4\dfrac{1}{3}$

47. $2\dfrac{2}{3} \diamond 3$

48. $^-5\dfrac{5}{6} \diamond ^-5$

49. $7 \diamond ^-8\dfrac{2}{3}$

50. $^-7 \diamond 8\dfrac{2}{3}$

51. $^-1\dfrac{1}{3} \diamond \dfrac{^-4}{3}$

52. $2\dfrac{5}{8} \diamond \dfrac{^-1}{2}$

53. $^-4 \diamond ^-4\dfrac{1}{3}$

54. $5\dfrac{5}{6} \diamond 5$

55. $^-2\dfrac{3}{4} \diamond ^-2\dfrac{5}{8}$

56. $5\dfrac{1}{4} \diamond 5\dfrac{2}{8}$

57. $^-6\dfrac{2}{3} \diamond ^-6\dfrac{3}{4}$

58. $8\dfrac{5}{8} \diamond 8\dfrac{1}{2}$

Problem Solving USING DATA

▶ **Solve. Use the daily stock quotations on page 210.**

59. Which stock went down $1 per share?

60. Which stock went up 25¢ a share?

61. Suppose that you owned 100 shares of duPont. By how much did the value of your shares increase?

62. Suppose that you owned 100 shares of Mobil. By how much did the value of your shares increase?

★63. If you owned 100 shares of Ford and 100 shares of Mobil, what would your total net change (in dollars) be for the day?

★64. If you owned 200 shares of TWA and 400 shares of Mobil, what would your total net change (in dollars) be for the day?

Algebra Using Rational Numbers | **211**

8-2 Writing Rational Numbers

OBJECTIVE: To write a rational number in decimal form.
To write a decimal in simplest fractional form.

A chili recipe calls for $\frac{3}{4}$ pound of ground
pork. You have 0.75 pound of ground pork. To
decide whether you have the proper amount
for the recipe, you can change $\frac{3}{4}$ to a decimal
or change 0.75 to a fraction.

EXAMPLES	Here's how to write a rational number in decimal form.

To write a rational number in decimal form, divide the numerator by the denominator.

A $\frac{3}{4} = ?$

$$\begin{array}{r} 0.75 \\ 4\overline{)3.00} \\ -28 \\ \hline 20 \\ -20 \\ \hline 0 \end{array}$$

$$\frac{3}{4} = 0.75$$

*Notice that the division came to an end, or terminated. For this reason, 0.75 is called a **terminating decimal**.*

B $\frac{^-4}{11} = ?$

$$\begin{array}{r} 0.3636 \\ 11\overline{)4.0000} \\ -33 \\ \hline 70 \\ -66 \\ \hline 40 \\ -33 \\ \hline 70 \\ -66 \\ \hline 4 \end{array}$$

$$\frac{^-4}{11} = {^-}0.\overline{36}$$

*Notice that the division did not come to an end. The bar over the "36" tells you that the 36 repeats. For this reason, $^-0.\overline{36}$ is called a **repeating decimal**.*

C $^-2\frac{2}{3} = ?$

First change $\frac{2}{3}$ to a decimal.

$$\begin{array}{r} 0.66 \\ 3\overline{)2.00} \\ -18 \\ \hline 20 \\ -18 \\ \hline 2 \end{array}$$

Since $\frac{2}{3} = 0.\overline{6}$

$$^-2\frac{2}{3} = {^-}2.\overline{6}$$

CHECK for Understanding

1. Look at Example A. Is $\frac{3}{4}$ pound the same as 0.75 pound?

EXAMPLES	Here's how to write a decimal in simplest fractional form.

	Read the decimal.	Write as a fraction or mixed number.	Write in simplest form.
D	0.75 ← 75 hundredths	$0.75 = \frac{75}{100}$	$0.75 = \frac{75}{100} = \frac{3}{4}$
E	$^-3.8$ ← negative 3 and 8 tenths	$^-3.8 = {^-}3\frac{8}{10}$	$^-3.8 = {^-}3\frac{8}{10} = {^-}3\frac{4}{5}$

EXERCISES

▶ **Write each rational number in decimal form.**
Here are scrambled answers for the next row of exercises: $2.\overline{6}$ $0.\overline{3}$ 0.5 $^-0.8$ $^-1.75$ $^-0.\overline{4}$

2. $\dfrac{1}{2}$ **3.** $\dfrac{^-7}{4}$ **4.** $\dfrac{1}{3}$ **5.** $\dfrac{^-4}{5}$ **6.** $\dfrac{8}{3}$ **7.** $\dfrac{^-4}{9}$

8. $\dfrac{^-15}{2}$ **9.** $\dfrac{7}{2}$ **10.** $\dfrac{^-9}{8}$ **11.** $\dfrac{15}{4}$ **12.** $\dfrac{^-8}{3}$ **13.** $\dfrac{11}{6}$

14. $\dfrac{^-8}{9}$ **15.** $\dfrac{5}{11}$ **16.** $\dfrac{^-13}{10}$ **17.** $\dfrac{17}{12}$ **18.** $\dfrac{^-19}{16}$ **19.** $\dfrac{25}{12}$

20. $2\dfrac{1}{2}$ ◀ | *Hint: First change $\dfrac{1}{2}$ to a decimal.* | **21.** $^-4\dfrac{1}{3}$ **22.** $3\dfrac{3}{5}$ **23.** $^-6\dfrac{5}{6}$

▶ **Write each decimal in simplest fractional form.**
Here are scrambled answers for the next row of exercises: $\dfrac{^-1}{4}$ $\dfrac{^-1}{8}$ $\dfrac{3}{5}$ $\dfrac{3}{4}$ $\dfrac{^-1}{2}$

24. 0.6 **25.** $^-0.25$ **26.** $^-0.5$ **27.** 0.75 **28.** $^-0.125$

29. 0.8 **30.** $^-0.24$ **31.** 0.48 **32.** $^-0.9$ **33.** 0.150

34. $^-0.35$ **35.** 0.375 **36.** $^-0.72$ **37.** $^-0.4$ **38.** 0.16

39. 2.25 **40.** $^-1.4$ **41.** 2.40 **42.** $^-5.5$ **43.** 9.35

Problem Solving USING RATIONAL NUMBERS

▶ **Solve.**

44. If it takes $\dfrac{3}{4}$ pound of pork for each batch of chili, how many batches can be made with 2 pounds of pork?

45. Suppose that a chili recipe calls for 1 pound of pork and $2\dfrac{1}{4}$ pounds of beef. How many pounds of beef would be needed for $1\dfrac{1}{2}$ pounds of pork?

Challenge! LOOK FOR A PATTERN

▶ **Copy and complete. A calculator may be helpful.**

46.

Fractional form	$\dfrac{1}{11}$	$\dfrac{2}{11}$	$\dfrac{3}{11}$	$\dfrac{4}{11}$	$\dfrac{5}{11}$	**e.**	**f.**	**g.**	**h.**
Decimal form	$0.\overline{09}$	**a.**	**b.**	**c.**	**d.**	$0.\overline{54}$	$0.\overline{63}$	$0.\overline{72}$	$0.\overline{81}$

OBJECTIVE: To add and subtract rational numbers.

Rational numbers can be written as decimals or as fractions. The sum $^-0.625 + 0.375$ can also be written as $\dfrac{^-5}{8} + \dfrac{3}{8}$.

EXAMPLES	Here's how to add rational numbers written as fractions.

To add rational numbers with a common denominator, write the sum of the numerators over the common denominator.

To add rational numbers with different denominators, find the least common denominator, change to equivalent rational numbers, and add.

$\boxed{A}$ $\dfrac{^-5}{8} + \dfrac{3}{8} = \dfrac{^-2}{8}$

$= \dfrac{^-1}{4}$ ← $\boxed{\textit{Simplest form}}$

$\boxed{B}$ $^-2\dfrac{1}{2} + {}^-1\dfrac{3}{4} = \dfrac{^-5}{2} + \dfrac{^-7}{4}$ ← $\boxed{\begin{array}{l}\textit{The LCD}\\\textit{of 2 and 4}\\\textit{is 4.}\end{array}}$

$= \dfrac{^-10}{4} + \dfrac{^-7}{4}$

$= \dfrac{^-17}{4}$

$= {}^-4\dfrac{1}{4}$ ← $\boxed{\textit{Simplest form}}$

EXAMPLES	Here's how to subtract rational numbers written as fractions.

To subtract a rational number, add the opposite of the rational number.

$\boxed{C}$ $\dfrac{5}{16} - \dfrac{13}{16} = \dfrac{5}{16} + \dfrac{^-13}{16}$

$= \dfrac{^-8}{16}$

$= \dfrac{^-1}{2}$ ← $\boxed{\textit{Simplest form}}$

$\boxed{D}$ $^-3\dfrac{2}{3} - {}^-1\dfrac{3}{4} = \dfrac{^-11}{3} - \dfrac{^-7}{4}$

$= \dfrac{^-11}{3} + \dfrac{7}{4}$ ← $\boxed{\begin{array}{l}\textit{The LCD}\\\textit{of 3 and 4}\\\textit{is 12.}\end{array}}$

$= \dfrac{^-44}{12} + \dfrac{21}{12}$

$= \dfrac{^-23}{12}$

$= {}^-1\dfrac{11}{12}$ ← $\boxed{\textit{Simplest form}}$

CHECK for Understanding

1. Look at Examples C and D. To subtract a rational number, add the __?__ of the rational number.

EXERCISES

▷ **Give each sum in simplest form.**
Here are scrambled answers for the next row of exercises: $\frac{^-7}{16}$ $\quad \frac{^-1}{8}$ $\quad 1\frac{1}{16}$ $\quad \frac{^-1}{5}$ $\quad \frac{^-13}{20}$

2. $\frac{2}{5} + \frac{^-3}{5}$ **3.** $\frac{^-1}{2} + \frac{3}{8}$ **4.** $\frac{3}{16} + \frac{7}{8}$ **5.** $\frac{^-1}{4} + \frac{^-2}{5}$ **6.** $\frac{5}{16} + \frac{^-3}{4}$

7. $\frac{^-2}{5} + \frac{^-3}{20}$ **8.** $\frac{1}{6} + \frac{^-5}{9}$ **9.** $\frac{7}{8} + \frac{5}{6}$ **10.** $\frac{^-9}{10} + \frac{7}{10}$ **11.** $\frac{5}{12} + \frac{^-5}{8}$

12. $-3\frac{1}{5} + 4\frac{2}{5}$ **13.** $^-4\frac{1}{2} + ^-3\frac{1}{4}$ **14.** $5\frac{7}{8} + ^-3$ **15.** $3\frac{2}{3} + ^-2\frac{2}{3}$ **16.** $^-8 + ^-6\frac{2}{5}$

17. $^-5\frac{1}{2} + 2\frac{3}{4}$ **18.** $3\frac{7}{8} + ^-4\frac{1}{6}$ **19.** $^-6\frac{7}{10} + 4\frac{3}{4}$ **20.** $^-8\frac{11}{12} + ^-2\frac{2}{3}$ **21.** $4\frac{5}{8} + ^-2\frac{2}{3}$

▷ **Give each difference in simplest form.**
Here are scrambled answers for the next row of exercises: $1\frac{1}{6}$ $\quad \frac{^-14}{15}$ $\quad \frac{1}{8}$ $\quad \frac{11}{12}$ $\quad \frac{^-13}{18}$

22. $\frac{2}{3} - \frac{^-1}{4}$ **23.** $\frac{^-3}{5} - \frac{1}{3}$ **24.** $\frac{^-3}{8} - \frac{1}{2}$ **25.** $\frac{^-1}{2} - \frac{2}{9}$ **26.** $\frac{5}{12} - \frac{^-3}{4}$

27. $\frac{5}{6} - \frac{^-7}{8}$ **28.** $\frac{11}{12} - \frac{5}{8}$ **29.** $4\frac{3}{5} - ^-2\frac{1}{5}$ **30.** $8\frac{3}{4} - 5\frac{1}{2}$ **31.** $^-9 - 4\frac{3}{4}$

32. $8\frac{5}{8} - ^-10\frac{2}{3}$ **33.** $6\frac{5}{6} - 4\frac{1}{8}$ **34.** $^-6\frac{3}{5} - ^-9\frac{1}{2}$ **35.** $^-8\frac{11}{12} - 9\frac{7}{8}$ **36.** $3\frac{5}{8} - ^-9\frac{5}{6}$

Group Project ANALYZING DATA *What's Your Reaction Time?*

▷ **Work in a small group.**

37. Each class member estimates the average reaction time for the class. Give your estimate to the nearest hundredth of a second.

38. a. First, all class members grasp hands to form a circle.
b. Next, with all students' eyes closed, your teacher will time how long it takes for a "squeeze" to go around the circle.
c. Compute the average reaction time for your class. Round your answer to the nearest hundredth of a second.

39. Who in your class had the closest estimate?

40. a. Repeat Exercise 38 five more times.
b. Did the average reaction time improve with practice?

41. Write a paragraph that summarizes the data.

OBJECTIVE: To multiply and divide rational numbers.

EXAMPLES	Here's how to multiply rational numbers.

To multiply rational numbers, multiply the numerators to get the numerator of the product and multiply the denominators to get the denominator of the product.

$\boxed{A}$ $\dfrac{^-3}{8} \cdot \dfrac{5}{4} = \dfrac{^-15}{32}$

$\boxed{B}$ $\dfrac{^-\overset{^-5}{10}}{3} \cdot \dfrac{5}{\underset{2}{4}} = \dfrac{^-25}{6}$

$= ^-4\dfrac{1}{6}$ ← $\boxed{Simplest\ form}$

$\boxed{C}$ $^-2\dfrac{2}{3} \cdot {}^-4\dfrac{1}{2} = \dfrac{\overset{^-4}{^-8}}{3} \cdot \dfrac{\overset{^-3}{^-9}}{\underset{1}{2}}$

$= 12$

CHECK for Understanding

1. Look at Example C.
   a. $^-2\dfrac{2}{3}$ and $^-4\dfrac{1}{2}$ were first written as $\dfrac{^-8}{3}$ and $\underline{\ ?\ }$.
   b. Next, the shortcut called $\underline{\ ?\ }$ was used.
   c. Finally, the product was written in $\underline{\ ?\ }$ form.

EXAMPLES	Here's how to divide rational numbers.

Remember that two numbers are reciprocals if their product is 1.
To divide by a rational number, multiply by its reciprocal.

$\boxed{D}$ $\dfrac{5}{3} \div \dfrac{^-2}{5} = \dfrac{5}{3} \cdot \dfrac{^-5}{2}$

$= \dfrac{^-25}{6}$

$= ^-4\dfrac{1}{6}$ ← $\boxed{Simplest\ form}$

$\boxed{E}$ $\dfrac{^-7}{8} \div \dfrac{3}{4} = \dfrac{^-7}{\underset{2}{8}} \cdot \dfrac{\overset{1}{4}}{3}$

$= \dfrac{^-7}{6}$

$= ^-1\dfrac{1}{6}$

$\boxed{F}$ $^-4\dfrac{2}{3} \div {}^-5\dfrac{3}{5} = \dfrac{^-14}{3} \div \dfrac{^-28}{5}$

$= \dfrac{\overset{^-1}{^-14}}{3} \cdot \dfrac{^-5}{\underset{2}{28}}$

$= \dfrac{5}{6}$

EXERCISES

▶ **Give each product in simplest form.**

Here are scrambled answers for the next row of exercises: $\quad ^-2\dfrac{2}{15} \quad \dfrac{^-3}{20} \quad 1 \quad \dfrac{1}{6} \quad 1\dfrac{1}{2}$

2. $\dfrac{^-3}{4} \cdot \dfrac{1}{5}$ 3. $\dfrac{^-1}{3} \cdot \dfrac{^-1}{2}$ 4. $\dfrac{1}{4} \cdot 4$ 5. $\dfrac{^-8}{3} \cdot \dfrac{4}{5}$ 6. $\dfrac{9}{10} \cdot \dfrac{5}{3}$

7. $\dfrac{5}{6} \cdot \dfrac{^-3}{4}$ 8. $\dfrac{^-7}{8} \cdot \dfrac{9}{2}$ 9. $\dfrac{6}{5} \cdot \dfrac{5}{6}$ 10. $\dfrac{^-7}{4} \cdot \dfrac{^-7}{4}$ 11. $^-8 \cdot \dfrac{5}{6}$

12. $2 \cdot {}^-1\dfrac{1}{2}$ 13. $2\dfrac{2}{3} \cdot 4$ 14. $^-1\dfrac{3}{4} \cdot 1\dfrac{3}{4}$ 15. $^-1\dfrac{5}{6} \cdot {}^-2\dfrac{1}{3}$ 16. $3\dfrac{1}{4} \cdot {}^-3\dfrac{3}{4}$

17. $4\dfrac{1}{6} \cdot 2\dfrac{1}{3}$ 18. $^-1\dfrac{1}{2} \cdot {}^-1\dfrac{3}{4}$ 19. $2\dfrac{2}{3} \cdot {}^-1\dfrac{3}{4}$ 20. $^-4\dfrac{1}{2} \cdot 2\dfrac{3}{8}$ 21. $3\dfrac{1}{8} \cdot {}^-3\dfrac{3}{4}$

▶ **Give each quotient in simplest form.**

Here are scrambled answers for the next row of exercises:

$$1\frac{1}{2} \qquad {}^-3 \qquad {}^-1\frac{1}{3} \qquad 1\frac{1}{4}$$

22. $\dfrac{{}^-3}{4} \div \dfrac{1}{4}$

23. $\dfrac{1}{2} \div \dfrac{1}{3}$

24. $\dfrac{{}^-5}{6} \div \dfrac{{}^-2}{3}$

25. $\dfrac{2}{3} \div \dfrac{{}^-1}{2}$

26. $\dfrac{{}^-3}{10} \div \dfrac{4}{5}$

27. $\dfrac{3}{4} \div \dfrac{1}{2}$

28. ${}^-6 \div \dfrac{3}{4}$

29. $\dfrac{5}{8} \div {}^-3$

30. $\dfrac{{}^-5}{6} \div {}^-4$

31. $\dfrac{5}{8} \div \dfrac{2}{3}$

32. $5 \div {}^-2\dfrac{1}{2}$

33. ${}^-2\dfrac{1}{2} \div 1\dfrac{1}{4}$

34. ${}^-2\dfrac{1}{3} \div {}^-1\dfrac{1}{4}$

35. $2\dfrac{7}{8} \div 3\dfrac{1}{4}$

36. $3\dfrac{5}{6} \div {}^-2\dfrac{1}{3}$

37. ${}^-5\dfrac{3}{4} \div {}^-2\dfrac{2}{3}$

38. $6\dfrac{2}{3} \div 5\dfrac{1}{3}$

39. ${}^-4\dfrac{5}{6} \div 4$

40. $4\dfrac{1}{4} \div {}^-3\dfrac{1}{8}$

41. ${}^-3\dfrac{1}{2} \div {}^-1\dfrac{3}{4}$

Problem Solving USING A FORMULA

Your reaction distance while driving a car is the distance that your car travels during the time it takes you to move your foot from the gas pedal to the brake pedal. This formula shows the relationship between reaction distance and speed.

$$d = \frac{11s}{10} \quad \leftarrow \begin{array}{l}\textit{speed in}\\ \textit{miles}\\ \textit{per hour}\end{array}$$

$\uparrow$
Reaction distance in feet

42. What would your approximate reaction distance be if you were driving
  a. 20 mph?   **b.** 30 mph?
  c. 40 mph?   **d.** 50 mph?

★**43.** How fast would you be driving if your reaction distance was
  a. 27.5 ft?   **b.** 49.5 ft?
  c. 60.5 ft?   **d.** 57.75 ft?

BUILD YOUR PROBLEM SOLVING POWER

You will use a variety of problem-solving strategies on these two pages. For some problems, you may wish to use a calculator.

1 MAKE A DRAWING

You are riding your bicycle from Santa Rosa to Sonoma. This sign is $1\frac{1}{2}$ miles ahead of you. How far are you from Geyserville?

✓ Problem Solving Tips

Make a drawing.* Use the data from the sign to fill in the missing information in this drawing. Then decide how to use the data to solve the problem.

GEYSERVILLE   SANTA ROSA   EL VERANO   SONOMA

*At least one more problem on these two pages can be solved by making a drawing.

EL VERANO	6 mi ►
SONOMA	8 mi ►
◄ SANTA ROSA	16 mi
◄ GEYSERVILLE	45 mi

2 ON THE ROAD AGAIN

Here is some information about a trip you plan to take this weekend:

- You are driving south from Amarillo to Lubbock, a distance of 120 miles.
- You will pass through Canyon, Tulia, and Plainview (in that order) on your trip.
- When you are 10 miles north of Plainview, you will be 55 miles from Lubbock.
- When you are 8 miles south of Canyon, you will be 24 miles from Amarillo.
- Tulia is 50 miles south of Amarillo.

How far is it from
a. Amarillo to Canyon?
b. Canyon to Tulia?
c. Tulia to Plainview?
d. Plainview to Lubbock

3 TENNIS ANYONE?

Scott, Bill, and Rick each have a sister. The girls' names are Heather, Beth, and Rose.

All six of them played in a mixed doubles tennis match. Scott teamed up with Bill's sister, and played against Rose's brother and Beth. Rick sat out with the other girl, who was not his sister.

Who is each boy's sister?

4 GEOMETRY—VISUAL THINKING

How many different triangles can you draw on a 3-by-3 dot grid? (Two triangles are the same if you can turn and/or flip one and make it fit exactly on the other.)

5 DOG POUNDS

How much does each dog weigh? *Hint: Which weighs more, Duffy or Ivory? How much more?*

6 McKAY'S DAYS

Dr. Thomas McKay lived $\frac{3}{13}$ of his life as a child and schoolboy. Then he spent $\frac{4}{39}$ of his life preparing for his outstanding medical career. For $\frac{1}{2}$ of his life, he was a successful surgeon in a famous children's hospital.

Since his retirement, he has lived 13 years on a comfortable pension.

How old is he now?

7 PEP IT UP!

Kevin, Larry, Monica, and Nancy each play a horn in the school band. Oliver, Polly, and Quinn each play a woodwind instrument. The band director wants to form a pep band using 3 horns and 2 woodwinds. Make a list of the possible combinations using these seven band members.

APPLICATIONS

8 How much did the whole cheesecake weigh?

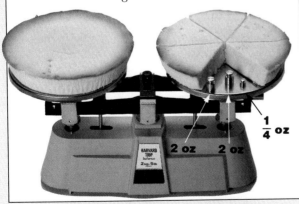

LOMBARDI'S World Famous CHEESECAKE

Size	Weight (ounces)	Number of Servings	Ounces per Serving
Large	$25\frac{1}{2}$	6	**a.**
Medium	$18\frac{3}{4}$	**b.**	$3\frac{3}{4}$
Small	**c.**	4	$3\frac{1}{2}$

9 Copy and complete the chart.

10 Lombardi's Deli charges $3.99 per pound for its World Famous Cheesecake. How much does a large cheesecake cost? Round your answer to the nearest cent.

Rate Your Problem-Solving Power: *6–7 correct = Good;   8–9 correct = Excellent;   10 correct = Exceptional*

OBJECTIVE: To use a sales-tax table for solving problems.

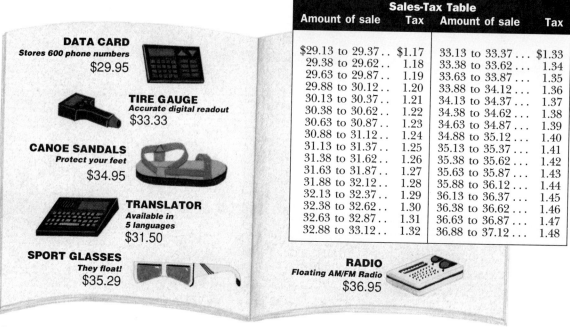

Sales-Tax Table			
Amount of sale	Tax	Amount of sale	Tax
$29.13 to 29.37..	$1.17	33.13 to 33.37 ...	$1.33
29.38 to 29.62..	1.18	33.38 to 33.62 ...	1.34
29.63 to 29.87..	1.19	33.63 to 33.87 ...	1.35
29.88 to 30.12..	1.20	33.88 to 34.12 ...	1.36
30.13 to 30.37..	1.21	34.13 to 34.37 ...	1.37
30.38 to 30.62..	1.22	34.38 to 34.62 ...	1.38
30.63 to 30.87..	1.23	34.63 to 34.87 ...	1.39
30.88 to 31.12..	1.24	34.88 to 35.12 ...	1.40
31.13 to 31.37..	1.25	35.13 to 35.37 ...	1.41
31.38 to 31.62..	1.26	35.38 to 35.62 ...	1.42
31.63 to 31.87..	1.27	35.63 to 35.87 ...	1.43
31.88 to 32.12..	1.28	35.88 to 36.12 ...	1.44
32.13 to 32.37..	1.29	36.13 to 36.37 ...	1.45
32.38 to 32.62..	1.30	36.38 to 36.62 ...	1.46
32.63 to 32.87..	1.31	36.63 to 36.87 ...	1.47
32.88 to 33.12..	1.32	36.88 to 37.12 ...	1.48

DATA CARD
Stores 600 phone numbers
$29.95

TIRE GAUGE
Accurate digital readout
$33.33

CANOE SANDALS
Protect your feet
$34.95

TRANSLATOR
Available in 5 languages
$31.50

SPORT GLASSES
They float!
$35.29

RADIO
Floating AM/FM Radio
$36.95

▶ **Use the information on the catalog page and the sales-tax table to solve the problem.**

1. The sales tax on a data card is $1.20. What is the sales tax on a radio?

2. Robert bought a translator and a pair of sport glasses. What was the total cost, including sales tax?

3. If you have $73, do you have enough money to buy a tire gauge and a radio? *Hint: Don't forget the sales tax.*

4. Molly bought 3 data cards and 2 pairs of canoe sandals. She paid $170 and received $3.86 change. What was the sales tax?

▶ **Write an equation and solve the problem.**

5. Rae bought 3 of the same catalog item. The sales tax was $3.78. The total cost, including the sales tax, was $98.28. What item did she buy?
(Let r = the cost of one of the items Rae bought.)

6. Stan bought 4 of the same catalog item. The sales tax was $5.32. The total cost, including the sales tax, was $138.64. What item did he buy?
(Let s = the cost of one of the items Stan bought.)

7. Sean bought a radio and some data cards. The sales tax was $6.28. How many data cards did he buy if the total cost, including the sales tax, was $163.03?
(Let d = the number of data cards Sean bought.)

8. Paige bought some pairs of canoe sandals and a radio. The sales tax was $4.28. How many pairs of canoe sandals did she buy if the total cost, including the sales tax, was $111.13?
(Let c = the number of pairs of canoe sandals Paige bought.)

▶ **Solve and check.** *(page 112)*

1. $^-7w + 15 = 1$ **2.** $^-6z - 4 = ^-10$ **3.** $4k - 9 = ^-9$ **4.** $5j + 33 = 18$

5. $\dfrac{k}{3} + 9 = ^-12$ **6.** $\dfrac{h}{-3} - 8 = 16$ **7.** $\dfrac{d}{-5} + 18 = 20$ **8.** $\dfrac{c}{9} - 24 = ^-24$

9. $^-5r - 7 = 8$ **10.** $^-3t + 9 = 27$ **11.** $\dfrac{x}{-4} + 21 = 30$ **12.** $\dfrac{y}{-8} - 12 = ^-40$

▶ **Write in scientific notation.** *(page 126)*

13. 14,000   **14.** 23,800   **15.** 120,000   **16.** 395,000

17. 5,800,000   **18.** 9,630,000   **19.** 18,000,000   **20.** 427,000,000

21. 0.0083   **22.** 0.052   **23.** 0.00084   **24.** 0.0009

25. 0.000178   **26.** 0.0000258   **27.** 0.005286   **28.** 0.0000718

▶ **Find the least common denominator.** *(page 142)*

29. $\dfrac{5}{6}, \dfrac{3}{4}$ **30.** $\dfrac{4}{9}, \dfrac{2}{3}$ **31.** $\dfrac{1}{8}, \dfrac{4}{3}$ **32.** $\dfrac{3}{10}, \dfrac{2}{5}$ **33.** $\dfrac{3}{8}, \dfrac{1}{6}$

34. $\dfrac{3}{a}, \dfrac{2}{3a}$ **35.** $\dfrac{a}{2b}, \dfrac{c}{b}$ **36.** $\dfrac{3}{8}, \dfrac{5}{64}$ **37.** $\dfrac{9}{2m^2}, \dfrac{7}{m}$ **38.** $\dfrac{1}{a^2b}, \dfrac{5}{ab^2}$

39. $\dfrac{a}{6b}, \dfrac{2}{3}$ **40.** $\dfrac{1}{2n}, \dfrac{t}{3n}$ **41.** $\dfrac{4}{d^2}, \dfrac{5}{d^3}$ **42.** $\dfrac{5}{c^2}, \dfrac{r}{3c^2}$ **43.** $\dfrac{e}{f^2}, \dfrac{e}{3f}$

▶ **Write in simplest form.** *(page 158)*

44. $\dfrac{24}{6}$ **45.** $\dfrac{30}{4}$ **46.** $\dfrac{39}{5}$ **47.** $\dfrac{21}{7}$ **48.** $\dfrac{40}{6}$ **49.** $\dfrac{42}{3}$

50. $\dfrac{4a}{6b}$ **51.** $\dfrac{12x}{16x}$ **52.** $\dfrac{7b^2}{4b}$ **53.** $\dfrac{10f}{4g}$ **54.** $\dfrac{28m}{8n}$ **55.** $\dfrac{5w}{3w^2}$

56. $\dfrac{3x}{2xy}$ **57.** $\dfrac{7pq}{3pq^2}$ **58.** $\dfrac{10m^2}{13mn^2}$ **59.** $\dfrac{5xy^2}{15xy}$ **60.** $\dfrac{24ab}{30a^2b^2}$ **61.** $\dfrac{18cd^2}{24c^2d}$

▶ **Add or subtract.** *(page 170)*

62. $\dfrac{4}{a} + \dfrac{2}{b}$ **63.** $\dfrac{3}{r} + \dfrac{4}{s}$ **64.** $\dfrac{2}{3} + \dfrac{6}{y}$ **65.** $\dfrac{2}{5} + \dfrac{2}{n}$ **66.** $\dfrac{2}{r} + \dfrac{r}{s}$

67. $\dfrac{5}{p} - \dfrac{6}{q}$ **68.** $\dfrac{2}{t} - \dfrac{8}{u}$ **69.** $\dfrac{3}{4} - \dfrac{2}{b}$ **70.** $\dfrac{3}{4} - \dfrac{7}{y}$ **71.** $\dfrac{4}{m} - \dfrac{m}{n}$

72. $\dfrac{a}{b} + \dfrac{7}{a}$ **73.** $\dfrac{2r}{t} - \dfrac{1}{2t}$ **74.** $\dfrac{1}{m} + \dfrac{n}{3}$ **75.** $\dfrac{1}{c} - \dfrac{c}{d}$ **76.** $\dfrac{2}{vw} - \dfrac{1}{v}$

OBJECTIVE: To solve equations having rational solutions.

There are no fractions in the equation
$3x - 2 = 0$. Nevertheless, the solution of the
equation is a fraction.

EXAMPLES | **Here's how to solve equations having rational solutions.**

A **Equation:**

	$3x - 2 = 0$	✓ **CHECK**
Add 2 to both sides.	$3x - 2 + 2 = 0 + 2$	$3x - 2 = 0$
Simplify.	$3x = 2$	$3 \cdot \dfrac{2}{3} - 2 \overset{?}{=} 0$
		$2 - 2 \overset{?}{=} 0$
Divide both sides by 3.	$\dfrac{3x}{3} = \dfrac{2}{3}$	$0 = 0$
		It checks!
Simplify.	$x = \dfrac{2}{3}$	

B **Equation:**

	$^-5x + 7 = 23$	✓ **CHECK**
Subtract 7 from both sides.	$^-5x + 7 - 7 = 23 - 7$	$^-5x + 7 = 23$
Simplify.	$^-5x = 16$	$^-5 \cdot {}^-3\dfrac{1}{5} + 7 \overset{?}{=} 23$
		$^-5 \cdot \dfrac{^-16}{5} + 7 \overset{?}{=} 23$
Divide both sides by $^-5$.	$\dfrac{^-5x}{^-5} = \dfrac{16}{^-5}$	$16 + 7 \overset{?}{=} 23$
		$23 = 23$
Simplify.	$x = {}^-3\dfrac{1}{5}$	*It checks!*

C **Equation:**

	$9 - 4x = {}^-5$	
To subtract, add the opposite. →	$9 + {}^-4x = {}^-5$	*Commutative property of addition*
	$^-4x + 9 = {}^-5$ ←	

Subtract 9 from both sides.	$^-4x + 9 - 9 = {}^-5 - 9$	✓ **CHECK**
Simplify.	$^-4x = {}^-14$	$9 - 4x = {}^-5$
		$9 - 4 \cdot 3\dfrac{1}{2} \overset{?}{=} {}^-5$
Divide both sides by $^-4$.	$\dfrac{^-4x}{^-4} = \dfrac{^-14}{^-4}$	$9 - \overset{2}{\cancel{4}} \cdot \dfrac{7}{\underset{1}{\cancel{2}}} \overset{?}{=} {}^-5$
		$9 - 14 \overset{?}{=} {}^-5$
Simplify.	$x = 3\dfrac{1}{2}$	$^-5 = {}^-5$
		It checks!

CHECK for Understanding

1. Complete these examples.

a.

$$5x - 4 = 8$$
$$5x - 4 + 4 = 8 + 4$$
$$5x = \underline{\ ?\ }$$
$$\frac{5x}{5} = \frac{12}{?}$$
$$x = 2\frac{2}{5}$$

b.

$$^-9y - 5 = 1$$
$$^-9y - 5 + 5 = 1 + \underline{\ ?\ }$$
$$^-9y = 6$$
$$\frac{^-9y}{^-9} = \frac{?}{^-9}$$
$$y = \frac{^-2}{3}$$

c.

$$6 - 8z = ^-10$$
$$6 + ^-8z = ^-10$$
$$^-8z + 6 = \underline{\ ?\ }$$
$$^-8z + 6 - 6 = ^-10 - \underline{\ ?\ }$$
$$^-8z = ^-16$$
$$\frac{^-8z}{^-8} = \frac{^-16}{?}$$
$$z = 2$$

EXERCISES

▶ Solve and check.
Here are scrambled answers for the next row of exercises: $\quad ^-1\frac{5}{7} \quad 3 \quad 1\frac{2}{3} \quad ^-3$

2. $3n + 8 = 13$ **3.** $4a - 3 = 9$ **4.** $^-8 - 5n = 7$ **5.** $9 + 7z = ^-3$

6. $^-5b - 12 = 7$ **7.** $^-12 + 5a = 8$ **8.** $^-7p + 11 = 15$ **9.** $^-10 - 6s = 18$

10. $6q + 13 = ^-9$ **11.** $9 - 8p = 0$ **12.** $10d - 5 = 0$ **13.** $^-10 + 12b = 8$

14. $16 + 7c = 10$ **15.** $^-12e - 18 = 5$ **16.** $^-8r + 16 = 20$ **17.** $18 - 7r = 12$

18. $10f - 7 = 11$ **19.** $^-15 - 6q = 15$ **20.** $14 + 9d = 13$ **21.** $15s + 10 = 12$

22. $12 - 8r = ^-3$ **23.** $^-7g - 16 = ^-12$ **24.** $^-16t + 50 = 0$ **25.** $^-10 + 8e = ^-2$

26. $12u + 18 = 32$ **27.** $18 + 12f = ^-6$ **28.** $8h - 10 = ^-19$ **29.** $16 - 4t = ^-20$

30. $^-21 + 11j = 21$ **31.** $^-15j - 9 = 15$ **32.** $^-20 - 11u = ^-10$ **33.** $^-18v + 12 = 12$

34. $24 - 15v = 16$ **35.** $25 + 4k = ^-17$ **36.** $20w + 24 = 6$ **37.** $20k - 12 = 26$

Problem Solving USING EQUATIONS

▶ **First choose the equation and solve the problem. Then reread the problem to see if your answer makes sense.**

> Equation A: $5n - 6 = 4$
> Equation B: $6 - 5n = 4$
> Equation C: $6n - 5 = 4$
> Equation D: $5 - 6n = 4$

38. If 5 is subtracted from 6 times a number, the difference is 4. What is the number?

39. If 6 times a number is subtracted from 5, the difference is 4. What is the number?

40. If 6 is subtracted from 5 times a number, the difference is 4. What is the number?

41. If you subtract the product of 5 times a number from 6, the difference is 4. What is the number?

OBJECTIVE: To simplify algebraic expressions by combining like terms.

In this lesson you will use the distributive property, the multiplying by 1 property, and the definition of subtraction to simplify expressions such as

$$^-5a - 3a \qquad\qquad 12c - c$$

EXAMPLES	Here's how to use the properties and the definition of subtraction to simplify expressions.

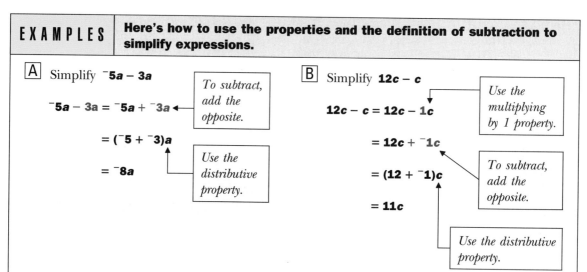

A Simplify $^-5a - 3a$

$^-5a - 3a = {}^-5a + {}^-3a$ ◄── *To subtract, add the opposite.*

$= (^-5 + {}^-3)a$

$= {}^-8a$ ◄── *Use the distributive property.*

B Simplify $12c - c$

$12c - c = 12c - 1c$ ── *Use the multiplying by 1 property.*

$= 12c + {}^-1c$

$= (12 + {}^-1)c$ ── *To subtract, add the opposite.*

$= 11c$

Use the distributive property.

CHECK for Understanding

1. Complete these examples.

   a. $^-5x + 3x = (^-5 + 3)x$
          $= \underline{\;?\;} x$

   b. $8z - 8z = 8z + {}^-8z$
          $= (8 + {}^-8)z$
          $= \underline{\;?\;} z$
          $= 0$

   c. $16w - w = 16w - 1w$
          $= 16w + {}^-1w$
          $= (16 + \underline{\;?\;})w$
          $= 15w$

EXAMPLES	Here's how to use a shortcut to simplify expressions.

To simplify these expressions, first change each subtraction to adding the opposite and then combine like terms (terms that have the same variable) by adding.

C Simplify $4a + 3 - 6a$

$4a + 3 - 6a = 4a + 3 + {}^-6a$

Like terms

$= {}^-2a + 3$

D Simplify $7b + 8 - b - 6$

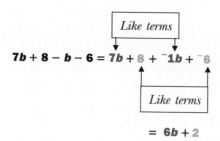

Like terms

$7b + 8 - b - 6 = 7b + 8 + {}^-1b + {}^-6$

Like terms

$= 6b + 2$

EXERCISES

▶ **Simplify by combining like terms.**
Here are scrambled answers for the next two rows of exercises: *11n − 6*   *9n*   *8n + 3*
   ⁻11n − 6   *7n*   *0*

2. $6n + 3n$

3. $8n - n$

4. $9n - 9n$

5. $8n + 3n - 6$

6. $^-8n - 3n - 6$

7. $8n - 3 + 6$

8. $3x + 11x$

9. $3x - 11x$

10. $^-3x - 11x$

11. $9y - 4y + 7$

12. $9y + 4y - 7$

13. $9y - 4 - 7$

14. $8 + 3w - 5$

15. $8w + 3 - 5$

16. $8 + 3w - 5w$

17. $10z - 12 - 6z$

18. $10z - 12z - 6$

19. $^-8r - r - 15$

20. $^-8r + r - 15$

21. $12t - 12t + 8$

22. $12t - 12 + 8t$

23. $5a + 3a + 9 - 8$

24. $5a + 3 + 9a - 8$

25. $11c - 4c + 12 - 12$

26. $11c - 4 + 10 - 12c$

27. $14j - 14j + 8 - 3$

28. $14j - 14 + 8j - 3$

29. $18m - 6 - m + 8$

30. $18m - 6m - 1 + 8$

31. $15p - 12 - 15p - 6$

32. $15p - 12p - 15 - 6$

33. $20t - 17 - 20t - 18$

34. $20t - 17t - 20 - 18$

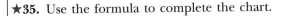 USING A FORMULA

Suppose that you throw a baseball upward at the rate of 64 feet per second (about 44 miles per hour). The formula below gives the distance that the ball would be above the ground in *t* seconds.

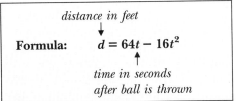

distance in feet

Formula: $d = 64t - 16t^2$

*time in seconds
after ball is thrown*

★**35.** Use the formula to complete the chart.

★**36.** How high would the ball be after 1 second? After 3 seconds? Why are the distances the same?

Time in seconds (*t*)	Distance in feet (*d*)
0	?
1	?
2	?
3	?
4	?

OBJECTIVE: To solve equations by combining like terms.

Sally bought 4 cassette tapes. Melissa bought 3 cassette tapes and a $7 storage box for her tapes. Each cassette costs the same. If they spent $63 together, how much did they spend for each cassette?

If you let c be the cost in dollars of each cassette, you can solve the problem by solving the equation

$$4c + 3c + 7 = 63$$

dollars Sally spent   dollars Melissa spent   total dollars spent

EXAMPLES	**Here's how to solve an equation that has the same variable in more than one term.**

A **Equation:** $4c + 3c + 7 = 63$

Combine like terms. $7c + 7 = 63$

Subtract 7 from both sides. $7c + 7 - 7 = 63 - 7$

Simplify. $7c = 56$

Divide both sides by 7. $\dfrac{7c}{7} = \dfrac{56}{7}$

Simplify. $c = 8$

✔ CHECK

$$4c + 3c + 7 = 63$$
$$4 \cdot 8 + 3 \cdot 8 + 7 \overset{?}{=} 63$$
$$32 + 24 + 7 \overset{?}{=} 63$$
$$63 = 63$$

It checks!

B
$$5y - 3y + 4 = 9$$
$$2y + 4 = 9$$
$$2y + 4 - 4 = 9 - 4$$
$$2y = 5$$
$$\dfrac{2y}{2} = \dfrac{5}{2}$$
$$y = 2\tfrac{1}{2}$$

C
$$16 + 4z - 6z = {}^-19$$
$${}^-2z + 16 = {}^-19$$
$${}^-2z + 16 - 16 = {}^-19 - 16$$
$${}^-2z = {}^-35$$
$$\dfrac{{}^-2z}{{}^-2} = \dfrac{{}^-35}{{}^-2}$$
$$z = 17\tfrac{1}{2}$$

CHECK for Understanding

1. Look at Example A. The first step in solving the equation was to __?__ like terms.

2. How much did each cassette cost? Does the answer seem reasonable?

3. Explain how to get the numbers printed in red in Examples B and C.

EXERCISES

▶ **Solve and check.**
Here are scrambled answers for the next two rows of exercises: $\quad 2\frac{3}{7} \quad {}^-6 \quad \frac{5}{9} \quad 3 \quad {}^-1\frac{3}{7} \quad {}^-4$

4. $7x + 4x = 33$

5. $9y - y = {}^-48$

6. $z - 5z = 16$

7. $7j + 2j - 5 = 0$

8. ${}^-5k - 2k - 4 = 6$

9. $7d - 2 + 5 = 20$

10. $4x + 10x = 28$

11. $4x - 10x = 28$

12. ${}^-4x - 10x = 28$

13. $8y - 3y + 6 = {}^-10$

14. $8y + 3y - 6 = {}^-10$

15. $8y - 3 - 6 = {}^-10$

16. $9 + 4w - 6 = 12$

17. $9w + 4 - 6 = 12$

18. $9 + 4w - 6w = 12$

19. $8z - 10 - 4z = {}^-20$

20. $8z - 10z - 4 = {}^-20$

21. $8 - 10z - 4z = {}^-20$

22. ${}^-6r - r - 13 = {}^-18$

23. ${}^-6r + r - 13 = {}^-18$

24. ${}^-6r + 1 - 13r = {}^-18$

25. $12t - t + 5 = 6$

26. $12t - 1 + 5t = 6$

27. $12 - t + 5t = 6$

28. $6a + 4a + 9 = 0$

29. $6a + 4 + 9a = 0$

30. $6a + 4 + 9 = 0$

31. $12c - 5c + 11 = {}^-4$

32. $12c - 5 + 11 = {}^-4$

33. $12c - 5 + 11c = {}^-4$

34. $15j - 10j + 8 = 11$

35. $15j - 10 + 8j = 11$

36. $15j - 10 + 8 = 11$

37. $18m - 6 - m = {}^-16$

38. $18m - 6m - 1 = {}^-16$

39. $18 - 6m - m = {}^-16$

Problem Solving USING EQUATIONS

▶ **First choose the equation and solve the problem. Then reread the problem to see if your answer makes sense.**

Equation A:	$4n + 1 = 27$
Equation B:	$4n - 1 = 27$
Equation C:	$4n + n = 27$
Equation D:	$4n - n = 27$

40. Agnes bought 4 cassette tapes and Barry bought 1 cassette tape. Together they spent $27. If each tape cost the same, how much did 1 tape cost?

41. Brenda bought 4 cassette tapes. One dollar was deducted from the total cost for her discount coupon. She paid $27 for the tapes. If each tape cost the same, how much did 1 tape cost?

42. Cliff bought 4 cassette tapes. Glenn bought 1 cassette tape. The difference between the amounts they spent was $27. If each tape cost the same, how much did 1 tape cost?

43. Kay bought 4 cassette tapes and a $1 bottle of tape cleaner. She paid a total of $27. If each tape cost the same, how much did 1 tape cost?

44. Write a problem that can be solved using Equation B. Then solve the problem.

45. Write a problem that can be solved using Equation D. Then solve the problem.

8-9 Solving Equations with a Variable on Each Side

OBJECTIVE: To solve equations that have a variable on each side of the equal sign.

The cheerleaders at Highland High School sold gym shorts, caps, and visors to raise money for new uniforms. Luis bought 3 pairs of gym shorts and a $2 visor. Joe bought 2 pairs of gym shorts and a $6 cap. Luis and Joe both paid the same amount. What was the cost of each pair of gym shorts?

If you let g be the cost in dollars of a pair of gym shorts, you can solve the problem by solving the equation.

$$\underbrace{3g + 2}_{\substack{Number\ of \\ dollars \\ Luis \\ spent}} = \underbrace{2g + 6}_{\substack{Number\ of \\ dollars \\ Joe \\ spent}}$$

EXAMPLE	**Here's how to solve an equation that has variables on both sides of the equation.**

Equation:	$3g + 2 = 2g + 6$
Subtract $2g$ from both sides to get the variables on one side.	$3g - 2g + 2 = 2g - 2g + 6$
Simplify.	$g + 2 = 6$
Subtract 2 from both sides.	$g + 2 - 2 = 6 - 2$
Simplify.	$g = 4$

✔ **CHECK**

$3g + 2 = 2g + 6$
$3 \cdot 4 + 2 \stackrel{?}{=} 2 \cdot 4 + 6$
$14 = 14$
It checks!

CHECK for Understanding

1. Look at the example above. The first step in solving the equation was to subtract __?__ from both sides to get the __?__ on one side of the equation.

2. What was the cost of each pair of gym shorts? Does the answer seem reasonable?

3. Complete these examples.

a.
$$4x = 11 - 2x$$
$$4x + \underline{\ ?\ } = 11 - 2x + 2x$$
$$6x = \underline{\ ?\ }$$
$$\frac{6x}{6} = \frac{11}{6}$$
$$x = \underline{\ ?\ }$$

b.
$$18 - 3y = 2y + 8$$
$$18 - 3y - \underline{\ ?\ } = 2y - 2y + 8$$
$$^-5y + 18 = 8$$
$$^-5y + 18 - 18 = 8 - \underline{\ ?\ }$$
$$^-5y = ^-10$$
$$\frac{^-5y}{^-5} = \frac{^-10}{^-5}$$
$$y = \underline{\ ?\ }$$

EXERCISES

▶ **Solve and check.**

Here are scrambled answers for the next two rows of exercises: $\quad ^-7\frac{1}{2} \quad ^-2\frac{1}{2} \quad ^-\frac{1}{2} \quad 1 \quad 1\frac{3}{5} \quad 4$

4. $3x = 5 - 2x$

5. $^-2k = 3 + 4k$

6. $^-4n = 5 - 2n$

7. $5m + 2 = 3m + 10$

8. $6n - 3 = 8n + 12$

9. $16 - 4y = y + 8$

10. $5y - 3 = 2y - 9$

11. $15 + 2c = c - 8$

12. $^-3a = 2a + 20$

13. $^-3k + 4 = 5k - 12$

14. $2r = ^-4r - 10$

15. $12 - 2c = 4 - 3c$

16. $6q - 3 = 8q + 7$

17. $9 + 8h = 6 - 2h$

18. $2j + 4 = 4j - 16$

19. $6 + 3y = y + 4$

20. $4y = 8 + 2y$

21. $^-7n - 6 = ^-3n + 8$

22. $9x + 3 = 4x - 8$

23. $12 - 3a = a + 6$

24. $18 - 3j = 2j + 2$

25. $5y + 3 = 2y - 21$

26. $3x - 4 = 5x - 4$

27. $8y = ^-3y + 12$

28. $6c = 9 - 3c$

29. $9m - 7 = 2m + 14$

30. $12 - 9k = 3k + 12$

31. $12 + 5j = j + 3$

32. $6w + 2 = 5w - 4$

33. $9g - 8 = 6g - 4$

34. $10 - 8f = 2f + 4$

35. $^-9n = 4 + 2n$

36. $2k + 3 = 4k - 6$

Challenge! **LOGICAL REASONING**

▶ **Solve.**

37. Janna is a cheerleader at Highland High School. She knows a number of different cheers. Study the clues to find out how many different cheers she knows.

Clues:

She knows fewer than 40 cheers.

If you divide the number of cheers by 5, the remainder is 1.

If you divide the number of cheers by 4, the remainder is 3.

If you divide the number of cheers by 3, the remainder is 1.

OBJECTIVE: To use formulas to solve problems.

When you push down or let up on the gas pedal or accelerator of a car, the car changes its speed. The change in speed that the car experiences during a period of time is called **acceleration.** Acceleration is positive when speed increases; it is negative when speed decreases. To find acceleration, you can use this formula:

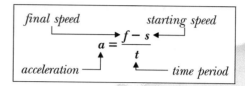

E X A M P L E S | **Here's how to use the acceleration formula to solve problems.**

A | A car's speed is increased from 30 miles per hour to 45 miles per hour in 10 seconds. Find the acceleration.

Write the formula.

$$a = \frac{f - s}{t}$$

Substitute 45 for f, 30 for s, and 10 for t.

$$a = \frac{45 - 30}{10}$$

Simplify.

$$= \frac{15}{10} \text{ or } 1\frac{1}{2}$$

The car accelerates $1\frac{1}{2}$ miles per hour each second.

B | A car's speed is decreased from 55 miles per hour to 30 miles per hour in 6 seconds. Find the acceleration.

Write the formula.

$$a = \frac{f - s}{t}$$

Substitute values for f, s, and t.

$$a = \frac{30 - 55}{6}$$

Simplify.

$$= \frac{^-25}{6} \text{ or } ^-4\frac{1}{6}$$

The acceleration is $^-4\frac{1}{6}$ miles per hour each second.

When acceleration is negative, it is called **deceleration.**

EXERCISES

▶ Use the formula $a = \dfrac{f - s}{t}$ to complete each statement.

1. If a car increases its speed from 10 miles per hour to 32 miles per hour in 5 seconds, its acceleration is ___?___ miles per hour each second.

2. If a car decreases its speed from 33 miles per hour to 22 miles per hour in 3 seconds, its acceleration is ___?___ miles per hour each second.

3. If a motorcycle goes from 40 miles per hour to 10 miles per hour in 4 seconds, its acceleration is ___?___ miles per hour each second.

4. If a bicycle goes from 4 miles per hour to 12 miles per hour in 9 seconds, its acceleration is ___?___ mile per hour each second.

5. If a car goes from a standstill to 40 miles per hour in 9 seconds, its acceleration is ___?___ miles per hour each second.

6. If a car traveling 50 miles per hour is stopped in 4 seconds, its acceleration is ___?___ miles per hour each second.

▶ Solve. Use the distance formula at the right.

7. A trucker drove 8.5 hours at an average speed of 44 miles per hour. How many miles did she drive?

8. A salesperson drove 6.25 hours at an average speed of 48 miles per hour. How far did he drive?

9. You drove 3.5 hours at an average speed of 40 miles per hour. How far did you drive?

10. If you drove 125 miles at an average speed of 50 miles per hour, how many hours would it take?

11. How many minutes does it take an airplane to travel 450 miles at 12 miles per minute?

12. A pilot wants to make a 350-mile trip in $2\frac{1}{2}$ hours. What average groundspeed must she maintain?

13. A runner wants to run a 10-kilometer race in 50 minutes. What average speed in km/h must he maintain?

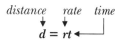

distance   rate   time
$$d = rt$$

The distance (d) traveled is equal to the rate (r) of speed times the time (t) it takes to travel the distance.

OBJECTIVE: To solve problems using a calculator when appropriate.

TV COMMERCIALS		
	Cost per minute	
Time slot	Local station	National network station
9:00–noon	$200	$35,000
12:30–3:30	360	60,000
3:30–5:30	440	75,000
7:00–10:00	800	150,000
10:30–midnight	480	90,000

▶ **Use the information in the chart to solve the problem.**

1. During the 9:00–noon time slot, one minute of commercial time on the local TV station costs $200. What is the cost for the same time on a national network TV station?

2. What does it cost to buy 4 minutes of commercial time on the local TV station in the 3:30–5:30 time slot?

▶ **Use the information in the chart. Fill in the facts that fit the calculator sequence.**

3. 1.5 ☒ 75000 ▭ (*112500*)

On national TV, ___**a.**___ minutes of
(number)
commercial time would cost $___**b.**___
(number)
in the ___**c.**___ time slot.
(time)

4. 2.25 ☒ 440 ▭ (*990*)

In the ___**a.**___ time slot on the ___**b.**___ sta-
(time) (type)
tion, ___**c.**___ minutes of commercial time
(number)
would cost $___**d.**___ .
(number)

When the local TV station televises a football game, it sells commercial time to advertisers. Suppose the game costs $4500 to televise and there are 15 minutes of commercial time. To find the profit the station will make from televising the game, you can use the formula at the right.

$$p = 15c - 4500$$

profit in dollars ↑   ↑ cost per minute in dollars

▶ **Use the formula. Decide when a calculator would be helpful.**

5. How much profit will the TV station make on the game if the cost for one minute of commercial time is $360?

6. Find the profit if the cost for one minute of commercial time is $1000.

7. If the cost for one minute of commercial time is $250, will the TV station make a profit or take a loss on the game? How much of a profit or a loss?

8. **Write a problem** that can be solved using the formula. Then solve the problem.

▶ **Give each product in simplest form.** *(page 184)*

1. $\dfrac{4}{3} \cdot \dfrac{4}{3}$
2. $\dfrac{3}{5} \cdot \dfrac{1}{3}$
3. $4 \cdot \dfrac{3}{2}$
4. $\dfrac{5}{12} \cdot \dfrac{3}{2}$
5. $\dfrac{7}{8} \cdot \dfrac{4}{5}$

6. $\dfrac{x}{5} \cdot \dfrac{1}{x}$
7. $12 \cdot \dfrac{y}{3}$
8. $\dfrac{3}{8} \cdot \dfrac{4y}{3}$
9. $\dfrac{2}{3d} \cdot \dfrac{3}{7f}$
10. $\dfrac{a}{b} \cdot \dfrac{2b}{5a}$

11. $\dfrac{a}{b^2} \cdot \dfrac{2b}{a}$
12. $\dfrac{x}{y} \cdot \dfrac{y}{x}$
13. $\dfrac{m}{n} \cdot \dfrac{n^2}{3}$
14. $\dfrac{2y}{z} \cdot \dfrac{z}{2w}$
15. $\dfrac{4d}{e} \cdot \dfrac{3e}{10d^2}$

▶ **Give each quotient in simplest form.** *(page 188)*

16. $\dfrac{3}{2} \div \dfrac{1}{2}$
17. $\dfrac{4}{5} \div \dfrac{2}{3}$
18. $\dfrac{5}{6} \div \dfrac{4}{3}$
19. $\dfrac{7}{8} \div 2$
20. $6 \div \dfrac{5}{2}$

21. $\dfrac{2}{a} \div \dfrac{3}{a}$
22. $\dfrac{4}{b} \div \dfrac{2}{c}$
23. $\dfrac{j}{4} \div \dfrac{k}{6}$
24. $\dfrac{2y}{z} \div \dfrac{y}{3}$
25. $\dfrac{a}{6} \div \dfrac{2b}{3}$

26. $\dfrac{2y}{z} \div \dfrac{4y}{z}$
27. $\dfrac{m}{p} \div \dfrac{3m}{2q}$
28. $\dfrac{d^2}{e} \div \dfrac{d}{e}$
29. $\dfrac{x}{9y} \div \dfrac{x^2}{6y}$
30. $\dfrac{9a}{2b} \div \dfrac{3a}{8b^2}$

▶ **Solve. Give the answer in simplest form.** *(page 196)*

31. $\dfrac{3}{8}$ of $40 = n$
32. $\dfrac{5}{6}$ of $120 = n$
33. $\dfrac{7}{8}$ of $96 = n$
34. $\dfrac{3}{2}$ of $150 = n$

35. $\dfrac{3}{5}$ of $12 = n$
36. $\dfrac{7}{2}$ of $15 = n$
37. $\dfrac{9}{4}$ of $22 = n$
38. $\dfrac{2}{5}$ of $34 = n$

▶ **Complete.** *(page 198)*

39. $2\dfrac{1}{2}$ ft = __?__ in.
40. $1\dfrac{2}{3}$ yd = __?__ ft
41. $3\dfrac{1}{4}$ yd = __?__ in.

42. $1\dfrac{1}{3}$ days = __?__ h
43. $2\dfrac{2}{3}$ h = __?__ min
44. $3\dfrac{1}{2}$ days = __?__ h

45. $1\dfrac{3}{4}$ min = __?__ s
46. $1\dfrac{5}{6}$ h = __?__ min
47. $2\dfrac{1}{5}$ min = __?__ s

48. $1\dfrac{1}{4}$ gal = __?__ qt
49. $2\dfrac{1}{2}$ qt = __?__ pt
50. $1\dfrac{1}{2}$ gal = __?__ pt

▶ **Solve. Give the answer in simplest form.** *(page 200)*

51. $\dfrac{3}{8}n = 36$
52. $\dfrac{5}{8}n = 45$
53. $\dfrac{5}{6}n = 75$
54. $\dfrac{4}{5}n = 48$

55. $\dfrac{2}{3}n = 33$
56. $\dfrac{5}{2}n = 42$
57. $\dfrac{7}{8}n = 50$
58. $\dfrac{7}{4}n = 65$

59. $\dfrac{5}{2}n = 12$
60. $\dfrac{4}{5}n = 15$
61. $\dfrac{2}{3}n = 25$
62. $\dfrac{1}{10}n = 10$

Here are scrambled answers for the review exercises:

⁻4	8	common	divide	numerators	repeating
⁻3	12	compare	equivalent	opposite	subtraction
5	addition	denominators	like	rational	terminating
7	combine	distributive	multiply	reciprocal	variable

1. A number that can be written as the quotient of two integers (divisor not 0) is called a _?_ number. To _?_ rational numbers with different denominators, you compare equivalent rational numbers with the same denominators. *(page 210)*

2. To write a rational number in decimal form, _?_ the numerator by the denominator. If the division comes to an end, the decimal is called a _?_ decimal. If the division does not come to an end, the decimal is called a _?_ decimal. *(page 212)*

3. To add rational numbers with different denominators, find the least _?_ denominator, change to _?_ rational numbers, and add. To subtract a rational number, add the _?_ of the rational number. *(page 214)*

4. To multiply rational numbers, multiply the _?_ to get the numerator of the product and multiply the _?_ to get the denominator of the product. To divide by a rational number, multiply by its _?_ . To find the quotient of $\frac{7}{8} \div \frac{^-3}{2}$, you would _?_ $\frac{7}{8}$ by $\frac{^-2}{3}$. *(page 216)*

5. To solve the equation $8 - 3y = 7$, you would first change the _?_ to adding the opposite. Next, you would use the commutative property of _?_ . Then you would subtract _?_ from both sides and finally divide both sides by _?_ . *(page 222)*

6. The next step in simplifying this expression is to use $15a - a = 15a - 1a$
   the _?_ property. $= 15a + {}^-1a$
   To simplify the expression $6b - 2b$, you can first change the subtraction to adding the opposite and then combine _?_ terms by adding. *(page 224)*

7. To solve the equation $8x - 3x + 7 = {}^-5$, you would first _?_ like terms. Next, you would subtract _?_ from both sides. Finally, you would divide both sides by _?_ . *(page 226)*

8. To solve the equation $8y - 12 = 12y + 3$, you would first subtract $12y$ from both sides to get the _?_ on one side of the equation. Next, you would add _?_ to both sides. Finally, you would divide both sides by _?_ . *(page 228)*

<, =, or >? *(page 210)*

1. $\dfrac{1}{4} \diamond \dfrac{^-5}{8}$

2. $\dfrac{^-2}{3} \diamond 0$

3. $^-3 \diamond ^-3\dfrac{3}{8}$

4. $\dfrac{^-7}{4} \diamond ^-1\dfrac{3}{4}$

Write each rational number in decimal form. *(page 212)*

5. $\dfrac{1}{2}$

6. $\dfrac{^-2}{3}$

7. $\dfrac{^-17}{8}$

8. $2\dfrac{5}{6}$

Write each decimal in simplest fractional form. *(page 212)*

9. 0.6

10. $^-0.25$

11. 3.75

12. 1.375

Give each sum or difference in simplest form. *(page 214)*

13. $\dfrac{3}{5} + \dfrac{^-1}{5}$

14. $\dfrac{^-5}{6} + \dfrac{^-5}{8}$

15. $\dfrac{^-5}{8} - \dfrac{1}{8}$

16. $\dfrac{7}{8} - \dfrac{^-2}{3}$

Give each product or quotient in simplest form. *(page 216)*

17. $\dfrac{^-1}{5} \cdot \dfrac{^-1}{4}$

18. $^-3\dfrac{1}{2} \cdot 2\dfrac{3}{4}$

19. $\dfrac{3}{8} \div \dfrac{1}{8}$

20. $^-4\dfrac{3}{4} \div 2\dfrac{1}{8}$

Solve and check. *(page 222)*

21. $5x + 3 = 11$

22. $^-6y - 2 = 17$

23. $8 + 9w = ^-20$

24. $7 - 10z = ^-1$

Simplify by combining like terms. *(page 224)*

25. $7a + 5a$

26. $6f - f - 3$

27. $15 - h + 10h - 4$

Solve and check. *(page 226)*

28. $2n + 3n - 6 = ^-8$

29. $7n - n + 10 = 16$

30. $3p + 8 - 4 = 0$

31. $3t + 6 - 5t = 1$

32. $7 + 4a - 10 = ^-2$

33. $r - 8 + r = ^-9$

Solve and check. *(page 228)*

34. $7d = 5 + 8d$

35. $^-6e - 3 = ^-2e + 8$

36. $p + 1 = 6 - 7p$

37. $4x + 3 = 2x - 11$

38. $2j - 7 = 5j + 3$

39. $14 - 6r = 4r - 16$

Write an equation and solve the problem. *(page 227)*

40. Four less than the product of 8 and a number n is $^-32$. What is the number?

41. The sum of 8 times a number n and 4 times a number n is $^-32$. What is the number?

42. The sum of 8 times a number n and 4 is $^-32$. What is the number?

43. Eight times a number n increased by 4 times a number n is $^-32$. What is the number?

▶ **Choose the correct letter.**

1. Solve.

$$24 = \frac{y}{-2} - 6$$

A. $^-60$ **B.** $^-36$

C. $^-15$ **D.** none of these

2. 0.000875 written in scientific notation is

A. $8.75 \cdot 10^4$

B. $0.875 \cdot 10^{-3}$

C. $8.75 \cdot 10^{-4}$

D. none of these

3. The least common denominator of $\frac{3x}{4y^2z}$ and $\frac{3}{yz}$ is

A. $4yz$ **B.** $4y^2z$

C. $4y^2z^2$ **D.** none of these

4. $\frac{5a}{10a^2b}$ written in simplest form is

A. $\frac{a}{2a^2b}$ **B.** $\frac{5}{10ab}$

C. $\frac{1}{2ab}$ **D.** none of these

5. Give the sum.

$$\frac{3a}{2} + \frac{b}{3}$$

A. $\frac{3a + b}{6}$ **B.** $\frac{9a + b}{6}$

C. $\frac{3a + 2b}{6}$ **D.** none of these

6. Give the difference.

$$\frac{2x}{3} - \frac{5y}{4}$$

A. $\frac{8x - 15y}{12}$ **B.** $\frac{2x - 5y}{12}$

C. $\frac{8x - 5y}{12}$ **D.** none of these

7. Give the product.

$$\frac{2c}{d} \cdot \frac{d}{2f}$$

A. $\frac{4cf}{d^2}$ **B.** $\frac{cd}{f}$

C. $\frac{c}{f}$ **D.** none of these

8. Give the quotient.

$$\frac{m}{3n} \div \frac{5m}{n^2}$$

A. $\frac{n}{15}$ **B.** $\frac{5m^2}{3n^2}$

C. $\frac{n^2}{15m}$ **D.** none of these

9. Solve.

$$\frac{2}{3} \text{ of } 20 = n$$

A. 20   **B.** $13\frac{1}{3}$

C. 30   **D.** none of these

10. $4\frac{2}{3}$ yd = __?__ ft

A. 13

B. 28

C. 56

D. none of these

11. Solve.

$$\frac{5}{2} \text{ of } n = 40$$

A. 16   **B.** 40

C. 100   **D.** none of these

12. Choose the equation.

You worked 5 hours on Friday and 4 hours on Saturday. After spending $9 you had $27 left. How much did you earn per hour?

A. $5n + 4n + 9 = 27$

B. $5n + 4n - 9 = 27$

C. $5n - 4n + 9 = 27$

D. $5n - 4n - 9 = 27$

Using Algebra in Ratio, Proportion, and Percent

9

$I = prt$

Interest equals principal times rate times time.
(page 260)

When you deposit money in your savings account, the bank pays you interest on the principal (the amount in your account). When you borrow money from the bank, you pay the bank interest on the principal (the amount you owe).

237

9-1 Equivalent Algebraic Ratios

OBJECTIVE: To give the ratio of two quantities. To change ratios to higher or lower terms.

You can use a **ratio** to compare two numbers. The table shows how many stamps John has from each of eight countries. The ratio of Austrian stamps to Canadian stamps is 24 to 96. Here are three ways to write the ratio:

$$24 \text{ to } 96 \qquad \frac{24}{96} \qquad 24:96$$

Read each ratio as "24 to 96."

Country	Number of stamps
Austria	24
Canada	96
Denmark	18
France	60
Great Britain	72
Soviet Union	64
United States	120
Germany	66

EXAMPLES | **Here's how to find equal ratios.**

To write a ratio as a fraction in lowest terms, divide both terms of the ratio by their greatest common factor.

A In arithmetic

$GCF = 24$

$$\frac{24}{96} = \frac{24 \div 24}{96 \div 24}$$

$$= \frac{1}{4}$$

B In algebra

$GCF = 2a$

$$\frac{6a^2}{4ab} = \frac{6a^2 \div 2a}{4ab \div 2a}$$

$$= \frac{3a}{2b}$$

If you multiply both terms of a ratio by the same number (not 0) or expression, you get an equal ratio.

C In arithmetic

$$\frac{9}{2} = \frac{9 \cdot 3}{2 \cdot 3}$$

$$= \frac{27}{6}$$

D In algebra

$$\frac{5x}{2xy} = \frac{5x \cdot 3x}{2xy \cdot 3x}$$

$$= \frac{15x^2}{6x^2y}$$

CHECK for Understanding

1. Look at Example A. What is the ratio of Austrian stamps to Canadian stamps written in lowest terms?

2. Look at Example B. To write $\frac{6a^2}{4ab}$ in lowest terms, both terms were divided by the GCF, which is __?__ .

3. If you multiply both terms by the same number (not 0) or expression, you get an __?__ ratio.

4. In Example D, both terms of $\frac{5x}{2xy}$ were multiplied by __?__ to get the equal ratio $\frac{15x^2}{6x^2y}$.

5. Complete these examples.

   a. $\dfrac{30}{40} = \dfrac{?}{4}$ b. $\dfrac{6}{5} = \dfrac{?}{15}$ c. $\dfrac{ab^2}{2ab} = \dfrac{?}{2}$ d. $\dfrac{3x}{5y} = \dfrac{?}{10y^2}$

EXERCISES

▶ **Give each ratio as a fraction in lowest terms.**
Here are scrambled answers for the next row of exercises: $\dfrac{4}{3}$ $\dfrac{1}{4}$ $\dfrac{5}{2}$ $\dfrac{1}{2}$ $\dfrac{4}{9}$

6. 4 to 8   **7.** 3 to 12   **8.** 15 to 6   **9.** 8 to 6   **10.** 16 to 36

11. $\dfrac{14}{21}$ **12.** $\dfrac{10}{4}$ **13.** $\dfrac{16}{6}$ **14.** $\dfrac{9}{45}$ **15.** $\dfrac{18}{27}$

16. $12:20$ **17.** $14:8$ **18.** $18:32$ **19.** $12:18$ **20.** $24:18$

21. $2a$ to $4b$ **22.** $9x$ to $6x$ **23.** cd to c^2 **24.** $2r^2$ to $8rs$ **25.** $6m^2$ to $8mn^2$

26. $\dfrac{8c}{24d}$ **27.** $\dfrac{24j}{16j}$ **28.** $\dfrac{5ab}{2a^2}$ **29.** $\dfrac{16yz}{12z^2}$ **30.** $\dfrac{18s^2t^2}{24st}$

31. $15m:25n$ **32.** $32x:28x$ **33.** $12cd^2:18d$ **34.** $24a:16ab^2$ **35.** $36jh^2:20j^2h$

▶ **Complete to get an equal ratio.**

36. $\dfrac{3}{4}=\dfrac{?}{8}$ **37.** $\dfrac{8}{5}=\dfrac{?}{15}$ **38.** $\dfrac{7}{8}=\dfrac{?}{24}$ **39.** $\dfrac{1}{3}=\dfrac{?}{18}$ **40.** $\dfrac{5}{2}=\dfrac{?}{16}$

41. $\dfrac{3}{2}=\dfrac{?}{24}$ **42.** $\dfrac{5}{4}=\dfrac{?}{12}$ **43.** $\dfrac{1}{2}=\dfrac{?}{18}$ **44.** $\dfrac{3}{4}=\dfrac{?}{16}$ **45.** $\dfrac{11}{2}=\dfrac{?}{20}$

46. $\dfrac{4}{5}=\dfrac{?}{40}$ **47.** $\dfrac{5}{6}=\dfrac{?}{30}$ **48.** $\dfrac{9}{4}=\dfrac{?}{24}$ **49.** $\dfrac{2}{5}=\dfrac{?}{50}$ **50.** $\dfrac{10}{3}=\dfrac{?}{18}$

51. $\dfrac{3}{b}=\dfrac{?}{7b}$ **52.** $\dfrac{5}{x}=\dfrac{?}{xy}$ **53.** $\dfrac{n}{8}=\dfrac{?}{8m}$ **54.** $\dfrac{12}{c}=\dfrac{?}{c^2}$ **55.** $\dfrac{w}{6}=\dfrac{?}{6z^2}$

56. $\dfrac{a}{b}=\dfrac{?}{2b^2}$ **57.** $\dfrac{j^2}{k}=\dfrac{?}{10k^2}$ **58.** $\dfrac{t^2}{s}=\dfrac{?}{4s^2}$ **59.** $\dfrac{5m}{3n}=\dfrac{?}{15mn}$ **60.** $\dfrac{8x^2y}{3x}=\dfrac{?}{18x}$

61. $\dfrac{3cd}{c^2d}=\dfrac{?}{c^2d^2}$ **62.** $\dfrac{2rs^2}{3rs}=\dfrac{?}{9r^2s}$ **63.** $\dfrac{9}{4a}=\dfrac{?}{12a^2b^3}$ **64.** $\dfrac{3m}{8n}=\dfrac{?}{24mn^2}$ **65.** $\dfrac{7x^2y}{3}=\dfrac{?}{15y}$

Problem Solving USING DATA

▶ **Solve. Use the chart on page 238.**

66. What is the ratio in lowest terms of French stamps to British stamps?

67. What is the ratio in lowest terms of British stamps to French stamps?

68. How many United States stamps are there for each French stamp?

69. How many British stamps are there for every 8 Soviet stamps?

70. For every 20 United States stamps, there are 3 stamps from which country?

71. What is the ratio in lowest terms of foreign stamps to United States stamps?

OBJECTIVE: To solve proportions.

Suppose that you decided to build a model of the *Douglas Chicago* that is $\frac{1}{10}$ the size of the real airplane. To decide how wide the wingspan should be, you could solve a proportion. A **proportion** is an equation that states that two ratios are equal.

Douglas Chicago
Wingspan: **50 feet**   Length: **35$\frac{1}{2}$ feet**

The *Douglas Chicago* was flown in 1924 by a team of U.S. Army pilots who were the first to fly around the world.

> Wingspan of model airplane $\longrightarrow \dfrac{n}{50} = \dfrac{1}{10} \longleftarrow$ Size of model airplane
>
> Wingspan of real airplane $\longrightarrow$ $\longleftarrow$ Size of real airplane

Note: *When setting up a proportion, make sure that the terms of the ratios are in the same order.*

1. Look at the proportion above. If you think about equal ratios, what must n equal?

2. Therefore, if your model is $\frac{1}{10}$ the size of the real airplane, what should the wingspan of the model be?

Lockheed Vega
Wingspan: **41 feet**   Length: **27$\frac{1}{2}$ feet**

The *Lockheed Vega* was flown in 1932 by Amelia Earhart who became the first woman to fly across the Atlantic solo and nonstop.

Suppose you decide to build a model of the *Lockheed Vega* that is $\frac{3}{32}$ the size of the real airplane. To decide how wide the wingspan should be, you could solve this proportion: $\frac{n}{41} = \frac{3}{32}$.

3. Is this proportion easy to solve by thinking about equal ratios?

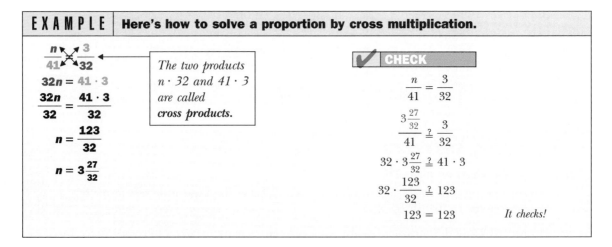

| E X A M P L E | Here's how to solve a proportion by cross multiplication. |

$$\frac{n}{41} \underset{\times}{\times} \frac{3}{32}$$

$$32n = 41 \cdot 3$$

$$\frac{32n}{32} = \frac{41 \cdot 3}{32}$$

$$n = \frac{123}{32}$$

$$n = 3\frac{27}{32}$$

The two products $n \cdot 32$ and $41 \cdot 3$ are called **cross products.**

✔ **CHECK**

$$\frac{n}{41} = \frac{3}{32}$$

$$\frac{3\frac{27}{32}}{41} \stackrel{?}{=} \frac{3}{32}$$

$$32 \cdot 3\frac{27}{32} \stackrel{?}{=} 41 \cdot 3$$

$$32 \cdot \frac{123}{32} \stackrel{?}{=} 123$$

$$123 = 123 \qquad \textit{It checks!}$$

CHECK for Understanding

4. Look at the example above. If your model is $\frac{3}{32}$ the size of the real *Lockheed Vega,* what should the wingspan of the model be?

5. Complete these examples.

a. $\dfrac{n}{8} \bowtie \dfrac{3}{5}$

$5n = 8 \cdot 3$
$5n = 24$
$\dfrac{5n}{5} = \dfrac{24}{5}$
$n = \underline{\ ?\ }$

b. $\dfrac{8}{6} \bowtie \dfrac{4}{n}$

$8n = 6 \cdot 4$
$8n = 24$
$\dfrac{8n}{8} = \dfrac{?}{8}$
$n = 3$

c. $\dfrac{5}{2} \bowtie \dfrac{n}{7}$

$2n = 5 \cdot \underline{\ ?\ }$
$2n = 35$
$\dfrac{2n}{2} = \dfrac{35}{2}$
$n = 17\dfrac{1}{2}$

d. $\dfrac{2}{n} \bowtie \dfrac{4}{1\frac{1}{2}}$

$\underline{\ ?\ } n = 2 \cdot 1\dfrac{1}{2}$
$4n = 3$
$\dfrac{4n}{4} = \dfrac{3}{4}$
$n = \dfrac{3}{4}$

EXERCISES

▶ **Solve each proportion by thinking about equal ratios or by cross multiplication. Give answers in simplest form.**

Here are scrambled answers for the next row of exercises: $2\dfrac{11}{12}$ 18 16 $1\dfrac{3}{5}$ $2\dfrac{1}{10}$

6. $\dfrac{1}{2} = \dfrac{8}{n}$ **7.** $\dfrac{2}{3} = \dfrac{12}{n}$ **8.** $\dfrac{n}{8} = \dfrac{2}{10}$ **9.** $\dfrac{7}{n} = \dfrac{12}{5}$ **10.** $\dfrac{3}{10} = \dfrac{n}{7}$

11. $\dfrac{9}{11} = \dfrac{n}{7}$ **12.** $\dfrac{13}{4} = \dfrac{10}{n}$ **13.** $\dfrac{n}{18} = \dfrac{6}{3}$ **14.** $\dfrac{11}{n} = \dfrac{18}{5}$ **15.** $\dfrac{16}{5} = \dfrac{20}{n}$

16. $\dfrac{n}{9} = \dfrac{14}{3}$ **17.** $\dfrac{20}{n} = \dfrac{14}{11}$ **18.** $\dfrac{9}{15} = \dfrac{n}{18}$ **19.** $\dfrac{16}{5} = \dfrac{30}{n}$ **20.** $\dfrac{18}{n} = \dfrac{9}{4}$

21. $\dfrac{1\frac{1}{2}}{3} = \dfrac{n}{2}$ **22.** $\dfrac{10}{7} = \dfrac{1\frac{1}{4}}{n}$ **23.** $\dfrac{4}{n} = \dfrac{8}{2\frac{3}{4}}$ **24.** $\dfrac{n}{2\frac{1}{2}} = \dfrac{4}{6}$ **25.** $\dfrac{5}{4} = \dfrac{1\frac{1}{4}}{n}$

26. $\dfrac{7}{8} = \dfrac{5\frac{2}{3}}{n}$ **27.** $\dfrac{4}{n} = \dfrac{2\frac{1}{3}}{5}$ **28.** $\dfrac{2}{3\frac{1}{4}} = \dfrac{n}{9}$ **29.** $\dfrac{n}{3} = \dfrac{8}{1\frac{1}{2}}$ **30.** $\dfrac{n}{5} = \dfrac{2}{6\frac{1}{5}}$

Problem Solving USING DATA

▶ **Solve. Use the information on page 240.**

31. Suppose you decide to make a model of the *Lockheed Vega* that is $\dfrac{1}{30}$ the size of the real airplane.
   a. What would the wingspan of the model be?
   b. How long should the model be?

32. Suppose you decide to make a model of the *Douglas Chicago* that is $\dfrac{3}{20}$ the size of the real airplane.
   a. What should the wingspan of the model be?
   b. How long should the model be?

33. A model of the *Lockheed Vega* has a wingspan of 3 feet. What should the length be?

34. A model of the *Douglas Chicago* has a length of $4\dfrac{7}{16}$ feet. What should the wingspan be?

OBJECTIVE: To solve rate problems using proportions.

A **rate** is a ratio of two unlike quantities.

You spent $11.13 for 8 gallons of gasoline.

$$\text{Rate: } \frac{\$11.13}{8 \text{ gal}}$$

Read as "$11.13 per 8 gallons."

EXAMPLES | **Here's how to use proportions to solve rate problems.**

A If you spent $11.13 for 8 gallons of gasoline, how much would you spend for 14 gallons?

$$dollars \longrightarrow \frac{\mathbf{11.13}}{\mathbf{8}} = \frac{\mathbf{n}}{\mathbf{14}} \longleftarrow dollars$$
$$gallons \longrightarrow \qquad\qquad \longleftarrow gallons$$

$$8n = 11.13 \cdot 14 \qquad 11.13 \boxed{\times} 14 \boxed{=} \boxed{155.82}$$

$$8n = 155.82 \longleftarrow \quad 155.82 \boxed{\div} 8 \boxed{=} \boxed{19.4775}$$

$$n \approx 19.48 \longleftarrow$$

Read ≈ as "is approximately equal to."

At the given rate, you would spend $19.48 for 14 gallons of gasoline.

> *Remember: When setting up a proportion, you must be sure that the terms of the ratios are in the same order!*

B You drive 196 miles in 4 hours. At that rate, how many hours will it take you to drive 340 miles?

$$miles \longrightarrow \frac{\mathbf{196}}{\mathbf{4}} = \frac{\mathbf{340}}{\mathbf{n}} \longleftarrow miles$$
$$hours \longrightarrow \qquad\qquad \longleftarrow hours$$

$$196n = 4 \cdot 340$$

$$196n = 1360$$

$$n \approx 6.94 \longleftarrow \quad 1360 \boxed{\div} 196 \boxed{=} \boxed{6.9387755}$$

At the given rate, it would take you about 6.94 hours to drive 340 miles.

EXERCISES

▶ **Solve by using proportions. If an answer does not come out evenly, round it to the nearest hundredth. Decide when a calculator would be helpful.**

1. You spend $18 for 12 gallons of gasoline. At that price,
   a. how many gallons could you buy for $13?

   Hint: $\dfrac{18}{12} = \dfrac{13}{n}$

   b. how many gallons could you buy for $7.50?
   c. how much would 6 gallons cost?
   d. how much would 11.4 gallons cost?

2. You drive 128 miles in 3 hours. At that speed,
   a. how many miles could you drive in 4 hours?

   Hint: $\dfrac{128}{3} = \dfrac{n}{4}$

   b. how many miles could you drive in 9 hours?
   c. how many hours would it take you to drive 300 miles?
   d. how many hours would it take you to drive 186 miles?

3. You drive 124 miles and use 4.8 gallons of gasoline. At that rate,
   a. how many miles could you drive on 10 gallons?
   b. how many miles could you drive on 13 gallons?
   c. how many gallons would you need for 400 miles?
   d. how many gallons would you need for 280 miles?

4. You spend $3.60 to drive 110 miles on a toll road. At that rate,
   a. how many miles could you drive for $1.80?
   b. how many miles could you drive for $3.00?
   c. how much would it cost to drive 180 miles?
   d. how much would it cost to drive 95 miles?

5. During the first 2 days of your trip, you spend $43 for meals. At that rate,
   a. how much will your meals cost for 7 days?
   b. how many days of meals could you buy for $200?

6. You drive 4 hours and use 8.2 gallons of gasoline. At that rate,
   a. how many hours could you drive on 10 gallons?
   b. how many gallons would you need to drive 9 hours?

★7. You want to be in Chicago by 12 noon. At 9:30 A.M., you are 127 miles from Chicago and you are traveling at the rate of 54 miles per hour. Will you be on time if you keep driving at the same rate?

★8. At 2:45 P.M., you are 117 miles from Tampa. You want to be in Tampa at 5:00 P.M. How many miles per hour must you average to be on time?

OBJECTIVE: To change a percent to a fraction. To change a percent to a decimal.

The circle graph shows how Alano budgeted his money. Notice that $12\frac{1}{2}\%$ $\left(12\frac{1}{2} \text{ percent}\right)$ of his money is spent for tapes. This means that $12\frac{1}{2}¢$ out of each $100¢$ is spent for tapes. *Percent* means "per 100" or "hundredths."

1. What percent is budgeted for lunches?

2. What percent is budgeted for recreation?

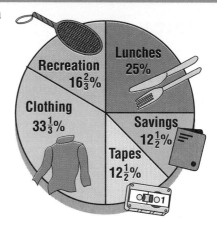

Recreation $16\frac{2}{3}\%$
Lunches 25%
Clothing $33\frac{1}{3}\%$
Savings $12\frac{1}{2}\%$
Tapes $12\frac{1}{2}\%$

| EXAMPLES | Here's how to change a percent to a fraction. |

To change a percent to a fraction, first write the percent as a fraction with a denominator of 100. Then write the fraction in simplest form.

Fraction spent for lunches

Fraction spent for recreation

$\boxed{A}$ $25\% = \dfrac{25}{100} = \dfrac{1}{4}$

$\boxed{B}$ $16\frac{2}{3}\% = \dfrac{16\frac{2}{3}}{100} = 16\frac{2}{3} \div 100 = \dfrac{\overset{1}{\cancel{50}}}{3} \cdot \dfrac{1}{\underset{2}{\cancel{100}}} = \dfrac{1}{6}$

CHECK for Understanding

3. Look at the examples above. What fraction of the budget is spent for lunches? What fraction is spent for recreation?

| EXAMPLES | Here's how to change a percent to a decimal. |

To change a percent to a decimal, move the decimal point two places to the left and remove the percent sign. Remember *percent* means "hundredths."

$\boxed{C}$ $25\% = 0.25$

$\boxed{D}$ $16\frac{2}{3}\% = 0.16\frac{2}{3}$

CHECK for Understanding

4. Complete these examples of changing a percent to a fraction.

   a. $75\% = \dfrac{?}{100} = \dfrac{3}{4}$ **b.** $83\frac{1}{3}\% = \dfrac{83\frac{1}{3}}{100} = 83\frac{1}{3} \div 100 = \dfrac{?}{3} \cdot \dfrac{1}{100} = \dfrac{5}{6}$

5. Complete these examples of changing a percent to a decimal.

   a. $47\% = \underline{\ ?\ }$ **b.** $133\frac{1}{3}\% = \underline{\ ?\ }$

EXERCISES

▶ **Change to a fraction, whole number, or mixed number. Give each answer in simplest form.**

Here are scrambled answers for the next row of exercises: $\frac{7}{10}$ 1 $\frac{1}{2}$ $\frac{1}{4}$ $1\frac{1}{4}$ $\frac{18}{25}$

6. 25%   **7.** 125%   **8.** 50%   **9.** 100%   **10.** 70%   **11.** 72%

12. 15%   **13.** 32%   **14.** 220%   **15.** 66%   **16.** 44%   **17.** 48%

18. 20%   **19.** 120%   **20.** 60%   **21.** 300%   **22.** 225%   **23.** 175%

24. $18\frac{3}{4}\%$ **25.** $8\frac{1}{3}\%$ **26.** $81\frac{1}{4}\%$ **27.** $66\frac{2}{3}\%$ **28.** $62\frac{1}{2}\%$ **29.** $116\frac{2}{3}\%$

30. $206\frac{1}{4}\%$ **31.** $33\frac{1}{3}\%$ **32.** $187\frac{1}{2}\%$ **33.** $106\frac{1}{4}\%$ **34.** $162\frac{1}{2}\%$ **35.** $233\frac{1}{3}\%$

▶ **Change to a decimal.**

Here are scrambled answers for the next row of exercises:
0.06   1.50   4.00   0.74   2.50   0.037

36. 74%   **37.** 6%   **38.** 150%   **39.** 400%   **40.** 250%   **41.** 3.7%

42. 5%   **43.** 75%   **44.** 10%   **45.** 3.25%   **46.** 200%   **47.** 40%

48. 350%   **49.** $7\frac{1}{2}\%$ **50.** $16\frac{2}{3}\%$ **51.** $166\frac{2}{3}\%$ **52.** $33\frac{1}{3}\%$ **53.** 21.2%

Problem Solving USING GRAPHS

▶ **Solve. Use the circle graph on page 244.**

54. For what two items did Alano budget half his money?

55. For which item did Alano budget the greatest percent of his money?

56. What fraction of Alano's money is spent for clothing?

57. What fraction of Alano's money is spent for tapes?

58. What fraction of Alano's money is budgeted for clothing and tapes?
Hint: Use your answers from exercises 56 and 57.

59. What fraction of Alano's money is not budgeted for savings?

★**60.** Suppose that Alano received $24. How much would be budgeted for

    a. lunches?   **b.** recreation?   **c.** clothing?
    d. tapes?   **e.** savings?

★**61.** Suppose that out of one paycheck Alano budgeted $16 for clothing. For what amount was the paycheck?

OBJECTIVE: To change a fraction to a percent. To change a decimal to a percent.

Look at the football helmets. Can you name the team for each helmet? This question was part of a football survey. The tally below shows the results.

Name	Number of teams named correctly			
Andrea	ⅿℍℍ	ℍℍ	ℍℍ	/
Rosa		ℍℍ	ℍℍ	////
Mel		ℍℍ	ℍℍ	//
Dan		ℍℍ	ℍℍ	ℍℍ ///
Tomas		ℍℍ	ℍℍ	ℍℍ //
Cheryl		ℍℍ	ℍℍ	ℍℍ ℍℍ

1. How many teams did Mel name correctly?

2. How many teams are there in all?

3. What fraction of the teams did Mel name?

4. What fraction of the teams did Andrea name?

EXAMPLES | **Here's how to change a fraction to a percent.**

Changing Mel's fraction to a percent

Use mental math

Change to an equivalent fraction with a denominator of 100. Then write as a percent.

$\boxed{A}$ $\dfrac{1}{2} = \dfrac{50}{100}$

So, $\dfrac{1}{2} = 50\%$

Changing Andrea's fraction to a percent

Use a proportion

$\boxed{B}$ $part \longrightarrow \dfrac{2}{3} = \dfrac{n}{100} \longleftarrow part$, whole $\longrightarrow 3 \quad 100 \longleftarrow whole$

$$3n = 200 \qquad n = 66\dfrac{2}{3}$$

So, $\dfrac{2}{3} = \dfrac{66\frac{2}{3}}{100}$ or $66\dfrac{2}{3}\%$.

EXAMPLES | **Here's how to change a decimal to a percent.**

To change a decimal to a percent, move the decimal point two places to the right and affix the percent sign.

$\boxed{C}$ $0.5 = 50\%$

$\boxed{D}$ $0.66\dfrac{2}{3} = 66\dfrac{2}{3}\%$

CHECK for Understanding

5. Look at the examples. What percent of the teams did Mel name correctly? What percent did Andrea name correctly?

6. Which method would you use to change $\frac{3}{10}$ to a percent? To change $\frac{1}{12}$ to a percent?

EXERCISES

▶ **Change to a percent.** *Hint: First change to an equivalent fraction with a denominator of 100.*
Here are scrambled answers for the next row of exercises: 225%   40%   60%   90%   50%   125%

7. $\frac{2}{5}$ 8. $\frac{9}{4}$ 9. $\frac{9}{10}$ 10. $\frac{3}{5}$ 11. $\frac{5}{4}$ 12. $\frac{1}{2}$

13. $\frac{1}{5}$ 14. $\frac{1}{4}$ 15. $\frac{4}{5}$ 16. 2   17. $\frac{5}{2}$ 18. $\frac{13}{4}$

▶ **Change to a percent.** *Hint: You may need to solve a proportion.*
Here are scrambled answers for the next row of exercises:

175% $16\frac{2}{3}\%$ $33\frac{1}{3}\%$ 120% $133\frac{1}{3}\%$ $83\frac{1}{3}\%$

19. $\frac{1}{3}$ 20. $\frac{1}{6}$ 21. $\frac{7}{4}$ 22. $\frac{6}{5}$ 23. $\frac{5}{6}$ 24. $\frac{4}{3}$

25. $\frac{9}{16}$ 26. $\frac{9}{25}$ 27. $\frac{5}{9}$ 28. $\frac{7}{2}$ 29. $\frac{2}{3}$ 30. $\frac{3}{8}$

▶ **Change to a percent.**
Here are scrambled answers for the next row of exercises: $33\frac{1}{3}\%$ 450%   300%   40%   3%   45%

31. 0.4   32. 0.45   33. 4.5   34. 0.03   35. $0.33\frac{1}{3}$ 36. 3

37. 2.5   38. $0.87\frac{1}{2}$ 39. 1   40. 0.8   41. 0.06   42. 0.39

Problem Solving USING DATA

▶ **Solve. Use the survey results on page 246.**

43. What percent of the teams did Tomas name correctly?

44. What percent of the teams did Rosa name correctly?

45. What percent of those surveyed knew more than 15 teams?

46. What percent of those surveyed knew 15 or fewer teams?

47. What percent of the teams did Cheryl miss?

48. Allison got $41\frac{2}{3}\%$ of the teams correct. What fraction did she name correctly?

★49. John named $87\frac{1}{2}\%$ of the teams correctly. How many teams did he miss?

50. **Write a percent problem** about the survey. Then solve the problem.

PROBLEM SOLVING POWER

BUILD YOUR

You will use a variety of problem-solving strategies on these two pages. For some problems, you may wish to use a calculator.

1 HOW TO PAY A LINK A DAY

In the days of the Old West, a family was stranded in a mining town waiting for the next stagecoach. It had no money, but the mother had a gold chain with 7 links. The hotel manager agreed to let the family stay for a link a day. How could the woman cut the chain only once, yet pay 1 link each night for up to 7 nights lodging?

2 NEWSY NICKELS

When I went to visit my uncle Wellington last summer, he made me this offer. "Casper," he said, "if you will be a good lad and bring me my newspaper each day, I'll give you 1 nickel the first day, 2 nickels the second day, and so on until you have received a total of $6.00. Then it will be time for you to go home."

I accepted his offer. How many days did I visit Uncle Wellington last summer?

3 GEOMETRY—VISUAL THINKING

These two triangles intersect in 3 points.

a. How many points of intersection are possible for two triangles?
b. How many points of intersection are possible for a triangle and a square?

4 DOUBLE TROUBLE

What number is twice the product of its digits? *Hint: 24 is 3 times the product of its digits.*

5 LEAP FROG PUZZLE

Use the rules to change the arrangement
of coins from this:

to this:

Rules:
- Pennies must always move to the right, nickels to the left.
- You can move to an adjacent space or jump one coin.

Hint: Use real coins or make a model.

6 PURSE-ONALLY SPEAKING!

How much money do I have in my coin
purse? Clues:

- I have only nickels, dimes, and quarters.
- I have twice as many dimes as quarters.
- I have 3 more nickels than dimes.
- The total amount is a little less than $3.00.

Hint: Make a table.

Number of quarters	1
Number of dimes	2
Number of nickels	5
Total value	$0.70

7 TALL TALES

Gary, Kevin, Ken, and Peter are all members of the high school band. Gary, at 213
centimeters, is the tallest band member. Kevin is 54 centimeters taller than Peter,
the shortest member of the band. Peter is 5 centimeters shorter than Ken, who is $\frac{2}{3}$
as tall as Gary. How tall are Kevin, Ken, and Peter?

Skin-Diving Cruises

3 days: $499*
7 days: $849*
12 days: $1395*
*per person

Make your reservation today!
- Deposit due when reservation is made: $\frac{1}{5}$ total fare.
- Balance due 2 weeks before sailing date.

APPLICATIONS

Complete.

8 Carla and three of her friends made reservations to go on a 7-day
skin-diving cruise. Each person paid a deposit of _?_ .

9 Their cruise leaves Miami on August 4. They will have to pay
the balance due on their cruise by _?_ .

10 The 12-day cruise costs about _?_ dollars per day less than the 3-day
cruise. (Round your answer to the nearest whole number of dollars.)

Rate Your Problem-Solving Power: *6–7 correct = Good;   8–9 correct = Excellent;   10 correct = Exceptional*

OBJECTIVE: To use a scale drawing (map) for solving problems.

A map is an example of a scale drawing. On this map, 1 centimeter stands for 130 kilometers. To find the air distance between cities, we can measure a distance on the map and solve a proportion.

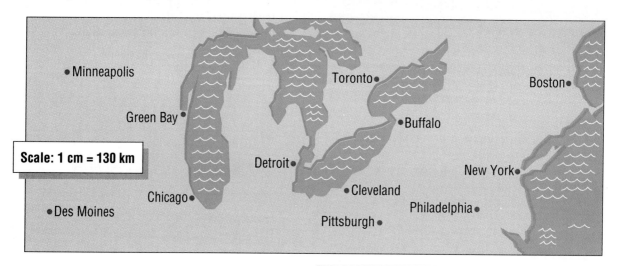

Scale: 1 cm = 130 km

Example The distance from Chicago to Boston on the map is 10.4 centimeters. What is the air distance from Chicago to Boston?

The air distance is 1352 kilometers.

$$cm\ on\ map \longrightarrow \frac{1}{130} = \frac{10.4}{n} \longleftarrow cm\ on\ map$$
$$actual\ km \longrightarrow \qquad \qquad \longleftarrow actual\ km$$
$$n = 10.4(130)$$
$$n = 1352$$

▶ **Find the air distance between the cities. The map distances are given.**

   1. Chicago to Des Moines, 3.9 cm

   2. New York to Chicago, 8.7 cm

   3. New York to Detroit, 5.8 cm

   4. Toronto to Minneapolis, 8.3 cm

   5. Buffalo to Des Moines, 9.4 cm

   6. Philadelphia to Buffalo, 3.3 cm

   7. Toronto to Philadelphia, 4.1 cm

   8. Detroit to Chicago, 2.9 cm

▶ **Use your answers to exercises 1–8 to solve.**

   9. How far is it from New York to Detroit to Chicago?

   10. How far is it from New York to Chicago to Des Moines?

   11. How much farther is it from Toronto to Minneapolis than from Toronto to Philadelphia?

   12. How much farther is it from Buffalo to Des Moines than from Buffalo to Philadelphia?

▶ **Solve. Use a ruler and the map.**

   13. How far is it from Detroit to Minneapolis?

   14. How far is it from Cleveland to Boston?

   15. Which city is about 1150 kilometers east of Des Moines?

   16. Which city is about 1750 kilometers west of Boston?

▶ **Solve and check.** *(page 112)*

1. $4y + y = {}^-21$ **2.** $6x - 5 = {}^-41$ **3.** $\dfrac{n}{4} + 3 = {}^-10$ **4.** $\dfrac{m}{7} - 12 = {}^-15$

5. $4 = {}^-5k + 24$ **6.** $16 = {}^-3j + 1$ **7.** ${}^-12 = 7z - 12$ **8.** $23 = 12c - 25$

9. ${}^-6 = \dfrac{a}{2} + 6$ **10.** $15 = \dfrac{d}{{}^-2} + 20$ **11.** $30 = \dfrac{f}{{}^-6} + 21$ **12.** ${}^-32 = \dfrac{b}{8} - 25$

13. ${}^-11s - 5 = 17$ **14.** ${}^-8t + 16 = 40$ **15.** $\dfrac{y}{{}^-9} + 15 = 23$ **16.** $\dfrac{x}{{}^-7} - 28 = {}^-36$

▶ **Simplify using laws of exponents.** *(page 124)*

17. $10^2 \cdot 10^3$ **18.** $7^3 \cdot 7^1$ **19.** $5^2 \cdot 5^4$ **20.** $8^5 \cdot 8^2$

21. $x^2 \cdot x^2$ **22.** $y^4 \cdot y^1$ **23.** $a^3 \cdot a^5$ **24.** $b^2 \cdot b^6$

25. $\dfrac{10^5}{10^3}$ **26.** $\dfrac{2^5}{2^3}$ **27.** $\dfrac{6^4}{6^1}$ **28.** $\dfrac{9^6}{9^3}$

29. $\dfrac{w^5}{w^2}$ **30.** $\dfrac{z^6}{z^4}$ **31.** $\dfrac{c^5}{c^1}$ **32.** $\dfrac{d^8}{d^3}$

▶ **Give the prime factorization using exponents or the algebraic factorization.**
(page 130)

33. 9   **34.** 4   **35.** 12   **36.** 20   **37.** 27   **38.** 24

39. 144   **40.** 32   **41.** 240   **42.** 180   **43.** 360   **44.** 200

45. $6a^2$ **46.** $4b^3$ **47.** $12c^2$ **48.** $3d^3$ **49.** $18e$ **50.** $16f^2$

51. $5ab^2$ **52.** $6a^2b$ **53.** $24ab$ **54.** $3ab^3$ **55.** $34a^2b^2$ **56.** $26ab^2$

▶ **Give the greatest common factor.** *(page 132)*

57. 3, 16   **58.** 18, 12   **59.** 25, 30   **60.** 60, 40

61. 48, 16   **62.** 4, $8c$ **63.** 24, $13d$ **64.** $9x$, $16x$

65. $12y$, y **66.** $6z$, $9z$ **67.** $24a$, $16a$ **68.** $10c$, $13c^2$

69. $5b^2$, $3b$ **70.** $24d$, $21d^2$ **71.** $36f$, $9f^2$ **72.** $15xy$, $18x^2$

73. $15w^2z$, $25wz^2$ **74.** $42m^2n$, $28n^2$ **75.** $32p^2q^2$, $24pq^2$ **76.** $40x^2y^3$, $25xy^2$

▶ **Write in lowest terms.** *(page 140)*

77. $\dfrac{9}{27}$ **78.** $\dfrac{15}{12}$ **79.** $\dfrac{21}{35}$ **80.** $\dfrac{48}{36}$ **81.** $\dfrac{28}{42}$ **82.** $\dfrac{25}{40}$

83. $\dfrac{z}{z^2}$ **84.** $\dfrac{9}{6y}$ **85.** $\dfrac{5n}{15m}$ **86.** $\dfrac{ab}{5b}$ **87.** $\dfrac{20mn}{4m^2}$ **88.** $\dfrac{21}{24h^2}$

89. $\dfrac{18x}{6y}$ **90.** $\dfrac{r^2s}{10s^2}$ **91.** $\dfrac{36p}{42pq}$ **92.** $\dfrac{17a^3}{3a^2}$ **93.** $\dfrac{35yz^2}{15y^2}$ **94.** $\dfrac{18s^2t}{48s^2t^2}$

9-7 Finding a Percent of a Number

OBJECTIVE: To find a percent of a number.

To find the sale price of an item, you can first compute the discount. The discount is the amount that is subtracted from the regular price.

1. Look at the ad. What is the regular price of the life vest?

2. The discount on the life vest is what percent of the regular price?

EXAMPLE | **Here's how to use mental math to find a percent of a number.**

[A] Finding discount on life vest: **20% of $45 = n**

Use mental math

Change the percent to a fraction and multiply.

$$\text{Think: } \frac{1}{\cancel{5}} \times \$\cancel{45}^{9} \rightarrow \textbf{20\% of \$45 = n}$$

The discount is $9.

CHECK for Understanding

3. Look at the ad. What is the regular price of the ski gloves?

4. The discount on the gloves is what percent of the regular price?

EXAMPLE | **Here's how to use a proportion or an equation to find a percent of a number.**

[B] Finding discount on ski gloves: **18% of $24 = n**

Use a proportion

18% of $24 = n

$part \rightarrow \dfrac{\textbf{18}}{\textbf{100}} = \dfrac{\textbf{n}}{\textbf{\$24}} \leftarrow part$
$whole \rightarrow \qquad\qquad \leftarrow whole$

$$100n = \$432$$
$$n = \$4.32$$

The discount is $4.32.

Use an equation

18% of $24 = n

Change the percent to a decimal and multiply.

$$0.18 \cdot \$24 = n$$
$$n = \$4.32$$

CHECK for Understanding

5. Look at the examples above.
   a. What is the discount on the life vest? What is the sale price of the life vest?
   b. What is the discount on the ski gloves? What is the sale price of the ski gloves?

EXERCISES

▶ **Solve using mental math. (Express the percent as a fraction.)**
Here are scrambled answers for the next row of exercises: 9   15   20

6. 50% of 30 = n

7. 25% of 36 = n

8. $33\frac{1}{3}$% of 60 = n

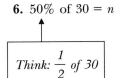
Think: $\frac{1}{2}$ *of 30*

Think: $\frac{1}{4}$ *of 36*

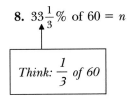
Think: $\frac{1}{3}$ *of 60*

9. 10% of 50 = n

10. 100% of 75 = n

11. 30% of 60 = n

12. 200% of 12 = n

13. 20% of 25 = n

14. $33\frac{1}{3}$% of 300 = n

▶ **Solve by using a proportion.**
Here are scrambled answers for the next row of exercises: 128.7   4.68   49.92

15. 32% of 156 = n

16. 9% of 52 = n

17. 78% of 165 = n

18. 56% of 61 = n

19. 87% of 46 = n

20. 115% of 32 = n

21. 4.6% of 35 = n

22. 82.5% of 8.8 = n

23. 0.75% of 40 = n

▶ **Solve by changing the percent to a decimal and multiplying.**
Here are scrambled answers for the next row of exercises: 38.4   9   21

24. 15% of 60 = n

25. 28% of 75 = n

26. 96% of 40 = n

27. 52% of 120 = n

28. 124% of 80 = n

29. 12% of 14 = n

30. 16.2% of 30 = n

31. 9.9% of 70 = n

32. 160% of 8.2 = n

▶ **Solve. *Hint: First try to decide which method would be the easiest.***

33. 25% of 80 = n

34. 14% of 32 = n

35. 70% of 20 = n

36. 10% of 125 = n

37. 16.5% of 80 = n

38. 50% of 40 = n

Problem Solving USING PERCENTS

▶ **Solve.**

39. Swimsuits that regularly sell for $36 are on sale for 25% off. How much is the discount?

40. Water skis are on sale for $33\frac{1}{3}$% off the regular price. What is the sale price if the regular price is $99?

41. The regular price of a wet suit is $120. You have $70. How much more money do you need to buy the wet suit if it is on sale for 20% off the regular price?

42. A nylon hat that regularly sold for $12 was put on sale for 30% off. A week later you bought the hat for 20% off the sale price. How much did you pay for the hat?

Using Algebra in Ratio, Proportion, and Percent | **253**

OBJECTIVE: To estimate a percent of a number using a conversion table.

The table below, which converts some sample percents to fractions, can help you estimate other percents.

1. Look at the table. What fraction is equivalent to 20%? To $33\frac{1}{3}$%?

2. What percent listed in the table is nearest 35%? Is nearest 73%?

Percent	Fraction	Percent	Fraction
10%	$\frac{1}{10}$	50%	$\frac{1}{2}$
20%	$\frac{1}{5}$	$66\frac{2}{3}$%	$\frac{2}{3}$
25%	$\frac{1}{4}$	75%	$\frac{3}{4}$
$33\frac{1}{3}$%	$\frac{1}{3}$		

Number of graduates up by 8%

EXAMPLES | Here's how to use the table to estimate a percent of a number.

A Estimate 35% of 24.

Find the nearest percent that is listed in the table. Then multiply by the equivalent fraction.

$$35\% \text{ of } 24 \approx \frac{1}{3} \cdot 24$$

$$= 8$$

So, 35% of 24 ≈ 8.

B Estimate 73% of 36.

Find the nearest percent that is listed in the table. Then multiply by the equivalent fraction.

$$73\% \text{ of } 36 \approx \frac{3}{4} \cdot \overset{9}{\underset{1}{36}}$$

$$= 27$$

So, 73% of 36 ≈ 27.

CHECK for Understanding

3. Look at the examples above.
  a. Is 35% of 24 greater than or less than 8?
  b. Is 73% of 36 greater than or less than 27?

4. Complete these examples.
  a. $19\% \text{ of } 40 \approx \frac{1}{5} \cdot 40$

$$= \underline{\ ?\ }$$

  b. $68\% \text{ of } 12 \approx \frac{2}{3} \cdot 12$

$$= \underline{\ ?\ }$$

EXERCISES

▶ **Use the table of percent-fraction equivalents. Estimate each percent of a number.**
Here are scrambled answers for the next row of exercises: 16   9   6   33

5. 21% of 45   **6.** 8% of 160   **7.** 26% of 24   **8.** 48% of 66

9. 76% of 40   **10.** 67% of 15   **11.** 32% of 90   **12.** 9% of 40

13. 19% of 30   **14.** 34% of 27   **15.** 11% of 70   **16.** 65% of 24

17. 35% of 15   **18.** 52% of 48   **19.** 73% of 12   **20.** 19% of 45

21. 9% of 480   **22.** 48% of 120   **23.** 24% of 400   **24.** 73% of 400

25. 12% of 360   **26.** 36% of 240   **27.** 51% of 660   **28.** 64% of 600

▶ **Choose greater than (>) or less than (<) for each ?.**

29. 10% of 60 is 6,
so 11% of 60 is _?_ 6.

30. $33\frac{1}{3}$% of 21 is 7,
so 32% of 21 is _?_ 7.

31. 25% of 80 is 20,
so 23% of 80 is _?_ 20.

32. 75% of 40 is 30,
so 77% of 40 is _?_ 30.

33. 50% of 340 is 170,
so 52% of 340 is _?_ 170.

34. $66\frac{2}{3}$% of 360 is 240,
so 64% of 360 is _?_ 240.

Group Project   USING PERCENT   *Your Favorite Things*

▶ **Work in a small group.**

35. Name your favorite
   a. recording artist or group.
   b. television program.
   c. brand name of athletic shoe.
   d. soft drink.

36. Ask your classmates to share their answers for Exercise 35. Each time a classmate agrees with one of your "favorite things," make a tally mark.

37. a. For each favorite thing, compute the percent of your classmates who agreed with you. Round your answers to the nearest tenth of a percent.
   b. In which category did the greatest percent of classmates agree with you?

38. Write a paragraph that summarizes your data.

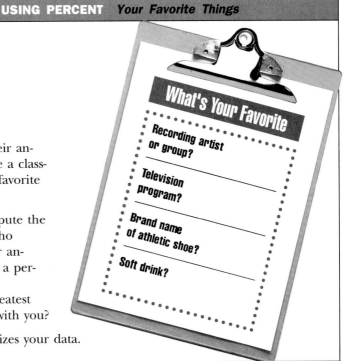

What's Your Favorite

Recording artist
or group?

Television
program?

Brand name
of athletic shoe?

Soft drink?

OBJECTIVE: To find a number when a percent of the number is know.

1. What percent of the questions did Carol get correct on her driver's license test?

2. How many questions did Carol get correct?

3. Were there more or fewer than 52 questions on the test?

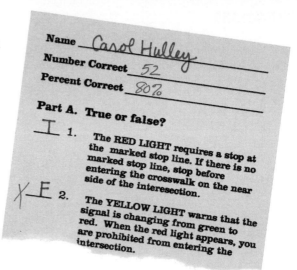

| EXAMPLE | Here's how to find the number when a percent is known. |

Finding the number of questions on the test:

$$80\% \text{ of what number is } 52?$$

Use a proportion

part ⟶ $\dfrac{80}{100} = \dfrac{52}{n}$ ⟵ part
whole ⟶   ⟵ whole

$$80n = 5200$$

$$\dfrac{80n}{80} = \dfrac{5200}{80}$$

$$n = 65$$

Use an equation

$$80\% \text{ of } n = 52$$

$$0.80\,n = 52$$

$$\dfrac{0.80n}{0.80} = \dfrac{52}{0.80}$$

$$n = 65$$

80% of 65 is 52.

CHECK for Understanding

4. Look at the above example. How many questions were on the test?

5. Complete these examples.

   a. 30% of $n = 60$

$$\dfrac{30}{100} = \dfrac{60}{n}$$

$$30n = 6000$$

$$n = \underline{\ ?\ }$$

   b. 42% of $n = 6.3$

$$\dfrac{42}{100} = \dfrac{6.3}{n}$$

$$42n = \underline{\ ?\ }$$

$$n = 15$$

   c. 10% of $n = 72$

$$0.10n = 72$$

$$\dfrac{0.10n}{0.10} = \dfrac{72}{0.10}$$

$$n = \underline{\ ?\ }$$

EXERCISES

▶ **Find the number.**

Here are scrambled answers for the next two rows of exercises: 62.5   60

6. 30% of what number is 18?

7. 80% of what number is 50?

8. 75% of what number is 80?

9. 150% of what number is 36?

10. 2.5% of what number is 24?

11. 12.5% of what number is 100?

▶ **Solve. Round each answer to the nearest tenth.**

Here are scrambled answers for the next row of exercises: 50.5   81.6   20.3

12. 12.5% of $n = 10.2$ **13.** 6.4% of $n = 1.3$ **14.** 9.3% of $n = 4.7$

15. 0.5% of $n = 0.9$ **16.** 0.8% of $n = 1.3$ **17.** 1.2% of $n = 4.2$

18. 125% of $n = 2.3$ **19.** 175% of $n = 12.4$ **20.** 150% of $n = 10.5$

Problem Solving USING PERCENTS

▶ **Solve.**

21. You took a test that had 72 questions. You got 18 questions wrong.
  a. What fraction of the questions did you get right?
  b. What percent of the questions did you get right?

22. You took a test that had 120 questions. You got 80% of the questions right. How many questions did you get right?
  Hint: 80% of 120 = n

23. You got 60 questions on a test right. You scored 75%. How many questions were on the test?
  Hint: 75% of n = 60.

24. You scored 90% on a test. You got 135 of the questions right. How many questions were on the test?

Challenge! USING A FORMULA

▶ **Solve.**

The approximate distance (d) in feet that it takes to stop a car traveling at the rate of r miles per hour is given by the formula:

> Formula: $d \approx \dfrac{6r^2}{100}$

25. About how many feet would it take to stop a car traveling at the rate of
  a. 20 mph?
  b. 35 mph?
  c. 45 mph?
  d. 55 mph?

OBJECTIVE: To find a percent of increase or decrease.

1. **a.** What was the balance of the account at month 1?
   b. Did the balance of the account increase or decrease from month 1 to month 2?
   c. By how many dollars did the balance increase or decrease from month 1 to month 2?

2. **a.** What was the balance of the account at month 4?
   b. Did the balance of the account increase or decrease from month 4 to month 5?
   c. By how many dollars did the balance increase or decrease from month 4 to month 5?

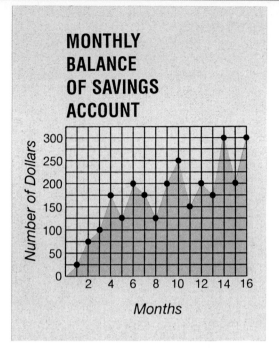

MONTHLY BALANCE OF SAVINGS ACCOUNT

EXAMPLES | **Here's how to find a percent of increase or decrease.**

A The percent of **increase** from month 1 to month 2

dollars of increase ⟶ $\dfrac{50}{25} = \dfrac{n}{100}$
*dollars **before** increase* ⟶

$$25n = 5000$$
$$\dfrac{25n}{25} = \dfrac{5000}{25}$$
$$n = 200$$

The monthly balance **increased** 200%.

B The percent of **decrease** from month 4 to month 5

dollars of decrease ⟶ $\dfrac{50}{175} = \dfrac{n}{100}$
*dollars **before** decrease* ⟶

$$175n = 5000$$
$$\dfrac{175n}{175} = \dfrac{5000}{175}$$
$$n \approx 28.6$$

The monthly balance **decreased** about 28.6%.

CHECK for Understanding

3. Complete.
   a. The percent of increase from 82 to 96 can be found by solving the proportion $\dfrac{?}{82} = \dfrac{n}{100}$.

   b. The percent of decrease from 120 to 108 can be found by solving the proportion $\dfrac{12}{?} = \dfrac{n}{100}$.

EXERCISES

▶ **Find the percent of increase or decrease. Use _i_ to indicate increase and _d_ to indicate decrease. Give answers to the nearest tenth of a percent.**

Here are scrambled answers for the next two rows of exercises:
50% i   123.7% i   100% i   22.5% d   25% d   36.2% d

4. from 24 to 36   **5.** from 80 to 60   **6.** from 38 to 85

7. from 116 to 74   **8.** from 32 to 64   **9.** from 160 to 124

10. from 28 to 37   **11.** from 100 to 81   **12.** from 38 to 106

13. from 135 to 99   **14.** from 208 to 187   **15.** from 120 to 169

16. from 212 to 90   **17.** from 178 to 256   **18.** from 160 to 80

19. from 80 to 160   **20.** from 50 to 125   **21.** from 125 to 50

22. from 220 to 153   **23.** from 150 to 220   **24.** from 184 to 0

Problem Solving USING GRAPHS

▶ **Solve. Refer to the graph on page 258.**

25. What was the percent of increase from month 2 to month 3?

26. What was the percent of decrease from month 7 to month 8?

27. a. By how many *dollars* did the account increase from month 3 to month 4?
b. By how many *dollars* did the account increase from month 5 to month 6?
c. Is the percent of increase from month 3 to month 4 the same as the percent of increase from month 5 to 6?

28. a. By how many *dollars* did the account decrease from month 10 to month 11?
b. By how many *dollars* did the account decrease from month 14 to month 15?
c. Is the percent of decrease from month 10 to month 11 the same as the percent of decrease from month 14 to 15?

29. Is the percent of decrease from month 14 to month 15 the same as the percent of increase from month 15 to month 16?

30. Is the percent of increase from month 1 to month 2 the same as the percent of increase from month 11 to month 12?

Challenge! USING PERCENTS

▶ **Solve.**

31. Ellen was earning $100 a week when she received a 10% increase in pay. Six months later she had to take a 10% decrease in pay. How much was she paid then?

32. Arthur was earning $100 a week when he had to take a 10% decrease in pay. Six months later he received a 10% increase in pay. How much was he paid then?

OBJECTIVE: To use formulas for solving problems.

When you borrow money from a bank, credit union, or loan company, you pay for the use of it. The amount you pay is called **interest.** The interest paid depends on the **principal** (the amount borrowed), the **rate** (percent of interest charged), and the **time** for which the money is borrowed.

To compute the interest on a loan, you can use this formula:

$$
\begin{array}{ccc}
& \textit{Interest} & \textit{rate} \\
& \downarrow & \downarrow \\
\textbf{Simple interest formula:} \quad I = & p \cdot & r \cdot t \\
& \uparrow & \uparrow \\
& \textit{principal} & \textit{time}
\end{array}
$$

E X A M P L E S	Here's how to use the simple interest formula to solve problems.

A In order to buy a car, you have to borrow $1250 for 3 years. The yearly interest rate is 14%. How much interest will you owe at the end of 3 years?

Write the formula. $I = prt$

Substitute 1250 for p, 0.14 for r, and 3 for t. $I = (1250)(0.14)(3)$

Simplify. $= 525$

The interest will be $525.

B Brian borrowed $400 for 6 months. He paid $32 in interest. What was the interest rate per year?

Write the formula. $I = prt$

Substitute values for I, p, and t. $32 = (400)(r)(0.5)$ ◀—

                                                   $= 200r$

Divide both sides by 200. $\dfrac{200r}{200} = \dfrac{32}{200}$

                                                   $r = 0.16$

                                                   $= 16\%$

The interest rate was 16% per year.

> *The time units must be the same. Since you are finding the yearly rate, use 0.5 of a year for* t.

EXERCISES

▶ **Use the formula $I = prt$ to solve each problem.**

1. Jill needs $1600. She borrows the money from a bank for 3 years. If the yearly interest rate is 14%, how much interest will she owe at the end of 3 years?

2. Matt needs $650. He borrows the money from a credit union for 2 years at 13% yearly interest. How much interest will Matt owe at the end of 2 years?

3. Earl borrows $5000 for 4 years to buy a car. The yearly interest rate is 11%. How much interest will he owe at the end of 4 years?

4. Harry borrows $120 from a friend. He agrees to pay 10% interest per year for 1.5 years. How much interest will Harry owe at the end of 1.5 years?

5. Bettina borrows $2000 from a loan company at 16% yearly interest. How much interest will she owe at the end of 30 months?
Hint: 30 months = 2.5 years

6. Miekka borrows $450. She agrees to pay 12% yearly interest. How much interest will she owe at the end of 6 months?
Hint: 6 months = 0.5 year

7. Seth borrowed $600 for 2 years. He paid $120 in interest. What was the interest rate per year?

8. Diane borrowed $1000 for 3 years. She paid $375 in interest. What was the interest rate per year?

▶ **Solve. Use the amount-owed formula below.**

> **Formula:** *Amount owed*   *principal*   *interest*
> $$A \quad = \quad p \quad + \quad prt$$
>
> When a loan is due, the amount (A) owed is equal to the principal (p) plus the interest (prt).

9. Troy borrows $500 at a yearly rate of 12%. What will be the amount he owes when the loan is due in 2 years?

10. Sonya borrows $3000 at a yearly rate of 15%. What will be the amount she owes when the loan is due in 4 years?

11. Chase borrows $300 at 16% interest per year. What will be the amount he owes when the loan is due in 1.5 years?

12. What is the amount you will have to repay if you borrow $8000 at 12% interest per year for 5 years?

13. What is the amount you will have to repay if you borrow $6000 at 12.5% interest per year for 8 years?

★14. You have to repay $6000 for the principal you borrowed at 12.5% for 8 years. How much was the principal?

15. **Write a problem** that can be solved using the formula above. Then solve the problem.

OBJECTIVE: To use a calculator for solving problems.

The interest on savings accounts is often compounded annually, semiannually, quarterly, or even daily. In this way, you earn interest on your interest. When the interest is compounded daily, interest is added to the account each day.

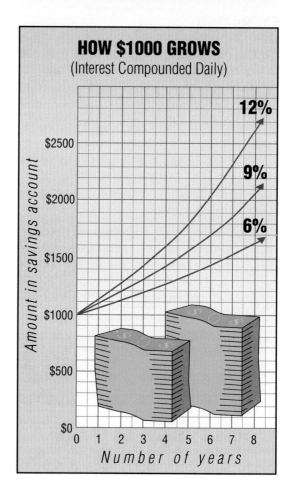

▶ **Use the graph to answer the questions.**

1. At 6% interest, about how much would a $1000 deposit be worth at the end of 3 years? 5 years? 7 years?

2. At 9% interest, about how much would a $1000 deposit be worth at the end of 3 years? 5 years? 7 years?

3. At 12% interest, about how much would a $1000 deposit be worth at the end of 3 years? 5 years? 7 years?

4. About how many years does it take to double a $1000 deposit at 9% interest?

5. About how many years does it take to double a $1000 deposit at 12% interest?

Here is how to use the formula below to find the amount in a savings account after 5 years when $1000 is deposited at 6% per year, compounded annually.

Formula: $A = P(1 + i)^n$
A = amount in the account
P = original deposit
i = interest rate per year, expressed as a decimal
n = number of years compounded

Write the formula: $A = P(1 + i)^n$

Substitute: $A = 1000(1 + 0.06)^5$

$A = 1000(1.06)^5$

Use a calculator: 1000 ☒ 1.06 y^x 5 ☰ 〔 *1338.2256* 〕

The amount is $1338.23.

▶ **Use the formula and a calculator to find the total savings.**

6. Find the amount for $1000 deposited for 5 years at 9% per year compounded annually.

7. Find the amount for $1000 deposited for 5 years at 12% per year compounded annually.

▶ **Give each sum or difference in simplest form.** *(page 214)*

1. $\dfrac{4}{9} + \dfrac{-3}{9}$

2. $\dfrac{-1}{4} + \dfrac{3}{8}$

3. $\dfrac{5}{12} + \dfrac{-5}{6}$

4. $\dfrac{-3}{4} + \dfrac{-1}{2}$

5. $\dfrac{-3}{10} + \dfrac{2}{5}$

6. $\dfrac{5}{7} + \dfrac{2}{7}$

7. $\dfrac{9}{32} + \dfrac{-11}{16}$

8. $\dfrac{-3}{16} + \dfrac{7}{8}$

9. $\dfrac{2}{3} - \dfrac{-3}{4}$

10. $\dfrac{-5}{8} - \dfrac{3}{4}$

11. $\dfrac{-7}{10} - \dfrac{1}{5}$

12. $\dfrac{2}{5} - \dfrac{9}{10}$

13. $\dfrac{-5}{6} - \dfrac{7}{8}$

14. $\dfrac{-8}{9} - \dfrac{-2}{3}$

15. $\dfrac{7}{12} - \dfrac{5}{8}$

16. $\dfrac{-11}{16} - \dfrac{7}{8}$

▶ **Give each product or quotient in simplest form.** *(page 216)*

17. $\dfrac{1}{3} \cdot \dfrac{1}{8}$

18. $\dfrac{-1}{4} \cdot \dfrac{2}{5}$

19. $\dfrac{5}{12} \cdot {-5}$

20. $\dfrac{-8}{5} \cdot \dfrac{3}{4}$

21. $\dfrac{-7}{8} \cdot \dfrac{-4}{5}$

22. $\dfrac{-5}{9} \cdot \dfrac{-3}{10}$

23. $8 \cdot \dfrac{-3}{4}$

24. $\dfrac{-8}{5} \cdot \dfrac{15}{16}$

25. $\dfrac{2}{3} \div \dfrac{-5}{6}$

26. $\dfrac{-4}{5} \div \dfrac{3}{10}$

27. $\dfrac{1}{2} \div \dfrac{3}{4}$

28. $8 \div \dfrac{2}{3}$

29. $\dfrac{-5}{6} \div 4$

30. $\dfrac{-7}{12} \div \dfrac{-14}{3}$

31. $\dfrac{7}{8} \div \dfrac{-21}{16}$

32. $\dfrac{-7}{8} \div \dfrac{5}{32}$

▶ **Simplify by combining like terms.** *(page 224)*

33. $5x + x$

34. $9y - y$

35. $4z - 4z$

36. $-3w + 5w - 8$

37. $-3w + 5 - 8$

38. $-3 + 5w - 8$

39. $-7r - r - 8$

40. $-7r + r - 18$

41. $-7r + 1 - 8$

42. $8 - 7t + 2t - 1$

43. $6t - 8 - 5t + 8$

44. $4 - 5 + 8t - t$

▶ **Solve and check.** *(page 226)*

45. $5f + f = {-17}$

46. $9h - h = 10$

47. $7e - 8 = {-8}$

48. $6x + x - 4 = 21$

49. $6x + 1 - 4x = 21$

50. $6 + x - 4x = 21$

51. $9y - 2y + 4 = {-3}$

52. $9 - 2y + 4y = {-3}$

53. $9y - 2 + 4y = {-3}$

54. $6t + 4 + 2t = 0$

55. $9 - t + 7t = {-1}$

56. $-8t + 3t - 9 = 6$

▶ **Solve and check.** *(page 228)*

57. $8y = 12 + 2y$

58. $-4k = 3 + 2k$

59. $8n + 8 = 3n + 18$

60. $6y + 2 = 2y - 4$

61. $-3x - 9 = 4x + 1$

62. $7w - 3 = {-2}w - 12$

63. $10 + 4g = g + 3$

64. $15 - 2k = 3k + 10$

65. $12 - 7c = 12 + 3c$

66. $5 - 2t = t + 6$

67. $9t - 8 = 4t + 7$

68. $-3t - 1 = 2 - 7t$

Here are scrambled answers for the review exercises:

4	25	decrease	estimate	increase	ratios
7	85	denominator	factor	left	right
12		divide	form	percent	simplify
		equivalent	fraction	rate	solve

1. To write a ratio as a fraction in lowest terms, divide both terms of the ratio by their greatest common __?__ . $\dfrac{3x^2}{4x}$ written as a fraction in lowest terms is $\dfrac{3x}{?}$.
   (*page 238*)

2. A proportion is an equation that states that two __?__ are equal. To finish solving this proportion, you would next __?__ both sides by 9 and then simplify both sides. (*page 240*)
   $$\frac{n}{7} = \frac{5}{9}$$
   $$9n = 35$$

3. A __?__ is a ratio of two unlike quantities. Suppose that you drive 203 miles in 5 hours. To find how many miles you would drive at that rate in 7 hours, you could solve the proportion $\dfrac{203}{5} = \dfrac{n}{?}$. (*page 242*)

4. To change a percent to a fraction, first write the percent as a fraction with a __?__ of 100. Then write the fraction in simplest __?__ . To change a fraction to a __?__ , change the fraction to an __?__ fraction with a denominator of 100. Then write as a percent. (*pages 244, 246*)

5. To change a percent to a decimal, move the decimal point two places to the __?__ and remove the percent sign. To change a decimal to a percent, move the decimal point two places to the __?__ and affix the percent sign. (*pages 244, 246*)

6. To solve the equation **25% of 38 = *n*,** you could change __?__ % to $\dfrac{1}{4}$ or 0.25 and multiply. (*page 252*)

7. To use a conversion table to __?__ a percent of a number, you would find the nearest percent listed in the table, then multiply by the equivalent __?__ .
   (*page 254*)

8. To find the number when a percent is known, you can change the percent to a fraction or decimal and __?__ the equation. To finish solving this example, you would divide both sides by 0.75 and then __?__ both sides.
   (*page 256*)
   75% of *n* = 51
   0.75*n* = 51

9. To find the percent of __?__ from **28 to 40,** you could solve the proportion $\dfrac{?}{28} = \dfrac{n}{100}$. To find the percent of __?__ from **85 to 68,** you could solve the proportion $\dfrac{17}{?} = \dfrac{n}{100}$. (*page 258*)

Complete to get an equal ratio. *(page 238)*

1. $\dfrac{2}{3} = \dfrac{?}{15}$
2. $\dfrac{35}{50} = \dfrac{?}{10}$
3. $\dfrac{2a}{b} = \dfrac{?}{3b}$
4. $\dfrac{4xy}{6x^2} = \dfrac{?}{3x}$
5. $\dfrac{2c}{9cd} = \dfrac{?}{18c^2d}$

Solve each proportion. *(page 240)*

6. $\dfrac{2}{5} = \dfrac{n}{20}$
7. $\dfrac{8}{n} = \dfrac{4}{13}$
8. $\dfrac{n}{8} = \dfrac{15}{4}$
9. $\dfrac{20}{11} = \dfrac{7}{n}$
10. $\dfrac{n}{21} = \dfrac{9}{16}$

Solve by using proportions. If your answer does not come out evenly, round it to the nearest hundredth. *(page 242)*

You drive 126 miles in 3 hours. At that speed, how many

11. miles could you drive in 4 hours?
12. miles could you drive in 5 hours?

13. hours would it take to drive 100 miles?
14. hours would it take to drive 240 miles?

Change to a fraction, whole number, or mixed number. Give each answer in simplest form. *(page 244)*

15. 75%
16. 16%
17. 150%
18. $37\frac{1}{2}\%$
19. $8\frac{1}{3}\%$
20. 200%

Change to a decimal. *(page 244)*

21. 37%
22. 250%
23. 8%
24. 5.5%
25. $8\frac{1}{4}\%$
26. $133\frac{1}{3}\%$

Change to a percent. *(page 246)*

27. $\dfrac{2}{5}$
28. $\dfrac{7}{4}$
29. $\dfrac{1}{3}$
30. $\dfrac{7}{8}$
31. $\dfrac{11}{6}$
32. 1

Solve. If necessary, round your answer to the nearest tenth. *(pages 252, 256)*

33. 20% of $65 = n$
34. 125% of $64 = n$
35. 24.5% of $75 = n$
36. 2.5% of $n = 31$

37. 10% of $n = 15$
38. 40% of $n = 24$
39. 1.5% of $n = 6$
40. 175% of $n = 9.6$

Find the percent of increase or decrease. *(page 258)*

41. from 18 to 27
42. from 30 to 40
43. from 56 to 42
44. from 72 to 62

Use the formula $I = prt$ to solve each problem. *(page 260)*

45. Arthur borrowed $2500 for 1.5 years. The yearly interest rate was 13%. How much interest did he owe at the end of 1.5 years?

46. Carol borrowed $1500 for 2 years. She paid $345 in interest. What was the interest rate per year?

▶ **Choose the correct letter.**

1. Solve.

$$-12 = \frac{x}{-3} + 5$$

A. 51

B. 21

C. −51

D. none of these

2. Simplify.

$$\frac{t^8}{t^2}$$

A. t^6

B. t^{10}

C. t^4

D. none of these

3. 8,400,000 written in scientific notation is

A. $84 \cdot 10^5$

B. $0.84 \cdot 10^7$

C. $8.4 \cdot 10^{-6}$

D. none of these

4. The prime factorization of 108 is

A. $2 \cdot 54$

B. $2^2 \cdot 27$

C. $2^2 \cdot 3^3$

D. none of these

5. The greatest common factor of $28cd^2$ and $42c^2d$ is

A. $7cd$

B. $14cd$

C. $14c^2d^2$

D. none of these

6. $\dfrac{28xy^2}{21y^2}$ written in lowest terms is

A. $\dfrac{28x}{21}$ **B.** $\dfrac{4x}{3}$

C. $\dfrac{4xy^2}{3y^2}$ **D.** none of these

7. Give the difference.

$$\frac{-5}{12} - \frac{-2}{3}$$

A. $\dfrac{-1}{3}$ **B.** $\dfrac{1}{3}$

C. $-1\dfrac{1}{12}$ **D.** none of these

8. Give the quotient.

$$\frac{-5}{6} \div \frac{20}{3}$$

A. $-5\dfrac{5}{9}$ **B.** -8

C. $\dfrac{-1}{8}$ **D.** none of these

9. Simplify by combining like terms.

$$3x - 5 - x + 3$$

A. $4x - 2$

B. $2x + 2$

C. $4x + 2$

D. none of these

10. Solve.

$$5y - 8 + y = -7$$

A. $\dfrac{1}{6}$

B. $-2\dfrac{1}{2}$

C. $\dfrac{1}{4}$

D. none of these

11. Solve.

$$4z + 9 = -2z - 6$$

A. $\dfrac{1}{2}$

B. $-2\dfrac{1}{2}$

C. $-7\dfrac{1}{2}$

D. none of these

12. Choose the equation.

You had $72. You bought 6 shirts yesterday and 3 shirts today. Each shirt cost the same. How much did 1 shirt cost?

A. $6n + 3n = 72$

B. $6n + 3 = 72$

C. $3n + 6 = 72$

D. $3n + 72 = 6n$

$A = \pi r^2$

The area of a circle equals π (approximately equal to 3.14) times the radius squared.
(page 296)

The water sprinkler shoots a stream of water out a distance of 30 feet while rotating on its axis. You can use these facts and the formula above to find the ground area that is watered by the sprinkler.

OBJECTIVE: To use geometric notation for line, ray, segment, and angle. To measure angles. To classify angles as acute, right, obtuse, congruent, complementary, or supplementary.

A **straight line,** or simply a **line,** extends indefinitely in two directions.

line *m* or line *AB* or $\overleftrightarrow{AB}$

A **ray** is part of a line that starts at a point and extends indefinitely in one direction.

ray *CD* or $\overrightarrow{CD}$

A **line segment,** or simply a **segment,** is part of a line that starts at one point and ends at another point.

segment *EF* or $\overline{EF}$

An **angle** is formed by two rays with a common endpoint. The common endpoint is called the **vertex** of the angle, and the rays are called the **sides** of the angle.

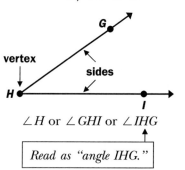

vertex

sides

$\angle H$ or $\angle GHI$ or $\angle IHG$

Read as "angle IHG."

EXAMPLE | **Here's how to use a protractor to measure an angle.**

Step 1. Place the center point of the protractor on the vertex of the angle.

Step 2. Place the 0 mark on one side of the angle.

Step 3. Read the measure of the angle where the other side crosses the protractor.

The measure of the angle is 35°.

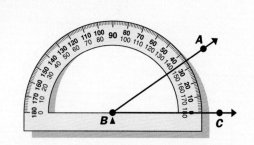

Measures of angles can be used to classify angles.
An **acute angle** measures between 0° and 90°.
A **right angle** measures 90°.
An **obtuse angle** measures between 90° and 180°.

Angles that have the same measure are **congruent angles.**

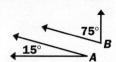

$\angle P$ is congruent to $\angle Q$ or $\angle P \cong \angle Q$.

Two angles whose measures have a sum of 90° are **complementary angles.**

$\angle A$ and $\angle B$ are complementary angles.

Two angles whose measures have a sum of 180° are **supplementary angles.**

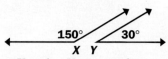

$\angle X$ and $\angle Y$ are supplementary angles.

| EXAMPLE | **Here's how to draw a 35° angle.** |

Step 1.

Draw one side. Place the protractor as you would for measuring and make a mark at 35.

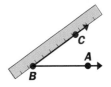

Step 2.

Draw the other side.

EXERCISES

▶ **Draw the figure.**

1. $\overleftrightarrow{RT}$ **2.** line segment AB **3.** $\overleftrightarrow{CD}$ **4.** right angle RST **5.** ray XY **6.** $\overline{PQ}$

▶ **Measure each angle.**

7.   **8.**   **9.**

▶ **Tell whether each angle is acute, right, or obtuse.**

10. $\angle ABD$ **11.** $\angle HBD$ **12.** $\angle HBF$

13. $\angle CBD$ **14.** $\angle GBA$ **15.** $\angle FBE$

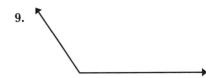

▶ **Complete. Use the figure at the right and the definitions on page 268.**

16. $\angle ABD$ and $\angle\ \underline{\ ?\ }$ are complementary.

17. $\angle ABC$ and $\angle\ \underline{\ ?\ }$ are congruent.

18. $\angle ABG$ and $\angle\ \underline{\ ?\ }$ are supplementary.

19. $\angle EBA$ is congruent to $\angle\ \underline{\ ?\ }$.

▶ **Draw angles having these measures.**

20. 45°   **21.** 120°   **22.** 75°   **23.** 150°   **24.** 20°   **25.** 135°

Problem Solving FINDING ANGLE MEASURES

26. Draw a pair of supplementary angles that are congruent. What is the measure of each angle?

27. Draw a pair of complementary angles that are congruent. What is the measure of each angle?

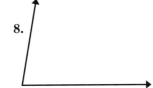

OBJECTIVE: To classify triangles as acute, right, obtuse, scalene, isosceles, or equilateral. To solve problems involving the sum of the angles of a triangle.

Triangles can be classified according to the types of angles or the number of congruent sides they contain.

| E X A M P L E S | Here's how to classify triangles using their angle measures. |

Every triangle has at least two acute angles. The third angle is used to classify the triangle.

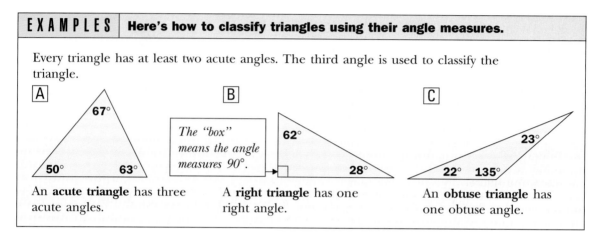

An **acute triangle** has three acute angles.

The "box" means the angle measures 90°.

A **right triangle** has one right angle.

An **obtuse triangle** has one obtuse angle.

CHECK for Understanding

1. Use the examples above to classify these triangles as acute, right, or obtuse.

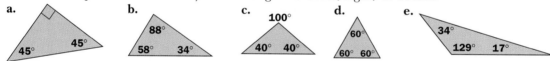

a.   b.   c.   d.   e.

2. Find the sum of the measures of the angles in each triangle.

> An important property of triangles: The sum of the measures of the angles in any triangle is 180°.

| E X A M P L E S | Here's how to classify triangles using the number of congruent sides. |

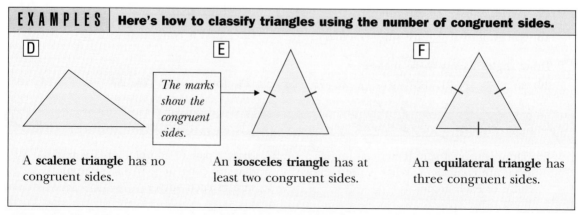

The marks show the congruent sides.

A **scalene triangle** has no congruent sides.

An **isosceles triangle** has at least two congruent sides.

An **equilateral triangle** has three congruent sides.

EXERCISES

▶ Choose two of the following terms to classify each triangle: acute, obtuse, right, scalene, isosceles, equilateral.

3.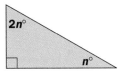

4.

5.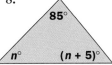

6.

▶ Write and solve an equation to find the measure of each angle.

7.
$2n°$
$n°$

8.
$85°$
$n°$ $(n + 5)°$

9.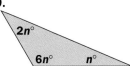
$2n°$
$6n°$ $n°$

10.
$n°$
$n°$ $n°$

11.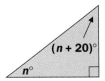
$(n + 20)°$
$n°$

12.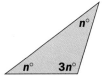
$n°$
$n°$ $3n°$

13. In right triangle *ABC*, the measure of acute ∠*A* is 4 times the measure of acute ∠*B*. Find the measure of each angle.

14. In triangle *DEF*, the measure of ∠*D* is 3 times the measure of ∠*E*, and the measure of ∠*F* is 5 times the measure of ∠*E*. Find the measure of each angle.

Group Project

FINDING A PATTERN   *Angles, Angles!*

▶ Work in a small group.

15. Think about drawing 10 rays from 1 vertex. Each member of the group estimates the number of angles formed.

16. Follow the steps to find out how many angles are formed by 10 rays.
    a. Copy and complete this table.

Number of rays from vertex	1	2	3	4	5	6
Number of angles formed	0	1	3			

   b. Look for a pattern in your completed table. Continue the pattern to find the number of angles formed by 10 rays.

17. Who in your group had the closest estimate?

18. How many angles are formed by 20 rays drawn from 1 vertex?

OBJECTIVE: To recognize parallel and perpendicular lines. To solve problems involving vertical angles and congruent corresponding angles.

E X A M P L E S	Here's how we define parallel and perpendicular lines.

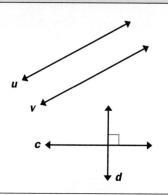

A. Two lines that do not intersect are **parallel lines.**
Line u is parallel to line v, or
line $u \parallel$ line v.

B. Two lines that form four right angles are
perpendicular lines.
Line c is perpendicular to line d,
or line $c \perp$ line d.

CHECK for Understanding

1. Look at Example A. Line u _?_ line v.

2. Look at Example B. Line c _?_ line d.

E X A M P L E S	Here's how we define vertical angles, a transversal, and corresponding angles.

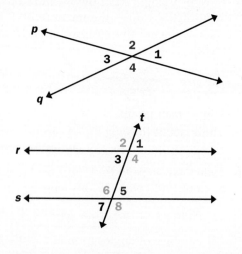

C. When two lines intersect, they form two
pairs of congruent angles called **vertical
angles.**
$\angle 1$ and $\angle 3$ are vertical angles; $\angle 1 \cong \angle 3$.
$\angle 2$ and $\angle 4$ are vertical angles; $\angle 2 \cong \angle 4$.

D. A **transversal** (line t in the diagram) is a
line that intersects two or more lines.

E. The pairs of angles that are color coded
($\angle 1$ and $\angle 5$, $\angle 2$ and $\angle 6$, $\angle 3$ and $\angle 7$,
and $\angle 4$ and $\angle 8$) are **corresponding
angles.**
When a transversal intersects two parallel
lines, the pairs of corresponding angles
are congruent.

CHECK for Understanding

3. Look at Example C. $\angle 3$ and _?_ are vertical angles.

4. Look at Example E. $\angle 3$ and _?_ are corresponding angles.

EXERCISES

▶ **True or false?**

5. Line $k \perp$ line m.   6. Line $j \parallel$ line k.

7. Line $p \parallel$ line k.   8. Line $m \perp$ line j.

9. Line $l \perp$ line n.   10. Line $j \parallel$ line p.

11. Line $n \parallel$ line p.   12. Lines l and m intersect.

13. There are 20 right angles in the drawing.

14. There are 16 acute angles in the drawing.

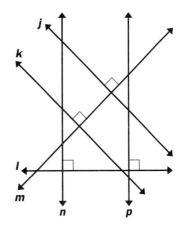

15. Find the measures of $\angle r$, $\angle s$, and $\angle t$.

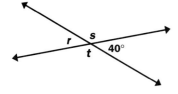

16. Find the measures of $\angle x$ and $\angle y$, if line $m \perp$ line n.

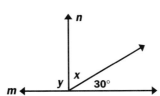

17. Find the measures of $\angle a$, $\angle b$, and $\angle c$, if line $p \parallel$ line q.

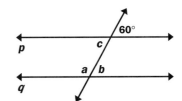

Problem Solving WRITING AND SOLVING EQUATIONS

▶ **Write and solve an equation to find the measure of the angle.**

18. Find the measure of $\angle ABC$.
Hint: Use supplementary angles.

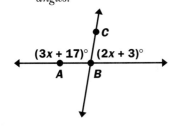

19. Find the measure of $\angle DEF$.
Hint: Use vertical angles.

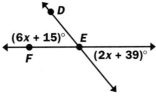

20. Find the measure of $\angle KLM$, if line $m \parallel$ line n.
Hint: Use congruent corresponding angles.

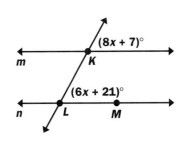

Using Algebra in Geometry | **273**

OBJECTIVE: To construct the perpendicular bisector of a segment, the bisector of an angle, and a triangle congruent to a given triangle.

In geometry, a compass and straightedge are used to **construct** figures. To **bisect a segment** is to divide it into two segments that are the same length.

| EXAMPLE | Here's how to construct the perpendicular bisector of a segment. |

Step 1.

Open the compass so that the setting is <u>more</u> than half the length of $\overline{CD}$. Use C as the center and draw arcs above and below $\overline{CD}$.

Step 2.

With the same setting, use D as the center and <u>draw</u> arcs above and below $\overline{CD}$.

Step 3.

Use a straightedge and draw a line through the points where the arcs cross.

Line XY is the perpendicular bisector of $\overline{CD}$.

CHECK for Understanding

1. Look at the example above. Line XY is the perpendicular bisector of which segment?

To **bisect an angle** is to divide it into two angles that have the same measure.

| EXAMPLE | Here's how to construct the bisector of an angle. |

Step 1.

Use V as the center and draw an arc that intersects both sides.

Step 2.

Use U and W as centers and draw intersecting arcs with the same compass setting.

Step 3.

Use a straightedge and draw the ray from V through the point where the arcs intersect.

Ray VP is the angle bisector of angle V.

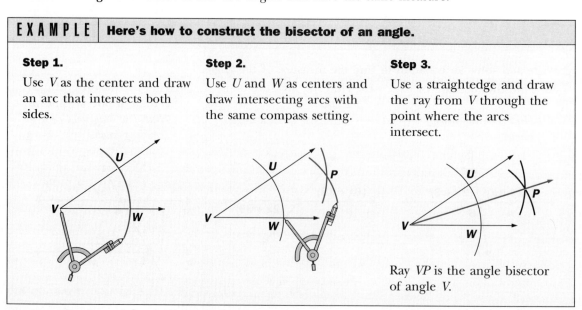

Congruent triangles have the same size and shape.

| E X A M P L E | **Here's how to construct a triangle congruent to a given triangle.** |

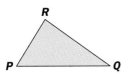

Step 1.

Use a straightedge to draw a ray and label its endpoint P'.

Step 2.

Set the compass legs on the endpoints of $\overline{PQ}$.

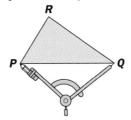

Step 3.

Use this setting and point P' as center to draw an arc that intersects the ray at point Q'.

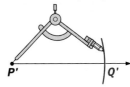

Step 4.

Set the compass legs on the endpoints of $\overline{PR}$. Use this setting and point P' as center to draw an arc.

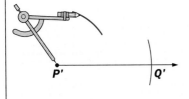

Step 5.

Set the compass legs on the endpoints of $\overline{QR}$. Use this setting and point Q' as center to draw an arc that intersects the arc drawn in Step 4 at point R'.

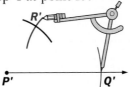

Step 6.

Draw $\overline{P'R'}$ and $\overline{Q'R'}$.

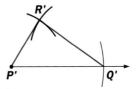

EXERCISES

▶ **Use a straightedge and compass.**

  2. Draw a segment. Construct its perpendicular bisector.

  3. Draw an angle. Construct its bisector.

  4. Draw a triangle. Construct a triangle congruent to it.

  5. Draw a large triangle. Construct the bisector of each angle. If you are careful, the three bisectors will all cross at the same point.

  6. Draw a large triangle. Construct the perpendicular bisector of each side. If you are careful, the three perpendicular bisectors will all cross at the same point.

OBJECTIVE: To classify polygons according to their sides and angles.

1. Look at the map. Which lot has 4 sides all the same length?

2. Which lot has exactly 1 pair of parallel sides?

3. Which lot has 2 pairs of parallel sides but no right angles?

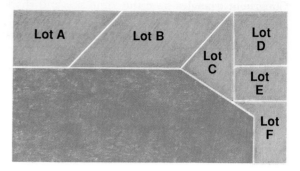

A **polygon** is a closed plane figure made up of segments.

| EXAMPLES | Here's how some polygons are named. |

Name of Polygon	Description	Examples
Quadrilateral	4 sides	
Square	4 sides the same length 4 right angles	
Rectangle	4 sides 4 right angles	A square is also a rectangle.
Parallelogram	4 sides 2 pairs of parallel sides	A rectangle is also a parallelogram.
Trapezoid	4 sides Exactly 1 pair of parallel sides	
Pentagon	5 sides	
Hexagon	6 sides	

CHECK for Understanding

4. Use the map and the chart above to answer these questions.
   a. Which lots are not quadrilaterals?
   b. Which lot is a square?
   c. Which 2 lots are rectangles?
   d. Which 3 lots are parallelograms?
   e. Which lot is a trapezoid?
   f. Which lot is a pentagon?

EXERCISES

▷ Use the chart on page 276 to name each polygon. Some shapes have more than one name.

5.

6.

7.

8.

9.

10.

11.

12.

 Problem Solving VISUAL THINKING

▷ Use the clues. Draw and name each polygon.

13. *Clues:*
This polygon has 4 sides.
It has 4 right angles.
Not all its sides are the same length.

14. *Clues:*
This polygon has no right angle.
It has 4 sides.
It has 2 pairs of parallel sides.

15. *Clues:*
This polygon has 4 sides.
It has 4 right angles.
All its sides are the same length.

16. *Clues:*
This polygon has no right angle.
It has 4 sides.
It has no pairs of parallel sides.

Challenge! USING SYMMETRY

 If you fold this figure along the dashed line, the two halves match exactly. We say that the figure has **symmetry,** and the fold line is a **line of symmetry.**

▷ Is the dashed line a line of symmetry? (You may want to trace the figures.)

17.

18.

19.

20.

21.

▷ Trace each figure and draw all lines of symmetry.

22. 23. 24. 25. 26.

Using Algebra in Geometry | 277

10-6 Metric Units of Length

OBJECTIVE: To become familiar with metric units of length.

In the metric system, the **meter** (m) is the basic unit of length. Here are the relationships between different metric units:

1 kilometer (km) = 1000 meters
1 hectometer (hm) = 100 meters
1 dekameter (dam) = 10 meters
1 meter (m) = 1 meter
1 decimeter (dm) = 0.1 meter
1 centimeter (cm) = 0.01 meter
1 millimeter (mm) = 0.001 meter

Note: The units listed in red are used most often.

| EXAMPLES | Here's how to estimate length in the metric system. |

A From the top of the Ferris wheel, a person can see about 5 city blocks— that's about 1 **kilometer.**

B The length of a seat on the Ferris wheel is about 1 **meter.**

C The width of an index fingernail is about 1 **centimeter.**

D The thickness of a dime is about 1 **millimeter.**

EXERCISES

▶ Choose mm, cm, m, or km.

1. The height of a Ferris wheel is 15 __?__ .

2. The height of a person is 185 __?__ .

3. The thickness of a dime is 1 __?__ .

4. The length of a river is 450 __?__ .

5. The length of a tennis court is 20 __?__ .

6. The length of a paper clip is 3 __?__ .

7. The width of a door is 0.9 __?__ .

8. The length of a new pencil is 190 __?__ .

9. The distance between two cities is 120 __?__ .

10. The length of a hiking trail is 12.5 __?__ .

11. The height of a kitchen counter is 0.95 __?__ .

12. The width of a dollar bill is 66 __?__ .

13. The height of a stepladder is 3.05 __?__ .

14. The width of a newspaper is 33 __?__ .

15. The distance across a room is 9.5 __?__ .

16. The length of a tennis racket is 655 __?__ .

Which measurement is reasonable?

17. Height of a ten-story building:
   a. 33 cm   **b.** 33 m   **c.** 33 km

18. Length of an automobile:
   a. 4.85 cm   **b.** 4.85 m   **c.** 4.85 mm

19. Length of a dollar bill:
   a. 16 mm   **b.** 16 cm   **c.** 16 m

20. Length of a baseball bat:
   a. 95 mm   **b.** 95 cm   **c.** 95 m

21. Height of a bicycle:
   a. 0.95 cm   **b.** 0.95 m   **c.** 0.95 km

22. Width of a thumb:
   a. 20 mm   **b.** 20 cm   **c.** 20 m

23. Thickness of a nickel:
   a. 2 mm   **b.** 2 cm   **c.** 2 m

24. Thickness of a dollar bill:
   a. 0.1 mm   **b.** 0.1 cm   **c.** 0.1 m

Problem Solving   USING A MAP

Solve. Use the map distances.

25. How far is it from
   a. Miami to Tallahassee through Orlando?
   b. Tampa to Jacksonville through Orlando?

26. How much farther is it from
   a. Orlando to Miami than from Orlando to Jacksonville?
   b. Tallahassee to Orlando than from Orlando to Miami?

Challenge!   USING A METRIC RULER

Use the code to answer the question.

CODE:

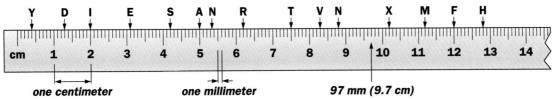

one centimeter   one millimeter   97 mm (9.7 cm)

27. *Question:* What is the height of the world's largest Ferris wheel?

   Answer:

| 4.2 cm | 20 mm | 10.2 cm | 75 mm | 0.4 cm | – | 7.5 cm | 128 mm | 6.2 cm | 31 mm | 3.1 cm |

| 50 mm | 53.5 mm | 13 mm |

| 120 mm | 2 cm | 8.3 cm | 3.1 cm | – | 75 mm | 31 mm | 5.35 cm | 7.5 cm | 128 mm | 4.2 cm |

| 11.2 cm | 3.1 cm | 7.5 cm | 31 mm | 62 mm | 42 mm |

OBJECTIVE: To make conversions between metric units of length.

Who Grew the Longer Beard?

Here are the facts you will need to know to change from one metric unit of length to another.

$$10 \text{ mm} = 1 \text{ cm}$$
$$100 \text{ cm} = 1 \text{ m}$$
$$1000 \text{ mm} = 1 \text{ m}$$
$$1000 \text{ m} = 1 \text{ km}$$

BEARD GROWING CONTEST
NAME _Charles_
LENGTH _0.13 M_

BEARD GROWING CONTEST
NAME _Larry_
LENGTH _125 MM_

EXAMPLES | **Here's how to change from one unit of length to another.**

To change units in the metric system, multiply or divide by 10, 100, or 1000.

Charles's beard

A $0.13 \text{ m} = \underline{\ ?\ } \text{ cm}$

Think: Since we are changing to a smaller unit, we should get a larger number. Therefore, we should multiply.

Remember: $1 \text{ m} = 100 \text{ cm}$
0.13 m = 13 cm
$\llcorner \times 100 \lrcorner$

Larry's beard

B $125 \text{ mm} = \underline{\ ?\ } \text{ cm}$

Think: Now we are changing to a larger unit, so we should get a smaller number. Therefore, we should divide.

Remember: $10 \text{ mm} = 1 \text{ cm}$
125 mm = 12.5 cm
$\llcorner \div 10 \lrcorner$

CHECK for Understanding

1. Look at the examples above.
   a. To change from meters to centimeters, multiply by __?__ .
   b. To change from millimeters to centimeters, divide by __?__ .

2. Who grew the longer beard, Charles or Larry?

3. Complete these examples.
   a. $1825 \text{ m} = \underline{\ ?\ } \text{ km}$

   Think: Changing smaller to larger units, so divide.

   Remember: $1000 \text{ m} = \underline{\ ?\ } \text{ km}$
   $1825 \text{ m} = \underline{\ ?\ } \text{ km}$
   $\llcorner \div 1000 \lrcorner$

   b. $9.65 \text{ km} = \underline{\ ?\ } \text{ m}$

   Think: Changing larger to smaller units, so multiply.

   Remember: $1 \text{ km} = \underline{\ ?\ } \text{ m}$
   $9.65 \text{ km} = \underline{\ ?\ } \text{ m}$
   $\llcorner \times 1000 \lrcorner$

EXERCISES

Copy and complete.

4. 6 cm = _?_ mm **5.** 4 m = _?_ cm **6.** 9 km = _?_ m **7.** 54 cm = _?_ m

8. 36 mm = _?_ cm **9.** 25 km = _?_ m **10.** 2485 m = _?_ mm **11.** 3.5 km = _?_ m

12. 2.9 cm = _?_ mm **13.** 58 m = _?_ cm **14.** 4.6 m = _?_ cm **15.** 250 cm = _?_ m

16. 83 mm = _?_ cm **17.** 2.75 km = _?_ m **18.** 2763 m = _?_ km **19.** 75 m = _?_ cm

20. 28 mm = _?_ cm **21.** 12.6 cm = _?_ mm **22.** 750 km = _?_ m **23.** 900 mm = _?_ m

24. 0.8 m = _?_ mm **25.** 8 cm + 4 mm = _?_ mm **26.** 40 cm + 4 mm = _?_ mm

27. 60 cm + 5 mm = _?_ mm **28.** 20 cm + 15 mm = _?_ cm **29.** 6 m + 25 cm = _?_ cm

30. 8 m + 175 cm = _?_ cm **31.** 5 m + 50 cm = _?_ m **32.** 6 m + 400 cm = _?_ m

Problem Solving USING METRIC UNITS

Solve.

33. Researchers say a man's beard grows an average of 0.038 of a centimeter a day. How many millimeters per day is that?

34. In one year, a man's beard grows an average of 138 millimeters. How many centimeters per year is that?

35. During the average lifetime, a man spends approximately 3350 hours removing 838 centimeters of whiskers. How many meters of whiskers is that?

36. The longest beard belonged to Hans Langseth of Barney, North Dakota. He let his whiskers grow to a length of 5.33 meters. How many centimeters is that?

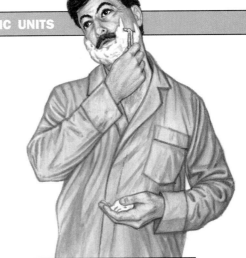

$10n$	$100n$	$1000n$
$\dfrac{n}{10}$	$\dfrac{n}{100}$	$\dfrac{n}{1000}$

Choose the expression that describes each of the following.

37. centimeters in n meters

38. millimeters in n meters

39. meters in n kilometers

40. millimeters in n centimeters

41. meters in n centimeters

42. kilometers in n meters

43. centimeters in n millimeters

44. meters in n millimeters

Using Algebra in Geometry | **281**

You will use a variety of problem-solving strategies on these two pages. For some problems, you may wish to use a calculator.

1 YOUR LICENSE, PLEASE

Tony's license plate says "4 KIX." Make a list of all 4-character license plates that can be made using 4, K, I, and X.

2 FRUIT FAVORITES

The favorite fruits of Brad, Robin, Joey, and Paige are banana, watermelon, pear, and pineapple. Use the clues to decide which fruit is the favorite of each person.

- No one's favorite fruit begins with the same letter as his or her name.
- Joey's favorite fruit and Robin's favorite fruit begin with the same letter.
- Joey's favorite fruit does not have seeds.

3 CIRCLE YOUR DIGITS!

Copy the diagram. Write the digits 3–9 in the circles so that the numbers in each row of circles (horizontal, vertical, and diagonal) have a sum of 18.

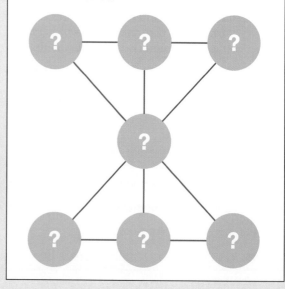

4 GEOMETRY—VISUAL THINKING

Which of these figures can you draw without lifting your pencil from the paper or retracing any lines?

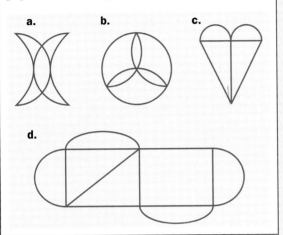

a. b. c.

d.

6 DISCOVERY PATTERN

Look for patterns. Without multiplying, find the missing products.

$1089 \times 1 = 1089$
$1089 \times 2 = 2178$
$1089 \times 3 = 3267$
$1089 \times 4 = \underline{\ ?\ }$
$1089 \times 5 = \underline{\ ?\ }$

$1089 \times 6 = \underline{\ ?\ }$
$1089 \times 7 = \underline{\ ?\ }$
$1089 \times 8 = \underline{\ ?\ }$
$1089 \times 9 = \underline{\ ?\ }$

5 IT WAS A VERY GREAT YEAR

Here are some important events in U.S. history. The year the event occurred is written as a Roman numeral.

Astronauts walk on the moon.
MCMLXIX

Great Depression begins.
MCMXXIX

Declaration of Independence signed.
MDCCLXXVI

Gold discovered in California.
MDCCCXLVIII

Civil War begins.
MDCCCLXI

Alaska and Hawaii become states.
MCMLIX

Use **Number Sense** to decide if the following statements could be true. Explain your answers.

a. My great-grandmother was born during the final stages of the Civil War. She watched Neil Armstrong walk on the moon on TV.

b. Fifty percent of the forty-niners who joined in the California gold rush lost their jobs in the Great Depression.

c. George Bush was President when Hawaii achieved statehood.

APPLICATIONS

7 Complete this box score from a Chicago Bulls basketball game.

8 Pippen and _?_ made 50% of their free-throw attempts.

9 Jordan made _?_ % of his field-goal attempts.

10 Together, Jordan and _?_ scored about 61% of the total points.

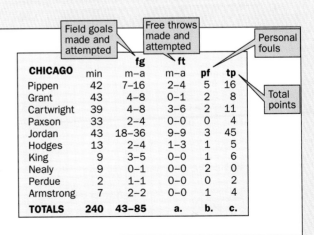

CHICAGO	min	fg m–a	ft m–a	pf	tp
Pippen	42	7–16	2–4	5	16
Grant	43	4–8	0–1	2	8
Cartwright	39	4–8	3–6	2	11
Paxson	33	2–4	0–0	0	4
Jordan	43	18–36	9–9	3	45
Hodges	13	2–4	1–3	1	5
King	9	3–5	0–0	1	6
Nealy	9	0–1	0–0	2	0
Perdue	2	1–1	0–0	0	2
Armstrong	7	2–2	0–0	1	4
TOTALS	**240**	**43–85**	a.	b.	c.

- Field goals made and attempted → fg
- Free throws made and attempted → ft
- Personal fouls → pf
- Total points → tp

OBJECTIVE: To read instruments and determine the precision of the measurements.

The scale markings on measurement instruments determine the **precision** of the measurements that can be made with them.

Ruler A measurements are precise to the nearest centimeter. (The marks are 1 cm apart.)

Ruler B measurements are precise to the nearest 0.5 centimeter. (The marks are 0.5 cm apart.)

Ruler C measurements are precise to the nearest 0.1 centimeter. (The marks are 0.1 cm apart.)

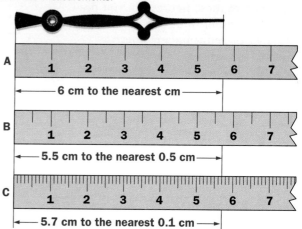

A

— 6 cm to the nearest cm —

B

— 5.5 cm to the nearest 0.5 cm —

C

— 5.7 cm to the nearest 0.1 cm —

▶ **Read these instruments and give the precision of the measurements.**

1.

Oil Pressure

__?__ lb/in² to the nearest __?__ lb/in²

2.

Fuel

__?__ gal to the nearest __?__ gal

3.

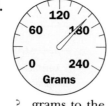

__?__ °C to the nearest __?__ °C

4.

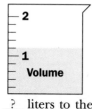

Volume

__?__ liters to the nearest __?__ liter

5.

Grams

__?__ grams to the nearest __?__ gram

6.

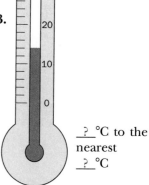

Voltage

__?__ volts to the nearest __?__ volt

▶ **Write and solve an equation to find the correct readings from these inaccurate gauges.**

7.

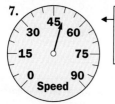

Speed

This reading is 10 mph greater than twice the actual speed.

8.

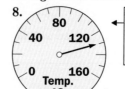

Temp. °C

This reading is 15° less than 3 times the actual temperature.

▶ **Simplify by combining like terms.** *(page 224)*

1. $7c + c$

2. $8x - x$

3. $5n - 5n$

4. $^-5d + 4d - 3$

5. $^-5d + 4 - 3d$

6. $^-5 + 4d - 3$

7. $^-10s - s - 8$

8. $^-10s + s - 8$

9. $^-10s + 1 - 8$

10. $12k - 10 + 6k - 7$

11. $12 - 10k + 6k - 7$

12. $12k - 10 + 6 - 7k$

▶ **Solve and check.** *(page 226)*

13. $4n + n = {}^-9$

14. $6r - r = 15$

15. $12v - 18 = {}^-18$

16. $11 + 5w - 6 = {}^-8$

17. $11 + 5w - 6w = {}^-8$

18. $11w + 5 - 6w = {}^-8$

19. $^-9c + 6 - 3c = 15$

20. $^-9 + 6c - 3 = 15$

21. $^-9 + 6c - 3c = 15$

▶ **Solve and check.** *(page 228)*

22. $5y = 8 - 3y$

23. $^-4n = 6 + 8n$

24. $18 - 8j = {}^-2j$

25. $4d + 3 = 2d + 17$

26. $5k - 8 = 3k + 10$

27. $9x - 3 = 4x - 11$

28. $12 + 2c = 10 + 6c$

29. $9 + 3f = 8 - 2f$

30. $15 - 7k = {}^-4k + 6$

31. $19 + 6j = {}^-2j - 5$

32. $8w - 3 = 7 + 2w$

33. $6t + 8 = {}^-6 - 3t$

▶ **Complete to get an equal ratio.** *(page 238)*

34. $\dfrac{5}{6} = \dfrac{?}{18}$

35. $\dfrac{3}{10} = \dfrac{?}{40}$

36. $\dfrac{3}{8} = \dfrac{?}{16}$

37. $\dfrac{4}{5} = \dfrac{?}{25}$

38. $\dfrac{9}{2} = \dfrac{?}{20}$

39. $\dfrac{4}{y} = \dfrac{?}{2y}$

40. $\dfrac{3}{w} = \dfrac{?}{4w}$

41. $\dfrac{j}{5} = \dfrac{?}{5k}$

42. $\dfrac{a}{b} = \dfrac{?}{b^2}$

43. $\dfrac{r}{s} = \dfrac{?}{3s^2}$

44. $\dfrac{a}{4b} = \dfrac{?}{12b^2}$

45. $\dfrac{6m}{2n} = \dfrac{?}{12n^2}$

46. $\dfrac{4w}{9xy} = \dfrac{?}{18xy^2}$

47. $\dfrac{w}{4} = \dfrac{?}{20xy}$

48. $\dfrac{5a}{6bc^2} = \dfrac{?}{18b^2c^2}$

▶ **Solve each proportion.** *(page 240)*

49. $\dfrac{n}{5} = \dfrac{2}{7}$

50. $\dfrac{8}{n} = \dfrac{4}{9}$

51. $\dfrac{8}{12} = \dfrac{n}{6}$

52. $\dfrac{9}{7} = \dfrac{8}{n}$

53. $\dfrac{n}{10} = \dfrac{12}{5}$

54. $\dfrac{11}{3} = \dfrac{n}{5}$

55. $\dfrac{15}{19} = \dfrac{10}{n}$

56. $\dfrac{n}{10} = \dfrac{12}{7}$

57. $\dfrac{20}{n} = \dfrac{16}{8}$

58. $\dfrac{18}{21} = \dfrac{n}{12}$

59. $\dfrac{4}{3\frac{1}{3}} = \dfrac{3}{n}$

60. $\dfrac{n}{2\frac{1}{2}} = \dfrac{8}{5}$

61. $\dfrac{12}{n} = \dfrac{6\frac{1}{8}}{4}$

62. $\dfrac{10}{2\frac{2}{3}} = \dfrac{n}{2}$

63. $\dfrac{7\frac{1}{2}}{5} = \dfrac{16}{n}$

OBJECTIVE: To use formulas to find the perimeters of polygons, given the lengths of their sides. To solve problems about perimeter.

The **perimeter** of a polygon is the distance around the polygon.

To find the perimeter of this tennis court, add the lengths of its sides.

$$11 + 23.8 + 11 + 23.8 = 69.6$$

The perimeter of the tennis court is 69.6 m.

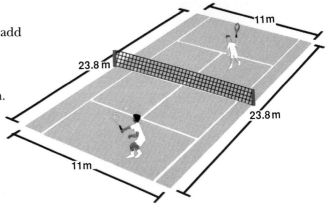

| EXAMPLES | Here's how to use formulas to find the perimeters of some polygons. |

A Triangle

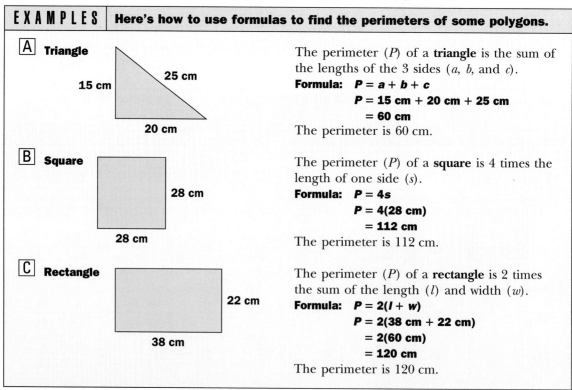

The perimeter (P) of a **triangle** is the sum of the lengths of the 3 sides (a, b, and c).
Formula: $P = a + b + c$
$P = 15\ cm + 20\ cm + 25\ cm$
$= 60\ cm$
The perimeter is 60 cm.

B Square

The perimeter (P) of a **square** is 4 times the length of one side (s).
Formula: $P = 4s$
$P = 4(28\ cm)$
$= 112\ cm$
The perimeter is 112 cm.

C Rectangle

The perimeter (P) of a **rectangle** is 2 times the sum of the length (l) and width (w).
Formula: $P = 2(l + w)$
$P = 2(38\ cm + 22\ cm)$
$= 2(60\ cm)$
$= 120\ cm$
The perimeter is 120 cm.

CHECK for Understanding

1. Look at the examples above. What is the formula for the perimeter of a square? What does the letter s represent?

2. What is the formula for the perimeter of a rectangle? What do the letters l and w represent?

EXERCISES

▶ **Find each perimeter.**

3.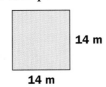
14 m

14 m

4.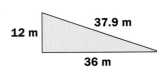
12 m   37.9 m

36 m

5.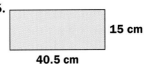
15 cm

40.5 cm

6.
19 cm

38.6 cm

7.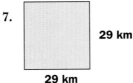
29 km

29 km

8.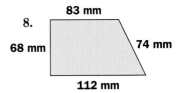
83 mm

68 mm   74 mm

112 mm

▶ **Substitute in the formula $P = a + b + c$, $P = 4s$, or $P = 2(l + w)$ to find the missing measure.**

9. $s = 6$ m
    $P = \underline{\ ?\ }$

10. $s = 7.5$ cm
    $P = \underline{\ ?\ }$

11. $s = \underline{\ ?\ }$
    $P = 48$ km

12. $l = 10$ cm
    $w = 7$ cm
    $P = \underline{\ ?\ }$

13. $l = 2.5$ m
    $w = 1.9$ m
    $P = \underline{\ ?\ }$

14. $l = 9$ km
    $w = \underline{\ ?\ }$
    $P = 26$ km

15. $a = 14$ m
    $b = 25$ m
    $c = 12$ m
    $P = \underline{\ ?\ }$

16. $a = 9.5$ cm
    $b = 10.2$ cm
    $c = 18.7$ cm
    $P = \underline{\ ?\ }$

17. $a = 250$ mm
    $b = 168$ mm
    $c = \underline{\ ?\ }$
    $P = 562$ mm

Problem Solving USING FORMULAS

▶ **Solve.**

18. How many meters of fence are needed to enclose a rectangular yard that is 25 meters by 18 meters?

19. How many centimeters of framing are needed to frame a painting that is 75 centimeters long and 55 centimeters wide?

20. A square photograph, 35 centimeters on each side, is to be framed. How many centimeters of framing are needed?

21. The perimeter of a rectangular pen is 120 meters. If the length is 45 meters, what is the width?

22. The perimeter of a square room is 52 meters. What is the length of each side of the room?

23. **Write a problem** that can be solved using one of the formulas $P = a + b + c$, $P = 4s$, or $P = 2(l + w)$

OBJECTIVE: To use formulas to find the circumference of a circle, given its diameter or radius. To solve problems about circumference.

The **radius** of the wheel is 10 inches. The **diameter** of the wheel is 20 inches. Notice that the diameter is twice the radius.

The distance around a circle is called the **circumference.** The circumference of any circle is a little more than 3 times the length of its diameter.

10 in. radius

20 in. diameter

EXAMPLES | **Here's how to use formulas to find the circumference of a circle.**

A To find the circumference (C), multiply π (read as "pi") by the diameter (d). We will use 3.14 as a decimal approximation for π.

20 in.

Formula: $C = \pi d$
$$C \approx 3.14 \ (20 \text{ in.})$$
$$\approx 62.8 \text{ in.}$$

B Since the diameter is twice the radius, we also can find the circumference by multiplying 2π by the radius (r).

10 in.

Formula: $C = 2\pi r$
$$C \approx 2(3.14)(10 \text{ in.})$$
$$\approx 62.8 \text{ in.}$$

CHECK for Understanding

1. Look at the examples above. A decimal approximation for π is __?__ .

2. One formula for finding circumference is $C = \pi d$. Another formula is $C =$ __?__ .

3. To compute in inches the circumference of a circle having a 36-inch diameter, you would multiply 3.14 by __?__ .

4. To compute in feet the circumference of a circle having a 4-foot radius, you would take 2 times 3.14 times __?__ .

EXERCISES

▶ **Find the circumference. Use 3.14 for π.**
Here are scrambled answers for the next row of exercises: *12.56 ft   21.98 ft   40.82 ft   25.12 ft*

5.

7 ft

6.

8 ft

7.

2 ft

8.

6.5 ft

9.

3 yd

10.

6 in.

11.
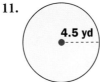
4.5 yd

12.

5 ft

▶ **Substitute in the formula $C = \pi d$ or $C = 2\pi r$ to find the missing measure.**

13. $r = 12$ in.
$\pi \approx 3.14$
$C \approx \underline{\ ?\ }$

14. $d = 7$ ft
$\pi \approx 3.14$
$C \approx \underline{\ ?\ }$

15. $r = 40$ yd
$\pi \approx 3.14$
$C \approx \underline{\ ?\ }$

16. $d = 4$ ft
$\pi \approx 3.14$
$C \approx \underline{\ ?\ }$

17. $r = 100$ in.
$\pi \approx 3.14$
$C \approx \underline{\ ?\ }$

18. $d = 2.5$ ft
$\pi \approx 3.14$
$C \approx \underline{\ ?\ }$

19. $r = \underline{\ ?\ }$
$\pi \approx 3.14$
$C \approx 9.42$ yd

20. $d = \underline{\ ?\ }$
$\pi \approx 3.14$
$C \approx 502.4$ ft

Problem Solving   CALCULATOR APPLICATION

▶ **Solve. Use 3.14 for π.**

21. a. The diameter of a wheel on a motorcycle is 26 inches. How far does the motorcycle travel during one revolution of a wheel?

b. Use your answer to Part a to compute how many revolutions a wheel would make in one mile. Round answer to the nearest whole number. *Hint: 1 mile = 5280 feet.*

c. It is 2794 miles from Los Angeles to New York City. Use your answer from Part b to compute how many revolutions a wheel would make riding from Los Angeles to New York City.

OBJECTIVE: To use formulas to find the areas of squares and other rectangles. To solve problems about areas of rectangles.

The **area** of a region is the number of square units that it takes to cover the region.

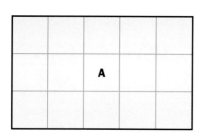

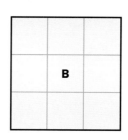

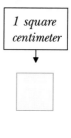

1 square centimeter

1. Count the squares. The area of rectangle A is _?_ square centimeters.

2. What is the area of square B?

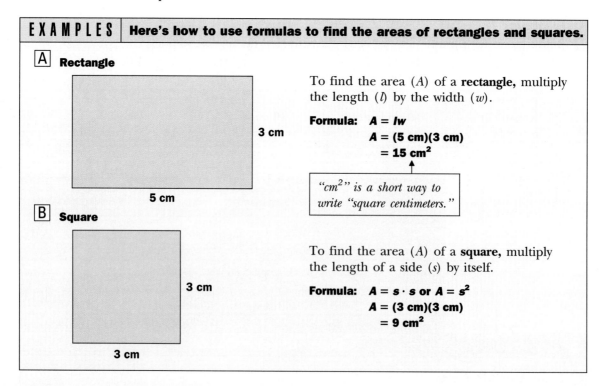

EXAMPLES | Here's how to use formulas to find the areas of rectangles and squares.

A Rectangle

3 cm

5 cm

To find the area (A) of a **rectangle,** multiply the length (l) by the width (w).

Formula: **A = lw**

$$A = (5 \text{ cm})(3 \text{ cm})$$
$$= 15 \text{ cm}^2$$

"cm^2" is a short way to write "square centimeters."

B Square

3 cm

3 cm

To find the area (A) of a **square,** multiply the length of a side (s) by itself.

Formula: **A = s · s or A = s²**

$$A = (3 \text{ cm})(3 \text{ cm})$$
$$= 9 \text{ cm}^2$$

CHECK for Understanding

3. Look at the examples above. What is the formula for the area of a rectangle? What does each letter represent?

4. If the length and width of a rectangle are 15 meters and 9 meters, the area is 135 _?_ meters.

5. If the side of a square is 12 centimeters, its area is _?_ square centimeters.

EXERCISES

▶ **Find the area.**

6.
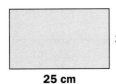
15 cm

25 cm

7.
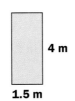
4 m

1.5 m

8.
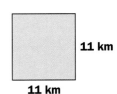
11 km

11 km

9.

9 m

6.5 m

10.

0.9 km

0.9 km

11.
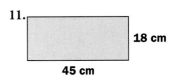
18 cm

45 cm

▶ **Substitute in the formula $A = lw$ or $A = s^2$ to find the missing measure.**

12. $l = 35$ cm
$w = 12$ cm
$A =$ _?_

13. $l = 2.6$ m
$w = 1.5$ m
$A =$ _?_

14. $l = 25$ km
$w = 12$ km
$A =$ _?_

15. $l = 10$ cm
$w = 8.4$ cm
$A =$ _?_

16. $l = 30$ m
$w = 20$ m
$A =$ _?_

17. $l = 2.5$ km
$w = 1.6$ km
$A =$ _?_

18. $s = 30$ m
$A =$ _?_

19. $s = 8.5$ cm
$A =$ _?_

20. $s = 2.2$ km
$A =$ _?_

21. $s = 3.4$ m
$A =$ _?_

22. $s = 0.8$ cm
$A =$ _?_

23. $s = 0.3$ km
$A =$ _?_

24. $l =$ _?_
$w = 14$ cm
$A = 126$ cm^2

25. $l =$ _?_
$w = 6.5$ m
$A = 78$ m^2

26. $l = 15$ km
$w =$ _?_
$A = 93$ km^2

27. $l = 0.5$ cm
$w =$ _?_
$A = 5$ cm^2

Problem Solving VISUAL THINKING

▶ **Tell whether the problem is about perimeter or area. Then solve the problem.**

28. How many 1-foot-square tiles are needed to cover an 18-foot by 12-foot family room?

29. How much molding is needed to go around a 24-foot by 16-foot ceiling?

30. How many yards of fencing are needed to enclose a 30-yard by 25-yard field?

31. How many square feet of sod are needed to cover an 80-foot by 105-foot lawn?

32. How much does it cost to paint a floor that is 15 feet by 25 feet? A $5.90 quart of paint covers about 125 square feet.

33. How much does it cost to frame a 25-foot by 2-foot poster? Framing costs $1.25 a foot.

OBJECTIVE: To use a formula to find the area of a parallelogram. To solve problems about areas of parallelograms.

To find the area of a parallelogram, we can cut the parallelogram into two pieces (**Step 1**) and rearrange the pieces to make a rectangle with the same area (**Step 2**).

Step 1.

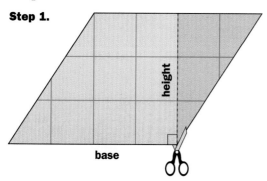

Step 2.

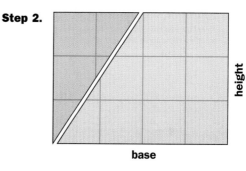

The area of the rectangle is 12 square centimeters. So the area of the parallelogram is also 12 square centimeters.

EXAMPLES | **Here's how to use a formula to find the area of a parallelogram.**

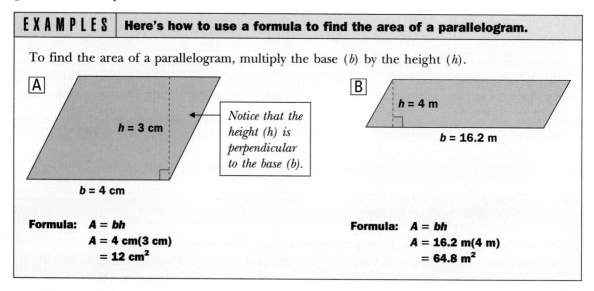

To find the area of a parallelogram, multiply the base (*b*) by the height (*h*).

A

h = 3 cm

Notice that the height (h) is perpendicular to the base (b).

b = 4 cm

Formula: $A = bh$
$A = 4 \text{ cm}(3 \text{ cm})$
$= 12 \text{ cm}^2$

B

h = 4 m

b = 16.2 m

Formula: $A = bh$
$A = 16.2 \text{ m}(4 \text{ m})$
$= 64.8 \text{ m}^2$

CHECK for Understanding

1. Look at the examples above. What is the formula for the area of a parallelogram? What does each letter represent?

2. If the base and height of a parallelogram are 25 centimeters and 20 centimeters, the area is 500 square __?__ .

3. If the base and height of a parallelogram are 8.4 meters and 3 meters, the area is __?__ square meters.

EXERCISES

▶ **Find the area.**

4.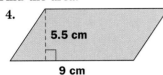
5.5 cm

9 cm

5.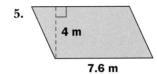
4 m

7.6 m

6.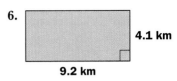
4.1 km

9.2 km

7.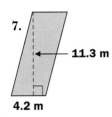
11.3 m

4.2 m

8.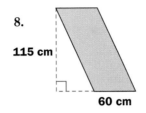
115 cm

60 cm

9.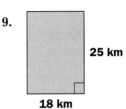
25 km

18 km

10.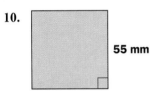
55 mm

55 mm

11.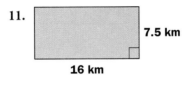
7.5 km

16 km

12.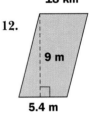
9 m

5.4 m

▶ **Substitute in the formula $A = bh$ to find the missing measure.**

13. $b = 25$ cm
 $h = 5$ cm
 $A = \underline{\ ?\ }$

14. $b = 30$ cm
 $h = 8.5$ cm
 $A = \underline{\ ?\ }$

15. $b = 10$ m
 $h = 5.4$ m
 $A = \underline{\ ?\ }$

16. $b = 2.1$ km
 $h = 5$ km
 $A = \underline{\ ?\ }$

17. $b = 100$ m
 $h = 100$ m
 $A = \underline{\ ?\ }$

18. $b = 65.2$ cm
 $h = 5.4$ cm
 $A = \underline{\ ?\ }$

19. $b = \underline{\ ?\ }$
 $h = 500$ m
 $A = 1500$ m^2

20. $b = 85$ cm
 $h = \underline{\ ?\ }$
 $A = 1020$ cm^2

21. $b = \underline{\ ?\ }$
 $h = 5.4$ km
 $A = 162$ km^2

Challenge! GUESS AND CHECK

▶ **Find the length of the side of the square.** *Hint: Estimate. Compare the area using a calculator. Repeat until you get the answer.*

22.

Area = 529 m^2

23.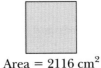

Area = 2116 cm^2

24.

Area = 7.29 km^2

OBJECTIVE: To use a formula to find the area of a triangle.

To find the area of a triangle, think about cutting a parallelogram into two triangles that have the same area.

Step 1.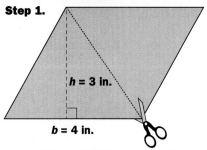
$h = 3$ in.
$b = 4$ in.

Step 2.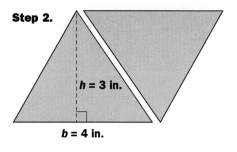
$h = 3$ in.
$b = 4$ in.

The area of the parallelogram is 12 square inches. The area of each triangle is half of 12 square inches or 6 square inches.

EXAMPLES | **Here's how to use a formula to find the area of a triangle.**

To find the area (A) of a triangle, multiply $\frac{1}{2}$ by the base (b) by the height (h).

A

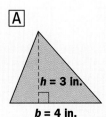

$h = 3$ in.
$b = 4$ in.

Formula: $A = \dfrac{1}{2}bh$

$A = \dfrac{1}{2}(4 \text{ in.})(3 \text{ in.})$

$= 6 \text{ in.}^2$

B

$h = 4\frac{1}{2}$ ft
$b = 10$ ft

Formula: $A = \dfrac{1}{2}bh$

$A = \dfrac{1}{2}(10 \text{ ft})(4\tfrac{1}{2} \text{ ft})$

$= 22\frac{1}{2} \text{ ft}^2$

CHECK for Understanding

1. Look at the examples above. For any triangle, the area is equal to one half the base times the __?__ .

2. If the base and height of a triangle are 5 yards and 7 yards, the area is $17\frac{1}{2}$ square __?__ .

3. If the base and height of a triangle are 12 inches and 9 inches, the area is __?__ square inches.

EXERCISES

▶ **Find the area.**

4.

9 in.

20 in.

5.

10 in.

12 in.

6.

15 in.

16 in.

7.

3 ft

10 ft

8.

12 yd

15 yd

9.

$8\frac{1}{2}$ ft

6 ft

10.

42 in.

15 in.

11.

6 ft

$18\frac{1}{2}$ ft

12.

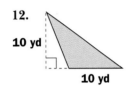

10 yd

10 yd

▶ **Substitute in the formula $A = \frac{1}{2}bh$ to find the missing measure.**

13. $b = 40$ yd

$h = 9$ yd

$A = \underline{?}$

14. $b = 28$ ft

$h = 20$ ft

$A = \underline{?}$

15. $b = 16$ yd

$h = 4\frac{1}{2}$ yd

$A = \underline{?}$

16. $b = 8$ in.

$h = \underline{?}$

$A = 20$ in.^{2}

17. $b = \underline{?}$

$h = 20$ yd

$A = 100$ yd^2

Challenge! MAKE A TABLE

18. Copy and complete this table.

19. Look for a pattern. What is the sum of the measures of the angles of a heptagon (7-sided polygon)? Of an octagon (8-sided polygon)?

Polygon	Triangle	Rectangle	Pentagon	Hexagon
Number of sides	3	4	5	6
Number of diagonals from a vertex	0	1		
Number of triangles formed	1	2		
Sum of the measures of the angles	$1(180°) = 180°$	$2(180°) = 360°$		

Using Algebra in Geometry | **295**

OBJECTIVE: To use a formula to find the area of a circle. To solve problems about areas of circles.

Here's a method you can use to estimate the area of a circle.

1. Look at the drawing at the right.
   a. The area of the red square is 18 square centimeters. What is the area of the green square?
   b. The circle is larger than the red square and smaller than the green square. You can estimate the area of the circle by finding the average of the areas of the squares. What is the average?

2. Estimate the area of the circle by counting squares. Is the area of the circle about 27 square centimeters?

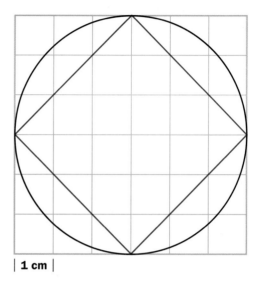

| 1 cm |

EXAMPLE | **Here's how to use a formula to find the area of a circle.**

To find the area (A) of a circle, multiply π (about 3.14) by the square of the radius (r).

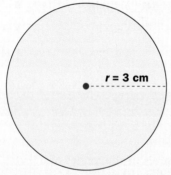

$r = 3$ cm

Formula: $A = \pi r^2$

$A \approx 3.14(3 \text{ cm})^2$

$\approx 3.14(9 \text{ cm}^2)$

$\approx 28.26 \text{ cm}^2$

CHECK for Understanding

3. Look at the example above. For any circle, the area equals π times the square of the __?__ .

4. In the formula $A = \pi r^2$, r^2 means __?__ times __?__ .

5. If the radius of a circle is 5 centimeters, its area is 3.14 $(5 \text{ cm})^2$, or 78.5 square __?__ .

6. If the radius of a circle is 10 meters, its area is 3.14 $(10 \text{ m})^2$, or __?__ square meters.

EXERCISES

▶ **Find the area. Use 3.14 for π.**
Here are scrambled answers for the next row of exercises: *50.24 cm²   12.56 cm²   0.785 m²   153.86 cm²*

7.
2 cm

8.
4 cm

9.
0.5 m

10.
7 cm

11.
12 cm

12.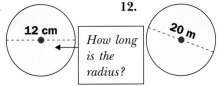
20 m *How long is the radius?*

13.
40 cm

14.
9 m

15.
1.2 m

16.
8 km

17.
0.8 m

18.
3.5 cm

▶ **Substitute in the formula $A = \pi r^2$ to find the missing area.**

19. $r = 5$ cm
$\pi \approx 3.14$
$A \approx \underline{\ ?\ }$

20. $r = 1.2$ m
$\pi \approx 3.14$
$A \approx \underline{\ ?\ }$

21. $r = 30$ km
$\pi \approx 3.14$
$A \approx \underline{\ ?\ }$

22. $r = 0.4$ m
$\pi \approx 3.14$
$A \approx \underline{\ ?\ }$

Problem Solving USING A FORMULA

▶ **Solve.**

23. A round tabletop has a radius of 80 centimeters. What is its area?

24. A round mirror has a radius of 35 centimeters. What is its area?

25. A circular platform has a diameter of 2 meters. What is its area?

26. A circular swimming pool has a diameter of 8.8 meters. What is its area?

Challenge! VISUAL THINKING

▶ **Add or subtract to find the area of each shaded part.**

27.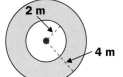
2 m
4 m

28.
6 m

★**29.**
4 m

OBJECTIVE: To solve problems by drawing diagrams. To use equations to solve problems.

In this lesson you will solve perimeter problems by writing equations. In solving each problem, it will be helpful to draw a diagram that shows the facts in the problem.

| E X A M P L E | **Here's how to use a diagram and an equation to solve a perimeter problem.** |

Problem: A rectangle is 12 centimeters longer than it is wide. What is the width of the rectangle if its perimeter is 96 centimeters?

Step 1. Choose a variable. Use it and a diagram to show the facts in the problem.

Let w = the width of the rectangle.
Let $w + 12$ = the length of the rectangle.

$$w + 12$$

w ⬚ w

$$w + 12$$

Step 2. Write an equation based on the facts.

$$w + w + 12 + w + w + 12 = 96$$

These are two equal expressions for the perimeter of the rectangle.

Step 3. Solve the equation.

$$w + w + 12 + w + w + 12 = 96$$
$$4w + 24 = 96$$
$$4w + 24 - 24 = 96 - 24$$
$$4w = 72$$
$$\frac{4w}{4} = \frac{72}{4}$$
$$w = 18$$

The width of the rectangle is 18 centimeters.

CHECK for Understanding

1. Look at the example above.
   a. In Step 1, we let w equal the width of the rectangle. Then $w +$ _?_ equals the length of the rectangle.
   b. In Step 2, the two equal expressions for the perimeter are _?_ and 96.
   c. An equation for the problem is $w + w + 12 + w + w + 12 =$ _?_

2. To check the solution, ask yourself if the answer fits the facts in the problem. If the width is 18 centimeters, what is the length? Is the sum of the two widths and two lengths equal to 96 centimeters?

EXERCISES

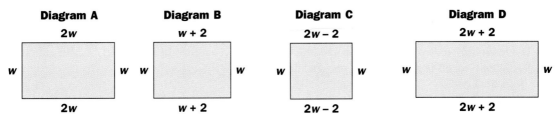

2w   w + 2   2w – 2   2w + 2

w   w   w   w   w   w   w   w

2w   w + 2   2w – 2   2w + 2

▶ **Decide which diagram above would be used to picture the facts in each problem. Then use the diagram to write an equation and solve the problem.**

3. The length of a rectangle is 2 centimeters more than its width (w). What is its width if its perimeter is 60 centimeters?

4. The length of a rectangle is 2 centimeters less than twice its width (w). What is its width if its perimeter is 50 centimeters?

5. The length of a rectangle is twice its width (w). What is its width if its perimeter is 120 centimeters?

6. The length of a rectangle is 2 centimeters more than twice its width (w). What is its width if its perimeter is 70 centimeters?

▶ **Draw a diagram that shows the facts. Then write an equation and solve the problem.**

7. The length of a rectangular tabletop is 30 centimeters more than its width (w). What is the width of the tabletop if its perimeter is 380 centimeters?

8. The length of a rectangular poster is 15 centimeters less than its width (w). What is the width of the poster if its perimeter is 230 centimeters?

9. The length of a rectangular garden is 3 times its width (w). What is the width of the garden if its perimeter is 240 meters?

10. The length of a rectangular field is 6 times its width (w). What is the width of the field if its perimeter is 280 meters?

11. The length of a rectangular mirror is 3 centimeters more than twice its width (w). What is the width of the mirror if its perimeter is 426 centimeters?

12. The length of a rectangular picture is 4 centimeters less than twice its width (w). What is the width of the picture if its perimeter is 436 centimeters?

★**13.** One side of a triangle is 3 centimeters longer than the shortest side (s), and the other side is 4 centimeters longer than the shortest side. How long is the shortest side if the perimeter is 67 centimeters?

★**14.** One side of a triangle is 5 centimeters longer than the shortest side (s), and the other side is 3 centimeters longer than the shortest side. How long is the shortest side if the perimeter is 59 centimeters?

OBJECTIVE: To solve problems using a calculator when appropriate.

▶ **Solve. Decide when a calculator would be helpful.**

1. How many feet of fencing are needed to enclose an 80-foot by 50-foot rectangular garden?

2. How many feet of fencing are needed to enclose a square garden that is 55 feet on a side?

3. How large a square can you fence in with 120 feet of fencing?

4. You use 240 feet of fencing to fence in a rectangular yard. If the length of the yard is 80 feet, how wide is the yard?

WEEKEND SPECIALS

Wire fence	$1.50 per foot
Steel posts	$5.50 each
Gates	$29.50 each

▶ **Use the weekend special prices. Find the total cost for each fencing project.**

5.

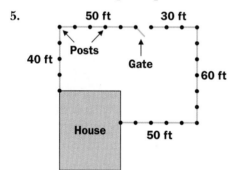

Cost of fence: _?_
Cost of posts: _?_
Cost of gate: _?_
Total cost: _?_

6.

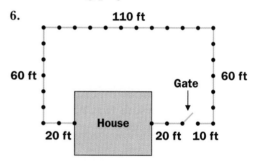

Cost of fence: _?_
Cost of posts: _?_
Cost of gate: _?_
Total cost: _?_

7.

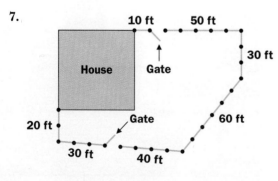

Total cost: _?_

8.

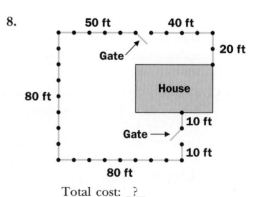

Total cost: _?_

▶ **Change to a fraction, whole number, or mixed number. Give each answer in simplest form.** *(page 244)*

1. 36% **2.** 85% **3.** 144% **4.** 5% **5.** 60% **6.** 1%

7. 300% **8.** 125% **9.** 250% **10.** $33\frac{1}{3}\%$ **11.** $37\frac{1}{2}\%$ **12.** $62\frac{1}{2}\%$

13. $87\frac{1}{2}\%$ **14.** $66\frac{2}{3}\%$ **15.** $162\frac{1}{2}\%$ **16.** $116\frac{2}{3}\%$ **17.** $233\frac{1}{3}\%$ **18.** $206\frac{1}{4}\%$

▶ **Change to a percent.** *(page 246)*

19. $\frac{1}{4}$ **20.** $\frac{3}{2}$ **21.** $\frac{1}{2}$ **22.** $\frac{3}{10}$ **23.** $\frac{3}{4}$ **24.** $\frac{9}{10}$

25. 1 **26.** 3 **27.** $\frac{3}{8}$ **28.** $\frac{4}{5}$ **29.** $\frac{5}{8}$ **30.** $\frac{5}{4}$

31. $\frac{13}{3}$ **32.** $\frac{15}{8}$ **33.** $\frac{11}{6}$ **34.** $\frac{4}{9}$ **35.** $\frac{7}{6}$ **36.** $\frac{8}{3}$

▶ **Solve.** *(page 252)*

37. 50% of $42 = n$ **38.** 30% of $20 = n$ **39.** 25% of $44 = n$

40. 31% of $80 = n$ **41.** 52% of $127 = n$ **42.** 79% of $174 = n$

43. 5.8% of $20 = n$ **44.** 9.2% of $235 = n$ **45.** 3.4% of $118 = n$

46. 80% of $45 = n$ **47.** 120% of $90 = n$ **48.** 200% of $64 = n$

▶ **Change each percent to a fraction or decimal and solve.** *(page 256)*

49. 50% of $n = 16$ **50.** 25% of $n = 8$ **51.** 75% of $n = 42$

52. 60% of $n = 39$ **53.** 150% of $n = 48$ **54.** 35% of $n = 56$

55. 12.5% of $n = 12$ **56.** 18% of $n = 27$ **57.** 37.5% of $n = 51$

58. 12.5% of $n = 18.2$ **59.** 275% of $n = 88$ **60.** 150% of $n = 24.6$

▶ **Find the percent of increase or decrease. Use *i* to indicate increase and *d* to indicate decrease. Give answers to the nearest tenth of a percent.** *(page 258)*

61. from 10 to 20 **62.** from 20 to 10 **63.** from 75 to 100

64. from 100 to 75 **65.** from 40 to 50 **66.** from 50 to 40

67. from 18 to 20 **68.** from 46 to 58 **69.** from 72 to 60

70. from 120 to 96 **71.** from 84 to 136 **72.** from 175 to 140

Here are scrambled answers for the review exercises:

0.001	area	congruent	height	obtuse	rectangle
5	base	construct	isosceles	perimeter	width
112	centimeters	decimal	kilometer	perpendicular	
113.04	circumference	diameter	meter	quadrilateral	
1000	complementary	divide	millimeters	radius	

1. An __?__ angle measures between 90° and 180°. Two angles whose measures total 90° are __?__ angles. *(page 268)*

2. An __?__ triangle has at least two congruent sides. *(page 270)*

3. Two lines that form four right angles are __?__ lines. The corresponding angles formed by parallel lines and a transversal are __?__. *(page 272)*

4. A compass and straightedge are used to __?__ figures. *(page 274)*

5. A parallelogram is also a __?__. A pentagon has __?__ sides. *(page 276)*

6. In the metric system, the __?__ is the basic unit of length. *(page 278)*
   1 kilometer = __?__ meters   1 millimeter = __?__ meter

7. To change units in the metric system, multiply or __?__ by 10, 100, or 1000. *(page 280)*
   100 __?__ = 1 meter   1000 __?__ = 1 meter   1000 meters = 1 __?__

8. The __?__ of a polygon is the distance around the polygon. Use the formula $P = 2(l + w)$ to find the perimeter of a __?__. *(page 286)*

9. In this drawing, the __?__ of the circle is 8 inches and the __?__ of the circle is 4 inches. The distance around a circle is called the __?__. To find the circumference, use the formula $C = \pi d$. A __?__ approximation for π is 3.14. *(page 288)*

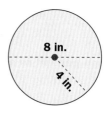

8 in.

4 in.

10. The __?__ of a region is the number of square units it takes to cover the region. To find the area of a rectangle, multiply the length by the __?__. To find the area of a parallelogram, multiply the base by the __?__. *(pages 290, 292)*

11. To find the area of a triangle, use the formula $A = \frac{1}{2}bh$, where b stands for the __?__ and h stands for the height. The area of this triangle is __?__ ft². *(page 294)*

14 ft

16 ft

12. To find the area of a circle, use the formula $A = \pi r^2$, where r stands for the radius. The area of this circle is __?__ in.² (Use 3.14 for π.) *(page 296)*

r = 6 in.

▶ Give the measure of each angle. Then tell whether it is acute, obtuse, or right. *(page 268)*

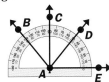

1. ∠BAE

2. ∠CAE

3. ∠DAE

▶ Write and solve an equation to find the measure of each angle. *(page 270)*

4.

5.
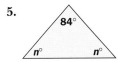

6.

▶ Write and solve an equation to find the measure of the angle. *(page 272)*

7.
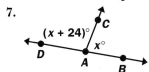

8.
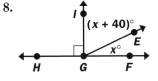

9.
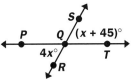

Find the measure of ∠CAD.   Find the measure of ∠EGI.   Find the measure of ∠PQR.

▶ Complete. *(page 280)*

10. 15.4 cm = _?_ m

11. 4.7 km = _?_ m

12. 8 cm + 3 mm = _?_ mm

▶ Find each perimeter or circumference. Use 3.14 for π. *(pages 286, 288)*

13.
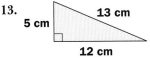

14.
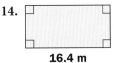

15.

6 cm

16.

0.5 m

▶ Find each area. *(pages 290, 292, 294, 296)*

17.
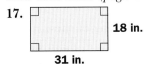

18.
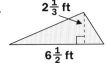

19.

20.

10 in.

▶ Write an equation and solve the problem. *(page 298)*

21. A rectangle has a length that is 3 feet more than its width. Its perimeter is 90 feet. What is its width?

22. A rectangle has a length that is 1 foot more than 3 times its width. Its perimeter is 90 feet. What is its width?

▶ **Choose the correct letter.**

---

1. Simplify by combining like terms.

$$5y - 8 - 3y + 1$$

A. $8y - 7$

B. $2y - 9$

C. $2y - 7$

D. none of these

---

2. Solve.

$$3w + 9 + 5w = {}^-11$$

A. $\dfrac{{}^-1}{4}$ B. $^-2\dfrac{1}{2}$

C. $2\dfrac{1}{2}$ D. none of these

---

3. Solve.

$$12 - 4n = 2n - 6$$

A. $^-9$ B. 3

C. $^-1$ D. none of these

---

4. Complete.

$$\dfrac{7a}{3b} = \dfrac{?}{12ab^2}$$

A. $28a^2b$ B. $28a^2$

C. $4ab$ D. none of these

---

5. Solve.

$$\dfrac{n}{5} = \dfrac{7}{1\frac{1}{2}}$$

A. $52\dfrac{1}{2}$ B. $1\dfrac{1}{14}$

C. $23\dfrac{1}{3}$ D. none of these

---

6. Change to a fraction.

$$62\dfrac{1}{2}\% = \underline{\ ?\ }$$

A. $\dfrac{3}{8}$ B. $\dfrac{1}{2}$

C. $\dfrac{7}{8}$ D. none of these

---

7. Change to a percent.

$$\dfrac{11}{6} = \underline{\ ?\ }$$

A. $87\dfrac{1}{2}\%$ B. $83\dfrac{1}{3}\%$

C. $62\dfrac{1}{2}\%$ D. none of these

---

8. Solve.

$$7.5\% \text{ of } 72 = n$$

A. 1.8

B. 3.6

C. 4.8

D. none of these

---

9. Solve.

$$125\% \text{ of } n = 38.5$$

A. 30.8

B. 38.5

C. 48.125

D. none of these

---

10. The percent of increase from 36 to 54 is

A. 18%

B. 33.3%

C. 50%

D. none of these

---

11. The percent of decrease from 50 to 36 is

A. 14%

B. 38.9%

C. 28%

D. none of these

---

12. Choose the equation.

Six more than the sum of 5 times a number and 3 times a number is 62. What is the number?

A. $5n + 3n + 6 = 62$

B. $5n - 3n + 6 = 62$

C. $5n - 3n - 6 = 62$

D. $5n + 3n - 6 = 62$

$$V = \frac{1}{3}Bh$$

The volume of a pyramid equals one third times the area of the base times the height of the pyramid.
(page 320)

American architect I. M. Pei designed this pyramid to serve as the entrance to the Louvre Museum in Paris, France. Its volume can be found by using the above formula.

305

OBJECTIVE: To classify space figures by their faces, vertices, and edges.

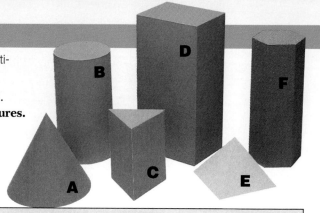

The models at the right are three-dimensional. Three-dimensional shapes are called **space figures.**

1. Which model has 9 edges, 6 vertices (corners), and 5 faces (sides)?

2. Which model has 4 triangular faces and 1 square face?

| E X A M P L E S | **Here's how space figures are named.** |

In the drawings below, the shaded faces are the **bases.** Prisms and pyramids are named according to the shapes of their bases. Notice that a cube is a rectangular prism whose faces are squares, all the same size.

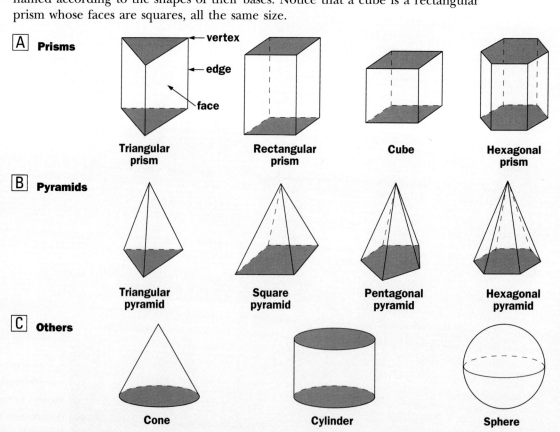

A Prisms

Triangular prism   Rectangular prism   Cube   Hexagonal prism

B Pyramids

Triangular pyramid   Square pyramid   Pentagonal pyramid   Hexagonal pyramid

C Others

Cone   Cylinder   Sphere

CHECK for Understanding

3. Use the models at the top of the page and the drawings in the examples to answer these questions.
   a. Which model is a square pyramid?
   b. Which model is a cylinder?
   c. Which model is a cone?
   d. Which 3 models are prisms?

EXERCISES

▶ Name the space figure in each exercise. A top view and a side view are given for each figure.

4.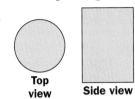

Top view   Side view

5.

Top view   Side view

6.

Top view

Side view

7.

Top view   Side view

8.

Top view   Side view

9.

Top view

Side view

Problem Solving VISUAL THINKING

▶ Use the clues and the drawings on page 306. Name each space figure.

10. *Clues:*
This space figure is a pyramid.
It has 6 vertices.

11. *Clues:*
This space figure is a prism.
All its faces are squares.

12. *Clues:*
This space figure has 8 edges.
It has 5 faces.

13. *Clues:*
This space figure has 2 bases.
It has 12 vertices.

Group Project AN INVESTIGATION *Faces, Edges, and Vertices*

▶ Work in a small group.

A **B**

C **D**

	Faces	Number of Edges	Vertices
Figure A	5	9	6
Figure B			
Figure C			
Figure D			

14. a. Copy this table.
    b. Look at the figures at the left and complete the table.
    c. Check to see if all the completed tables in your group are the same.

15. Look for a pattern in your completed table. Let F = number of faces, E = number of edges, and V = number of vertices. Copy and complete this formula: $F =$ _____ .

OBJECTIVE: To visualize how triangular, square, and rectangular faces are used to build space figures.

The pattern below was made from some of the pieces that are shown at the right. When the pattern is folded, it forms a triangular prism.

Square
30¢

Triangle A
14¢

Triangle B
29¢

Rectangle
60¢

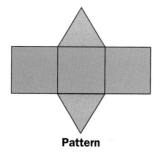

Pattern

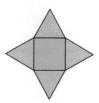

Fold

Prism

1. Look at the pattern.
   a. How many pieces were used to make the pattern?
   b. Which shape was used for the bases (top and bottom)? What is the total cost of the bases?
   c. Which shape was used for the other faces? What is their total cost?

2. What is the total cost of the triangular prism?

EXERCISES

▶ **Find the total cost of each pattern.**

3.

4.

5.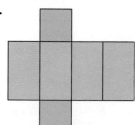

6.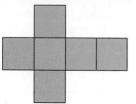

7.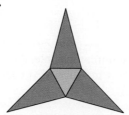

8.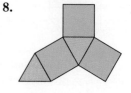

Find the total cost of each space figure.

9.

10.

11.

12.

13.

14.

15.

16.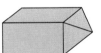

17.

18.

19.

20.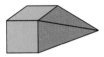

21.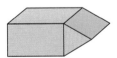

22.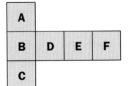

23.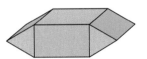

Challenge! VISUAL THINKING

Imagine each pattern folded to make a cube.

24. Which face is opposite

   a. face A?

   b. face B?

   c. face D?

25. Which face is opposite

   a. face A?

   b. face B?

   c. face F?

26. Which face is opposite

   a. face A?

   b. face B?

   c. face C?

OBJECTIVE: To find the surface areas of cubes and other rectangular prisms.

Alex and Jason used 1-inch-square pictures to cover the 6 faces of their photo boxes.

1. Which boy used 52 pictures to cover all 6 faces of his photo box?

The sum of the areas of the 6 faces of a rectangular prism or cube is called the **surface area.**

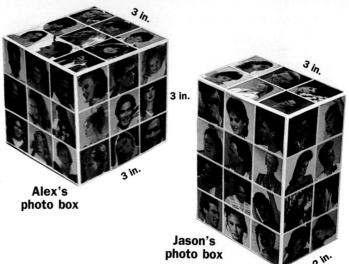

3 in.

3 in.

3 in.

Alex's
photo box

3 in.

3 in.

4 in.

2 in.

Jason's
photo box

E X A M P L E | **Here's how to find the surface area of a rectangular prism.**

Think about unfolding Jason's photo box. To find the surface area, compute the area of each face by multiplying its length by its width. Then add all six areas.

Top
(2" × 3")

Left face
(2" × 4")

Front
(3" × 4")

Right face
(2" × 4")

Back
(3" × 4")

Bottom
(2" × 3")

Area of front .... 12 in.^{2}
back .... 12 in.^{2}
top .... 6 in.^{2}
bottom .... 6 in.^{2}
left face .... 8 in.^{2}
right face .... 8 in.^{2}
Surface area = 52 in.^{2}

CHECK for Understanding

2. Look at the example above. The area of the front is the same as the area of the __?__ . The area of the top is the same as the area of the __?__ . The area of the left face is the same as the area of the __?__ face.

3. To find the surface area of a rectangular prism, you could first find the total area of its front, top, and left faces and then multiply by __?__ .

4. Look at Alex's photo box. If the area of each face is 9 square inches, then the surface area of the cube is __?__ square inches.

5. Explain how to find the surface area of a cube if you know the area of one face.

EXERCISES

▶ **Find the surface area of each rectangular prism.**
Here are scrambled answers for the next row of exercises: 162 in.²   216 in.²   158 in.²

6.
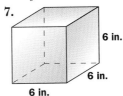
3 in.
5 in.
8 in.

7.
6 in.
6 in.
6 in.

8.
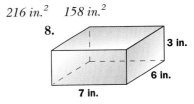
3 in.
6 in.
7 in.

9.
2 ft
5 ft
2 ft

10.
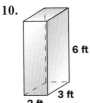
6 ft
3 ft
2 ft

11.
3 ft
3 ft
3 ft

12.
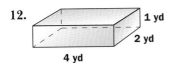
1 yd
2 yd
4 yd

13.
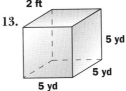
5 yd
5 yd
5 yd

14.
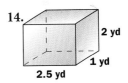
2 yd
1 yd
2.5 yd

Problem Solving   USING SURFACE AREA

▶ **Solve.**

15. How many 1-inch-square pictures are needed to cover a photo box 4 inches long, 3 inches wide, and 5 inches high?

16. How many 1-inch-square pictures are needed to cover a photo cube that is 4 inches on an edge?

Challenge!   GUESS AND CHECK

▶ **Use the area clues to find the missing length (*l*), width (*w*), and height (*h*).**

17.
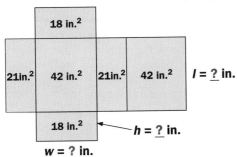
18 in.²
21 in.²   42 in.²   21 in.²   42 in.²   *l* = ? in.
18 in.² ← *h* = ? in.
w = ? in.

18.
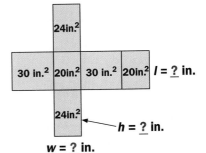
24 in.²
30 in.²   20 in.²   30 in.²   20 in.²   *l* = ? in.
24 in.² ← *h* = ? in.
w = ? in.

You will use a variety of problem-solving strategies on these two pages. For some problems, you may wish to use a calculator.

1 WORK BACKWARD

The plaza guide said to her guests, "It is now 4:50 P.M. and I have heard the clock chime 39 times since I entered the plaza today." If the clock strikes the hours and half hours, at what time did the guide start counting chimes?

✓ Problem Solving Tips

Work Backward. Put the data into a table and then work backward.

Time	4:30	4:00	3:30	3:00
Number of chimes	1	4	1	?
Total number of chimes	39	35	34	?

2 GEOMETRY—MULTIPLE SOLUTIONS

All the circles around the edge of this diagram are red. All the inner circles are white. There are more red circles than white circles. Can you draw a diagram in which the red circles around the edge and the white inner circles are equal in number? *Hint: Two figures are possible.*

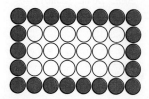

3 INTEGER PUZZLE

How many equations can you complete? Use each puzzle number only once.

PUZZLE NUMBERS

a. $10 \div \boxed{?} + \boxed{?} = {}^-1$ **b.** $5 \times (\boxed{?} + \boxed{?}) = 15$

c. $\boxed{?} \times \boxed{?} + 6 = {}^-2$ **d.** $(\boxed{?} - 1) \times \boxed{?} = 4$

e. $(\boxed{?} - \boxed{?}) \div 7 = 1$

4 ALPHAMETICS

In this sum, each different letter stands for a different digit. What digit can each letter stand for? *Hint: Two solutions are possible.*

$$
\begin{array}{r}
I\ C\ E \\
+\ H\ E\ A\ T \\
\hline
W\ A\ T\ E\ R
\end{array}
$$

5 A PATTERNED WORKOUT

A long-distance cyclist rode 100 miles in 5 days, each day riding 4 miles more than the day before.

a. How many miles were ridden each day?
b. If the cyclist continued the same pattern for 10 days, how many miles would be ridden altogether?

6 ARMADILLO ARITHMETIC

How can you put 21 armadillos into 4 pens and have an odd number in each pen?
Hint: Make a drawing.

7 TAKE NOTE

Five boys filled 5 math notebooks in 5 weeks during the last grading period, and 3 girls filled 3 math notebooks in 3 weeks.

At these rates, how many notebooks will 15 boys and 15 girls fill in 15 weeks?

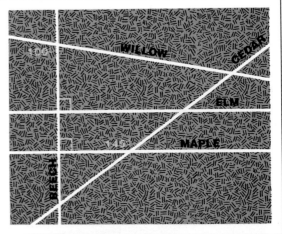

GEOMETRY—APPLICATIONS

Complete each sentence.
You may wish to use a protractor.

8 The acute angles formed by the intersection of Elm Street and Cedar Street measure _?_ degrees.

9 A 55° angle is formed by the intersection of _?_ Street and _?_ Street.

10 The smallest obtuse angle shown on the map measures _?_ degrees and is formed by the intersection of _?_ Street and _?_ Street.

Rate Your Problem-Solving Power: *6–7 correct = Good; 8–9 correct = Excellent; 10 correct = Exceptional*

OBJECTIVE: To recognize reflections, rotations, and translations of geometric figures.

Here are three ways to move geometric figures without changing their size and shape.

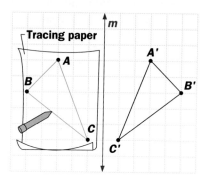

Reflection (flip)

If you trace △ABC and flip the tracing over, it will fit exactly on △A'B'C'. So △ABC ≅ △A'B'C'.

You can also think about placing a mirror along line *m*. The mirror image of △ABC is △A'B'C'. We say that △A'B'C' is a **reflection** of △ABC.

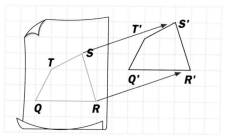

Translation (slide)

If you trace figure QRST, you can slide the tracing to fit exactly on figure Q'R'S'T'. Notice that corresponding sides of the two figures are parallel. We say that Q'R'S'T' is a **translation** of QRST.

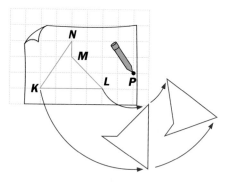

Rotation (turn)

If you trace figure KLMN, place your pencil point at P, and turn the tracing, you can rotate the tracing to fit exactly on the two red figures. Each of the red figures is a **rotation** of KLMN.

▶ **Use the definitions to answer these questions.**

1. Compare figures B, C, and D with A. Which figure is a
   a. reflection?   **b.** translation?
   c. rotation?

2. Copy these figures on grid paper and draw a reflection, translation, and rotation for each figure.

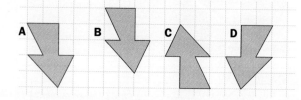

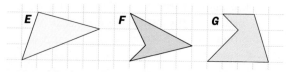

▶ **Give each sum or difference in simplest form.** *(page 168)*

1. $\dfrac{1}{8} + \dfrac{7}{8}$ 2. $\dfrac{1}{2} + \dfrac{5}{6}$ 3. $\dfrac{5}{8} + \dfrac{1}{6}$ 4. $\dfrac{7}{8} - \dfrac{3}{8}$ 5. $\dfrac{5}{6} - \dfrac{3}{10}$

6. $\dfrac{a}{2} + \dfrac{b}{6}$ 7. $\dfrac{r}{2} + \dfrac{s}{3}$ 8. $\dfrac{x}{6} + \dfrac{2y}{3}$ 9. $\dfrac{3m}{4} + \dfrac{n}{3}$ 10. $\dfrac{2r}{3} + \dfrac{6s}{5}$

11. $\dfrac{p}{4} - \dfrac{q}{6}$ 12. $\dfrac{t}{3} - \dfrac{u}{12}$ 13. $\dfrac{5a}{4} - \dfrac{b}{8}$ 14. $\dfrac{3x}{5} - \dfrac{2y}{7}$ 15. $\dfrac{5m}{6} - \dfrac{5n}{6}$

▶ **Add or subtract.** *(page 170)*

16. $\dfrac{1}{a} + \dfrac{2}{b}$ 17. $\dfrac{b}{a} - \dfrac{3}{ab}$ 18. $\dfrac{b}{3a} + \dfrac{2}{a}$ 19. $\dfrac{a}{2} - \dfrac{a}{b}$ 20. $\dfrac{3}{b} + \dfrac{a}{b^2}$

21. $\dfrac{y}{x} - \dfrac{2}{y}$ 22. $\dfrac{1}{xy} + \dfrac{y}{x}$ 23. $\dfrac{2}{x} - \dfrac{y}{4x}$ 24. $\dfrac{3}{x} + \dfrac{x}{2}$ 25. $\dfrac{5}{y^2} - \dfrac{7}{y}$

▶ **Give each product in simplest form.** *(page 184)*

26. $\dfrac{4}{5} \cdot \dfrac{1}{4}$ 27. $3 \cdot \dfrac{5}{6}$ 28. $\dfrac{7}{12} \cdot \dfrac{4}{5}$ 29. $\dfrac{5}{6} \cdot \dfrac{6}{5}$ 30. $\dfrac{10}{3} \cdot \dfrac{18}{5}$

31. $\dfrac{x}{5} \cdot \dfrac{1}{x}$ 32. $10 \cdot \dfrac{y}{2}$ 33. $\dfrac{5}{6} \cdot \dfrac{3y}{10}$ 34. $\dfrac{2}{3d} \cdot \dfrac{3}{9f}$ 35. $\dfrac{a}{b} \cdot \dfrac{3b}{2a}$

36. $\dfrac{a}{b^2} \cdot \dfrac{3b}{a}$ 37. $\dfrac{r}{s} \cdot \dfrac{s}{r}$ 38. $\dfrac{n^2}{5} \cdot \dfrac{m}{n}$ 39. $\dfrac{3y}{z} \cdot \dfrac{z}{3w}$ 40. $\dfrac{5d}{e} \cdot \dfrac{4e}{15d^2}$

▶ **Give each quotient in simplest form.** *(page 188)*

41. $\dfrac{5}{2} \div \dfrac{1}{2}$ 42. $\dfrac{4}{5} \div \dfrac{2}{9}$ 43. $\dfrac{7}{8} \div \dfrac{7}{2}$ 44. $\dfrac{5}{8} \div 2$ 45. $4 \div \dfrac{3}{2}$

46. $\dfrac{3}{a} \div \dfrac{4}{a}$ 47. $\dfrac{6}{b} \div \dfrac{2}{c}$ 48. $\dfrac{j}{6} \div \dfrac{k}{4}$ 49. $\dfrac{3y}{z} \div \dfrac{y}{6}$ 50. $\dfrac{a}{b} \div \dfrac{3b}{4}$

51. $\dfrac{3y}{z} \div \dfrac{9y}{z}$ 52. $\dfrac{m}{p} \div \dfrac{5m}{3q}$ 53. $\dfrac{2d^2}{e} \div \dfrac{d}{e}$ 54. $\dfrac{x}{12y} \div \dfrac{x^2}{8y}$ 55. $\dfrac{9a}{2b} \div \dfrac{3a}{4b^2}$

▶ **Solve. Give answers in simplest form.** *(page 196)*

56. $\dfrac{1}{3}$ of $27 = n$ 57. $\dfrac{3}{4}$ of $40 = n$ 58. $\dfrac{7}{8}$ of $72 = n$ 59. $\dfrac{3}{2}$ of $28 = n$

60. $\dfrac{3}{8}$ of $56 = n$ 61. $\dfrac{5}{6}$ of $60 = n$ 62. $\dfrac{9}{4}$ of $15 = n$ 63. $\dfrac{2}{5}$ of $27 = n$

64. $\dfrac{5}{9}$ of $38 = n$ 65. $\dfrac{5}{3}$ of $41 = n$ 66. $\dfrac{4}{5}$ of $49 = n$ 67. $\dfrac{3}{8}$ of $2 = n$

OBJECTIVE: To use a formula to find the volumes of cubes and other rectangular prisms.

The **volume** of a space figure is the number of cubic units that it takes to fill it.

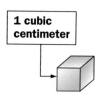

1 cubic centimeter

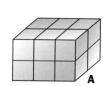
A

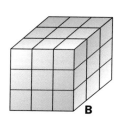
B

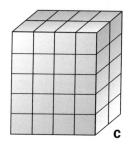
C

1. Count the cubes. Which prism has a volume of 12 cubic centimeters?

2. Which prism has a volume of 27 cubic centimeters?

3. What is the volume of prism C?

EXAMPLE | **Here's how to use a formula to find the volume of a prism.**

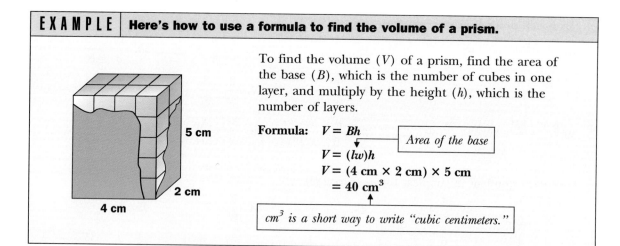

5 cm

2 cm

4 cm

To find the volume (V) of a prism, find the area of the base (B), which is the number of cubes in one layer, and multiply by the height (h), which is the number of layers.

Formula: $V = Bh$ → *Area of the base*

$V = (lw)h$

$V = (4 \text{ cm} \times 2 \text{ cm}) \times 5 \text{ cm}$

$= 40 \text{ cm}^3$

cm^3 is a short way to write "cubic centimeters."

CHECK for Understanding

4. Look at the example above. What is the formula for the volume of a rectangular prism? What do the letters V, l, w, and h represent?

5. Complete these examples.

a.

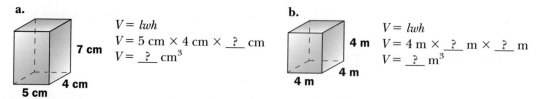

7 cm

4 cm

5 cm

$V = lwh$

$V = 5 \text{ cm} \times 4 \text{ cm} \times \underline{\ ?\ } \text{ cm}$

$V = \underline{\ ?\ } \text{ cm}^3$

b.

4 m

4 m

4 m

$V = lwh$

$V = 4 \text{ m} \times \underline{\ ?\ } \text{ m} \times \underline{\ ?\ } \text{ m}$

$V = \underline{\ ?\ } \text{ m}^3$

EXERCISES

▷ **Find the volume.**

6.

5 cm
4 cm
9 cm

7.
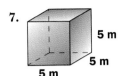
5 m
5 m
5 m

8.
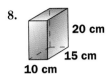
20 cm
15 cm
10 cm

9.
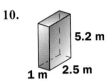
4 cm
10 cm
31.5 cm

10.
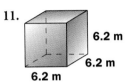
5.2 m
1 m   2.5 m

11.
6.2 m
6.2 m
6.2 m

▷ **Substitute in the formula $V = lwh$ to find the missing volume.**

12. $l = 6$ m
    $w = 7$ m
    $h = 10$ m
    $V = \underline{\ ?\ }$

13. $l = 12$ cm
    $w = 6$ cm
    $h = 20$ cm
    $V = \underline{\ ?\ }$

14. $l = 8$ m
    $w = 8$ m
    $h = 8$ m
    $V = \underline{\ ?\ }$

15. $l = 13$ cm
    $w = 5$ cm
    $h = 12$ cm
    $V = \underline{\ ?\ }$

16. $l = 2.5$ m
    $w = 6$ m
    $h = 4$ m
    $V = \underline{\ ?\ }$

17. $l = 12.4$ cm
    $w = 8$ cm
    $h = 20$ cm
    $V = \underline{\ ?\ }$

18. $l = 7.5$ cm
    $w = 6.2$ cm
    $h = 8$ cm
    $V = \underline{\ ?\ }$

19. $l = 2.1$ m
    $w = 3.2$ m
    $h = 4$ m
    $V = \underline{\ ?\ }$

Problem Solving APPLICATIONS

▷ **Tell whether the question involves perimeter, area, or volume.**

20. How much sand is needed to fill a sandbox?

21. How much fence is needed to fence a garden?

22. How much sod is needed to cover a lawn?

23. How much paper is needed to wrap a box?

24. How much water is needed for a swimming pool?

25. How many flowers are needed to border a patio?

Challenge! VISUAL THINKING

▷ **Each space figure is made from 1-centimeter cubes. Find each volume.**

26.

27.

28.
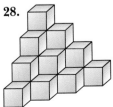

OBJECTIVE: To use a formula to find the volume of a cylinder.

The volume of the rectangular container is 96 cubic inches. Is the volume of the cylindrical container more or less than 96 cubic inches?

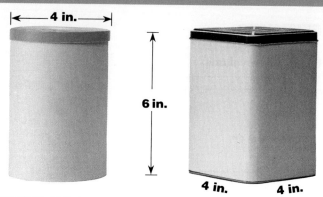

4 in.

6 in.

4 in.   **4 in.**

EXAMPLE | **Here's how to use a formula to find the volume of a cylinder.**

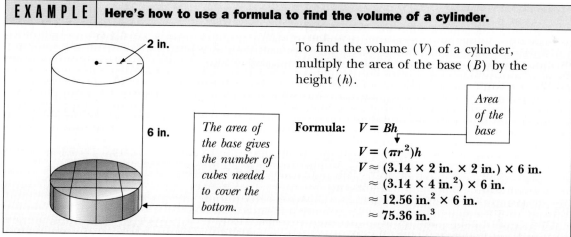

2 in.

6 in.

The area of the base gives the number of cubes needed to cover the bottom.

To find the volume (V) of a cylinder, multiply the area of the base (B) by the height (h).

Area of the base

Formula: $V = Bh$

$V = (\pi r^2)h$

$V \approx (3.14 \times 2 \text{ in.} \times 2 \text{ in.}) \times 6 \text{ in.}$

$\approx (3.14 \times 4 \text{ in.}^2) \times 6 \text{ in.}$

$\approx 12.56 \text{ in.}^2 \times 6 \text{ in.}$

$\approx 75.36 \text{ in.}^3$

CHECK for Understanding

1. Look at the example above. What is the formula for the volume of a cylinder? What do the letters V, π, r, and h represent?

2. In the formula $V = \pi r^2 h$, the r^2 means __?__ times __?__ .

3. Complete these examples.

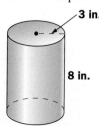

3 in.

8 in.

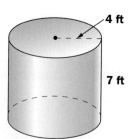

4 ft

7 ft

a. $V = \pi r^2 h$

$V \approx (3.14 \times 3 \text{ in.} \times 3 \text{ in.}) \times 8 \text{ in.}$

$\approx (3.14 \times \underline{\ ?\ } \text{ in.}^2) \times 8 \text{ in.}$

$\approx 28.26 \text{ in.}^2 \times 8 \text{ in.}$

$\approx \underline{\ ?\ } \text{ in.}^3$

b. $V = \pi r^2 h$

$V \approx (3.14 \times 4 \text{ ft} \times \underline{\ ?\ } \text{ ft}) \times 7 \text{ ft}$

$\approx (3.14 \times \underline{\ ?\ } \text{ ft}^2) \times 7 \text{ ft}$

$\approx 50.24 \text{ ft}^2 \times \underline{\ ?\ } \text{ ft}$

$\approx \underline{\ ?\ } \text{ ft}^3$

EXERCISES

▶ **Find the volume. Use 3.14 as an approximation for π.**

4.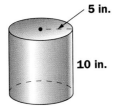
5 in.
10 in.

5.
6 ft
4 ft

6.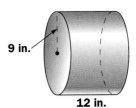
9 in.
12 in.

7.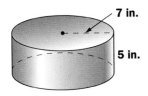
7 in.
5 in.

8.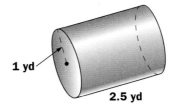
1 yd
2.5 yd

9.
1.5 in.
4 in.

▶ **Substitute in the formula $V = \pi r^2 h$ to find the missing measure.**

10. $r = 4$ in.
$\pi \approx 3.14$
$h = 9$ in.
$V \approx$ __?__

11. $r = 6$ in.
$\pi \approx 3.14$
$h = 2$ in.
$V \approx$ __?__

12. $r = 2$ yd
$\pi \approx 3.14$
$h = 3$ yd
$V \approx$ __?__

13. $r = 5$ ft
$\pi \approx 3.14$
$h = 2$ ft
$V \approx$ __?__

14. $r = 8$ in.
$\pi \approx 3.14$
$h = 2.5$ in.
$V \approx$ __?__

15. $r = 10$ in.
$\pi \approx 3.14$
$h = 20.5$ in.
$V \approx$ __?__

16. $r = 9$ ft
$\pi \approx 3.14$
$h = 3.25$ ft
$V \approx$ __?__

17. $r = 0.5$ yd
$\pi \approx 3.14$
$h = 4$ yd
$V \approx$ __?__

18. $r = 1$ yd
$\pi \approx 3.14$
$h =$ __?__
$V \approx 6.28$ yd^3

19. $r = 3$ in.
$\pi \approx 3.14$
$h =$ __?__
$V \approx 56.52$ in.^{3}

20. $r = 20$ in.
$\pi \approx 3.14$
$h =$ __?__
$V \approx 12{,}560$ in.^{3}

21. $r = 5$ ft
$\pi \approx 3.14$
$h =$ __?__
$V \approx 549.5$ ft^3

Challenge! VISUAL THINKING

22. Here are two ways to roll a sheet of paper to make a cylinder. Do you think the two cylinders have the same volume? Find a way to decide.

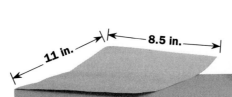

11 in.
8.5 in.
11 in.
8.5 in.

OBJECTIVE: To use formulas to find the volumes of
pyramids and cones.

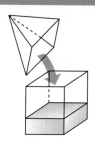
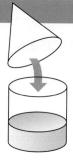

1. It takes 3 pyramids of sand to fill the prism.
So, the volume of a pyramid is __?__ the volume of a prism having the same base and height.

2. It takes 3 cones of sand to fill the cylinder.
So, the volume of a cone is __?__ the volume of a cylinder having the same base and height.

EXAMPLES | **Here's how to use a formula to find the volume of a pyramid or cone.**

A **Pyramid**

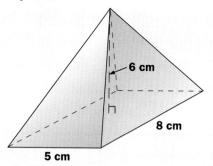

6 cm

8 cm

5 cm

The volume (V) of a pyramid is $\frac{1}{3}$ times the area of the base (B) times the height (h).

Formula: $V = \frac{1}{3}Bh$ — *Area of the rectangular base*

$$V = \frac{1}{3}(lw)h$$

$$V = \frac{1}{\overset{1}{\underset{1}{3}}}(8 \text{ cm} \cdot 5 \text{ cm})(\overset{2}{6} \text{ cm})$$

$$= 80 \text{ cm}^3$$

B **Cone**

9 cm

4 cm

The volume (V) of a cone is $\frac{1}{3}$ times the area of the base (B) times the height (h).

Formula: $V = \frac{1}{3}Bh$ — *Area of the circular base*

$$V = \frac{1}{3}(\pi r^2)h$$

$$V \approx \frac{1}{3}(3.14)(4 \text{ cm})^2 \cdot 9 \text{ cm}$$

$$\approx \frac{1}{\underset{1}{3}}(3.14)(16 \text{ cm}^2) \cdot \overset{3}{9} \text{ cm}$$

$$\approx 150.72 \text{ cm}^3$$

CHECK for Understanding

3. Look at the examples above.

a. If the area of the base of a pyramid is 40 square centimeters and the height is 6 centimeters, the volume is __?__ cubic centimeters.

b. What is the formula for the volume of a cone? What do the letters V, π, r, and h represent?

EXERCISES

▶ **Find the volume of each pyramid or cone. Use 3.14 for π.**

4.
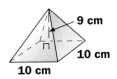
9 cm
10 cm
10 cm

5.
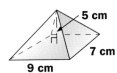
5 cm
7 cm
9 cm

6.

12 cm
8 cm
5 cm

7.
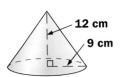
12 cm
9 cm

8.

10 cm
3 cm

9.
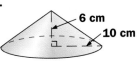
6 cm
10 cm

▶ **Each pyramid described below has a rectangular base. Find the volume.**

10. $l = 5$ m
$w = 6$ m
$h = 3$ m

11. $l = 12$ cm
$w = 7$ cm
$h = 4$ cm

12. $l = 1$ m
$w = 9$ m
$h = 5$ m

13. $l = 24$ cm
$w = 10$ cm
$h = 2$ cm

14. $l = 4$ cm
$w = 11$ cm
$h = 12$ cm

15. $l = 2.5$ m
$w = 4$ m
$h = 6$ m

16. $l = 10$ m
$w = 4.4$ m
$h = 3$ m

17. $l = 4.2$ cm
$w = 6.5$ cm
$h = 15$ cm

▶ **Find the volume of each cone. Use 3.14 for π. Round the answer to the nearest tenth.**

18. $r = 2$ m
$h = 6$ m

19. $r = 5$ m
$h = 9$ m

20. $r = 3$ m
$h = 10$ m

21. $r = 4$ m
$h = 3$ m

22. $r = 6$ m
$h = 20$ m

23. $r = 8$ m
$h = 12$ m

24. $r = 30$ m
$h = 8.5$ m

25. $r = 2.2$ m
$h = 3$ m

Challenge! USING A GRAPH

The cone in the cup is 8 centimeters high and holds 80 cubic centimeters of juice when full. The graph shows the volume of juice in the cone at different heights.

26. About how many centimeters high is the juice in the cone when the cone is
  a. $\frac{3}{4}$ full?
  b. $\frac{1}{2}$ full?

27. About how many cubic centimeters of juice are in the cone when the height is
  a. $\frac{3}{4}$ as high?
  b. $\frac{1}{2}$ as high?

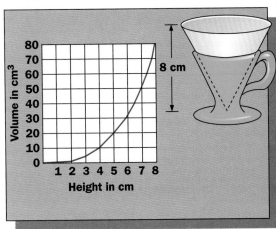

11-8 Metric Units of Capacity and Weight

OBJECTIVE: To make conversions between metric units of capacity. To make conversions between metric units of weight.

The **liter** (L) and **milliliter** (mL) are metric units of liquid measure or capacity.

$$1 \text{ L} = 1000 \text{ mL}$$

The **gram** (g) and **kilogram** (kg) are metric units of weight.

$$1 \text{ kg} = 1000 \text{ g}$$

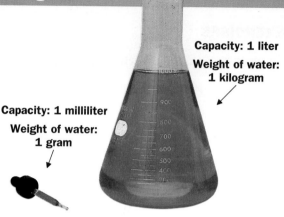

Capacity: 1 liter
Weight of water:
1 kilogram

Capacity: 1 milliliter
Weight of water:
1 gram

EXAMPLES	Here's how to change from one unit of capacity to another.

A

350 mL = ? L

Think: Changing from smaller units to larger units, so divide.

Remember: 1000 mL = 1 L

350 mL = 0.350 L
⌐ ÷ 1000 ↱

B

4.2 L = ? mL

Think: Changing from larger units to smaller units, so multiply.

Remember: 1 L = 1000 mL

4.2 L = 4200 mL
⌐ × 1000 ↱

CHECK for Understanding

1. Look at Examples A and B. Explain how to change from milliliters to liters. Explain how to change from liters to milliliters.

EXAMPLES	Here's how to change from one unit of weight to another.

C

8275 g = ? kg

Think: Changing from smaller units to larger units, so divide.

Remember: 1000 g = 1 kg

8275 g = 8.275 kg
⌐ ÷ 1000 ↱

D

0.625 kg = ? g

Think: Changing from larger units to smaller units, so multiply.

Remember: 1 kg = 1000 g

0.625 kg = 625 g
⌐ × 1000 ↱

CHECK for Understanding

2. Look at Examples C and D. Explain how to change from grams to kilograms. Explain how to change from kilograms to grams.

EXERCISES

▶ **Which capacity seems reasonable?**

3. A tablespoon:

    a. 15 mL   **b.** 150 mL

4. A bathtub:

    a. 3 L   **b.** 300 L

5. A soft-drink can:

    a. 4 mL   **b.** 400 mL

6. A thermos bottle:

    a. 800 mL   **b.** 8000 mL

7. A fruit-juice pitcher:

    a. 0.1 L   **b.** 1 L

8. A car's gas tank:

    a. 6 L   **b.** 60 L

▶ **Which weight seems reasonable?**

9. A dime:

    a. 3 g   **b.** 300 g

10. An orange:

    a. 4 g   **b.** 400 g

11. A can of peaches:

    a. 46.4 g   **b.** 464 g

12. A bicycle:

    a. 1.2 kg   **b.** 12 kg

13. A straight pin:

    a. 0.13 g   **b.** 13 g

14. An automobile:

    a. 200 kg   **b.** 2000 kg

▶ **Copy and complete.**

15. 9 L = ? mL

16. 25 L = ? mL

17. 125 L = ? mL

18. 7000 mL = ? L

19. 2875 mL = ? L

20. 1400 mL = ? L

21. 6.4 L = ? mL

22. 0.65 L = ? mL

23. 25.75 L = ? mL

24. 17,000 mL = ? L

25. 870 mL = ? L

26. 25 mL = ? L

27. 7 kg = ? g

28. 13 kg = ? g

29. 28 kg = ? g

30. 8000 g = ? kg

31. 1250 g = ? kg

32. 14,288 g = ? kg

33. 4.6 kg = ? g

34. 12.75 kg = ? g

35. 0.33 kg = ? g

Problem Solving USING METRIC UNITS

▶ **Solve.**

36. A recipe calls for 0.25 liter of cream. How many milliliters is that?

37. A bag of peanuts weighs 575 grams. How many kilograms is that?

38. A marshmallow weighs about 5 grams. How many marshmallows are there in a 0.45-kilogram bag of marshmallows?

39. How many 240-milliliter glasses can be filled from a 1.2-liter bottle of orange juice?

40. A teaspoon has a capacity of 5 milliliters. How many teaspoons of vanilla are there in a liter bottle?

41. When empty, a jelly jar weighs 85 grams. When full, it weighs 0.35 kilogram. How many grams of jelly does the jar hold?

OBJECTIVE: To relate metric units of volume, capacity, and weight.

The metric units of volume, capacity, and weight have a special relationship.

Look at the small container. It holds 1 cubic centimeter of water. Its capacity is 1 milliliter. The weight of the water is 1 gram.

The large container holds 1000 cubic centimeters of water. Its capacity is 1 liter (1000 milliliters). The weight of the water is 1 kilogram (1000 grams).

1. A container that has a volume of 75 cubic centimeters can hold ? milliliters of water. The weight of the water is ? grams.

2. A container that has a volume of 3000 cubic centimeters can hold ? liters of water. The weight of the water is ? kilograms.

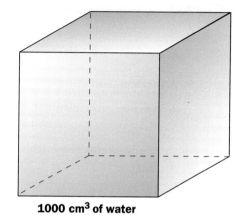

1 cm³ of water

1000 cm³ of water

EXAMPLE | **Here's how to find the amount and weight of water that a container can hold.**

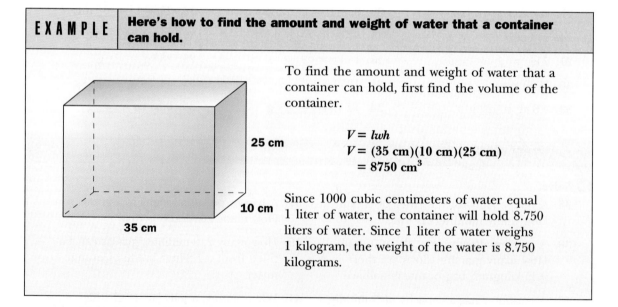

25 cm

10 cm

35 cm

To find the amount and weight of water that a container can hold, first find the volume of the container.

$$V = lwh$$
$$V = (35 \text{ cm})(10 \text{ cm})(25 \text{ cm})$$
$$= 8750 \text{ cm}^3$$

Since 1000 cubic centimeters of water equal 1 liter of water, the container will hold 8.750 liters of water. Since 1 liter of water weighs 1 kilogram, the weight of the water is 8.750 kilograms.

CHECK for Understanding

3. Look at the example above. A container that has a volume of 8750 cubic centimeters will hold ? liters of water. The weight of the water is ? kilograms.

EXERCISES

▶ Find the number of liters of water each container will hold.

4.

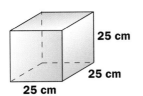

15 cm
10 cm
30 cm

5.

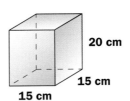

20 cm
15 cm
15 cm

6.

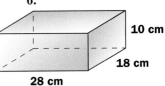

10 cm
18 cm
28 cm

7.

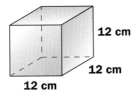

25 cm
25 cm
25 cm

8.

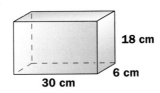

18 cm
6 cm
30 cm

9.

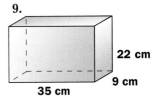

22 cm
9 cm
35 cm

▶ Find the number of kilograms of water each container will hold.

10.

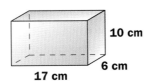

12 cm
12 cm
12 cm

11.

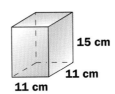

10 cm
6 cm
17 cm

12.

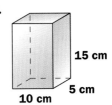

15 cm
5 cm
10 cm

13.

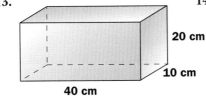

20 cm
10 cm
40 cm

14.

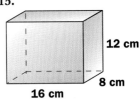

15 cm
11 cm
11 cm

15.

12 cm
8 cm
16 cm

Challenge! GUESS AND CHECK

▶ **Find the length of an edge of each cube.** *Hint: Estimate. Compute the volume using a calculator. Repeat until you get the answer.*

16.

?

This cube holds
2197 milliliters
of water.

17.

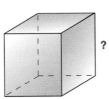

?

This cube holds
13,824 milliliters
of water.

18.

?

This cube holds
3.375 milliliters
of water.

OBJECTIVE: To use formulas to solve problems about area and volume.

Earlier you used formulas to solve area and volume problems. In this lesson you will solve problems by combining formulas.

Area Formulas	Volume Formulas
Rectangle: $A = lw$ Square: $A = s^2$ Circle: $A = \pi r^2$	Rectangular prism: $V = lwh$ Cylinder: $V = \pi r^2 h$ Cone: $V = \dfrac{1}{3}\pi r^2 h$

E X A M P L E S	Here's how to combine formulas to solve problems.

 A

Area problem: Find the area of the lawn.

Notice that the lawn has the shape of a rectangle with a circle removed. To find the grass area, you need to subtract the area of the circle from the area of the rectangle.

Write a formula.

$$A = lw - \pi r^2$$

Substitute and simplify.

$$A \approx 40 \text{ ft}(30 \text{ ft}) - 3.14(10 \text{ ft})^2$$
$$\approx 1200 \text{ ft}^2 - 314 \text{ ft}^2$$
$$\approx 886 \text{ ft}^2$$

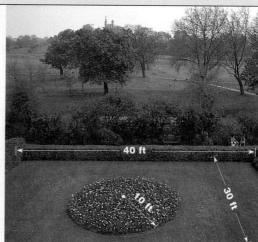

The area of the lawn is about 886 ft².

 B

Volume problem: Find the volume of the water tank.

Notice that the tank is shaped like a cylinder with a cone on top. To find the total volume, you need to add the volume of the cone to the volume of the cylinder.

Write a formula.

$$V = \pi r^2 h + \frac{1}{3}\pi r^2 h$$

Substitute and simplify.

$$V \approx 3.14(6 \text{ ft})^2(4 \text{ ft}) + \frac{1}{3}(3.14)(6 \text{ ft})^2(1 \text{ ft})$$
$$\approx 452.16 \text{ ft}^3 + 37.68 \text{ ft}^3$$
$$\approx 489.84 \text{ ft}^3$$

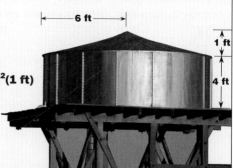

The volume of the water tank is about 490 ft³.

EXERCISES

▶ **Find the area of each shaded region. Use 3.14 for π.**

1.

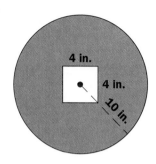

4 in.

4 in.

10 in.

2.

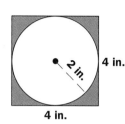
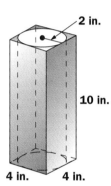

4 in.

2 in.

4 in.

4 in.

3.

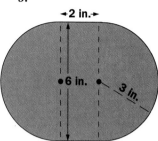

2 in.

6 in.

3 in.

▶ **Find the volume of each shaded space figure. Use 3.14 for π.**

4.

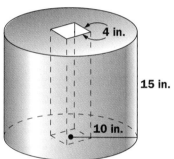

4 in.

15 in.

10 in.

5.

2 in.

10 in.

4 in.   4 in.

6.

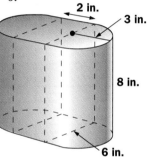

2 in.

3 in.

8 in.

6 in.

▶ **Make a drawing to picture the facts. Then write a formula and solve the problem. Use 3.14 for π.**

7. A rectangular room is 12 feet by 15 feet. A circular rug is in the center of the floor. The radius of the rug is 4 feet. What is the area of the uncovered portion of the floor?

8. A tabletop is shaped like a rectangle with half circles at each of the two longer sides of the rectangle. The rectangular part of the tabletop is 40 inches by 30 inches. The radius of each half circle is 20 inches. What is the area of the tabletop?

9. A storage bin is shaped like a cylinder attached to a cone. The height of the cylindrical part is 12 feet. The height of the conical part is 4 feet. The radius is 5 feet. What is the volume of the storage bin?

10. A cylindrical can is packed in a box 6 inches by 6 inches by 8 inches. The height of the can is 8 inches, and its radius is 3 inches. What is the volume of the space remaining in the box?

11. Write a problem that can be solved using the area formulas on page 326.

12. Write a problem that can be solved using the volume formulas on page 326.

OBJECTIVE: To solve problems using a calculator when appropriate.

Computers are used to draw "blueprints" of an architect's design. The computer can be programmed to display a floor plan on a screen or printout.

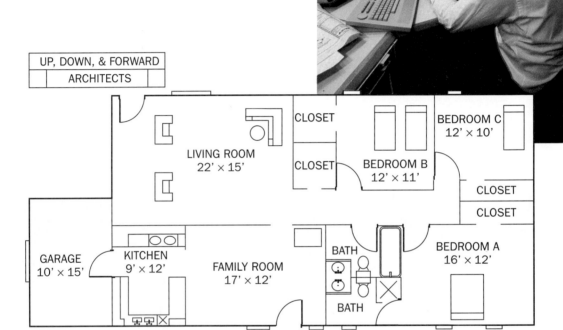

UP, DOWN, & FORWARD
ARCHITECTS

CLOSET

BEDROOM C
12' × 10'

LIVING ROOM
22' × 15'

CLOSET

BEDROOM B
12' × 11'

CLOSET

CLOSET

GARAGE
10' × 15'

KITCHEN
9' × 12'

FAMILY ROOM
17' × 12'

BATH

BEDROOM A
16' × 12'

BATH

▶ **Solve. Decide when a calculator would be helpful.**

1. Which room is 17 feet by 12 feet?

2. Which room is 16 feet by 12 feet?

3. What are the length and width of the living room?

4. What are the dimensions of the kitchen?

5. Which room has an area of 132 square feet?

6. What is the area of the smallest bedroom?

7. What is the area of the largest bedroom?

8. Carpet costs $1.50 per square foot. How much will it cost to carpet Bedroom B?

9. At $1.75 per square foot, how much will it cost to carpet the living room?

10. A floor tile 1 foot by 1 foot costs $.49. How much will it cost to tile the family-room?

11. A gallon of paint covers 600 square feet. What fraction of a gallon of paint will it take to cover the garage floor?

12. **Write a problem** that can be solved using the blueprint. Then solve the problem.

▶ **Solve. Give answers in simplest form.** *(page 200)*

1. $\frac{1}{3}n = 11$ **2.** $\frac{1}{4}n = 15$ **3.** $\frac{2}{3}n = 22$ **4.** $\frac{3}{4}n = 21$ **5.** $\frac{5}{8}n = 35$

6. $\frac{5}{6}n = 40$ **7.** $\frac{3}{2}n = 39$ **8.** $\frac{5}{2}n = 41$ **9.** $\frac{2}{3}n = 19$ **10.** $\frac{5}{4}n = 32$

11. $\frac{7}{8}n = 40$ **12.** $\frac{9}{4}n = 20$ **13.** $\frac{4}{9}n = 20$ **14.** $\frac{5}{9}n = 37$ **15.** $\frac{9}{5}n = 37$

▶ **Solve and check.** *(page 226)*

16. $3y + 5y = {}^-5$

17. $9x - x = 11$

18. $w - 6w = {}^-8$

19. $5n - 8n - 7 = 24$

20. $5n - 8 - 7n = 24$

21. $5 - 8n - 7n = 24$

22. $9y + 13 - 4y - 11 = {}^-18$

23. $9y + 13 - 4 - 11y = {}^-18$

24. ${}^-7t + 8 + t - 5 = 6$

25. ${}^-7 + 8t + t - 5 = 6$

▶ **Solve each proportion.** *(page 240)*

26. $\frac{n}{3} = \frac{5}{6}$ **27.** $\frac{7}{n} = \frac{5}{9}$ **28.** $\frac{9}{12} = \frac{n}{6}$ **29.** $\frac{8}{5} = \frac{5}{n}$ **30.** $\frac{n}{10} = \frac{12}{5}$

31. $\frac{5}{8} = \frac{n}{13}$ **32.** $\frac{16}{12} = \frac{4}{n}$ **33.** $\frac{n}{6} = \frac{14}{9}$ **34.** $\frac{13}{n} = \frac{5}{9}$ **35.** $\frac{11}{4} = \frac{n}{5}$

36. $\frac{n}{31} = \frac{5}{8}$ **37.** $\frac{17}{n} = \frac{3}{8}$ **38.** $\frac{9}{10} = \frac{n}{30}$ **39.** $\frac{3}{1\frac{1}{4}} = \frac{5}{n}$ **40.** $\frac{n}{2\frac{2}{3}} = \frac{7}{4}$

▶ *Solve.* *(pages 252, 256)*

41. 25% of 32 = n **42.** $33\frac{1}{3}$% of 48 = n **43.** 200% of 42 = n **44.** 15% of 80 = n

45. 12.5% of 64 = n **46.** 0.75% of 60 = n **47.** 20% of n = 30 **48.** 60% of n = 15

49. 150% of n = 72 **50.** 125% of n = 150 **51.** 8.5% of n = 2.72 **52.** 0.6% of n = 0.72

▶ **Find the percent of increase or decrease. Use *i* to indicate increase and *d* to indicate decrease. Give answers to the nearest tenth of a percent.** *(page 258)*

53. from 24 to 30

54. from 60 to 30

55. from 30 to 60

56. from 50 to 20

57. from 112 to 138

58. from 138 to 112

59. from 16 to 56

60. from 39 to 91

61. from 91 to 39

62. from 50 to 100

63. from 100 to 150

64. from 150 to 200

65. from 20 to 18

66. from 18 to 16

67. from 16 to 14

Here are scrambled answers for the review exercises.

$\frac{1}{3}$	200	area	divide	milliliter	space	triangular
1	1000	bases	height	multiply	sphere	volume
13.5	2512	cone	kilogram	pyramid	square	weight
20	add	cylinder	liter	rectangular	surface	width

1. Three-dimensional shapes are called _?_ figures. Prisms and pyramids are named according to the shapes of their _?_. *(page 306)*

   _?_ prism   **_?_ prism**   **_?_ pyramid**   **_?_**   **_?_**   **_?_**

2. The sum of the areas of the 6 faces of a rectangular prism or cube is called the _?_ area. To find the surface area of this rectangular prism, you would compute the area of each face by multiplying its length by its _?_. Then you would _?_ all six areas. *(pages 308, 310)*

3. The _?_ of a space figure is the number of cubic units that it takes to fill it. To find the volume of a prism, you can use the formula **V = Bh,** where *B* stands for the area of the base and *h* stands for the _?_. The volume of this prism is _?_ ft³. *(page 316)*

4. To find the volume of a cylinder, you can use the formula **V = Bh,** where *B* stands for the _?_ of the base and *h* stands for the height. The volume of this cylinder is _?_ in.³. Use 3.14 as an approximation for π. *(page 318)*

5. The volume of a _?_ or cone is _?_ times the area of the base times the height. The volume of this pyramid is _?_ ft³. *(page 320)*

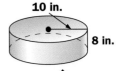

6. The _?_ and the milliliter are metric units of liquid measure or capacity. The gram and the _?_ are metric units of weight.
1 liter = _?_ milliliters   _?_ kilogram = 1000 grams   To change from milliliters to liters, you would _?_ by 1000. To change from kilograms to grams, you would _?_ by 1000. *(page 322)*

7. The metric units of volume, capacity, and _?_ have a special relationship. A container that will hold 1 cm³ of water has a capacity of 1 _?_. A container that will hold 1000 cm³ of water has a capacity of 1 liter. The weight of 1 liter of water is 1 kilogram. The container shown at the right holds _?_ liters of water, and the weight of the water would be 13.5 kilograms. *(page 324)*

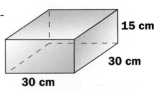

▶ **Match each space figure with its name.** *(page 306)*

1. Cone
2. Cube
3. Cylinder
4. Pentagonal pyramid
5. Rectangular prism
6. Sphere
7. Square pyramid
8. Triangular prism
9. Triangular pyramid

A.   B.   C.   D.

E.   F.   G.   H. 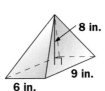 I.

▶ **Find the surface area.** *(pages 308, 310)*

10.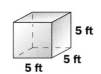
5 ft, 5 ft, 5 ft

11.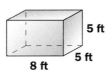
5 ft, 5 ft, 8 ft

12.
12 ft, 6 ft, 4 ft

13.
6 ft, 3 ft, 7 ft

▶ **Find the volume. Use 3.14 as an approximation for π.** *(pages 316, 318, 320)*

14.
6 in., 4 in., 4 in.

15.
2 in., 8 in.

16.
8 in., 9 in., 6 in.

17.
10 in., 3 in.

▶ **Complete.** *(page 322)*

18. 45 L = ? mL
19. 2500 mL = ? L
20. 0.65 L = ? mL
21. 856 mL = ? L
22. 25 kg = ? g
23. 1650 g = ? kg
24. 0.46 kg = ? g
25. 37 g = ? kg

▶ **Complete.** *(page 324)*

26. The container will hold ? liters of water.
27. The container will hold ? kilograms of water.

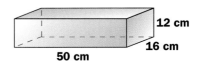
12 cm, 16 cm, 50 cm

▶ **Choose the formula and find the area of each shaded region. Use 3.14 for π.** *(page 326)*

A. $A = s^2 - \pi r^2$ B. $A = S^2 - s^2$
C. $A = \pi r^2 - lw$ D. $A = \pi R^2 - \pi r^2$

28.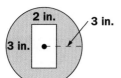
2 in., 3 in., 3 in.

29.
3 in.

30.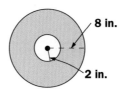
8 in., 2 in.

31.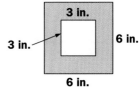
3 in., 3 in., 6 in., 6 in.

▶ **Choose the correct letter.**

---

1. Change to a decimal.

$$\frac{5}{6} = ?$$

A. 1.2

B. $8.33\frac{1}{3}$

C. $0.83\frac{1}{3}$

D. none of these

---

2. Give the sum.

$$\frac{2x}{3} + \frac{y}{4}$$

A. $\frac{2x + y}{12}$

B. $\frac{2x + y}{7}$

C. $\frac{8x + 3y}{12}$

D. none of these

---

3. Give the difference.

$$\frac{2r}{3} - \frac{s}{6}$$

A. $\frac{2r - 3s}{6}$

B. $\frac{2r - s}{6}$

C. $\frac{r - s}{3}$

D. none of these

---

4. Give the product.

$$\frac{5c}{6d} \cdot \frac{5d}{2c^2}$$

A. $\frac{25}{12c}$

B. $\frac{25c}{6c^2}$

C. $\frac{12c}{25}$

D. none of these

---

5. Give the quotient.

$$\frac{a}{b} \div \frac{7a}{4d}$$

A. $\frac{7a^2}{4bd}$

B. $\frac{7b}{4d}$

C. $\frac{4d}{7a}$

D. none of these

---

6. Solve.

$$\frac{5}{6} \text{ of } 33 = n$$

A. $39\frac{3}{5}$

B. $27\frac{1}{2}$

C. $5\frac{1}{2}$

D. none of these

---

7. Solve.

$$\frac{5}{8}n = 31$$

A. $49\frac{3}{5}$

B. $43\frac{3}{5}$

C. $19\frac{3}{8}$

D. none of these

---

8. Solve.

$$5k - 6 - 3k = {}^-9 - 6$$

A. $^-4\frac{1}{2}$

B. $^-1\frac{3}{8}$

C. $2\frac{1}{2}$

D. none of these

---

9. Solve.

$$\frac{n}{5} = \frac{3}{2\frac{2}{3}}$$

A. $4\frac{5}{9}$

B. $5\frac{5}{8}$

C. 40

D. none of these

---

10. Solve.

$$6.5\% \text{ of } 108 = n$$

A. 6.50

B. 16.62

C. 7.02

D. none of these

---

11. The percent of increase from 21 to 28 is

A. $33\frac{1}{3}\%$

B. 25%

C. 7%

D. none of these

---

12. Choose the equation.

You sold 7 tickets on Monday and 5 tickets on Tuesday. After deducting $6 for selling the tickets, you turned in $36. What was the cost of each ticket?

A. $7n + 5n + 6 = 36$

B. $7n - 5n + 6 = 36$

C. $7n + 5n - 6 = 36$

D. $7n - 5n - 6 = 36$

Graphing Equations and Inequalities

12

$$y = \frac{1}{4}x + 40$$

The temperature in degrees Fahrenheit (y) equals one-fourth the number of times a cricket chirps per minute (x) plus 40.
(page 347)

Y ou can estimate the temperature by counting the chirps of a cricket. If a cricket chirps 60 times per minute, the temperature is about 55°F.

OBJECTIVE: To write an inequality for a graph on the number line. To graph the solution of an inequality on the number line.

Here are some examples of inequalities:

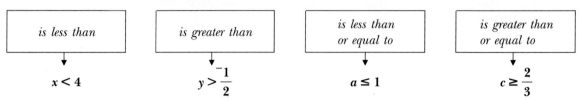

| *is less than* | *is greater than* | *is less than or equal to* | *is greater than or equal to* |

$$x < 4 \qquad y > \frac{^-1}{2} \qquad a \le 1 \qquad c \ge \frac{2}{3}$$

Look at the first inequality. There are many numbers that are solutions of $x < 4$. In fact, every number less than 4 is a solution.

1. Is 3 a solution of $x < 4$? Is $^-7$? Is $3\frac{5}{8}$?

2. Look at the second inequality. Is 0 a solution? Is 12? Is $\frac{^-1}{3}$?

EXAMPLES **Here's how to graph the solutions of inequalities on the number line.**

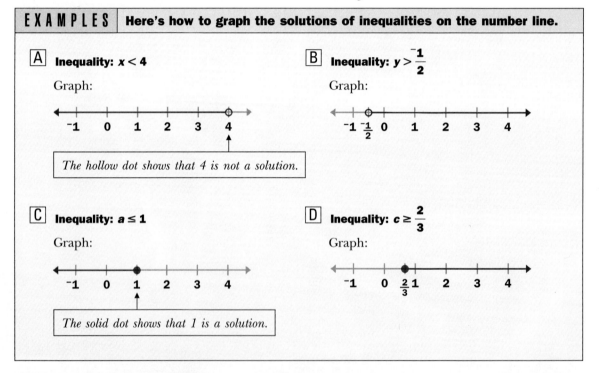

A **Inequality:** $x < 4$

Graph:

The hollow dot shows that 4 is not a solution.

B **Inequality:** $y > \frac{^-1}{2}$

Graph:

C **Inequality:** $a \le 1$

Graph:

The solid dot shows that 1 is a solution.

D **Inequality:** $c \ge \frac{2}{3}$

Graph:

CHECK for Understanding

3. Look at the graph in Example A. Is 4 a solution?

4. Look at the graph in Example B. Is $\frac{^-1}{2}$ a solution?

5. Complete.
   a. A hollow dot tells you that the number ___?___ a solution.
   is/is not

   b. A solid dot tells you that the number ___?___ a solution.
   is/is not

EXERCISES

▶ Match each inequality with its graph.

6. $n < 2$

a.

7. $n > 2$

b.

8. $n \leq 2$

c.

9. $n \geq 2$

d.

10. $n < {}^-2$

e.

11. $n > {}^-2$

f.

12. $n \leq {}^-2$

g.

13. $n \geq {}^-2$

h.

▶ Write an inequality for each graph.

14.

15.

16.

17.

18.

19.

20.

21.

22.

23.

▶ Graph the solutions of each inequality.

24. $x < 3$

25. $m \geq 4$

26. $a > {}^-3$

27. $f \leq {}^-1$

28. $b > 0$

29. $g \leq 5$

30. $y < \frac{{}^-1}{2}$

31. $p \geq \frac{3}{4}$

32. $n < 1\frac{1}{2}$

33. $q \geq {}^-3\frac{2}{3}$

34. $c > {}^-3\frac{1}{4}$

35. $h \leq 4\frac{3}{4}$

OBJECTIVE: To solve inequalities.

Solving an inequality is much like solving an equation. In fact, the only difference is that **when you multiply or divide both sides of an inequality by a negative number, you must reverse the inequality sign.**

EXAMPLES	Here's how to solve inequalities.

A **Inequality:** $\qquad x - 3 < {}^-4$

Add 3 to both sides. $\qquad x - 3 + 3 < {}^-4 + 3$

Simplify. $\qquad x < {}^-1$

B **Inequality:** $\qquad y + 4 > {}^-5$

Subtract 4 from both sides. $\qquad y + 4 - 4 > {}^-5 - 4$

Simplify. $\qquad y > {}^-9$

C **Inequality:** $\qquad \dfrac{c}{6} \geq {}^-2$

Multiply both sides by 6. $\qquad 6 \cdot \dfrac{c}{6} \geq 6 \cdot {}^-2$

Simplify. $\qquad c \geq {}^-12$

D **Inequality:** $\qquad {}^-2n \leq {}^-11$

Divide both sides by ${}^-2$ and reverse the inequality sign. $\qquad \dfrac{{}^-2n}{{}^-2} \geq \dfrac{{}^-11}{{}^-2}$

$\qquad n \geq 5\dfrac{1}{2}$

Simplify.

Notice that since we multiplied both sides by a **positive** number, the inequality sign was not reversed.

Notice that since we divided both sides by a **negative** number, the inequality sign was reversed.

CHECK for Understanding

1. In Example C, both sides of the inequality were __?__ by a **positive** number, so the inequality sign was not reversed.

2. In Example D, both sides of the inequality were __?__ by a **negative** number, so the inequality sign was reversed.

3. Complete.

a. $\quad z - 8 < {}^-3$

$\quad z - 8 + 8 < {}^-3 + 8$

$\qquad z < \underline{\ ?\ }$

b. $\quad \dfrac{h}{{}^-5} \geq 3$

$\quad {}^-5 \cdot \dfrac{h}{{}^-5} \leq \underline{\ ?\ } \cdot 3$

$\qquad h \leq {}^-15$

c. $\quad 4t \leq 18$

$\quad \dfrac{4t}{4} \leq \dfrac{18}{4}$

$\qquad t \leq \underline{\ ?\ }$

EXERCISES

 Would you reverse the inequality sign when solving the inequality?

4. $5a \le 3$ **5.** $^-2z > 9$ **6.** $n + 6 \ge {}^-3$ **7.** $\dfrac{j}{-4} \le 3$

8. $\dfrac{n}{6} > {}^-7$ **9.** $y - 8 < {}^-5$ **10.** $^-8w \le 5$ **11.** $\dfrac{t}{9} \ge {}^-10$

 Solve.

12. $^-4n > {}^-9$ **13.** $n + 9 \ge {}^-2$ **14.** $\dfrac{n}{-2} < {}^-6$ **15.** $n - 4 \le {}^-12$

16. $n - 9 < 16$ **17.** $6n \le 15$ **18.** $n + 12 \ge 5$ **19.** $\dfrac{n}{-5} > 3$

20. $n - 12 \ge {}^-5$ **21.** $\dfrac{n}{8} < {}^-10$ **22.** $^-9n > 30$ **23.** $n + 13 \le {}^-5$

24. $n - 21 \le {}^-6$ **25.** $^-12n > 34$ **26.** $\dfrac{n}{11} < {}^-9$ **27.** $n + 15 \ge 23$

28. $\dfrac{n}{-12} > 0$ **29.** $n + 18 \ge {}^-6$ **30.** $^-10n \le {}^-42$ **31.** $n - 12 < 11$

 First solve the inequality. Then match the inequality with the graph of its solutions.

32. $^-2y \le 7$

a.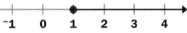

33. $y - 6 > {}^-4$

b.

34. $\dfrac{y}{2} \ge 1$

c.

35. $y + 3 < 4$

d.

36. $4y < 14$

e.

37. $y - 4 \ge {}^-3$

f.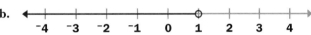

38. $\dfrac{y}{-3} \ge {}^-1$

g.

39. $y + 2 < {}^-1$

h.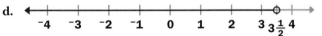

 First solve the inequality. Then graph its solutions.

40. $^-4n > {}^-3$ **41.** $n + 8 \ge 6$ **42.** $\dfrac{n}{-3} < {}^-1$ **43.** $n - 6 \le {}^-10$

44. $n - 7 < 11$ **45.** $5n \le 7$ **46.** $n + 10 \ge 7$ **47.** $\dfrac{n}{-3} > 2$

48. $n - 19 \ge {}^-15$ **49.** $\dfrac{n}{-8} < 0$ **50.** $^-8n > 20$ **51.** $n + 12 \le {}^-7$

12-3 Solving Two-Step Inequalities

OBJECTIVE: To solve two-step inequalities.

In this lesson, you will learn how to solve two-step inequalities.
Again, solving an inequality is much like solving an equation.

| EXAMPLES | Here's how to solve two-step inequalities. |

A Inequality: $\qquad 4y - 3 \le {}^-8$

Add 3 to both sides. $\qquad 4y - 3 + 3 \le {}^-8 + 3$

Simplify. $\qquad 4y \le {}^-5$

Divide both sides by 4. $\qquad \dfrac{4y}{4} \le \dfrac{{}^-5}{4}$

Simplify. $\qquad y \le {}^-1\frac{1}{4}$

B Inequality: $\qquad \dfrac{z}{{}^-6} + 7 > 9$

Subtract 7 from both sides. $\qquad \dfrac{z}{{}^-6} + 7 - 7 > 9 - 7$

Simplify. $\qquad \dfrac{z}{{}^-6} > 2$

Multiply both sides by $^-6$ and reverse the inequality sign. $\qquad {}^-6 \cdot \dfrac{z}{{}^-6} < {}^-6 \cdot 2$

Simplify. $\qquad z < {}^-12$

CHECK for Understanding

1. Look at Example B. In the next to the last step, both sides of the equation were multiplied by __?__. Was the inequality sign reversed?

2. Complete.

   a. $4a + 5 < {}^-1$

   $4a + 5 - 5 < {}^-1 - \underline{\ ?\ }$

   $4a < {}^-6$

   $\dfrac{4a}{4} < \dfrac{{}^-6}{4}$

   $a < {}^-1\frac{1}{2}$

   b. $\quad {}^-3j - 8 \ge {}^-4$

   ${}^-3j - 8 + 8 \ge {}^-4 + 8$

   ${}^-3j \ge \underline{\ ?\ }$

   $\dfrac{{}^-3j}{{}^-3} \underset{?}{\quad} \dfrac{4}{{}^-3}$

   $j \le {}^-1\frac{1}{3}$

   c. $\quad 2n + 6 \le 7$

   $2n + 6 - 6 \le 7 - 6$

   $2n \le \underline{\ ?\ }$

   $\dfrac{2n}{2} \underset{?}{\quad} \dfrac{1}{2}$

   $n \le \dfrac{1}{2}$

3. Would you reverse the inequality sign when solving the inequality?

   a. $2y + 6 \le 4$ **b.** $\dfrac{r}{3} - 2 \ge {}^-4$ **c.** ${}^-3a - 5 > 2$ **d.** $\dfrac{t}{{}^-2} + 5 \le 0$

   e. $\dfrac{j}{6} - 3 > {}^-1$ **f.** ${}^-7b + 6 < 5$ **g.** $\dfrac{c}{{}^-8} + 9 \le 11$ **h.** $5j - 4 \ge 6$

   i. $4y + 8 \ge 16$ **j.** $\dfrac{m}{{}^-4} + 2 \le 5$ **k.** ${}^-4a + 1 \ge 5$ **l.** $\dfrac{t}{{}^-3} + 4 \ge 0$

EXERCISES

 Solve.

   4. $5x + 3 < 3$ **5.** $\dfrac{x}{3} + 2 > 1$ **6.** ${}^-7x - 4 > 3$ **7.** $\dfrac{x}{{}^-5} - 3 < 2$

8. $\dfrac{x}{-4} + 5 \leq {}^-3$ **9.** ${}^-3x + 5 \geq {}^-4$ **10.** $\dfrac{x}{6} - 4 \geq 14$ **11.** $9x - 3 \leq 5$

12. ${}^-6x + 5 \geq {}^-2$ **13.** $\dfrac{x}{2} - 7 > 6$ **14.** $\dfrac{x}{-10} + 6 < 11$ **15.** $4x - 6 \leq 6$

16. $9x + 7 > {}^-5$ **17.** $\dfrac{x}{-7} + 9 \leq 12$ **18.** ${}^-8x - 9 < 4$ **19.** $\dfrac{x}{9} - 8 \geq 7$

20. $\dfrac{x}{10} - 12 > {}^-12$ **21.** $10x + 3 \leq 8$ **22.** $\dfrac{x}{-12} + 10 < 16$ **23.** ${}^-2x - 7 \geq {}^-3$

24. ${}^-12x + 5 \leq 6$ **25.** $\dfrac{x}{-8} + 13 \leq 15$ **26.** ${}^-16x - 6 \geq 4$ **27.** $\dfrac{x}{16} - 15 \leq {}^-11$

▶ **First solve the inequality. Then match the inequality with the graph of its solution.**

28. $2n + 3 > 4$ **a.**

29. $\dfrac{n}{2} - 6 \leq {}^-5$ **b.**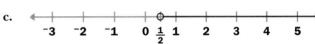

30. $\dfrac{n}{-4} + 9 < 8$ **c.**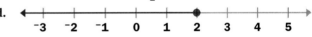

31. $4n - 3 \geq {}^-1$ **d.**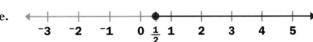

32. ${}^-4n - 5 > {}^-1$ **e.**

▶ **First solve the inequality. Then graph its solutions.**

33. $6a - 2 \leq 4$ **34.** ${}^-9c + 6 > {}^-21$ **35.** $\dfrac{r}{3} - 5 < {}^-6$ **36.** $\dfrac{t}{-2} + 5 \geq 7$

37. ${}^-3y + 5 > 7$ **38.** $\dfrac{a}{5} - 8 < {}^-8$ **39.** $5j + 9 \geq 4$ **40.** ${}^-2k - 10 \leq {}^-13$

Challenge!   GRAPHING INEQUALITIES

▶ **Graph on a number line**

41. all numbers less than 2 *and* greater than ${}^-3$.

42. all numbers less than 4 *and* greater than or equal to ${}^-1$.

43. all numbers less than or equal to ${}^-1$ *and* greater than ${}^-4$.

44. all numbers less than or equal to 3 *and* greater than or equal to ${}^-1$.

45. all numbers not equal to ${}^-2$.

BUILD YOUR
PROBLEM·SOLVING·POWER

You will use a variety of problem-solving strategies on these two pages. For some problems, you may wish to use a calculator.

1 USE RESOURCES

Thomas Jefferson was born on Saturday, April 13, 1743, and died on July 4, 1826. How many of Jefferson's birthdays were on Saturday?

✓ *Problem Solving Tips*

Use Resources. You may wish to use a reference book (encyclopedia, almanac, mathematics handbook, etc.) that has a perpetual calendar to count the Saturdays,

or

Use logical reasoning and make a list. Check today's date on last year's calendar and next year's calendar. How do the days in the week compare? What happens during a leap year? (The year 1800 was not a leap year.)

2 GEOMETRY—DRAW DIAGRAMS

What is the maximum number of regions that can be formed by 4 circles? The circles can overlap.

3 MINUTE BY MINUTE

The sum of the digits shown on this digital watch is 10. How many minutes before the sum of the digits will be 20?

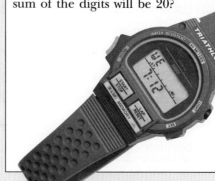

The number 808 looks like its mirror image. What other numbers less than 1000 look like their mirror image?

Add together the numbers for the days in April before Mindy's birthday. The sum will be four times the number for her birthday. When is Mindy's birthday?

APRIL						
S	M	T	W	TH	F	S
	1	2	3	4	5	6
7	8	9	10	11	12	13
14	15	16	17	18	19	20
21	22	23	24	25	26	27
28	29	30				

What is the area of the red region? *Hint: Think about the area of the uncolored region.*

a. 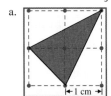   b.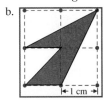

If a coach bought 12 of these boxes of baseballs, how many boxes would be left?

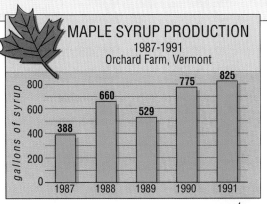

MAPLE SYRUP PRODUCTION
1987–1991
Orchard Farm, Vermont

gallons of syrup

Year	Gallons
1987	388
1988	660
1989	529
1990	775
1991	825

8 It takes about 10 gallons of sap to make a quart of maple syrup. About how many gallons of sap were required in 1991?

9 In its maple grove, Orchard Farm has 3200 taps (a small hole drilled 3 inches deep into the trunk of a maple tree). About how many fluidounces of syrup per tap did it average in 1988? (1 quart = 32 fluidounces)

10 What was the percentage increase in production between 1987 and 1991? Round your answer to the nearest whole percent.

Rate Your Problem-Solving Power: *6–7 correct = Good; 8–9 correct = Excellent; 10 correct = Exceptional*

OBJECTIVE: To use budgets for solving problems.

A **budget** is a plan for using one's money.

Danielle uses a budget to keep track of her spending so that she won't run out of money between paychecks. Alvita keeps a budget to make sure she puts some money into savings each week.

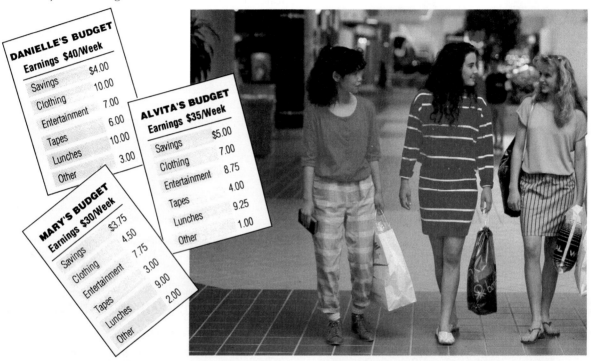

DANIELLE'S BUDGET
Earnings $40/Week

Savings	$4.00
Clothing	10.00
Entertainment	7.00
Tapes	6.00
Lunches	10.00
Other	3.00

ALVITA'S BUDGET
Earnings $35/Week

Savings	$5.00
Clothing	7.00
Entertainment	8.75
Tapes	4.00
Lunches	9.25
Other	1.00

MARY'S BUDGET
Earnings $30/Week

Savings	$3.75
Clothing	4.50
Entertainment	7.75
Tapes	3.00
Lunches	9.00
Other	2.00

▶ **Use the budgets to solve each problem.**

1. Which person budgets $\frac{1}{4}$ of her total earnings for clothing?

2. Who budgets $\frac{1}{8}$ of her total earnings for savings?

3. What fraction of Danielle's total earnings does she save? Give the answer in lowest terms.

4. What fraction of her total earnings does Alvita budget for entertainment? Give the answer in lowest terms.

5. Who spends 15% of her total earnings for clothing?

6. Who spends 50% of her total earnings for clothing and lunches?

7. What percent of Alvita's total earnings does she budget for clothing?

8. **Write a problem** that can be solved using one of the budgets given above.

▶ **Study the clues. Use the budgets to name each person.**

9. *Clues:*
   This person budgets more than 19% of her total earnings for clothing.
   She budgets less than 15% of her total earnings for tapes.

10. *Clues:*
    This person budgets less than 13% of her total earnings for savings.
    She budgets less than 21% of her total earnings for clothing.

▶ **Give the prime or algebraic factorization.** *(page 130)*

1. 56　　　　**2.** 98　　　　**3.** 16　　　　**4.** 72　　　　**5.** 81　　　　**6.** 68

7. $9f$　　　**8.** $7e$　　　**9.** gh　　　**10.** $10m$　　**11.** p^3　　**12.** $8qv$

13. $12x^2$　　**14.** $42ab$　　**15.** $38y$　　**16.** $3r^2t$　　**17.** $40n^3$　　**18.** c^3d

▶ **Give the greatest common factor.** *(page 132)*

19. 6, 24　　　　**20.** 12, 18　　　　**21.** 20, 10　　　　**22.** 36, 24

23. 40, 30　　　　**24.** $4, 8x$　　　　**25.** $18y, 27$　　　**26.** $f, 9f$

27. $6c, 9c$　　　**28.** $32j, 8j$　　　**29.** $3w^2, 14w$　　**30.** $18y, 20y^2$

31. $6n^2, 54n$　　**32.** $12u^2v, 18uv$　　**33.** $18ab, 21b^2$　　**34.** $36c^2d^2, 24cd^2$

▶ **Give each sum or difference in simplest form.** *(page 168)*

35. $\dfrac{1}{6} + \dfrac{1}{3}$　　**36.** $\dfrac{1}{2} + \dfrac{1}{3}$　　**37.** $\dfrac{3}{4} + \dfrac{5}{6}$　　**38.** $\dfrac{5}{6} - \dfrac{5}{8}$　　**39.** $\dfrac{7}{8} - \dfrac{2}{3}$

40. $\dfrac{a}{4} + \dfrac{b}{8}$　　**41.** $\dfrac{m}{16} + \dfrac{n}{8}$　　**42.** $\dfrac{5x}{12} + \dfrac{y}{8}$　　**43.** $\dfrac{w}{5} + \dfrac{4z}{15}$　　**44.** $\dfrac{2x}{21} + \dfrac{3y}{7}$

45. $\dfrac{x}{3} - \dfrac{y}{6}$　　**46.** $\dfrac{a}{4} - \dfrac{b}{5}$　　**47.** $\dfrac{n}{12} - \dfrac{3m}{8}$　　**48.** $\dfrac{2v}{9} - \dfrac{u}{12}$　　**49.** $\dfrac{5p}{24} - \dfrac{3q}{16}$

▶ **Give each product in simplest form.** *(page 184)*

50. $\dfrac{5}{8} \cdot \dfrac{8}{5}$　　**51.** $\dfrac{1}{8} \cdot \dfrac{3}{4}$　　**52.** $4 \cdot \dfrac{5}{8}$　　**53.** $\dfrac{9}{10} \cdot \dfrac{5}{4}$　　**54.** $\dfrac{12}{5} \cdot \dfrac{10}{3}$

55. $\dfrac{x}{3} \cdot \dfrac{1}{x}$　　**56.** $8 \cdot \dfrac{y}{4}$　　**57.** $\dfrac{5}{8} \cdot \dfrac{4y}{15}$　　**58.** $\dfrac{2}{5d} \cdot \dfrac{5}{11e}$　　**59.** $\dfrac{u}{v} \cdot \dfrac{3v}{2u}$

60. $\dfrac{n}{m^2} \cdot \dfrac{4m}{n}$　　**61.** $\dfrac{f}{g} \cdot \dfrac{g}{f}$　　**62.** $\dfrac{j^2}{4} \cdot \dfrac{k}{j}$　　**63.** $\dfrac{6y}{z} \cdot \dfrac{z}{3w}$　　**64.** $\dfrac{8d}{e} \cdot \dfrac{4e}{24d^2}$

▶ **Give each quotient in simplest form.** *(page 188)*

65. $\dfrac{5}{8} \div \dfrac{1}{8}$　　**66.** $\dfrac{7}{8} \div \dfrac{3}{2}$　　**67.** $\dfrac{7}{8} \div \dfrac{3}{4}$　　**68.** $\dfrac{3}{2} \div 2$　　**69.** $8 \div \dfrac{5}{4}$

70. $\dfrac{5}{c} \div \dfrac{3}{d}$　　**71.** $\dfrac{8}{n} \div \dfrac{2}{m}$　　**72.** $\dfrac{d}{12} \div \dfrac{e}{8}$　　**73.** $\dfrac{2y}{z} \div \dfrac{y}{6}$　　**74.** $\dfrac{8}{q} \div \dfrac{4q}{5}$

75. $\dfrac{4z}{y} \div \dfrac{3x}{y}$　　**76.** $\dfrac{a}{b} \div \dfrac{3a}{4b}$　　**77.** $\dfrac{3d^2}{e} \div \dfrac{d}{e}$　　**78.** $\dfrac{x}{12y} \div \dfrac{x^2}{9y}$　　**79.** $\dfrac{12m}{5n} \div \dfrac{4m}{10n^2}$

12-5 Graphing Ordered Pairs

OBJECTIVE: To give the coordinates of a point on a graph.
To graph an ordered pair.

Look at the picture at the right.

1. What is the horizontal number line called?

2. What is the vertical number line called?

3. Notice that the two axes intersect. What is the point of intersection called?

The Coordinate Plane

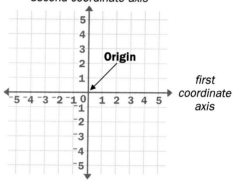

EXAMPLES | **Here's how to graph ordered pairs on the coordinate plane.**

A (3, 4) is read as "the ordered pair three four."
3 is called the first coordinate and 4 is called the second coordinate.

To graph (3, 4):
  a. Start at the origin.
  b. Move along the first coordinate axis to the first coordinate of the ordered pair, 3.
  c. Move parallel to the second coordinate axis to the second coordinate of the ordered pair, 4.

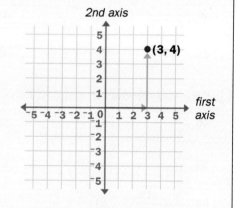

B **To graph** $\left(^-2, ^-3\frac{1}{2}\right)$:
  a. Start at the origin.
  b. Move along the first coordinate axis to the first coordinate of the ordered pair, $^-2$.
  c. Move parallel to the second coordinate axis to the second coordinate of the ordered pair, $^-3\frac{1}{2}$.

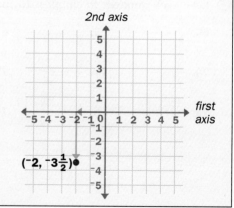

EXERCISES

▶ **Give the ordered pair for each point.**

4. A

5. B

6. C

7. D

8. E

9. F

10. G

11. H

12. I

13. J

14. K

15. L

16. M

17. N

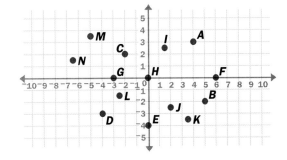

▶ **Graph these ordered pairs. Label each point with its ordered pair.**

18. $(2, 5)$

19. $(^-4, 6)$

20. $(5, ^-3)$

21. $(6, 0)$

22. $(0, 0)$

23. $(0, 8)$

24. $(^-4, ^-1)$

25. $\left(^-4, 5\frac{1}{2}\right)$

26. $\left(^-2\frac{1}{2}, ^-5\frac{1}{2}\right)$

27. $\left(5\frac{1}{2}, ^-1\frac{1}{2}\right)$

▶ **Solve.**

28. Here are some ordered pairs in which the second coordinate is 3 more than the first coordinate.

$(^-4, ^-1), (^-3, 0), (^-2, 1)$

a. List 3 more such ordered pairs.
b. Graph all 6 ordered pairs.
c. Do the points "line up"?

29. Here are some ordered pairs in which the second coordinate is the square of the first coordinate.

$(^-3, 9), (^-2, 4), (^-1, 1)$

a. List 3 more such ordered pairs.
b. Graph all 6 ordered pairs.
c. Do the points "line up"?

Group Project REFLECTIONS AND TRANSLATIONS *Triangle Tricks*

▶ **Work in a small group.**

30. On graph paper, draw the triangle with vertices at $(0, 1)$, $(3, 4)$, and $(6, 1)$. Then draw on the same graph paper the triangle you get when you
   a. multiply the first coordinate of the ordered pairs $(0, 1)$, $(3, 4)$, and $(6, 1)$ by $^-1$ and keep the same second coordinate.
   b. multiply the second coordinate of the ordered pairs $(0, 1)$, $(3, 4)$, and $(6, 1)$ by $^-1$ and keep the same first coordinate.

31. Compare your drawings with the drawings of other members of your group. About which axis would you flip the original triangle to get the new triangle in Exercise 30, Part a? In Exercise 30, Part b?

32. Predict what effect each of the following would have on the original triangle.
   a. Adding 2 to each first coordinate
   b. Subtracting 3 from each second coordinate

33. Verify your predictions in Exercise 32.

OBJECTIVE: To make a table of values that are solutions of equations in two variables.

Thus far you have worked with equations that have only one variable. In this lesson you will work with equations that have two variables. Here is an example:

$$y = 2x + 3$$

Notice that for $x = 0$ and $y = 3$ you get a true equation. Therefore, the ordered pair

(0, 3) ← | *Notice that the value of x is written first and the value of y is written second.*

is one solution of the equation. Actually such equations have an infinite number of solutions.

1. Substitute 3 for x and 0 for y. Is $(3, 0)$ a solution?

2. Is $(^-1, 1)$ a solution?

3. Is $\left(\dfrac{^-1}{2}, 2 \right)$ a solution?

E X A M P L E | **Here's how to find solutions of equations with two variables.**

To find an ordered pair that is a solution of an equation, choose a value for x and solve the equation to find the value for y.

Find three solutions for $y = 3x - 5$.

If you choose
2 for x, then

$$y = 3 \cdot 2 - 5$$
$$= 1$$

So $(2, 1)$ is a solution.

If you choose
$^-3$ for x, then

$$y = 3 \cdot {}^-3 - 5$$
$$= {}^-14$$

So $(^-3, {}^-14)$ is a solution.

If you choose
$\dfrac{2}{3}$ for x, then

$$y = 3 \cdot \dfrac{2}{3} - 5$$
$$= 2 - 5$$
$$= {}^-3$$

So $\left(\dfrac{2}{3}, {}^-3 \right)$ is a solution.

CHECK for Understanding

4. Look at the example. If you choose 2 for x and solve the equation for y, you get the solution $(2, \underline{\ ?\ })$.

5. What two other ordered pairs were found to be solutions of the equation?

6. Often a table of values is used to show solutions of an equation with two variables. Complete this table of values for the solutions found above.

$y = 3x - 5$	
x	y
2	1
$^-3$	$^-14$
$\dfrac{2}{3}$	?

EXERCISES

▶ Copy and complete each table of values.

7.

y = 2x	
x	y
⁻2	?
⁻1	?
0	?
1	?
2	?

8.

y = ⁻3x	
x	y
⁻3	?
⁻2	?
⁻1	?
0	?
1	?

9.

y = 5x + 1	
x	y
⁻1	?
0	?
1	?
2	?
3	?

10.

$y = \dfrac{1}{2}x - 1$	
x	y
⁻2	?
0	?
2	?
4	?
6	?

▶ Copy the equation and make a table of values showing four solutions.

11. $y = {}^-2x$

12. $y = 5x$

13. $y = 5x - 8$

14. $y = 3x + 7$

15. $y = 4x + 3$

16. $y = {}^-2x - 6$

17. $y = \dfrac{1}{4}x + 2$

18. $y = \dfrac{{}^-3}{4}x - 1$

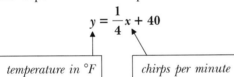
Problem Solving   USING A FORMULA

▶ Use the equation below to solve the problems.

Nature provides us with many interesting number relationships. For example, the number of times a cricket chirps per minute depends on the temperature.

$$y = \frac{1}{4}x + 40$$

temperature in °F   chirps per minute

19. What is the temperature when a cricket chirps 20 times per minute?

20. How many chirps per minute should a cricket make when the temperature is 60°F?

Challenge!   LOOK FOR A PATTERN

▶ Write an equation for the table of values.

21.

?	
x	y
1	4
2	8
3	12
4	16

22.

?	
x	y
⁻2	⁻3
⁻1	⁻1
0	1
1	3

23.

?	
x	y
⁻1	2
0	0
1	⁻2
2	⁻4

24.

?	
x	y
0	4
1	2
2	0
3	⁻2

Graphing Equations and Inequalities | **347**

OBJECTIVE: To graph linear equations.

An equation of the form of

$$y = 2x + 1$$

is called a **linear equation,** because if you graph the solutions on the coordinate plane, the points all lie on a straight line. (Notice that the first four letters in the word *linear* spell the word *line.*)

$y = 2x + 1$	
x	y
$^-2$	$^-3$
$^-1$	$^-1$
$\frac{1}{2}$	2
$1\frac{1}{2}$	4
2	5

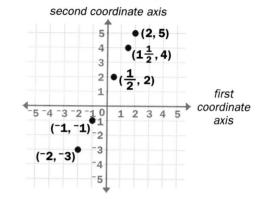

E X A M P L E | **Here's how to graph a linear equation.**

Graph the equation $y = {}^-3x + 2$.

Step 1. Make a table of values.

$y = {}^-3x + 2$	
x	y
$^-1$	5
0	2
1	$^-1$
2	$^-4$

Step 2. Label the first coordinate axis x and the second coordinate axis y.

Step 3. Graph the ordered pairs given in the table of values.

Step 4. Draw a line through the points.

Step 5. Label the graph with the equation.

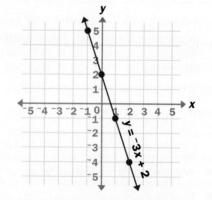

CHECK for Understanding

1. Look at the example. When a linear equation is graphed, the first coordinate axis is labeled x and the second coordinate axis is labeled __?__ .

2. How many points were graphed before the line was drawn?

3. Every ordered pair on the line is a solution of the equation $y = {}^-3x + 2$. By looking at the graph, do you think that $\left(\frac{^-1}{3}, 3 \right)$ could be a solution? Is it a solution?

EXERCISES

▶ First copy and complete the table of values. Then graph the equation.

4.

$y = x$	
x	y
$^-2$	?
$^-1$	?
0	?
1	?

5.

$y = x - 4$	
x	y
$^-1$	?
0	?
1	?
2	?

6.

$y = {}^-2x$	
x	y
$^-2$	?
$^-1$	?
0	?
1	?

7.

$y = 2x - 3$	
x	y
0	?
1	?
2	?
3	?

8.

$y = {}^-3x + 1$	
x	y
$^-1$	?
0	?
1	?
2	?

9.

$y = 3x + 2$	
x	y
$^-2$	?
$^-1$	?
0	?
1	?

10.

$y = \dfrac{1}{2}x + 2$	
x	y
$^-4$	?
$^-2$	?
0	?
2	?

11.

$y = \dfrac{2}{3}x - 3$	
x	y
$^-3$	?
0	?
3	?
6	?

▶ First make a table of values showing three solutions. Then graph the equation.

12. $y = 3x$ **13.** $y = 4x$ **14.** $y = 2x + 3$ **15.** $y = 3x - 2$

16. $y = {}^-2x + 1$ **17.** $y = 3x + 1$ **18.** $y = \dfrac{1}{3}x - 2$ **19.** $y = \dfrac{{}^-1}{3}x - 2$

20. Refer to your graphs for exercises 16 through 19. At what point does the graph cross the y-axis?

LOOK FOR A PATTERN

▶ Find the equation for each graph. *Hint: List some ordered pairs and look for a pattern.*

21.

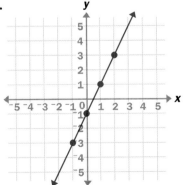

22.

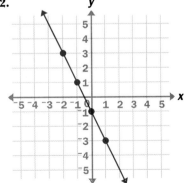

OBJECTIVE: To find the slope and *y*-intercept of a line. To write the equation of a line, given the slope and *y*-intercept.

The **slope** of a line is the ratio of the change in *y* to the change in *x* between any two points on the line.

$$\text{slope} = \frac{\text{change in } y\text{-variable}}{\text{change in } x\text{-variable}}$$

E X A M P L E | **Here's how to find the slope of a line using two points on the line.**

A Find the slope of the line with points *A* and *B*.

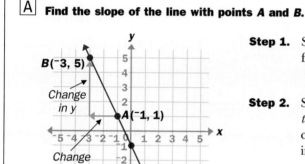

Step 1. Subtract the *y* coordinates to find the change in *y*.

$$5 - 1 = 4$$

Step 2. Subtract the *x* coordinates in *the same order* as the *y* coordinates to find the change in *x*.

$$^-3 - {^-1} = {^-2}$$

Step 3. Write the ratio. slope $= \dfrac{4}{^-2} = {^-2}$

CHECK for Understanding

1. Look at Example A. The slope of the line is __?__ .

2. At what point does the line cross the *y*-axis?

The **y-intercept** of a line is the *y*-value at the point where the line crosses the *y*-axis. The *y*-intercept of the line in the example is $^-1$.

E X A M P L E | **Here's how to use the slope *(m)* and the *y*-intercept *(b)* to write the equation of a line.**

B Find the equation of the line shown in Example A.

For any line, $y = mx + b$ where *m* is the slope and *b* is the *y*-intercept.
We have computed the slope to be $^-2$, so $m = {^-2}$.
The graph crosses the *y*-axis at $^-1$, so $b = {^-1}$.
So the equation of the line is $y = mx + b$

$$y = {^-2}x + {^-1}$$
$$y = {^-2}x - 1$$

CHECK for Understanding

3. For any line with equation $y = mx + b$, m is the __?__ of the line and b is the y-intercept.

4. The equation of the line in the examples is __?__ .

EXERCISES

▶ **Write the slope (m) and y-intercept (b) of each line.**

   5. $y = 2x + 1$ **6.** $y = {}^-3x + 4$ **7.** $y = 4x + 1$ **8.** $y = 4x - 1$

   9. $y = x + 6$ **10.** $y = x - 6$ **11.** $y = 3x$ **12.** $y = 4$

▶ **For each slope (m) and y-intercept (b), write the equation of the line.**

   13. $m = 4$; $b = 3$ **14.** $m = 3$; $b = 1$ **15.** $m = {}^-2$; $b = 4$ **16.** $m = {}^-3$; $b = 2$

   17. $m = 2$; $b = {}^-1$ **18.** $m = {}^-4$; $b = {}^-2$ **19.** $m = 1$; $b = 5$ **20.** $m = 0$; $b = {}^-6$

▶ **First find the slope and y-intercept. Then write the equation of the line.**

21.

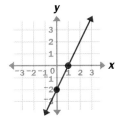

22.

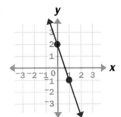

23.

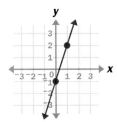

24.

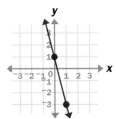

25.

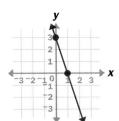

26.

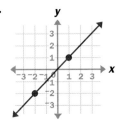

Challenge!   WRITING EQUATIONS

▶ **Find the equation of the line containing each pair of points.** *Hint: One of the ordered pairs gives the y-intercept.*

   27. $(2, 5)$, $(0, 1)$

   28. $(0, 4)$, $(2, {}^-6)$

   29. $(0, 0)$, $({}^-1, 4)$

Graphing Equations and Inequalities | **351**

OBJECTIVE: To write an inequality in two variables for a graph on the coordinate plane. To graph an inequality in two variables on the coordinate plane.

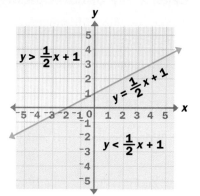

The graph of a linear equation separates the coordinate plane into two regions. The **region above** the graph of $y = \frac{1}{2}x + 1$ is the graph of the linear inequality $y > \frac{1}{2}x + 1$.

1. Are the following ordered pairs solutions of the linear inequality $y > \frac{1}{2}x + 1$? *Hint: Substitute and simplify.*

    a. $(1, 2)$ **b.** $(^-2, 1)$ **c.** $(^-4, 0)$ **d.** $(0, 3)$

2. Do the graphs of the ordered pairs lie in the yellow region?

The **region below** the graph of $y = \frac{1}{2}x + 1$ is the graph of the linear inequality $y < \frac{1}{2}x + 1$.

3. Are the following ordered pairs solutions of the linear inequality $y < \frac{1}{2}x + 1$?

    a. $(2, 0)$ **b.** $(3, 2)$ **c.** $(^-1, 0)$ **d.** $(^-4, ^-3)$

4. Do the graphs of the ordered pairs lie in the blue region?

EXAMPLES | Here's how to graph a linear inequality.

A   **Graph $y \leq x + 1$.**

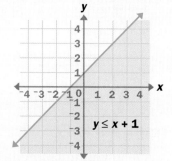

Step 1. Graph $y = x + 1$.

Step 2. Shade in the region **below** the graph of $y = x + 1$.

Step 3. Label the graph $y \leq x + 1$.

Note: a solid line for the graph of $y = x + 1$ shows that it is included in the graph of $y \leq x + 1$.

B   **Graph $y > \frac{1}{3}x - 2$.**

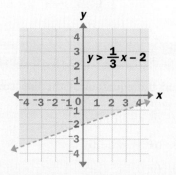

Step 1. Graph $y = \frac{1}{3}x - 2$.

Step 2. Shade in the region **above** the graph of $y = \frac{1}{3}x - 2$.

Step 3. Label the graph $y > \frac{1}{3}x - 2$.

Note: A dashed line for the graph of $y = \frac{1}{3}x - 2$ shows

*that it is **not** included in the graph of $y > \frac{1}{3}x - 2$.*

CHECK for Understanding

5. Look at Example A. The graph of $y \leq x + 1$ includes the graph of $y = x + 1$ and the region __?__ the graph of $y = x + 1$.

6. Look at Example B. The graph of $y > \frac{1}{3}x - 2$ is the region __?__ the graph of $y = \frac{1}{3}x - 2$.

EXERCISES

▶ In each exercise you are given the graph of a linear inequality. First study the equation of the line. Then give the inequality.

7.

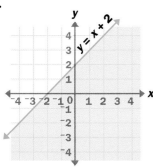

8.

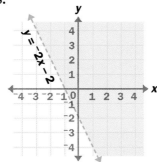

9.

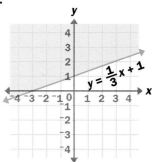

10.

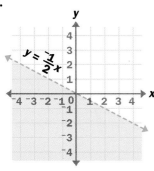

11.

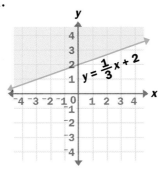

12.

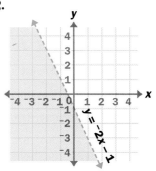

▶ Graph each linear equation or linear inequality.

13. $y = 3x - 5$

14. $y = {}^-2x + 8$

15. $y = \frac{-1}{2}x + 3$

16. $y = \frac{2}{3}x - 3$

17. $y < 2x$

18. $y > {}^-3x$

19. $y < x + 4$

20. $y > {}^-2x - 5$

21. $y \geq 3x + 1$

22. $y \leq {}^-2x - 3$

23. $y \geq \frac{1}{2}x - 1$

24. $y \leq \frac{2}{3}x + 2$

Graphing Equations and Inequalities | **353**

OBJECTIVE: To solve a system of linear equations by graphing.

Notice that the graphs of

$$y = {}^-2x - 6 \text{ and } y = \frac{1}{2}x - 1$$

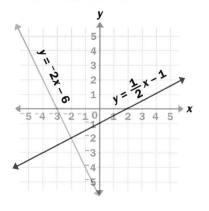

are shown on the same coordinate plane.

1. The ordered pair $({}^-3, 0)$ is a solution of which equation?

2. The ordered pair $({}^-4, {}^-3)$ is a solution of which equation?

3. What ordered pair appears to be a solution of both equations? *Hint: Where do the graphs intersect?*

The ordered pair that is a solution of both equations is the solution of the **system of equations**

$$y = {}^-2x - 6$$

$$y = \frac{1}{2}x - 1$$

EXAMPLE | **Here's how to solve a system of equations by graphing.**

Solve the system of equations

$$y = {}^-2x + 5$$

$$y = 3x$$

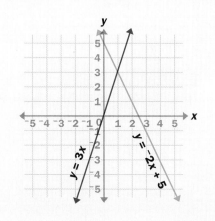

Step 1. Graph both equations on the same coordinate plane.

Step 2. It appears that $(1, 3)$ is a solution of both equations. To be sure that $(1, 3)$ is a solution of the system of equations, we check to see whether $(1, 3)$ is a solution of both equations.

$$y = {}^-2x + 5 \qquad\qquad y = 3x$$

$$3 \stackrel{?}{=} {}^-2 \cdot 1 + 5 \qquad 3 \stackrel{?}{=} 3 \cdot 1$$

$$3 \stackrel{?}{=} {}^-2 + 5$$

$$3 = 3 \qquad\qquad\qquad 3 = 3$$

So $(1, 3)$ is the solution of the system of equations.

EXERCISES

▶ Determine whether the ordered pair is the solution of the system of equations by substituting in both equations.

4. $(1, 2)$

$y = {}^-3x + 5$

$y = 2x$

5. $(2, {}^-1)$

$y = x - 3$

$y = 3x - 7$

6. $(0, 1)$

$y = 2x + 1$

$y = 3x - 1$

7. $({}^-1, {}^-1)$

$y = x$

$y = {}^-2x - 3$

8. $({}^-2, 6)$

$y = \dfrac{{}^-1}{3}x$

$y = 4x + 2$

▶ First study the graphs and list the ordered pair that appears to be the solution of the system of equations. Then check to see if the ordered pair is a solution by substituting in both equations.

9.

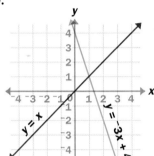

10.

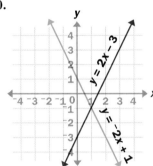

11.

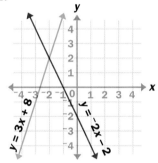

12.

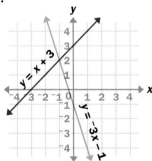

13.

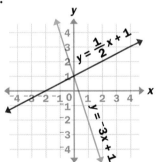

14.

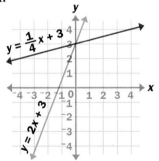

▶ Solve each system of equations by graphing.

15. $y = x$

$y = 3x - 4$

16. $y = 2x$

$y = x - 1$

17. $y = x + 3$

$y = 2x + 1$

18. $y = 3x - 2$

$y = {}^-4x - 2$

19. $y = 2x + 3$

$y = \dfrac{1}{2}x + 2$

20. $y = {}^-3x - 3$

$y = \dfrac{{}^-1}{2}x + 2$

Graphing Equations and Inequalities | **355**

12-11 Problem Solving—Using Systems of Linear Equations

OBJECTIVE: To use systems of linear equations to solve problems.

In this lesson you will use what you learned about systems of linear equations to solve problems.

| EXAMPLE | Here's how to use systems of linear equations to solve problems. |

Problem: Cassie and Daniel caught a total of 12 fish. Cassie caught 3 times as many fish as Daniel. How many fish did each person catch?

Step 1. Choose variables. Use them and the facts to write two equations.

Let c = the number of fish Cassie caught. Let d = the number of fish Daniel caught.

Together they caught 12 fish.

$c + d = 12$

Cassie caught 3 times as many fish as Daniel.

$c = 3d$

Step 2. Solve the system of equations.

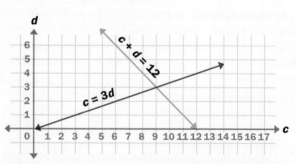

It appears that (9, 3) is the solution to both equations.
So Cassie caught 9 fish and Daniel caught 3.

CHECK for Understanding

1. Look at the example above.
   a. In Step 1, we let c equal the number of fish __?__ caught and __?__ equal the number of fish Daniel caught. The two equations for the problem are __?__ and __?__ .
   b. In Step 2, the ordered pair (__?__ , __?__) is the solution to both equations. So Cassie caught __?__ fish and Daniel caught __?__ fish.

2. To check the solution, ask yourself whether the answer fits the facts in the problem. Does 9 plus 3 equal 12? Does 3 times 3 equal 9?

EXERCISES

▶ **Read the facts. Then complete the steps to answer the question.**

3. *Facts:* Michelle and Nathan earned a total of $20. Michelle earned $8 more than Nathan.

   Question: How many dollars did each earn?

   a. If you let *m* equal the number of dollars Michelle earned and let *n* equal the number of dollars Nathan earned, then two equations for the problem are $m + n =$ _?_ and $n +$ _?_ $= m$.

   b. Look at the graph. The ordered pair (_?_ , _?_) is the solution to both equations. So Michelle earned _?_ dollars and Nathan earned _?_ dollars.

4. *Facts:* Stuart and Tony scored a total of 17 points. Tony scored 7 fewer points than Stuart.

   Question: How many points did each score?

   a. If you let *s* equal the number of points Stuart scored and let *t* equal the number of points Tony scored, then two equations for the problem are $s + t =$ _?_ and $s -$ _?_ $= t$.

   b. Look at the graph. The ordered pair (_?_ , _?_) is the solution to both equations. So Stuart scored _?_ points and Tony scored _?_ points.

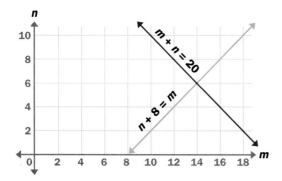

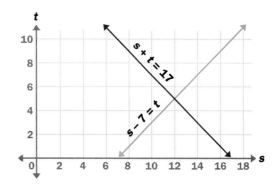

▶ **Solve each problem. First write two equations. Then solve the system of equations by graphing.**

5. The sum of two numbers is 22. The first number (*f*) is 8 more than the second number (*s*). What are the two numbers?

6. The sum of two numbers is 23. The first number (*f*) is 5 less than the second number (*s*). What are the two numbers?

7. The sum of two numbers is 24. The first number (*f*) is 3 times the second number (*s*). What are the two numbers?

8. The sum of two numbers is 18. The first number (*f*) is 2 times the second number (*s*). What are the two numbers?

9. Jim and Kelly sold a total of 26 tickets. The number of tickets Jim sold (*j*) was 8 more than the number of tickets Kelly sold (*k*). How many tickets did each sell?

10. Fran and Greg spent a total of $10. The number of dollars Fran spent (*f*) was 3 times as great as the number of dollars Greg spent (*g*). How many dollars did each spend?

Graphing Equations and Inequalities | **357**

OBJECTIVE: To solve problems about direct and indirect variation.

Think about riding a bicycle 8 miles per hour. In 1 hour, you would travel 8 miles, in 2 hours you would travel 16 miles, and so on. The equation $d = 8t$ or $y = 8x$ can be used to describe the situation. A table of values is given. Notice that as the value of x increases, the value of y increases. This is an example of **direct variation.**

$y = 8x$	
x	y
1	8
2	16
3	24
4	32

A variable y varies directly with x if there is a positive number k such that $y = kx$. What is the value of k in the example above?

Think about riding a bicycle a distance of 24 miles. At the rate of 24 miles per hour, it would take you 1 hour; at the rate of 12 miles per hour, it would take 2 hours; and so on. The equation $rt = 24$, or $xy = 24$ can be used to describe the situation. A table of values is given. Notice that as the value of x decreases, the value of y increases. This is an example of an **indirect** or **inverse variation.**

$xy = 24$	
x	y
24	1
12	2
8	3
6	4

A variable y **varies inversely with** x if there is a positive number k such that $xy = k$.

What is the value of k in the example above?

▶ **First tell whether y varies directly or inversely with x. Then give the value of k.**

1.

x	y
1	3
2	6
3	9
4	12

2.

x	y
1	12
2	6
3	4
4	3

3.

x	y
1	5
2	10
3	15
4	20

4.

x	y
1	6
2	3
3	2
4	$\dfrac{3}{2}$

▶ **Decide whether the situation is a direct or indirect variation.**

5. A sports car averages 32 miles on a gallon of gasoline. How does the total miles traveled vary with the total number of gallons used?

6. A motorcyclist is planning a 120-mile trip. How will his total time traveling vary with his average speed?

▶ **Change to a fraction, whole number, or mixed number. Give each answer in simplest form.** *(page 244)*

1. 40%   **2.** 75%   **3.** 5%   **4.** 80%   **5.** 1%   **6.** 50%

7. 18%   **8.** 65%   **9.** 132%   **10.** 100%   **11.** 64%   **12.** 150%

13. 400%   **14.** 175%   **15.** 350%   **16.** $12\frac{1}{2}\%$ **17.** $33\frac{1}{3}\%$ **18.** $208\frac{1}{3}\%$

▶ **Change to a percent.** *(page 246)*

19. 0.25   **20.** 1   **21.** 0.6   **22.** $0.16\frac{2}{3}$ **23.** 0.05   **24.** 1.36

25. 0.1   **26.** $0.87\frac{1}{2}$ **27.** 2   **28.** 0.62   **29.** 1.50   **30.** 0.3

31. 0.29   **32.** 0.09   **33.** $1.33\frac{1}{3}$ **34.** 5   **35.** 0.06   **36.** 2.75

▶ **Solve.** *(page 252)*

37. 25% of 64 = n **38.** 60% of 80 = n **39.** 120% of 65 = n

40. 43% of 75 = n **41.** 89% of 27 = n **42.** 61% of 98 = n

43. 6.2% of 30 = n **44.** 5.9% of 45 = n **45.** 0.5% of 82 = n

46. 0.9% of 46 = n **47.** 125.8% of 100 = n **48.** 87.5% of 120 = n

▶ **Solve.** *(page 256)*

49. 20% of n = 17   **50.** 50% of n = 23   **51.** 75% of n = 42

52. 80% of n = 36   **53.** 150% of n = 48   **54.** 125% of n = 65

55. 37.5% of n = 18   **56.** 1% of n = 2.25   **57.** 87.5% of n = 84

58. 17.5% of n = 50.75   **59.** 225% of n = 96.75   **60.** 275% of n = 49.5

▶ **Write and solve an equation to find the measure of each angle.** *(page 270)*

61.

62.

63.
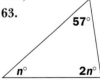

Here are scrambled answers for the review exercises:

⁻1	both	graphing	not	ratio
1	coordinate	line	ordered	reverse
2	crosses	linear	origin	second
	divide	negative	plane	slope

1. **Inequality:** $x < \dfrac{^{-}1}{2}$ The graph shows that $\dfrac{^{-}1}{2}$ is __?__ a solution.
   (page 334)

   Graph: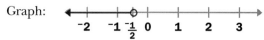

2. If you multiply or divide both sides of an inequality by a __?__ number, you must reverse the inequality sign. To solve the inequality $^{-}3n < 8,$ you would __?__ both sides by $^{-}3$ and __?__ the inequality sign. *(page 336)*

3. To graph the ordered pair $\left(2,\ ^{-}1\dfrac{1}{2}\right),$ you would start at the __?__ and move along the first coordinate axis to the first __?__ of the ordered pair, 2. Then you would move parallel to the __?__ coordinate axis to the second coordinate, $^{-}1\dfrac{1}{2}.$
   (page 344)

4. To find an __?__ pair that is a solution of $y = 2x - 7,$ choose a value for x and solve the equation for $y.$ *(page 346)*

5. An equation of the form $y = 2x + 1$ is called a __?__ equation. If you graph the equation, you will get a straight __?__ . *(page 348)*

6. The slope of a line is the __?__ of the change in y to the change in x between any two points. The slope of the line shown at the right is __?__ .
   The y-intercept is the y-value at the point where the line __?__ the y-axis.
   The y-intercept of the line shown at the right is __?__ .
   For any line, $y = mx + b,$ m is the __?__ and b is the y-intercept. The equation of the line shown at the right is $y = 2x - 1.$ *(page 350)*

   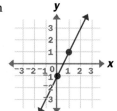

7. To solve a system of equations by __?__ , you would first graph both equations on the same coordinate __?__ . In the example, it appears that $(\underline{\ ?\ }, 0)$ is a solution of both equations. The last step would be to check to see whether $(1, 0)$ is a solution of __?__ equations. *(page 354)*

   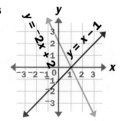

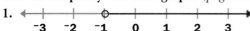
▶ **Write an inequality for each graph.** *(page 334)*

1. 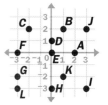 ← -3 -2 -1 0 1 2 3 →

2. ← -3 -2 -1 0 1 2 3 →

▶ **Solve.** *(pages 336, 338)*

3. $n + 8 \leq {}^-3$ **4.** $\dfrac{n}{{}^-3} \geq 8$ **5.** ${}^-5n + 8 \leq {}^-6$ **6.** $\dfrac{n}{{}^-6} + 2 \geq 0$

▶ **Give the ordered pair of each point.** *(page 344)*

7. A   **8.** B   **9.** C   **10.** D

11. E   **12.** F   **13.** G   **14.** H

15. I   **16.** J   **17.** K   **18.** L

▶ **First copy and complete each table of values. Then graph the equation.**
(pages 346, 348)

19. $y = {}^-2x$ **20.** $y = x + 8$ **21.** $y = 3x - 7$ **22.** $y = \dfrac{1}{2}x + 2$

x	y
0	?
1	?
2	?

x	y
⁻3	?
⁻2	?
⁻1	?

x	y
⁻1	?
0	?
1	?

x	y
⁻2	?
0	?
2	?

▶ **For each slope (m) and y-intercept (b), write the equation of the line.** *(page 350)*

23. $m = 2; b = 3$ **24.** $m = {}^-3; b = {}^-1$ **25.** $m = 0; b = 4$

▶ **Graph each linear inequality.** *(page 352)*

26. $y \geq 2x - 3$ **27.** $y \leq {}^-3x + 6$ **28.** $y > \dfrac{1}{3}x - 5$ **29.** $y < \dfrac{{}^-3}{4}x + 7$

▶ **Solve each system of equations by graphing.** *(page 354)*

30. $y = x$ **31.** $y = 2x$ **32.** $y = x + 5$ **33.** $y = {}^-3x + 1$

     $y = 2x - 1$ $y = x - 3$ $y = 2x - 1$ $y = \dfrac{1}{2}x + 1$

▶ **Write two equations for each problem. Then solve each system of equations
by graphing.** *(page 356)*

34. The sum of two numbers is 10. The
first number (f) is 6 more than the
second number (s). What are the two
numbers?

35. The sum of two numbers is 12. The
first number (f) is 2 less than the second number (s). What are the two
numbers?

Graphing Equations and Inequalities | **361**

▶ **Choose the correct letter.**

1. The prime factorization of 72 is

A. $8 \cdot 9$

B. $2 \cdot 3 \cdot 2 \cdot 6$

C. $2^3 \cdot 3^2$

D. none of these

2. The greatest common factor of

$12x^2y$ and $20xy^2$ is

A. $4x$ B. $4y$

C. $4x^2y^2$ D. none of these

3. Give the sum.

$$\frac{2u}{3} + \frac{v}{4}$$

A. $\dfrac{2u+v}{7}$ B. $\dfrac{8u+3v}{12}$

C. $\dfrac{4y+3v}{12}$ D. none of these

4. Give the difference.

$$\frac{c}{6} - \frac{3d}{4}$$

A. $\dfrac{2c-9d}{12}$ B. $\dfrac{c-3d}{2}$

C. $\dfrac{3c-2d}{12}$ D. none of these

5. Give the product.

$$\frac{2a}{b} \cdot \frac{b^2}{4a}$$

A. $\dfrac{2ab}{4}$ B. $\dfrac{2b}{4a}$

C. $\dfrac{b}{2}$ D. none of these

6. Give the quotient.

$$\frac{m}{3n} \div \frac{4m}{n^2}$$

A. $\dfrac{n}{12}$ B. $\dfrac{12}{n}$

C. $\dfrac{n^2}{12}$ D. none of these

7. Change to a decimal.

$87\frac{1}{2}\% = ?$

A. $8.7\frac{1}{2}$ B. $87\frac{1}{2}$

C. $0.87\frac{1}{2}$ D. none of these

8. Change to a percent.

$0.4 = ?$

A. 4% B. 400%

C. 0.004% D. none of these

9. Solve.

4.5% of $30 = n$

A. 1.35

B. 13.5

C. 6.67

D. none of these

10. Solve.

1% of $n = 4.75$

A. 0.0475

B. 47.5

C. 475

D. none of these

11. The percent of decrease from 75 to 60 is

A. 15%

B. 20%

C. 25%

D. none of these

12. Choose the equation.

Al bought 3 posters and Amy bought 2. Each poster cost the same. Amy also spent $6 for film. Together they spent $50. How much did 1 poster cost?

A. $3n + 2n + 6 = 50$

B. $3n + 2n - 6 = 50$

C. $3n - 2n + 6 = 50$

D. $3n - 2n - 6 = 50$

$$P(4, 6 \text{ or } 6, 4) = \frac{1}{18}$$

The probability of rolling a 4 and a 6, or a 6 and a 4 is $\frac{1}{18}$.

(page 371)

When you roll two dice, there are 36 possible outcomes. Therefore, the probability of rolling a 4 and a 6, or a 6 and a 4, so that two of the green Parcheesi tokens can reach "home" is $\frac{2}{36}$ or $\frac{1}{18}$.

OBJECTIVE: To find the total number of outcomes by using a tree diagram. To find the total number of outcomes by using a basic counting principle.

To make a baseball T-shirt, you need a T-shirt and a baseball decal. A store has red and green T-shirts and 3 kinds of decals.

1. How many different baseball T-shirts can be made with the red T-shirt?

2. How many different baseball T-shirts can be made with the green T-shirt?

3. How many different baseball T-shirts can be made in all?

  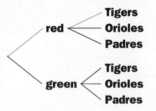

EXAMPLE | **Here's how to find the total number of baseball T-shirts that can be made.**

You can find how many different baseball T-shirts can be made by drawing a **tree diagram.**

```
              Tigers
    red  <--- Orioles
              Padres

              Tigers
    green <-- Orioles
              Padres
```

The blue "branch" represents a red T-shirt and a Tigers decal.

You can compute how many different baseball T-shirts can be made by using a **basic counting principle:**

If there are **2** ways to choose a T-shirt and **3** ways to choose a baseball decal, then the total number of ways to make a baseball T-shirt is **2 × 3**, or **6**.

CHECK for Understanding

4. Look at the tree diagram. What baseball T-shirt is represented by the top branch? The bottom branch?

5. How many different branches are there in the tree diagram?

6. Does the number of branches in the tree diagram equal the number of ways to make a baseball T-shirt?

EXERCISES

▶ **Solve.**

7. Suppose you have a supply of blue T-shirts and yellow T-shirts and a supply of the baseball decals shown at the right.

a. How many choices of T-shirts do you have? How many choices of baseball decals do you have?

b. Draw a tree diagram to show all possible baseball T-shirts you could make.

c. How many different baseball T-shirts could you make?

d. How many different baseball T-shirts could you make with the yellow T-shirt?

e. How many different baseball T-shirts could you make if you did not use a Cardinals decal?

f. Suppose instead that you have 5 colors of T-shirts and the 4 baseball decals. How many different baseball T-shirts could you make?

8. A snack bar serves 6 kinds of sandwiches and 4 kinds of juice. How many different orders of 1 sandwich and 1 juice could you place?

★9. How many different outfits could you make from 3 pairs of pants, 5 shirts, and 2 sweaters if each outfit consists of pants, a shirt, and a sweater?

★10. In an election of class officers, 3 students were running for president, 2 for vice president, and 2 for secretary. How many different ways could you vote?

★11. You decide to buy a stereo system. You can choose from 5 amplifiers, 4 speakers, and 3 turntables. How many different systems could you buy?

★12. Suppose you decide to buy a car and can choose any of the options shown on the list. How many choices would you still have if you decided that you wanted only

a. a red car?
b. a white interior?
c. a blue car with a black interior?
d. a yellow car with manual transmission?
e. a brown interior with automatic transmission?

CAR OPTIONS		
6 Paint colors	**3 Interior colors**	**2 Transmissions**
red	brown	manual
yellow	black	automatic
blue	white	
brown		
black		
white		

OBJECTIVE: To compute the number of permutations
(possible arrangements of things in a definite order).

Candice, Douglas, and Aileen are seated on a
toboggan.

1. Who is in the front? The middle? The back?

2. The letters *CDA* may be used to describe the
order in which they are seated. Use the three
letters to list all possible orders in which they
can be seated.

3. How many ways can the three people
be seated on the toboggan?

A **permutation**
is a
possible
arrangement
of things
in a
definite order.

Douglas

Aileen

Candice

EXAMPLE	Here's how to compute the number of permutations.

Number of people to choose from for first position		Number left to choose from for second position		Number left to choose from for third position		Number of possible arrangements
↓		↓		↓		↓
3	×	2	×	1	=	6

CHECK for Understanding

4. Look at the example above. To compute the number of permutations (possible
arrangements) of 3 things, you would multiply what three numbers?

5. Think about 4 people on a toboggan.
   a. How many people would there be to choose from for the front position?
   b. How many people would there be left to choose from for the second position? The third position? The back position?
   c. What four numbers would you multiply to compute the number of permutations of 4 things?
   d. How many ways can the 4 people be seated on a toboggan?

LIFTS
DOUBLE CHAIR LIFT
TRIPLE CHAIR LIFT
GONDOLA
T-BAR
ROPE TOW

EXERCISES

▶ Solve.

6. Candice, Douglas, and Aileen decided to go skiing. How many ways could they line up to purchase their lift tickets?

7. The sign shows the different lifts available at the ski area.
   a. How many lifts are there?
   b. In how many different orders could Candice take the 5 lifts?

8. Each gondola car seats 4 people. In how many ways can 4 people be seated in a gondola car?

9. Each seat on the triple-chair holds 3 people. In how many ways can 6 people be seated in 2 chairs?

10. There were 8 skiers in a race. In how many different orders could the 8 skiers finish?

★11. Each gondola car seats 4 people. If Candice, Douglas, and Aileen were the only people in the gondola car, in how many ways could they be seated?

12. Douglas decided to buy a sandwich and a drink for lunch. He could choose from 8 sandwiches and 5 drinks. How many different lunches could he buy?

13. Candice decided to buy some soup and a sandwich for lunch. She could choose from 6 soups and 8 sandwiches. How many different lunches could she buy?

14. There were 6 soups, 5 drinks, and 3 desserts. How many soup-drink-dessert lunches could be ordered?

15. How many soup-sandwich-drink-dessert lunches could be ordered?
    Hint: See exercises 13 and 14.

Challenge! **CALCULATOR APPLICATION**

▶ Solve.

16. Suppose that you and 9 of your friends are to be seated in a row of 10 seats at a theater.
    a. How many ways can the group be seated?
    b. Suppose that the group can get into a new seating order every 5 seconds. How many seconds would it take for the group to sit in all possible orders?
    c. How many minutes would it take? How many hours? How many days?

OBJECTIVE: To determine the probability of a simple event.

1. Suppose that you rolled the die. How many different outcomes would be possible? (How many different numbers could possibly land faceup?)

2. Would the possible outcomes be equally likely? (Would all numbers have the same chance of landing faceup?)

EXAMPLES | **Here's how to find the probability (the chance) of a particular event when all the outcomes are equally likely.**

$$\textbf{Probability of a particular event} = \frac{\text{Number of ways that the particular event can occur}}{\text{Number of possible outcomes}}$$

A Probability of rolling a 2

$$P(2) = \frac{\text{Number of ways of rolling a 2}}{\text{Number of possible outcomes}}$$

$$P(2) = \frac{1}{6}$$

B Probability of rolling either a 3 or a 4

$$P(3 \text{ or } 4) = \frac{2}{6}$$ *There are two ways to get either a 3 or a 4.*

$$P(3 \text{ or } 4) = \frac{1}{3}$$

CHECK for Understanding

3. Look at Example A. What is the probability of rolling a 2?

4. Look at Example B. What is the probability of rolling a 3 or a 4?

EXERCISES

▶ **Give each probability in simplest form. Think about rolling a die.**

  5. $P(4)$ **6.** $P(1 \text{ or } 3)$ **7.** $P(\text{odd number})$ **8.** $P(4, 5, \text{ or } 6)$

  9. $P(4 \text{ or less})$ **10.** $P(\text{not } 6)$ **11.** $P(\text{prime number})$ **12.** $P(\text{factor of } 12)$

13. $P(8) = \dfrac{0}{6}$ *There are no, or 0, ways that an outcome of 8 can occur.*

14. Rolling an 8 is an example of an impossible event. Look at your answer to Exercise 13. What is the probability of an impossible event?

15. $P(\text{a number less than } 9) = \dfrac{6}{6}$ *All six outcomes are less than 9.*

16. Rolling a number less than 9 is an example of an event that is certain to occur. Look at your answer to Exercise 15. What is the probability of an event that is certain to occur?

Give each probability in simplest form.

Think about spinning this spinner.

17. *P*(green)

18. *P*(not red)

19. *P*(blue)

20. *P*(not blue)

21. *P*(yellow)

22. *P*(not yellow)

23. *P*(red)

24. *P*(not green)

25. *P*(black)

26. *P*(not black)

27. *P*(brown or red)

28. *P*(yellow or blue)

29. *P*(yellow or green)

30. *P*(red or green)

31. *P*(blue or green)

32. *P*(black or yellow)

33. *P*(yellow, blue, or green)

34. *P*(brown, blue, or red)

35. *P*(red, blue, or yellow)

Think about shuffling these cards and then turning one of the cards faceup.

P	R	O	B	A	B	I	L	I	T	Y		A	N	D

S	T	A	T	I	S	T	I	C	S

36. *P*(A)

37. *P*(T)

38. *P*(N)

39. *P*(S)

40. *P*(I)

41. *P*(B)

42. *P*(L)

43. *P*(not L)

44. *P*(A or I)

Challenge! THINKING ABOUT OUTCOMES

45. When you toss a thumbtack, there are these two possible outcomes:

point up   **point down**

a. Do you think that the outcomes are equally likely?

b. Toss a thumbtack 60 times and keep a record of the outcomes.

c. From the results of your experiments, do you think a thumbtack has a greater chance of landing point up or point down?

46. When you toss a paper cup, there are these three possible outcomes:

bottom down   **top down**   **side down**

a. Do you think that the outcomes are equally likely?

b. Toss a paper cup 60 times and keep a record of the outcomes.

c. From the results of your experiments, predict which outcome is most likely. Predict which outcome is least likely.

13-4 Sample Space

OBJECTIVE: To use a tree diagram to show the sample space. To compute the probability of an event using the sample space.

Every customer who buys a cassette at the Sensational Sounds store gets a Cassette Match card. When you rub each box, a picture of a red or blue cassette is equally likely to appear.

1. What is the prize for a winning card?

2. Is the card shown a winning card?

To find the probability of getting a winning card, it is helpful to first list the **sample space.** The sample space is the set of all possible outcomes.

SENSATIONAL SOUNDS CASSETTE MATCH

Rub Each Box

If all cassettes are the same color, present this card and get **$1⁰⁰ off !**

EXAMPLE | **Here's how to use a tree diagram to show the sample space.**

Note: R stands for red.
B stands for blue.

1st cassette	2nd cassette	3rd cassette	Sample space
R	R	R	RRR
		B	RRB
	B	R	RBR
		B	RBB
B	R	R	BRR
			BRB
	B	R	BBR
		B	BBB

CHECK for Understanding

3. Look at the example. How many outcomes are in the sample space?

4. How many of the outcomes are winners (that is, have all pictures the same color)?

5. What is the probability (in simplest form) of getting a winning card?

EXERCISES

▶ **Use the sample space above. Give each probability in simplest form.**

What is the probability of getting

6. a card having exactly 1 red cassette?

7. a card having exactly 2 blue cassettes?

8. a card having a blue cassette in the first box?

9. a card having a red cassette in the middle box?

10. a card having a blue cassette in the first two boxes?

11. a losing card?

12. Think about tossing
this coin 3 times.
Make a tree diagram
to show the sample space.

13. Refer to your sample space in Exercise 12 to find the following probabilities:

    a. P(all heads)
        b. P(no heads)

    c. P(1 head and 2 tails)
        d. P(2 heads and 1 tail)

    e. P(less than 3 heads)
        f. P(more than 3 tails)

14. Think about rolling these dice. Copy and complete the table to show the
sample space. The outcome shown on the dice is entered in the table.

Number on green die

	1	2	3	4	5	6
1						
2						
3				(3, 4)		
4						
5						
6						

Number on red die

15. Refer to your sample space in Exercise 14 to find the following probabilities:

    a. P(sum of 12)
      b. P(not sum of 12)
      c. P(sum of 3)

    d. P(sum of 7)
      e. P(sum less than 7)
      f. P(sum greater than 7)

    g. P(doubles)
      h. P(not doubles)
      i. P(sum of 1)

Challenge! **USING A TREE DIAGRAM**

⊘ Solve.

16. Think about putting these four marbles
in a bag. Then imagine taking two marbles out of the bag one after the other
without looking.

    a. Draw a tree diagram of the sample
space.

    b. What is the probability that the red mar-
ble will be picked?

    c. What is the probability that the blue
marble will not be picked?

    d. What is the probability that at least 1
marble will be green?

13-5 Probability—Independent Events

OBJECTIVE: To compute the probability of two or more independent events.

Think about first rolling a die and then tossing a coin. Since the outcome on the die in no way affects the outcome on the coin, two events such as rolling a 1 and tossing heads are called **independent events.**

The tree diagram shows all possible outcomes of first rolling a die and then tossing a coin.

1. First think about rolling a die. What is the probability of rolling a 1?

2. Next think about tossing a coin. What is the probability of tossing heads (H)?

3. Now look at the tree diagram. What is the probability of first rolling a 1 and then tossing heads (H)?

Sample Space

1	H --->	1, H
	T	1, T
2	H	2, H
	T	2, T
3	H	3, H
	T	3, T
4	H	4, H
	T	4, T
5	H	5, H
	T	5, T
6	H	6, H
	T	6, T

E X A M P L E | **Here's how to compute the probability of two or more independent events.**

The probability of the first event followed by the second event is the product of the probabilities of the individual events.

$$P(1) \qquad P(H)$$

Probability of rolling a 1 and then tossing heads → $P(1, H) = \dfrac{1}{6} \times \dfrac{1}{2}$

$$= \dfrac{1}{12}$$

CHECK for Understanding

4. Look at the example. What is $P(1)$? What is $P(H)$? What is $P(1, H)$?

5. What is $P(4, T)$?

EXERCISES

▶ **Give each probability as a fraction in simplest form.**
   Think about first rolling a die and then tossing a coin.

6. **a.** $P(5)$
   b. $P(T)$
   c. $P(5, T)$

7. **a.** $P(\text{odd number})$
   b. $P(H)$
   c. $P(\text{odd number, H})$

8. $P(\text{number less than 4, T})$

9. $P(\text{not 6, T})$

▶ **Give each probability as a fraction in simplest form.**

Think about first rolling the die and then
spinning the spinner.

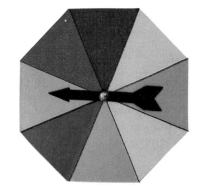

10. P(1, green)

11. P(5, red)

12. P(odd number, yellow)

13. P(not 4, blue)

14. P(6, not yellow)

15. P(not 3, not red)

16. P(number greater than 3, green)

Think about spinning the spinner above once and then spinning it again.

17. P(red, red)

20. P(red, yellow)

23. P(yellow, not yellow)

18. P(yellow, yellow)

21. P(yellow, green)

24. P(not green, not blue)

19. P(brown, blue)

22. P(not red, green)

25. P(not green, not yellow)

Think about placing the 6 marbles in a bag and
thoroughly mixing them up. Suppose that, without
looking, you picked out a first marble, **put it back into
the bag,** and then picked out a second marble.

26. P(green, blue)

28. P(blue, red)

30. P(not yellow, green)

27. P(yellow, green)

29. P(green, not yellow)

31. P(blue, not blue)

Problem Solving FINDING PROBABILITIES

▶ **Solve.**

32. If you toss a coin 3 times, what is the
probability that you will get 3 heads?

33. If you toss a coin 3 times, what is the
probability in simplest form that you
will get either all heads or all tails?

Challenge! FINDING PROBABILITIES

▶ **Solve.**

34. If you toss a coin 4 times, what is the
probability that you will not get all
heads?

35. If you toss a coin 6 times, what is the
probability in simplest form that all outcomes will be alike?

OBJECTIVE: To compute the probability of two or more dependent events.

Suppose that these cards were thoroughly shuffled and spread out facedown in front of you.

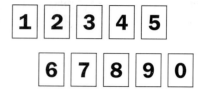

1. If you picked a card, what is the probability of picking the 3?

2. Suppose that you picked the 3 on your first draw, **did not replace the card,** and drew a second card. What is the probability that your second card would be the 8?

Since the outcome of picking the first card affects the outcome of picking the second card from the remaining cards, two events such as drawing a 3 and then drawing an 8 are called **dependent events.**

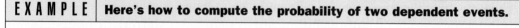

| E X A M P L E | **Here's how to compute the probability of two dependent events.** |

The probability of the first event followed by the second event is the product of the probabilities of the individual events.

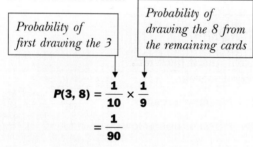

Probability of first drawing the 3

Probability of drawing the 8 from the remaining cards

$$P(3, 8) = \frac{1}{10} \times \frac{1}{9}$$

$$= \frac{1}{90}$$

CHECK for Understanding

3. Look at the example. To find $P(3, 8)$, __?__ the probability of first drawing the 3 by the probability of drawing the 8 from the remaining cards.

4. What is $P(3, 8)$?

EXERCISES

▶ **Give each probability as a fraction in simplest form.**

Think about drawing a card from the cards shown above, then without replacement drawing a second card.

5. $P(6, 1)$

6. $P(\text{even}, 5)$

7. $P(8, \text{odd})$

8. $P(\text{odd}, \text{even})$

9. $P(3, \text{prime})$

10. $P(\text{prime}, \text{composite})$

Suppose that these cards were thoroughly shuffled and spread out facedown in front of you. Think about drawing cards without replacement.

A B C D E F G H I

 Give each probability as a fraction in simplest form.

11. $P(A, H)$

12. $P(B, \text{not } A)$

13. $P(\text{not } I, I)$

14. $P(\text{vowel}, H)$

15. $P(E, \text{consonant})$

16. $P(\text{not } E, \text{not } E)$

17. $P(\text{not vowel}, E)$

18. $P(\text{consonant}, \text{vowel})$

19. $P(\text{not vowel}, \text{vowel})$

20. $P(\text{consonant}, \text{consonant})$

21. $P(A, I, D)$

22. $P(\text{vowel}, \text{vowel}, \text{vowel})$

★**23.** $P(H, I, D, E)$

★**24.** $P(\text{not } E, \text{not } E, \text{not } E, E)$

★**25.** $P(\text{consonant}, \text{consonant}, \text{consonant}, \text{consonant})$

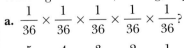 **CALCULATOR APPLICATION**

A container has 36 Ping-Pong balls numbered 1–36.

Look at the rules
for these two games:

Game 1

- Guess five numbers. (You can choose the same number more than once.)
- Mix the balls and draw one without looking.
- Replace the ball, mix, and draw another ball.
- Repeat until five balls have been drawn.

Game 2

- Guess five different numbers.
- Mix the balls.
- Without looking, draw five balls *without* replacement.

You win if you guessed all five numbers that were drawn.

26. Which game has $P(\text{winning})$ equal to

   a. $\dfrac{1}{36} \times \dfrac{1}{36} \times \dfrac{1}{36} \times \dfrac{1}{36} \times \dfrac{1}{36}$?

   b. $\dfrac{5}{36} \times \dfrac{4}{35} \times \dfrac{3}{34} \times \dfrac{2}{33} \times \dfrac{1}{32}$?

27. What is $P(\text{winning})$

   a. Game 1?

   b. Game 2?

28. Which game would you be more likely to win?

You will use a variety of problem-solving strategies on these two pages. For some problems, you may wish to use a calculator.

1 TENNIS TIME-OUT

Erica and Karen played two games of tennis and decided to take a break. They have one 12-ounce can of juice and want to split it evenly. They have an empty 7-ounce cup and an empty 5-ounce cup. How can they divide the 12 ounces of juice evenly?

✓ Problem Solving Tips

Make a model. Use a 12-ounce can (or other 12-ounce container), a 7-ounce cup, and a 5-ounce cup. Pour from one container to another until you have measured exactly 6 ounces. Keep a record of what you do,

or

Make a chart.

CONTAINER		
12-ounce	7-ounce	5-ounce
12	0	0
5	7	0
5	2	5

2 FACULTY FINAGLE

The principal, the ESL teacher, the algebra teacher, and the librarian of our school are, in no particular order: Mr. McNulty, Ms. McDonald, Ms. McInerney, and Mr. McCoy. But I can never remember who does what! However, I do know that

- Ms. McDonald is taller than the algebra teacher and the librarian.
- The principal lunches alone.
- Ms. McInerney always has lunch with the algebra teacher and the librarian.
- Mr. McCoy is older than the algebra teacher.

What is each person's job?

3 GEOMETRY—VISUAL THINKING

This is a map of Fitzgerald's Farm. Can you divide it into three identical parts by drawing only two straight lines?
Hint: Trace the figure and draw the lines.

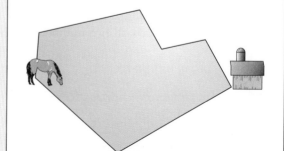

4 INTEGER PUZZLES

Use the number cards to get the answers.
Use each card only once in each puzzle.

a. $(\boxed{?} \div \boxed{?}) + (\boxed{?} \div \boxed{?}) = {}^{+}5$
b. $(\boxed{?} + \boxed{?}) \div (\boxed{?} + \boxed{?}) = {}^{-}3$
c. $(\boxed{?} \div \boxed{?}) \times (\boxed{?} + \boxed{?}) = {}^{+}8$
d. $(\boxed{?} - \boxed{?}) \div (\boxed{?} + \boxed{?}) = {}^{+}7$

5 GEOMETRY—SPACE FIGURES

This cube measures 3 inches on each side. Suppose that the cube was cut up into 1-inch cubes as shown.

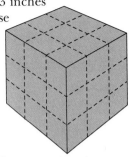

How many 1-inch cubes are painted on

a. 4 sides?   **b.** 3 sides?   **c.** 2 sides?
d. 1 side?   **e.** 0 sides?

6 TIME FOR CHANGE

Looking at the coins in my pocket, I noticed that I could pay the exact price for any item from one cent up to and including one dollar without receiving any change.

What is the smallest number of coins I could have in my pocket? What coins are they? *Hint: Make an organized list.*

7 TRAIN TRAVEL

It is 60 miles from Atlantic City to Philadelphia. Two trains leave at 10:00 A.M., one from Philadelphia at 40 miles an hour and the other from Atlantic City at 50 miles an hour. When they meet, are they nearer to Philadelphia or to Atlantic City?

APPLICATIONS

Ed's Golf Shop is having a sale on golf packages. Each package has a set of clubs, a bag, and a pull cart. Customers choose from the following.

Golf Package Sale!

4 BAGS TO CHOOSE FROM		SELECT FROM 3 CLUB SETS		YOUR CHOICE OF 2 CARTS	
Weekender	$19.19	Starter Set	$159.59	Carry All	$39.39
Lightweight	35.35	Shot Makers	199.99	EZ Roller	75.75
Par Maker	79.79	Ultimates	229.99		
Pro's Choice	149.49				

8 How many different packages are possible?

9 Bonnie has decided on Ultimates for clubs and the EZ Roller cart.
a. How many packages can she make?   **b.** What is the total cost of each package?

10 Ted made up a package that cost $319.17. Which options did Ted choose?

Rate Your Problem-Solving Power: *6–7 correct = Good;   8–9 correct = Excellent;   10 correct = Exceptional*

OBJECTIVE: To use postal rates for solving problems.

FIRST-CLASS STAMP FACTS

Before 1885, the first-class letter rate (cost for the first ounce) depended on distance traveled. On July 1, 1885, the national rate became 2 cents. Since then the rate has been changed several times. The following list shows these changes.

DATE	RATE
	3 cents
November 3, 1917	2 cents
July 1, 1919	3 cents
July 6, 1932	4 cents
August 1, 1958	5 cents
January 7, 1963	8 cents
May 16, 1971	10 cents
March 2, 1974	13 cents
December 11, 1975	15 cents
May 29, 1978	18 cents
March 22, 1981	20 cents
November 1, 1981	22 cents
February 17, 1985	25 cents
April 3, 1988	29 cents
February 3, 1991	

▶ **Use the information in the article to solve each problem.**

1. What was the total cost of postage for mailing 15 one-ounce letters on July 5, 1971?

2. On July 6, 1932, a roll of 50 stamps cost $1.50. How much did a similar roll of 50 stamps cost 40 years later?

3. How much more did it cost to mail a Father's Day card in 1975 than it did in 1950? *Hint: Father's Day is always in June.*

4. How much more did it cost to mail 24 Valentine cards in 1980 than it did in 1950? *Hint: Valentine's Day is February 14.*

5. What was the percent of decrease in the first-class letter rate from November 10, 1917, to November 10, 1919?

6. What was the percent of increase in the first-class letter rate from March 15, 1974, to March 15, 1976?

▶ **Write an equation and solve the problem.**

7. The cost to mail a 7.5-ounce letter first-class is $1.65. That is $.55 less than twice the cost to mail the same letter third-class. How much does it cost to mail the letter third-class?
(Let c = the cost to mail the letter third-class.)

★8. The cost to mail a 10.5-ounce package third-class is $1.30. That is $.55 more than one-third the cost to mail the same package first-class. How much does it cost to mail the package first-class?
(Let p = the cost to mail the package first-class.)

Cumulative Algebra Practice

▶ **Give the greatest common factor.** *(page 132)*

1. 5, 10   **2.** 16, 24   **3.** 28, 21   **4.** 72, 24   **5.** 84, 63

6. $3, 6c$ **7.** $18, 12y$ **8.** $5w, 13w$ **9.** $9j, 4j^2$ **10.** $35g, 7g^2$

11. $21xy, 18y^2$ **12.** $14a^2b, 21ab^2$ **13.** $56cd^2, 42c^2$ **14.** $48r^2s^2, 32rs^2$ **15.** $50x^2y^3, 15xy^3$

▶ **Write in lowest terms.** *(page 140)*

16. $\dfrac{8}{24}$ **17.** $\dfrac{42}{36}$ **18.** $\dfrac{21}{42}$ **19.** $\dfrac{60}{45}$ **20.** $\dfrac{27}{45}$ **21.** $\dfrac{35}{50}$

22. $\dfrac{n}{n^2}$ **23.** $\dfrac{12}{9x}$ **24.** $\dfrac{6x}{18y}$ **25.** $\dfrac{rs}{3s}$ **26.** $\dfrac{32ab}{8a^2}$ **27.** $\dfrac{40}{56c^2}$

28. $\dfrac{36j}{9k}$ **29.** $\dfrac{c^2d}{9d^2}$ **30.** $\dfrac{45s}{72st}$ **31.** $\dfrac{19k^3}{4k^2}$ **32.** $\dfrac{45pq^2}{25p^2}$ **33.** $\dfrac{16y^2z}{40y^2z^2}$

▶ **Give each sum or difference in simplest form.** *(page 168)*

34. $\dfrac{1}{2}+\dfrac{1}{8}$ **35.** $\dfrac{1}{4}+\dfrac{1}{5}$ **36.** $\dfrac{2}{3}+\dfrac{3}{4}$ **37.** $\dfrac{7}{8}+\dfrac{5}{6}$ **38.** $\dfrac{5}{9}+\dfrac{4}{5}$

39. $\dfrac{a}{3}+\dfrac{b}{6}$ **40.** $\dfrac{j}{24}+\dfrac{k}{8}$ **41.** $\dfrac{5y}{12}+\dfrac{z}{8}$ **42.** $\dfrac{j}{6}+\dfrac{5k}{18}$ **43.** $\dfrac{3a}{35}+\dfrac{2b}{7}$

44. $\dfrac{5}{6}-\dfrac{1}{3}$ **45.** $\dfrac{3}{4}-\dfrac{1}{2}$ **46.** $\dfrac{5}{6}-\dfrac{4}{5}$ **47.** $\dfrac{8}{9}-\dfrac{3}{4}$ **48.** $\dfrac{13}{6}-\dfrac{5}{8}$

49. $\dfrac{c}{5}-\dfrac{d}{10}$ **50.** $\dfrac{m}{3}-\dfrac{n}{4}$ **51.** $\dfrac{j}{8}-\dfrac{k}{12}$ **52.** $\dfrac{y}{15}-\dfrac{3z}{10}$ **53.** $\dfrac{5q}{18}-\dfrac{5r}{12}$

▶ **Give each product in simplest form.** *(page 184)*

54. $\dfrac{1}{6}\cdot\dfrac{1}{5}$ **55.** $\dfrac{3}{4}\cdot\dfrac{1}{3}$ **56.** $5\cdot\dfrac{7}{10}$ **57.** $\dfrac{7}{8}\cdot\dfrac{8}{7}$ **58.** $\dfrac{5}{6}\cdot\dfrac{3}{10}$

59. $\dfrac{y}{4}\cdot\dfrac{1}{y}$ **60.** $9\cdot\dfrac{c}{3}$ **61.** $\dfrac{5}{6}\cdot\dfrac{4k}{15}$ **62.** $\dfrac{3}{4d}\cdot\dfrac{4}{13e}$ **63.** $\dfrac{a}{b}\cdot\dfrac{7b}{3a}$

64. $\dfrac{r}{s^2}\cdot\dfrac{6s}{r}$ **65.** $\dfrac{m}{n}\cdot\dfrac{n}{m}$ **66.** $\dfrac{t^2}{5}\cdot\dfrac{s}{t}$ **67.** $\dfrac{9x}{y}\cdot\dfrac{y}{6x}$ **68.** $\dfrac{12c}{d}\cdot\dfrac{3d}{30c^2}$

▶ **Give each quotient in simplest form.** *(page 188)*

69. $\dfrac{5}{6}\div\dfrac{1}{6}$ **70.** $\dfrac{6}{7}\div\dfrac{2}{7}$ **71.** $\dfrac{9}{8}\div\dfrac{3}{4}$ **72.** $\dfrac{4}{5}\div5$ **73.** $8\div\dfrac{3}{5}$

74. $\dfrac{7}{n}\div\dfrac{3}{n}$ **75.** $\dfrac{10}{a}\div\dfrac{5}{b}$ **76.** $\dfrac{r}{16}\div\dfrac{s}{12}$ **77.** $\dfrac{3s}{t}\div\dfrac{5s}{t}$ **78.** $\dfrac{2j^2}{k}\div\dfrac{j}{k}$

OBJECTIVE: To compute expectation.

Here is a game that was played at a school carnival.

1. How much did it cost to spin the wheel?

2. What is the probability that the wheel will stop on green?

3. What is the value of the prize?

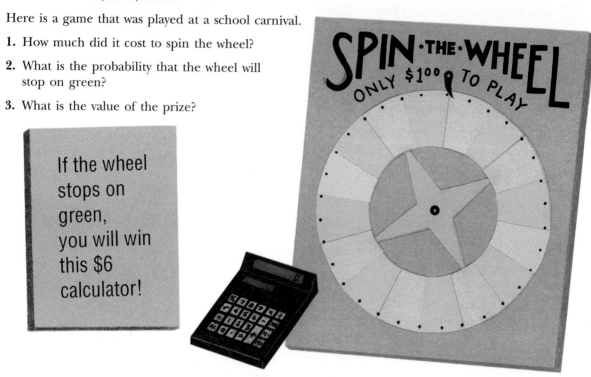

If the wheel stops on green, you will win this $6 calculator!

SPIN·THE·WHEEL
ONLY $1⁰⁰ TO PLAY

| E X A M P L E | Here's how to compute the expectation for a game of Spin-the-Wheel. |

To find the expectation, multiply the probability of winning the prize by the value of the prize.

	P(winning)		Value of Prize		
Expectation =	$\dfrac{1}{8}$	×	**$6**	=	**$.75**

The expectation for a game of Spin-the-Wheel is $.75.

CHECK for Understanding

4. Look at the example. What is the expectation for a game of Spin-the-Wheel?

5. To decide whether such a game is a good deal for the player, you compare the cost of playing with the expectation. Is the expectation less than or greater than the cost of playing?

6. If the expectation is less than the cost of playing, then such a game is a "bad deal" for the player. Is the game Spin-the-Wheel a bad deal for the player?

7. How much would the expectation have to be for the game to be considered a "good deal" for the player?

EXERCISES

Buy a chance on a $120 guitar!

▶ **Solve.**

A merchant donated this guitar to the school carnival. The carnival committee decided to sell 180 chances on it, at $.75 each.

8. What is the cost of a chance?

9. If you bought one chance, what would be your probability of winning?

10. What is the value of the guitar?

11. What is the expectation rounded to the nearest cent?

12. Is the expectation greater than or less than the cost of a chance?

13. Is buying a chance a good deal or a bad deal?

Suppose that the same wheel pictured on page 380 was used for the following games.

14. Pay $2; if the wheel stops on green, you win a $17 camera.
  a. What is the probability of winning?
  b. What is the expectation rounded to the nearest cent?
  c. Is the game a good deal?

15. Pay $3; if the wheel stops on yellow, you win a school sweatshirt worth $6.25.
  a. What is the probability of winning?
  b. What is the expectation rounded to the nearest cent?
  c. Is the game a good deal?

16. Pay $2; if the wheel stops on blue, you win a school T-shirt worth $5. Is the game a good deal?

17. Pay $3; if the wheel stops on yellow or green, you win a $6 gift certificate to a record store. Is the game a good deal?

Challenge! **GUESS AND CHECK**

▶ **Solve.**

18. Laura bought a chance on a camera and a chance on a radio. She spent $2.25. The chance on the radio cost $.25 more than the chance on the camera. How much did she pay for each chance?

19. Adam bought 2 chances on a watch and 1 chance on a radio. He spent $4.00. Joan bought 1 chance on the watch and 2 chances on the radio. She spent $3.50. How much did they pay for each chance?

OBJECTIVE: To analyze data using line plots.

The study of collecting, organizing, and interpreting data is called **statistics**.

F. Roosevelt

T. Roosevelt

U.S. Presidents Inaugurated 1901–1963		
President	Inauguration Year / Age	Age at death
T. Roosevelt	1901 / 42	60
W. H. Taft	1909 / 51	72
W. Wilson	1913 / 56	67
W. Harding	1921 / 55	57
C. Coolidge	1923 / 51	60
H. Hoover	1929 / 54	90
F. Roosevelt	1933 / 51	63
H. Truman	1945 / 60	88
D. Eisenhower	1953 / 62	78
J. Kennedy	1961 / 43	46
L. Johnson	1963 / 55	64

Line plots are a quick, simple way to organize data. They allow you to see how the data are grouped.

EXAMPLE │ **Here's how to make a line plot of the presidents' inaugural ages.**

Draw a number line and write an X to represent each age.

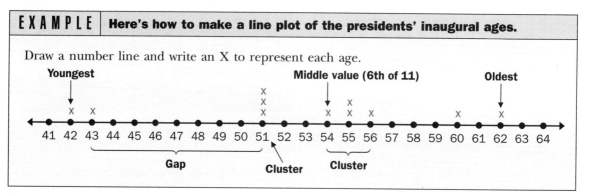

CHECK for Understanding

1. How many X's are below the middle value of 54? How many are above?

2. What was the greatest number of presidents inaugurated at the same age?

A line plot helps you find the median, mode, and range of a set of numbers because it shows the order of the numbers.

The **median** is the middle number of an ordered set of numbers. The median of the inaugural ages is 54. If there is an even number of numbers, the median is the average of the two middle numbers.

The **mode** is the number that occurs most often. The mode of the inaugural ages is 51. Not all sets of data have a mode.

The **range** is the difference between the largest and smallest numbers. The range of the inaugural ages is $62 - 42 = 20$.

The **mean** is the sum of the numbers divided by the number of them. The average of the inaugural ages is $(42 + 43 + 51 + 51 + \cdots + 56 + 60 + 62) \div 11 = 52.\overline{72}$. The mean age rounded to the nearest year is 53.

EXERCISES

▶ **Find the mean. Round each answer to the nearest tenth.**

3. 12, 4, 14, 9

4. 35, 39, 27, 28

5. 195, 176, 183, 178, 181

6. 253, 276, 248, 281, 265

7. 29.6, 32.7, 31.9, 28.7

8. 17.5, 13.8, 12.4, 11.7, 12.4, 17.0

▶ **Find the median.**

9. 13, 12, 11, 6, 7, 12, 9

10. 12, 9, 8, 14, 10, 11

11. 14, 7, 12, 7, 7, 10, 13

12. 13, 14, 9, 11, 13, 10, 9

13. 183, 182, 183, 180, 187, 182

14. 12, 15, 13, 19, 13, 15, 13

▶ **Find the mode and range. (It is possible for a list to have more than one mode.)**

15. 5, 7, 9, 7, 5, 9, 7

16. 11, 15, 12, 15, 10, 14, 15

17. 20, 27, 22, 23, 21, 24, 22

18. 66, 82, 69, 73, 68, 66, 76, 66

Problem Solving ANALYZING DATA

19. a. Use the data from the chart on page 382 to make a line plot of the presidents' ages at death. *Hint: Number the line from 45 to 90.*

b. Compare your line plot with the one in the example. Which one has the larger gap? In which one are the ages more clustered?

c. Use your line plot to make an ordered list of the presidents' ages at death.

d. Find the median, mean, mode, and range of the presidents' ages at death.

20. a. Look at the U.S. Presidents chart on page 382. Notice that President Harding lived only 2 years after his inauguration. Compute the number of years each president lived after his inauguration.

b. Make a line plot of your data. *Hint: Number the line from 0 to 40.*

c. Find the median, mean, mode, and range of your data.

d. **Write a description** of your data using the information from Parts b and c.

OBJECTIVE: To interpret frequency tables, bar graphs, broken-line graphs, picture graphs, and circle graphs.

A **frequency table** shows the number of times different events or responses occur.

▶ **Use the frequency table to answer each question.**

1. How many students attended 4 movies last month?

2. How many didn't attend a movie?

3. How many attended 2 or fewer movies?

4. How many attended 3 or more movies?

5. How many students were in the survey?

6. What percent of the students attended only 1 movie?

7. What percent of the students attended fewer than 3 movies?

8. What percent attended at least 1 movie?

Number of Movies Attended Last Month		
Number of movies	**Number of students**	
	Tally	**Frequency**
4	III	3
3	⊬⊬ I	6
2	⊬⊬ ⊬⊬	10
1	II	2
0	III	3

▶ **Use the bar graph to answer each question.**

9. How many hours did Bert spend doing homework?

10. Who spent the most time doing homework?

11. How many more hours did Alma spend doing homework than Claire?

12. By what percent would Claire have to increase her homework time to spend as much time as Bert?

13. Who spent 10% less time on homework than Alma?

Homework During the Week

▶ **Use the broken-line graph to answer each question.**

14. By the end of the second week, Sally had saved $8. How much had she saved by the end of the third week?

15. During which week did Sally save the most?

16. How much did Sally save during the six weeks?

17. Sally is saving to buy a $90 radio-tape player. What percent of the money will she have by the end of the sixth week?

18. If Sally continues to save at the rate of her first six weeks, how many weeks must she save for the $90 radio-tape player?

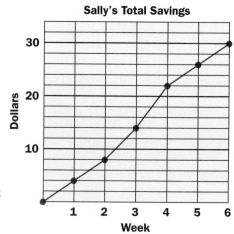

Sally's Total Savings

EXERCISES

▶ **Use the picture graph to answer each question.**

19. Who has the most tapes? The fewest?

20. How many tapes does Elliot have?

21. How many tapes does Ernie have?

22. Elliot has how many more tapes than Cindy?

23. Who has 40% more tapes than Fred?

24. Who has about 37% fewer tapes than Elliot?

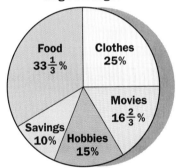

Number of Tapes in Collection

Elliot	⊕ ⊕ ⊕ ⊕ ⊕ ⊕
Cindy	⊕ ⊕ ⊕ ⊕ ⊖
Ernie	⊕ ⊕ ⊕ ⊕ ⊕ ▹
Fred	⊕ ⊕ ⊕ ⊖

Each ⊕ *stands for 4 tapes.*

▶ **Use the circle graph to answer the following questions.**

25. For which category does Hugo allow the most money? The least money?

26. How much does Hugo spend on food each week?

27. How much more does he spend a week on clothes than on hobbies?

28. On which two categories does he spend half his budget?

29. How much would Hugo spend on movies during a year (52 weeks)?

Hugo's Budget

Food $33\frac{1}{3}$ %

Clothes 25%

Movies $16\frac{2}{3}$ %

Savings 10%

Hobbies 15%

Weekly earnings: $60

▶ **Use the broken-line graphs to answer the following questions.**

30. What percent took French in 1985? In 1989?

31. In what year did the percent of students taking algebra decrease?

32. In what year were there more students taking biology than algebra?

33. What was the first year that more students were enrolled in French than in Spanish?

34. Suppose that there were 300 ninth grade students in 1988. Estimate how many students took each elective.

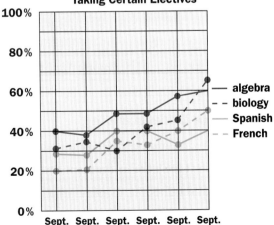

Percent of Ninth Grade Students Taking Certain Electives

— algebra
- - biology
— Spanish
- - French

Sept. 1985 Sept. 1986 Sept. 1987 Sept. 1988 Sept. 1989 Sept. 1990

OBJECTIVE: To analyze data using stem-and-leaf plots.

Statisticians use stem-and-leaf plots to quickly organize and display data. A stem-and-leaf plot can help in comparing the January snowfalls of these vacation areas.

Vacation Area Statistics—January Averages					
Vacation area	**Snowfall (in.)**	**P.M. temp (°F)**	**Vacation area**	**Snowfall (in.)**	**P.M. temp (°F)**
Adirondack Mtn., NY	28	25	Land O'Lakes, WI	15	21
Anchorage, AK	12	19	Mount Hood, OR	62	36
Aspen, CO	26	34	Pocono Mtn., PA	14	32
Black Hills, SD	10	34	Redwoods, CA	48	40
Blue Ridge Mtn., VA	11	36	Reno, NV	46	38
Bryce Canyon, UT	20	36	Sun Valley, ID	33	31
Cascade Mtn., OR	72	33	Traverse City, MI	21	26
Catskill Mtn., NY	15	30	Wasatch Range, UT	65	31
Glacier Ntl. Park, MT	35	28	White Mtn., NH	35	26
Grand Canyon, AZ	31	36	Wisconsin Dells, WI	10	24
Green Mtn., VT	23	27	Yosemite, CA	25	48
Klamath Falls, OR	15	39			

EXAMPLE | **Here's how to make a stem-and-leaf plot.**

The January snowfalls are 2-digit numbers. The tens digits range from 1 to 7, so choose these digits as stems. The ones digits are the leaves.

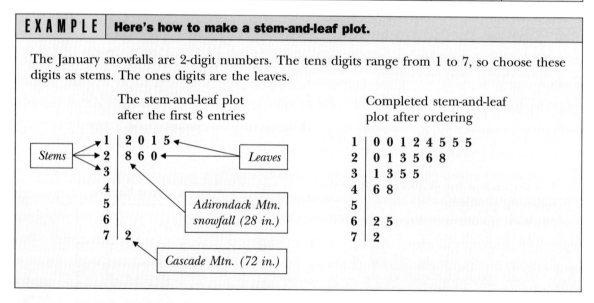

The stem-and-leaf plot after the first 8 entries

Completed stem-and-leaf plot after ordering

1	0 0 1 2 4 5 5 5
2	0 1 3 5 6 8
3	1 3 5 5
4	6 8
5	
6	2 5
7	2

CHECK for Understanding

1. How many vacation areas have a January snowfall greater than 59 inches?

2. How many leaves are in the stem-and-leaf plot?

3. The middle leaf is the median. What snowfall does it represent?

4. The leaf that occurs most often (3 times) is the mode. What snowfall does it represent?

EXERCISES

5. Make a stem-and-leaf plot of the January temperatures for the vacation areas in the chart on page 386. Order the leaves from smallest to largest. Use the stems 1, 2, 3, and 4.

6. Use your stem-and-leaf plot from Exercise 5.
   a. What is the lowest afternoon temperature? The highest?
   b. How many areas had an afternoon temperature that was below freezing (32°F)? That was above freezing?
   c. What is the median, mode, and range of the temperatures?

Vacation Areas with Wide Open Spaces			
West of the Mississippi River	**People per square mile**	**East of the Mississippi River**	**People per square mile**
Sun Valley, ID	27	Adirondack Mtn., NY	34
Crater Lake, OR	25	Eastern Shore, VA	65
Spokane, WA	38	Myrtle Beach, SC	73
Anchorage, AK	11	White Mtn., NH	29
Glacier National Park, MT	9	Land O'Lakes, WI	17
Redwoods, CA	17	Bar Harbor, ME	39
Coos Bay, OR	28	Door County, WI	51
Aspen, CO	11	Outer Banks, NC	53
Cascade Mtn., OR	11	Mackinac Island, MI	18
Lake Tahoe, CA	33	Rangeley Lakes, ME	5

7. Make a **back-to-back stem-and-leaf plot** of the data in the chart. (The first four entries have been made.)

Write the stems in the center of the plot.

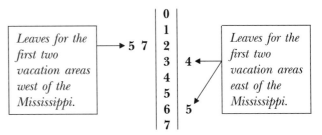

Leaves for the first two vacation areas west of the Mississippi. → 5 7

| 0 |
| 1 |
| 2 |
| 3 | 4 ←
| 4 |
| 5 |
| 6 | 5 |
| 7 |

Leaves for the first two vacation areas east of the Mississippi.

8. a. What is the range of the densities of the western vacation areas? Of the eastern vacation areas?
   b. Which range is greater? How is this difference shown by the stem-and-leaf plot?
9. a. What is the median of the densities of the western areas? Of the eastern areas?
   b. Which median is greater? How is this difference shown by the stem-and-leaf plot?

OBJECTIVE: To analyze data using box-and-whisker plots.

Box-and-whisker plots are used by statisticians to compare sets of data.

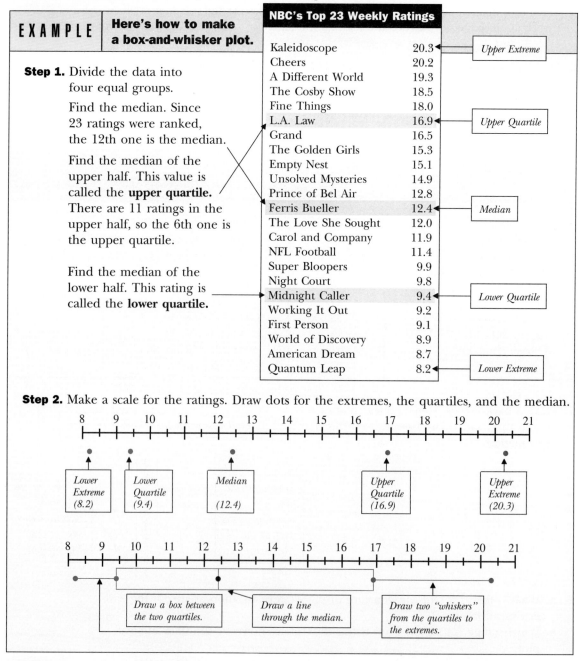

EXAMPLE	Here's how to make a box-and-whisker plot.

Step 1. Divide the data into four equal groups.

Find the median. Since 23 ratings were ranked, the 12th one is the median.

Find the median of the upper half. This value is called the **upper quartile.** There are 11 ratings in the upper half, so the 6th one is the upper quartile.

Find the median of the lower half. This rating is called the **lower quartile.**

NBC's Top 23 Weekly Ratings

Kaleidoscope	20.3	Upper Extreme
Cheers	20.2	
A Different World	19.3	
The Cosby Show	18.5	
Fine Things	18.0	
L.A. Law	16.9	Upper Quartile
Grand	16.5	
The Golden Girls	15.3	
Empty Nest	15.1	
Unsolved Mysteries	14.9	
Prince of Bel Air	12.8	
Ferris Bueller	12.4	Median
The Love She Sought	12.0	
Carol and Company	11.9	
NFL Football	11.4	
Super Bloopers	9.9	
Night Court	9.8	
Midnight Caller	9.4	Lower Quartile
Working It Out	9.2	
First Person	9.1	
World of Discovery	8.9	
American Dream	8.7	
Quantum Leap	8.2	Lower Extreme

Step 2. Make a scale for the ratings. Draw dots for the extremes, the quartiles, and the median.

Lower Extreme (8.2)   Lower Quartile (9.4)   Median (12.4)   Upper Quartile (16.9)   Upper Extreme (20.3)

Draw a box between the two quartiles.   Draw a line through the median.   Draw two "whiskers" from the quartiles to the extremes.

CHECK for Understanding

1. Look at the example. Was the program *Cheers* above or below the upper quartile?

2. How many of the 23 ratings are *between* the quartiles?

EXERCISES

▶ Here is a box-and-whisker plot of the ratings for 23 CBS programs that competed with the NBC programs shown in the example.

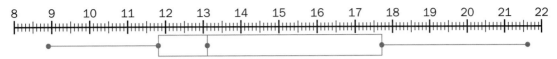

3. Use the box-and-whisker plot to estimate the
  a. upper extreme. **b.** upper quartile. **c.** median.
  d. lower quartile. **e.** lower extreme.

4. Compare the CBS plot with the NBC plot. Which network had the higher
  a. upper extreme? **b.** upper quartile? **c.** median?
  d. lower quartile? **e.** lower extreme?

5. Which network had the better ratings?

▶ Here are box-and-whisker plots of the average scores for the 19 top-scoring players in the National Basketball Association (NBA).

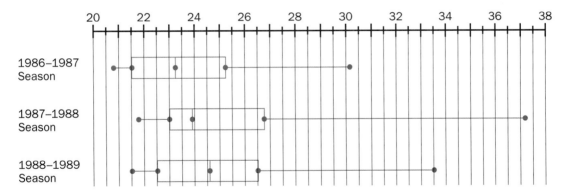

6. Compare the box-and-whisker plots for the 1986–1987 and the 1987–1988 seasons. In which season were the scores higher?

7. Compare the box-and-whisker plots for the 1987–1988 and the 1988–1989 seasons. Which season had the higher
  a. upper extreme? **b.** upper quartile? **c.** median?
  d. lower quartile? **e.** lower extreme?

8. Compare the box-and-whisker plots for the 1987–1988 and the 1988–1989 seasons.
  a. During which season did the highest scorer have the better score?
  b. During which season did the middle scorer have the better score?
  c. During which season did the top quarter of the players appear to have the better scores?
  d. During which season did the bottom quarter of the players appear to have the better scores?

OBJECTIVE: To recognize how statistics can be misleading.

These six students were asked if they lived more than three blocks from school. The results of this "sample" were printed in the school newspaper:

"Five Out of Six Students Live More Than Three Blocks from School."

1. How many students were in the sample?

2. Do you think that the students in the sample were representative of the **population** (the entire student body)? Why or why not?

3. Do you think that the statement in the school newspaper could be misleading?

A sample that is *not representative* of the population being sampled is called a **biased sample.** The sample above is likely to be biased because students who ride in a car to school probably live farther than three blocks from the school.

4. Suppose instead that you asked the first 40 students in a lunch line if they lived more than 3 blocks from school. Would the sample be biased if all students must eat lunch at school?

Often misleading statements result from the wrong interpretation of facts. Explain why each statement could be misleading.

5. *Fact:* Fifty-seven percent of the automobile accidents in Lake City involved male drivers.

6. *Fact:* Sixty-four percent of the people in Lake City that had an accident did so within 25 miles of their homes.

> *"Male drivers are worse drivers then female drivers!"*

> *"The most unsafe place to drive is within 25 miles of where you live!"*

EXERCISES

If not studied carefully, some graphs can be misleading.

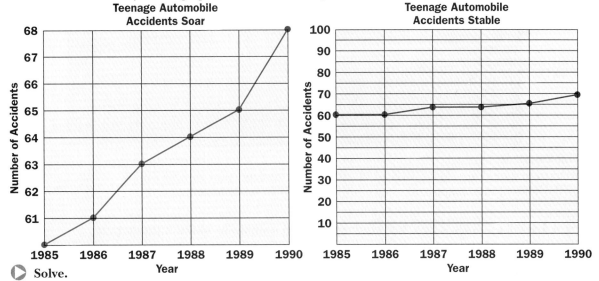

Teenage Automobile Accidents Soar

Teenage Automobile Accidents Stable

▶ Solve.

7. Look at the first graph. How many accidents occurred in each of the years?

8. Does the second graph appear to show the same information?

9. Which graph would you use if you were trying to
   a. raise auto-insurance rates for teenagers?
   b. lower auto-insurance rates for teenagers?

10. Look at the first graph. What was the percent of increase in accidents
    a. from 1985 to 1990?
    b. from 1989 to 1990?
    Round your answers to the nearest tenth percent.

Group Project   ANALYZING DATA   *Are You an Average Listener?*

▶ Work in a small group.

11. a. For each day of last week, list the number of hours that you listened to the radio.
    b. Find the total number of hours for the week and round to the nearest hour.

12. a. List the number of hours each class member listened to the radio last week.
    b. Order the numbers from least to greatest.

13. a. Find the median, mode, and range of the data above.
    b. Compute the mean to the nearest tenth hour.

14. Did you listen to the radio more or less than most class members?

15. **Write a paragraph** that summarizes the data.

OBJECTIVE: To use sampling to make predictions.

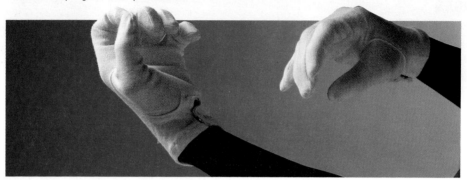

Some students were asked to find what percent of the student body could identify what these hands are doing. Instead of asking each student, they asked a sample of the student population and estimated the percent from their sample.

| E X A M P L E | Here's how the students used a sample to make a prediction. |

Step 1. For their sample, they chose the first 40 students to enter the school. The results were:

| Students who knew the hands are playing a violin | HHT HHT HHT HHT HHT I |
| Students who did not know the hands are playing a violin | HHT HHT IIII |

Step 2. From their sample they predicted the percent of the student body that would know what the hands are doing. To do this, they solved a proportion.

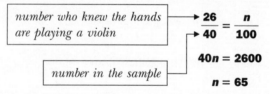

$$\frac{26}{40} = \frac{n}{100}$$

$$40n = 2600$$

$$n = 65$$

From the sample of 40, they could predict that 65% of the student body would know that the hands are playing a violin.

CHECK for Understanding

1. Look at the example above.
   a. How many students were in the sample?
   b. How many students in the sample could identify what the hands are doing?
   c. What percent of the students sampled knew that the hands are playing a violin?

2. On the basis of the sample, how many students of a student population of 1000 would know what the hands are doing?

EXERCISES

▷ **Use the sample to solve.**

3. *Sample question:*
What are these
hands doing?

Sample results:

Number who know what the hands are doing	̶H̶H̶ ̶H̶H̶ ̶H̶H̶ |||
Number who do not know	̶H̶H̶ ||

a. How many people in the sample
know that the hands are holding a
golf club?

b. How many people are in the sample?

c. What percent of the sample know
what the hands are doing?

d. On the basis of the sample, how
many people out of 600 would you
predict would know that the hands
are holding a golf club?

4. *Sample question:*
What are these
hands doing?

Sample results:

Number who know what the hands are doing	̶H̶H̶ ̶H̶H̶ ̶H̶H̶ ̶H̶H̶ ̶H̶H̶ ̶H̶H̶ ||
Number who do not know	̶H̶H̶ ̶H̶H̶ ̶H̶H̶ |||

a. How many people in the sample
know that the hands are threading a
needle?

b. How many people are in the sample?

c. What percent of the sample know
what the hands are doing?

d. On the basis of the sample, how
many people out of 800 would you
predict would know that the hands
are threading a needle?

▷ **Use a proportion to solve each problem. Round each answer to the nearest
whole number.**

5. In a sample, 15 out of 40 people said
their favorite color is red. Based on the
sample, how many people out of 1000
would you predict would pick red as
their favorite color?

6. In a sample, 45 out of 80 people said
that they sleep at least 8 hours a night.
Based on the sample, how many people
out of 500 would you predict sleep at
least 8 hours a night?

7. In a sample, 38 out of 60 people said
vanilla is their favorite ice cream. Based
on the sample, how many people out of
1500 would pick vanilla as their favorite
ice cream?

8. In a sample, 24 out of 30 people see at
least one movie a year. Based on the
sample, how many people out of 200
see at least one movie a year?

9. In a sample, 38 out of 70 families have
2 cars. Suppose 200 families are sampled. How many do you predict would
have 2 cars?

10. Write a sampling problem that can be
solved using a proportion.

OBJECTIVE: To draw bar graphs, broken-line graphs, and picture graphs.

Here are the results of a television survey:

Name	Hours of television watched last week
Alex	12
Beth	9
Joan	13
Mark	19

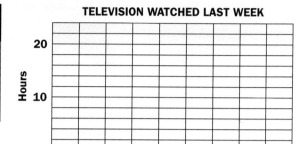

TELEVISION WATCHED LAST WEEK

1. Copy and complete the bar graph. Be sure to include the title of the graph.

2. Take a similar survey. Show the results on a bar graph.

The daily attendance in John's mathematics class is given below.

Day	Students present
Monday	21
Tuesday	23
Wednesday	19
Thursday	20
Friday	22

3. Copy and complete the broken-line graph.

4. Make a similar graph of the weekly attendance in your math class for a week.

ATTENDANCE IN JOHN'S MATH CLASS

Here are the results of a cassette tape survey:

Name	Number of cassette tapes in collection
Jacob	16
Elsa	21
Lee	24
Matthew	19
Kate	26

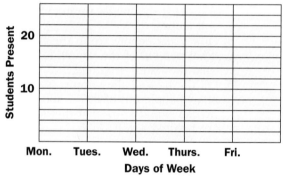

NUMBER OF TAPES IN COLLECTION

Jacob

Elsa

Lee

Matthew

Kate

Each ● stands for 4 tapes

5. Copy and complete the picture graph.

6. Take a similar survey. Show the results on a picture graph.

▶ **Solve. Give each answer in simplest form.** *(page 196)*

1. $\frac{1}{3}$ of $39 = n$

2. $\frac{1}{5}$ of $55 = n$

3. $\frac{2}{3}$ of $48 = n$

4. $\frac{3}{4}$ of $52 = n$

5. $\frac{5}{8}$ of $64 = n$

6. $\frac{4}{5}$ of $65 = n$

7. $\frac{5}{6}$ of $42 = n$

8. $\frac{3}{2}$ of $80 = n$

9. $\frac{3}{5}$ of $27 = n$

10. $\frac{7}{3}$ of $41 = n$

11. $\frac{5}{4}$ of $42 = n$

12. $\frac{4}{5}$ of $51 = n$

▶ **Complete.** *(page 198)*

13. $1\frac{1}{3}$ ft = __?__ in.

14. $2\frac{3}{4}$ ft = __?__ in.

15. $2\frac{1}{3}$ yd = __?__ ft

16. $1\frac{1}{2}$ days = __?__ h

17. $2\frac{2}{3}$ h = __?__ min

18. $1\frac{1}{3}$ min = __?__ s

19. $1\frac{1}{3}$ yd = __?__ in.

20. $2\frac{1}{4}$ yd = __?__ in.

21. $1\frac{1}{2}$ gal = __?__ qt

22. $2\frac{1}{2}$ qt = __?__ pt

23. $2\frac{3}{4}$ gal = __?__ pt

24. $3\frac{1}{2}$ pt = __?__ c

▶ **Solve. Give answers in simplest form.** *(page 200)*

25. $\frac{1}{3}n = 14$

26. $\frac{1}{2}n = 21$

27. $\frac{1}{5}n = 33$

28. $\frac{2}{3}n = 36$

29. $\frac{3}{4}n = 42$

30. $\frac{5}{8}n = 50$

31. $\frac{5}{3}n = 65$

32. $\frac{7}{4}n = 56$

33. $\frac{5}{8}n = 18$

34. $\frac{4}{5}n = 41$

35. $\frac{7}{5}n = 60$

36. $\frac{5}{6}n = 62$

▶ **Simplify by combining like terms.** *(page 224)*

37. $9y + y$

38. $5z - z$

39. $7j - 7j$

40. $^-4n + 7n - 11$

41. $^-4n + 7 - 11n$

42. $^-4 + 7n - 11n$

43. $^-12t - t - 15$

44. $^-12t + t - 15$

45. $^-12t + 1 - 15$

46. $8a - 7 - 10a$

47. $^-8 - 7a + 10$

48. $^-8a + 7 + 8a$

▶ **Solve and check.** *(page 228)*

49. $^-3n = 4 + n$

50. $6y = 8 - 2y$

51. $^-5w = ^-3w + 10$

52. $2y + 9 = 4y - 3$

53. $4w - 3 = w - 1$

54. $5x - 11 = 2x + 11$

55. $15 - 3k = k + 8$

56. $12 + 2n = n - 9$

57. $16 - 3c = 9 + c$

58. $^-2r + 4 = 6r$

59. $^-7 - d = ^-3d$

60. $5t + 3 = t$

Here are scrambled answers for the review exercises:

8	120	mean	outcomes	sample
12	box-and-whisker	median	permutation	stem-and-leaf
15	dependent	mode	probability	value
48	independent	multiply	range	

1. To compute the total number of ways that several decisions can be made, _?_ the number of choices for each of the decisions. If you have 6 pairs of slacks and 8 sweaters, you would have _?_ slacks-sweater combinations. *(page 364)*

2. An arrangement of things in a definite order is called a _?_ . When 5 students line up for lunch, they can line up in _?_ different ways. *(page 366)*

3. If you roll a die,

$$P(\text{3 or less}) = \frac{\textbf{number of ways to roll a 3 or less}}{\textbf{number of possible } \underline{\ ?\ }}$$

   A _?_ space is the set of all possible outcomes. If you list the sample space for tossing a coin 3 times, the sample space will have _?_ outcomes.
   (pages 368, 370)

4. Since an event such as rolling a 2 does not affect an event such as tossing heads, the two events are called _?_ events.

$$P(\text{2, H}) = \frac{1}{?}$$

   If the outcome of a first event can affect the outcome of a second event, the events are called _?_ events. If you draw two cards without replacement,

$$P(\text{B, vowel}) = \frac{1}{?}$$

   $$\boxed{\text{A}}\ \boxed{\text{B}}\ \boxed{\text{C}}\ \boxed{\text{D}}\ \boxed{\text{E}}\ \boxed{\text{F}}$$

   (pages 372, 374)

5. To find the expectation, multiply the _?_ of winning by the _?_ of the prize.
   (page 380)

6. The average of a set of data is called the _?_ . The _?_ is the score in the middle. The _?_ is the number that occurs most often. The _?_ is the difference between the largest number and the least number. *(page 382)*

7. Statisticians use _?_ plots and _?_ plots to organize and display data.
   (pages 386, 388)

▶ **Solve.** *(pages 364, 366)*

1. How many different outfits can you make with 6 blouses and 5 skirts?

2. In how many ways can 6 people line up at a ticket booth?

▶ **Give each probability in simplest form.** *(page 368)*

Think about drawing a card, without looking, from the five cards shown at the right.

$$\boxed{1}\ \boxed{2}\ \boxed{3}\ \boxed{4}\ \boxed{5}$$

3. $P(3)$

4. $P(\text{odd})$

5. $P(\text{even})$

6. $P(\text{not } 4)$

▶ **Solve.** *(page 370)*

Think about drawing a card, without looking, from the five cards shown above, replacing it, and drawing a second card without looking.

7. List the sample space.

8. What is the probability that each card will have an even number?

9. What is the probability that each card will have an odd number?

10. What is the probability that the sum of the numbers on the cards will be 6?

▶ **Give each probability in simplest form.** *(pages 372, 374)*

Think about drawing a card, without looking, from the five cards shown above, then tossing a coin.

11. $P(4, \text{heads})$

12. $P(\text{even, tails})$

13. $P(\text{not } 2, \text{heads})$

Think about drawing a card, without looking, from the five cards shown above, then without replacement drawing a second card without looking.

14. $P(5, 4)$

15. $P(\text{even}, 1)$

16. $P(\text{odd, even})$

▶ **Solve.** *(page 380)*

17. **a.** Eighty chances were sold on binoculars worth $120. Each chance cost $2. If you bought one chance, what would be your expectation?
    b. Is buying a chance a good deal?

▶ **Solve using the scores.** *(page 382)*

Scores: 35, 43, 31, 43, 45, 52, 50, 43, 36

18. Make a line plot.

19. Find the mean.

20. Find the median.

21. Find the mode.

22. Find the range.

▶ **Use a proportion to solve.** *(page 392)*

23. In a sample, 12 out of 40 people said their favorite color is blue. Based on the sample, how many people out of 500 would you predict to pick blue as their favorite color?

▶ **Choose the correct letter.**

1. The greatest common factor of $16ab^2$ and $20a^2b$ is

A. $4a$

B. $4b$

C. $4a^2b^2$

D. none of these

2. $\dfrac{21u^2v}{15uv^2}$ written in lowest terms is

A. $\dfrac{7u^2}{5uv}$

B. $\dfrac{7uv}{5v^2}$

C. $\dfrac{7u}{5v}$

D. none of these

3. Give the sum.

$$\dfrac{3c}{4} + \dfrac{d}{5}$$

A. $\dfrac{15c + 4d}{20}$

B. $\dfrac{15c + d}{20}$

C. $\dfrac{3c + d}{20}$

D. none of these

4. Give the product.

$$\dfrac{7x}{y^2} \cdot \dfrac{y}{14x}$$

A. $\dfrac{7}{2xy}$

B. $\dfrac{1}{2y}$

C. $\dfrac{7y}{2}$

D. none of these

5. Give the quotient.

$$\dfrac{a}{4b} \div \dfrac{5a}{b^2}$$

A. $\dfrac{b}{20}$

B. $\dfrac{20}{b}$

C. $\dfrac{b^2}{20}$

D. none of these

6. Solve.

$$\dfrac{4}{3} \text{ of } 72 = n$$

A. 54

B. 72

C. 96

D. none of these

7. $2\dfrac{1}{3}$ yd = ___?___ in.

A. 7

B. 28

C. 48

D. none of these

8. Solve.

$$\dfrac{5}{3}n = 33$$

A. 55

B. 33

C. $19\dfrac{4}{5}$

D. none of these

9. Simplify by combining like terms.

$$9x + 7 - x - 17$$

A. $10x - 10$

B. $8x + 10$

C. $8x - 10$

D. none of these

10. Solve.

$$12 - 3c = 9 + c$$

A. $^-5\dfrac{1}{4}$

B. $\dfrac{3}{4}$

C. $^-1\dfrac{1}{2}$

D. none of these

11. Find the volume. Use 3.14 for π.

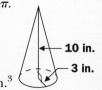

10 in.

3 in.

A. 62.8 in.^{3}

B. 282.6 in.^{3}

C. 94.2 in.^{3}

D. none of these

12. Choose the equation.

Adela worked 5 hours yesterday and 3 hours today. She earned a total of $52. How much did she earn per hour?

A. $5n + 3n = 52$

B. $5n - 3n = 52$

C. $3n + 52 = 5n$

D. $5n - 52 = 3n$

Polynomials

$F = 1.8C + 32$

This formula converts degrees Celsius to degrees Fahrenheit.
(page 418)

The thermometer shows that the temperature for the swimmers was 30°C. You can use the formula to convert this temperature to degrees Fahrenheit.

399

OBJECTIVE: To identify a polynomial by the number of terms.

1. Look at the tiles.

   a. What is the area of the *n*-by-*n* square?

   b. What is the area of the 1-by-*n* rectangle?

A **monomial** is a variable, a number, or a product of variables and numbers. For example, n^2, 1, and $3n$ are monomials.

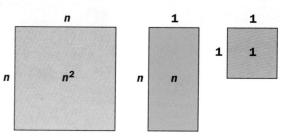

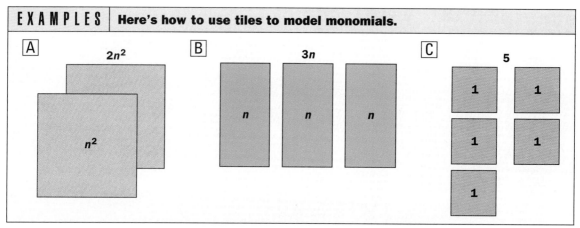

| EXAMPLES | Here's how to use tiles to model monomials. |

A $2n^2$

B $3n$

C 5

CHECK for Understanding

2. Look at Examples A, B, and C. Which monomial is represented by

   a. the two large squares?   **b.** the five small squares?

Expressions such as $2n^2 + 3n$ and $n^2 + 2n + 3$ are called polynomials. A **polynomial** is a monomial or the sum or difference of monomials. The monomials are called the **terms** of the polynomial.

| EXAMPLES | Here's how to use tiles to model polynomials. |

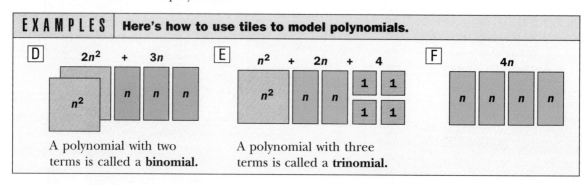

D $2n^2$ + $3n$

A polynomial with two terms is called a **binomial.**

E n^2 + $2n$ + 4

A polynomial with three terms is called a **trinomial.**

F $4n$

CHECK for Understanding

3. Look at Examples D, E, and F. Which of the polynomials is

   a. a monomial?   **b.** a binomial?   **c.** a trinomial?

EXERCISES

▶ In each exercise, a polynomial is modeled by tiles. Give the polynomial.

4.

5.

6.

▶ Tell whether each polynomial is a monomial, a binomial, or a trinomial.

7. $6y + 2x$ **8.** $3b$ **9.** $m^2 + 6$ **10.** $4a + b + 3c$

11. $9m^2$ **12.** $8n^2 + {}^-5$ **13.** $x^2 + y + 4$ **14.** ab

15. $r^2 + 3r + 1$ **16.** $6ab$ **17.** ${}^-2n$ **18.** $11m^2 + 1$

19. $ab^2 + a + 4$ **20.** ${}^-a^2$ **21.** $x^2 + 4$ **22.** $xy^2 + xy + 3$

▶ Use your mental math skills. Evaluate each polynomial for $a = 1$, $b = 2$, $c = {}^-2$, and $d = {}^-3$.

23. $a^2 + b$ **24.** $b^2 + d$ **25.** $3ab$ **26.** $a + c + d$

27. $2a + 2b$ **28.** abc **29.** $2d + c$ **30.** $a^2 + b^2$

31. $a^2 + d^2$ **32.** $4b^2 + 3d$ **33.** $a + b + d$ **34.** $2a + 3b + 2c$

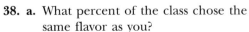 **ANALYZING DATA** *Favorite Flavor*

▶ Work in a small group.

35. Which one of these ice-cream flavors would you choose—vanilla, chocolate, strawberry, butter pecan, or coffee?

36. Make a ''favorite flavor'' tally for the members of your class by recording their answers to the question in Exercise 35.

37. Compute the percent of your class that chose each flavor.

38. a. What percent of the class chose the same flavor as you?
b. Which flavor was the most popular?
c. Which flavor was the least popular?

39. Write a paragraph that summarizes the data.

14-2 Adding Polynomials

OBJECTIVE: To add polynomials.

You can use tiles to help you understand addition.

1. **a.** What polynomial is being added to $2n^2 + 3n$?

   b. What is the sum?

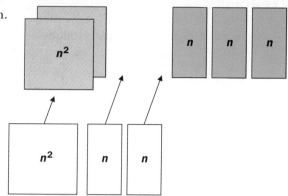

EXAMPLE	Here's how to use number properties to add polynomials.

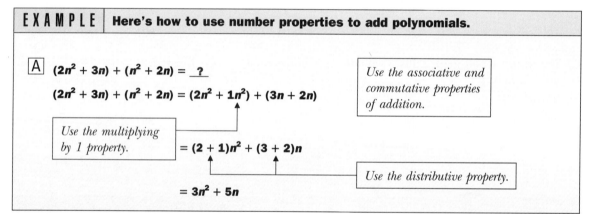

A $(2n^2 + 3n) + (n^2 + 2n) = $ _?_

$(2n^2 + 3n) + (n^2 + 2n) = (2n^2 + 1n^2) + (3n + 2n)$

Use the associative and commutative properties of addition.

Use the multiplying by 1 property.

$= (2 + 1)n^2 + (3 + 2)n$

Use the distributive property.

$= 3n^2 + 5n$

CHECK for Understanding

2. Complete these examples.

   a. $(5n^2 + 4n) + (2n^2 + 3n) = (5 + 2)n^2 + (4 + 3)n$

   $= $ _?_ $n^2 + $ _?_ n

   b. $(7n + 3) + (2n + {}^-2) = (7 + 2)n + 3 + {}^-2$

   $= $ _?_ $n + $ _?_

EXAMPLE	Here's how to use a shortcut to add polynomials.

First write the sum without grouping symbols and then combine like terms.

B $(4r^2 + 3s) + (2r^2 + 6s) = $ _?_

Like terms

$(4r^2 + 3s) + (2r^2 + 6s) = 4r^2 + 3s + 2r^2 + 6s$

Like terms

$= 6r^2 + 9s$

EXERCISES

 Give each sum.
Here are scrambled answers for the next two rows of exercises:
$5x^2 + 7x$ $5r + 3s$ $6x + 13$ $8x^2 + 4$

3. $(4x + 6) + (2x + 7)$

4. $(3x^2 + 3x) + (2x^2 + 4x)$

5. $(6x^2 + 7) + (2x^2 + {}^-3)$

6. $(3r + 2s) + (2r + s)$

7. $(7a^2 + b) + (2a^2 + 2b)$

8. $(8m + n) + (2m + 3n)$

9. $(5t + {}^-7u) + (6t + 8u)$

10. $(7c + 2d) + (4c + {}^-3d)$

11. $(3m + {}^-2n) + (9m + {}^-4n)$

12. $(11xy + 4y) + (8xy + 2y)$

13. $(3c + 3) + ({}^-6c + 8)$

14. $(12a + 6b) + (9a + 2b)$

15. $(2n + 3m) + ({}^-4n + 2m)$

16. $(3x^2 + 2x) + (4x^2 + {}^-3x)$

17. $(4c + 5d) + ({}^-2c + {}^-3d)$

18. $(a^2 + b^2) + (2a^2 + 2b^2)$

19. $(3a + {}^-7c) + ({}^-2a + 9c)$

20. $(6r + 8s) + ({}^-5r + {}^-9s)$

21. $(2a + 3b + 4c) + (3a + b + 2c)$

22. $(3m + 4n + 2) + (2m + {}^-3n + 5)$

23. $(6r + 5s + 7t) + (5r + 2t)$

24. $(4x + 2y + 5z) + (2x + 3z)$

25. $(5x^2 + 2x + 4) + (2x^2 + {}^-2x)$

26. $(18y^2 + 8y + 5) + (9y^2 + 6y + {}^-7)$

 Use your mental math skills. Combine like terms and evaluate for
$r = 10$ **and** $s = 100.$

27. $(3r + 7) + (2r + 2)$

28. $(4r + 3) + (2r + 3)$

29. $(6r + 5) + (3r + 3)$

30. $(2s + 2r) + (3s + 3r)$

31. $(4s + 5) + (3s + 2)$

32. $(2s + 3r) + (5s + 6r)$

33. $(3s + 2r + 4) + (6s + 5r + 3)$

34. $(4s + 4r + 3) + (5s + 2r + 4)$

35. $(7s + 3r + 7) + (s + 4r + 2)$

36. $(2s + 5r + 3) + (3s + r + 1)$

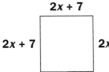 **WRITING EXPRESSIONS**

 Write and simplify an expression for the perimeter of each figure.

37.

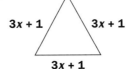

38.

39.

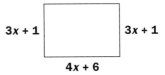

14-3 Subtracting Polynomials

OBJECTIVE: To subtract polynomials.

In this lesson, you will learn two ways to subtract polynomials.

EXAMPLE | **Here's how to use tiles to model subtracting polynomials.**

$(3n^2 + 5n) - (n^2 + 2n) = \underline{\ ?\ }$

Step 1.

Lay out tiles to represent $3n^2 + 5n$.

Step 2.

Subtract (remove) the tiles that represent $n^2 + 2n$. The tiles that are left represent the difference.

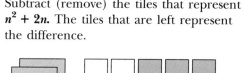

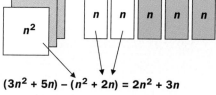

$(3n^2 + 5n) - (n^2 + 2n) = 2n^2 + 3n$

CHECK for Understanding

1. Look at the example above.

    a. What polynomial is being subtracted from $3n^2 + 5n$?   **b.** What is the difference?

EXAMPLE | **Here's how to use opposites to subtract polynomials.**

To subtract a polynomial, add the opposite of the polynomial.

$(3n^2 + 5n) - (n^2 + 2n) = \ ?$

$(3n^2 + 5n) - (n^2 + 2n) = (3n^2 + 5n) + (^-1)(n^2 + 2n)$

> *Multiply by ($^-1$) to get the opposite of $(n^2 + 2n)$.*

Like terms

$= 3n^2 + 5n + {}^-1n^2 + {}^-2n$

$= 2n^2 + 3n$ *Like terms*

CHECK for Understanding

2. Complete this example.

    $(4r + 3s) - (5r + 2s) = 4r + 3s + {}^-5r + {}^-2s = \underline{\ ?\ } r + \underline{\ ?\ } s$

EXERCISES

▶ **Subtract by thinking about removing tiles.**
Here are scrambled answers for the next row of exercises: $3n^2 + 1$ $4n + 1$ $n^2 + n$

3. $(3n^2 + 4n) - (2n^2 + 3n)$

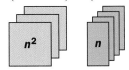

4. $(4n^2 + 3) - (n^2 + 2)$

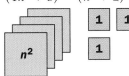

5. $(6n + 5) - (2n + 4)$

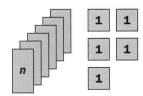

6. $(2n^2 + 5n) - (2n^2 + 3n)$

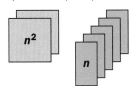

7. $(3n + 4) - (2n + 4)$

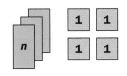

8. $(2n^2 + 3n + 5) - (2n + 4)$

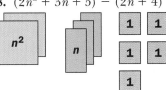

▶ **Subtract by adding the opposite.**
Here are scrambled answers for the next row of exercises: $3x + y$ $x^2 + y$

9. $(6x + 3y) - (3x + 2y)$

10. $(3x^2 + 2y) - (2x^2 + y)$

11. $(4x^2 + 3) - (x^2 + 2)$

12. $(6r + 5) - (2r + 4)$

13. $(6a^2 + 5b) - (2a^2 + 3b)$

14. $(6m + 2n) - (3m + n)$

15. $(7t^2 + 2y) - (6t^2 + 2y)$

16. $(8d + 6e) - (5d + 3e)$

17. $(3a + 5b) - (2a + 3b)$

18. $(13xy + 4) - (8xy + 2)$

19. $(3e + 4f) - (3e + 2f)$

20. $(12a + 2b) - (3a + b)$

21. $(3a + 2b + 4c) - (2a + b + 3c)$

22. $(2m + 5n + 6p) - (m + 2n + 5p)$

23. $(5x^2 + 3x + 4) - (3x^2 + 2x + 3)$

24. $(12y^2 + 6y + 3) - (8y^2 + 2)$

Problem Solving USING EXPRESSIONS

▶ **Solve. Use the signpost.**

25. Which city is $(2n + 2)$ miles from Norris?

26. Which city is $(3n + 1)$ miles from Joplin?

27. Which city is $(7n + 6)$ miles from Sun River?

28. If $n = 25$, how many miles is it from Shelby to Sun River?

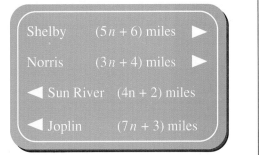

Polynomials | **405**

You will use a variety of problem-solving strategies on these two pages. For some problems, you may wish to use a calculator.

1 SUBJECT SURVEY

Here are the results of a survey about math, English, and science taken in Homeroom 114:

Everyone likes at least one of the subjects. 18 like math.   8 like both math and English.   32 like English.   16 like both English and science.   25 like science.   7 like both math and science.   3 like all three subjects.

How many students are in Homeroom 114?

✓ *Problem Solving Tips*

Draw a diagram.   You may find the diagram useful. Each region in the diagram represents part of the survey.

Hint: The shaded region represents the 7 students who like math and science.

math   **English**

science

2 ALPHAMETICS

In this sum, each different letter stands for a different digit. What digit does each letter stand for?

Hint 1: What digit does M have to be?

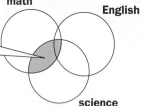

```
  S E N D
+ M O R E
---------
M O N E Y
```

Hint 2: After you decide on M, think about S and O.

3 TAKE ME OUT TO THE BALL GAME

Two fathers and two sons went to a baseball game. When they totaled their expenses, they found that they had spent $57 altogether. Each spent the same amount, and each spent a whole number of dollars. How was this possible?

4 MYSTERY DIGITS

Copy each multiplication problem and write in the missing mystery digits.

a.
```
      6 ? ? 9
    ×   ? 7
    4 3 8 1 3
  5 ? ? 7 2
  5 ? ? 5 3 3
```

b.
```
      ? ? ?
    ×   3 7
    6 7 7 6
  ? ? ? ?
  ? ? ? ? 6
```

5 WELLINGTON'S CHALLENGE

My Uncle Wellington said, "If you can show me how to use 21 coins to make $1, I'll give you $1. If you can show me three different ways to solve the problem, I'll give you $100."

Can you help me earn $100?

6 SWAP SPOTS

Each set of cards has a different total number of diamonds. Can you move one card from one set to another set so that each set will have the same total number of diamonds?

7 CHECKERBOARD COINS

Place the coins in the squares in such a way that each row, column, and diagonal contains each type of coin.

GEOMETRY—APPLICATIONS

The Jackson family is landscaping its backyard. This drawing shows the lawn and patio dimensions in feet.

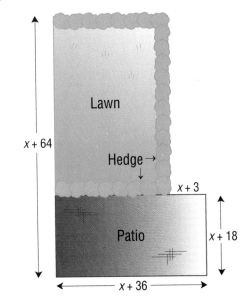

8 Write and simplify an expression for the perimeter of the patio.

9 What is the area of the lawn?

10 Mrs. Jackson wants to plant a hedge as shown in the drawing. How long will the hedge be?

Rate Your Problem-Solving Power: *6–7 correct = Good;   8–9 correct = Excellent;   10 correct = Exceptional*

14-4 Problem Solving—Applications

OBJECTIVE: To use shipping costs for solving problems.

SHIPPING COSTS FROM CHICAGO				
Shipping Weight*	**Zone 1** Up to 150 miles**	**Zone 2** 151 to 300 miles	**Zone 3** 301 to 600 miles	**Zone 4** 601 to 1000 miles
1 oz to 8 oz	$ .50	$ .53	$ .56	$ .60
9 oz to 15 oz	.83	.86	.90	.94
1 lb to 2 lb	1.22	1.34	1.57	1.83
2 lb 1 oz to 3 lb	1.32	1.46	1.69	1.98
3 lb 1 oz to 4 lb	1.46	1.63	1.90	2.25
4 lb 1 oz to 5 lb	1.60	1.78	2.09	2.51

*Weight rounded to the nearest ounce   **Distance rounded to the nearest mile

Zone 4

Bismarck • 　Duluth•

Zone 3
Toronto •

Zone 2

Milwaukee •

Chicago　Toledo •　Pittsburgh •　New York

Zone 1

Denver•

Kansas City•　St. Louis

▶ **Use the map and the shipping information to solve these problems.**

1. How much does it cost to ship a 20-ounce package to Zone 4?

2. What is the cost to ship a 4-pound package 500 miles?

3. Can you ship a 40-ounce package from Chicago to Pittsburgh for less than $1.50?

4. How much does it cost to ship a 30-ounce package 800 miles?

5. It cost Danny $1.90 to ship a 3.5-pound box to his sister. Does his sister live in Kansas City or Toledo?

6. How much more does it cost to ship two 20-ounce packages to St. Louis than it costs to ship one 40-ounce package to St. Louis?

7. How much more does it cost to ship a 5-pound package from Chicago to New York than from Chicago to Kansas City?

8. What is the total cost to ship a 2-pound 6-ounce package and a 1-pound 4-ounce package from Chicago to Milwaukee?

▶ **Write an equation and solve the problem.**

9. The weight of a box is 20 ounces more than the weight of a package. If the box weighs 60 ounces, what is the weight of the package? (Let p = the weight in ounces of the package.)

10. A package weighs 14 ounces. It weighs 2 ounces less than 4 times the weight of a letter. What is the weight of the letter? (Let w = the weight in ounces of the letter.)

▶ Write each decimal in simplest fractional form. *(page 213)*

1. 0.4

2. ⁻0.20

3. ⁻0.6

4. 0.25

5. ⁻0.375

6. ⁻0.24

7. 0.72

8. ⁻0.16

9. 0.150

10. ⁻0.875

11. 2.25

12. ⁻1.8

13. 2.40

14. ⁻9.6

15. 4.55

16. ⁻7.2

17. 3.08

18. ⁻5.125

19. 6.40

20. ⁻1.625

▶ Give each sum or difference in simplest form. *(page 214)*

21. $\frac{2}{3} + \frac{^-1}{3}$

22. $\frac{3}{8} + \frac{1}{2}$

23. $\frac{^-5}{16} + \frac{3}{8}$

24. $\frac{^-3}{5} + \frac{^-1}{4}$

25. $2\frac{5}{8} + {^-4}\frac{1}{4}$

26. $^-9\frac{5}{8} + 4\frac{2}{3}$

27. $\frac{3}{5} - \frac{1}{10}$

28. $\frac{^-3}{8} - \frac{1}{2}$

29. $\frac{1}{3} - \frac{3}{5}$

30. $\frac{5}{6} - \frac{^-7}{8}$

31. $\frac{^-9}{10} - \frac{^-3}{4}$

32. $4\frac{3}{5} - {^-2}\frac{1}{10}$

33. $^-6 - 3\frac{7}{8}$

34. $5\frac{1}{2} - {^-1}\frac{3}{4}$

35. $^-6\frac{7}{8} - {^-7}\frac{2}{3}$

36. $^-9 - 5\frac{5}{6}$

▶ Give each product or quotient in simplest form. *(page 216)*

37. $\frac{^-2}{3} \cdot \frac{1}{5}$

38. $\frac{^-1}{4} \cdot \frac{^-1}{3}$

39. $\frac{1}{8} \cdot 8$

40. $\frac{^-5}{3} \cdot \frac{3}{10}$

41. $\frac{^-4}{5} \cdot \frac{^-5}{4}$

42. $3 \cdot {^-1}\frac{1}{2}$

43. $^-2\frac{7}{8} \cdot {^-3}$

44. $^-2\frac{1}{2} \cdot 2\frac{1}{2}$

45. $^-6 \div \frac{3}{4}$

46. $\frac{9}{5} \div \frac{3}{10}$

47. $\frac{^-5}{12} \div \frac{15}{4}$

48. $^-10 \div 2\frac{1}{2}$

49. $^-1\frac{1}{4} \div {^-2}\frac{1}{3}$

50. $2\frac{5}{6} \div 1\frac{3}{4}$

51. $4\frac{2}{5} \div {^-1}\frac{3}{10}$

52. $^-4\frac{2}{3} \div 2\frac{5}{6}$

▶ Solve and check. *(page 222)*

53. $6n + 7 = 19$

54. $^-5j - 14 = 31$

55. $^-6 - 12n = 9$

56. $14k - 13 = 23$

57. $8 - 7q = 0$

58. $12d + 8 = 8$

59. $20m - 24 = 26$

60. $^-11 + 11j = 31$

61. $^-20 - 14p = {^-6}$

▶ Simplify by combining like terms. *(page 224)*

62. $12j + j$

63. $15y - y$

64. $16x - 16x$

65. $3v + v + 12$

66. $10n - 6 + 3n$

67. $^-12j + 9 + 5j$

68. $15v - 11v - 3 + 5$

69. $15v - 11 - 3v + 5$

70. $15 - 11v - 3 + 5v$

OBJECTIVE: To multiply a polynomial by a monomial.

You can use tiles to help you understand multiplication.

| **EXAMPLE** | **Here's how to multiply 2n and n + 2.** |

Step 1.

Use the edges of the tiles below to mark off a rectangle that has the dimensions **2n** and **n + 2.**

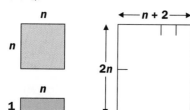

Step 2.

Fill in the rectangle with tiles to find its area.

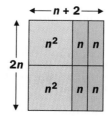

CHECK for Understanding

1. To find the area of the rectangle, you could multiply its length $(2n)$ by its width (_?_).

2. Count the tiles it takes to fill the rectangle. Is $2n(n + 2) = 2n^2 + 4n$?

| **EXAMPLE** | **Here's how to use the distributive property to multiply a polynomial by a monomial.** |

A $2n(n + 2) = $ _?_

$2n(n + 2) = 2n \cdot n + 2n \cdot 2$

$\qquad = 2n^2 + 4n$

B $3r\,(r + 2s + {}^-4) = $ _?_

$3r\,(r + 2s + {}^-4) = 3r \cdot r + 3r \cdot 2s + 3r \cdot {}^-4$

$\qquad = 3r^2 + 6rs + {}^-12r$

CHECK for Understanding

3. Complete these examples.

a. $8x(2x + 7) = 8x \cdot 2x + 8x \cdot 7$
$\qquad = \underline{\ ?\ } x^2 + \underline{\ ?\ } x$

b. ${}^-3y(x + y + {}^-4) = {}^-3y \cdot x + {}^-3y \cdot y + {}^-3y \cdot {}^-4$
$\qquad = \underline{\ ?\ } + \underline{\ ?\ } + \underline{\ ?\ }$

c. $t({}^-6t + 5) = t \cdot {}^-6t + t \cdot 5$
$\qquad = \underline{\ ?\ } + \underline{\ ?\ }$

d. ${}^-7(2c^2 + c + {}^-1) = {}^-7 \cdot 2c^2 + {}^-7 \cdot c + {}^-7 \cdot {}^-1$
$\qquad = \underline{\ ?\ } + \underline{\ ?\ } + \underline{\ ?\ }$

EXERCISES

▶ **Multiply.**
Here are scrambled answers for the first row of exercises:
$6n + 15$ $6n + 24$ $4n^2 + 12n$

4. $6(n + 4)$ **5.** $3(2n + 5)$ **6.** $4n(n + 3)$

7. $2n(2n + 3)$ **8.** $5(x + 3)$ **9.** $5y(y + 3)$

10. $3y(y + 8)$ **11.** $4z(z + 2)$ **12.** $6a(a + 4)$

13. $3b(b + 2)$ **14.** $5r(r + 2)$ **15.** $t(4t + 1)$

16. $7(x + {}^-2)$ **17.** $n(n + {}^-3)$ **18.** $6d(d + 3)$

19. $3a(a + {}^-2)$ **20.** $3a(a + 1)$ **21.** $4b(b + 2)$

22. $c(4c + d)$ **23.** $2t(r + s)$ **24.** $7a(2b + c)$

25. $5x(3x + {}^-2)$ **26.** $x({}^-3x + y)$ **27.** ${}^-2(3ab + 4)$

28. $4(x^2 + 2x + 3)$ **29.** $6(y^2 + 3y + 2)$ **30.** $5(2a^2 + 3a + {}^-4)$

31. $2a(3a + 4b + c)$ **32.** $2r(3r + 4s + {}^-3)$ **33.** $4(3c^2 + c + {}^-3)$

34. ${}^-2a(3a + 4b + c)$ **35.** ${}^-4x(2x + {}^-3y + 2z)$ **36.** $3({}^-12x^2 + 4x + 3)$

▶ **Use your mental math skills. Evaluate each expression for $a = 2$, $b = 3$.**

37. $2(a + 4)$ **38.** $3(b + 2)$ **39.** $2(b + a)$ **40.** $2(a + b)$

41. $a(a + 1)$ **42.** $b(b + 3)$ **43.** $a(a + 2)$ **44.** $b(a + 2)$

45. $a(a + b)$ **46.** $b(a + b)$ **47.** $2a(a + b)$ **48.** $2b(a + b)$

Problem Solving VISUAL THINKING

▶ **The areas of the shaded regions are given at the right. Match the shaded region with the area.**

Areas of shaded regions
$4n^2 + 6n$
$2n^2 + 6n$
$6n^2 + 6n$

49.

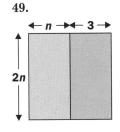

50.

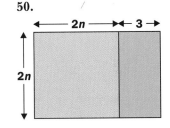

51.

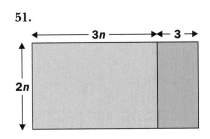

OBJECTIVE: To divide a polynomial by a monomial.

You have already divided monomials by factoring and canceling.

$$\frac{6x^2}{2x} = \frac{3 \cdot \overset{1}{\cancel{2}} \cdot \overset{1}{\cancel{x}} \cdot x}{\underset{1}{\cancel{2}} \cdot \underset{1}{\cancel{x}}} = 3x$$

In this lesson, you will use factoring and canceling to divide polynomials.

E X A M P L E S | **Here's how to divide a polynomial by a monomial.**

To divide a polynomial by a monomial, divide each term of the polynomial by the monomial.

A $\dfrac{10x + 2}{2} = \dfrac{10x}{2} + \dfrac{2}{2}$

$= \dfrac{5 \cdot \overset{1}{\cancel{2}} \cdot x}{\underset{1}{\cancel{2}}} + \dfrac{\overset{1}{\cancel{2}}}{\underset{1}{\cancel{2}}}$

$= 5x + 1$

B $\dfrac{9n^3 + 6n}{3n} = \dfrac{9n^3}{3n} + \dfrac{6n}{3n}$

$= \dfrac{3 \cdot \overset{1}{\cancel{3}} \cdot \overset{1}{\cancel{n}} \cdot n \cdot n}{\underset{1}{\cancel{3}} \cdot \underset{1}{\cancel{n}}} + \dfrac{3 \cdot 2 \cdot \overset{1}{\cancel{n}}}{\underset{1}{\cancel{3}} \cdot \underset{1}{\cancel{n}}}$

$= 3n^2 + 2$

C $\dfrac{12a + 4}{4} = \dfrac{\overset{1}{\cancel{4}} \cdot 3 \cdot a}{\underset{1}{\cancel{4}}} + \dfrac{\overset{1}{\cancel{4}}}{\underset{1}{\cancel{4}}}$

$= 3a + 1$

D $\dfrac{8c^2 + {}^-6c}{2c} = \dfrac{4 \cdot \overset{1}{\cancel{2}} \cdot \overset{1}{\cancel{c}} \cdot c}{\underset{1}{\cancel{2}} \cdot \underset{1}{\cancel{c}}} + \dfrac{{}^-3 \cdot \overset{1}{\cancel{2}} \cdot \overset{1}{\cancel{c}}}{\underset{1}{\cancel{2}} \cdot \underset{1}{\cancel{c}}}$

$= 4c + {}^-3$

CHECK for Understanding

1. Look at Example A. The red marks indicate that both numerators and denominators were divided by __?__ .

2. Look at Example B. The red marks indicate that both numerators and denominators were divided by __?__ .

3. Explain how to get the numbers that are printed in red in Examples C and D.

4. Complete these examples.

a. $\dfrac{24n + 8}{4} = \dfrac{6 \cdot \overset{1}{\cancel{4}} \cdot n}{\underset{1}{\cancel{4}}} + \dfrac{\overset{1}{\cancel{4}} \cdot 2}{\underset{1}{\cancel{4}}}$

$= \underline{\;?\;} + 2$

b. $\dfrac{6y^3 + 3y^2}{3y} = \dfrac{\overset{1}{\cancel{3}} \cdot 2 \cdot \overset{1}{\cancel{y}} \cdot y \cdot y}{\underset{1}{\cancel{3}} \cdot \underset{1}{\cancel{y}}} + \dfrac{\overset{1}{\cancel{3}} \cdot \overset{1}{\cancel{y}} \cdot y}{\underset{1}{\cancel{3}} \cdot \underset{1}{\cancel{y}}}$

$= \underline{\;?\;} + \underline{\;?\;}$

c. $\dfrac{6a^2 + {}^-2a}{2a} = \dfrac{? \cdot \overset{1}{\cancel{2}} \cdot \overset{1}{\cancel{a}} \cdot a}{\underset{1}{\cancel{2}} \cdot \underset{1}{\cancel{a}}} + \dfrac{{}^-\overset{1}{\cancel{2}} \cdot \overset{1}{\cancel{a}}}{\underset{1}{\cancel{2}} \cdot \underset{1}{\cancel{a}}} = \underline{\;?\;} + \underline{\;?\;}$

EXERCISES

▶ **Divide.**

Here are scrambled answers for the next row of exercises:
$n + 1$ $2n^2 + 1$ $3n + 1$ $4n + 1$

5. $\dfrac{12n^2 + 6}{6}$

6. $\dfrac{8n + 2}{2}$

7. $\dfrac{21n^2 + 7n}{7n}$

8. $\dfrac{n^2 + n}{n}$

9. $\dfrac{3n^3 + 3n^2}{3n}$

10. $\dfrac{24c + 12}{12}$

11. $\dfrac{15x^2 + 5x}{5x}$

12. $\dfrac{6d + {}^-3}{3}$

13. $\dfrac{d^3 + d^2}{d}$

14. $\dfrac{20x^2 + 10x}{5x}$

15. $\dfrac{16a^2 + {}^-8a}{4}$

16. $\dfrac{18e^3 + 12e^2}{6e}$

17. $\dfrac{15a + 30b}{5}$

18. $\dfrac{6m + {}^-3n}{3}$

19. $\dfrac{7x^2 + 14y}{7}$

20. $\dfrac{{}^-20a + 15b}{5}$

21. $\dfrac{9ab^2 + 3b}{3b}$

22. $\dfrac{6a^2b + {}^-4b}{2b}$

23. $\dfrac{12x^2y^2 + 6xy}{6xy}$

24. $\dfrac{12n^2 + 10n + 6}{2}$

25. $\dfrac{25b^2 + 15b + 10}{5}$

26. $\dfrac{16y^2 + {}^-8y + 12}{4}$

27. $\dfrac{20x^3 + 15x^2 + 10x}{5x}$

28. $\dfrac{3a^3 + {}^-6a^2 + 9a}{3a}$

29. $\dfrac{12n^3 + 16n^2 + {}^-20n}{4n}$

Challenge! USING EXPRESSIONS

30. Follow the instructions for this number trick.

> Instructions:
> - Pick a number.
> - Add 4.
> - Multiply by 3.
> - Subtract 6.
> - Divide by 3.
> - Subtract the number you picked.
>
> Your answer is __?__ .

Pick other numbers and try the number trick again. Do you always get the same answer?

31. To see why the number trick works, complete the chart.

Instructions	Expression	Simplified Expression
Pick a number.	n ———————→	n
Add 4.	$n + 4$ ———————→	$n + 4$
Multiply by 3.	$3(n + 4)$ ———→	$3n + 12$
Subtract 6.	$3n + 12 - 6$ ——→	__?__
Divide by 3.	$\dfrac{3n + 6}{3}$ ————→	__?__
Subtract the number you picked.	$n + 2 - n$ ———→	__?__

Look at the simplified expression. Does its value depend on the value of n?

OBJECTIVE: To solve equations containing polynomials.

Clara bowled three games. Her score in the second game was 28 lower than her first game score. Her score in the third game was twice her second game score. Her total score was 436. What was Clara's score in her first game?

If you let s be Clara's score in her first game, you can solve the problem by solving the equation:

$$s + (s - 28) + 2(s - 28) = 436$$

↑	↑	↑	↑
score in game 1	score in game 2	score in game 3	total score

E X A M P L E	**Here's how to solve an equation containing polynomials.**

Equation:	$s + (s - 28) + 2(s - 28) = 436$
Simplify to remove grouping symbols.	$s + s - 28 + 2s - 56 = 436$
Combine like terms.	$4s - 84 = 436$
Add 84 to both sides.	$4s - 84 + 84 = 436 + 84$
Simplify.	$4s = 520$
Divide both sides by 4.	$\dfrac{4s}{4} = \dfrac{520}{4}$
Simplify.	$s = 130$

 CHECK

$$s + (s - 28) + 2(s - 28) = 436$$
$$130 + (130 - 28) + 2(130 - 28) \stackrel{?}{=} 436$$
$$130 + 102 + 204 \stackrel{?}{=} 436$$
$$436 = 436$$

It checks.

CHECK for Understanding

1. Look at the example above. To find s, first simplify to remove grouping _?_ and then combine _?_ terms. Next add _?_ to both sides of the equation and then divide both sides by _?_ .

2. Check the solution. What were Clara's three scores?

EXERCISES

▶ **Copy and finish solving each example. Check your solution.**

3. $n + (n + 3) + 3(n + 5) = 8$

$n + n + 3 + 3n + \underline{} = 8$

$5n + 18 = 8$

$5n + 18 - \underline{} = 8 - 18$

$5n = \underline{}$

$$\frac{5n}{5} = \frac{^-10}{?}$$

$n = \underline{}$

4. $^-2n + 2(n - 2) + 4(n + 3) = 40$

$^-2n + 2n - \underline{} + 4n + 12 = 40$

$\underline{} n + 8 = 40$

$4n + 8 - \underline{} = 40 - 8$

$4n = \underline{}$

$$\frac{4n}{?} = \frac{32}{4}$$

$n = \underline{}$

▶ **Solve and check.**

Here are scrambled answers for the next two rows of exercises: $^-1$ 2 3 $^-5$

5. $4x + 3(x + 1) + 7 = 24$

6. $2m + 2(m - 1) + 9 = 19$

7. $r + 3(r + 2) + 6 = 8$

8. $m + 2(m + 7) + 3(m + 6) = 2$

9. $a + 3(a - 1) + 2(a + 4) = 5$

10. $^-2c + 2(c + 1) + 6c = 26$

11. $2(m - 1) + 3m + 4(m + 1) = 29$

12. $2(t - 1) + 3(t + 2) + 6t = 15$

13. $2r + 3(r + 2) + 4r = {}^-12$

14. $9(x - 1) + 3x + 2(x + 1) = 21$

15. $7(d + 2) + {}^-5d + 6(d + 1) = 44$

16. $^-6c + 3(c + 1) + 7c = 43$

17. $4(c + 1) + 2c + 3(c - 1) = 37$

18. $5(t - 1) + 2(t + 3) + 7 = 85$

19. $3(n + 2) + 3(n - 2) + 1 = 13$

20. $^-6r + 3(r + 1) + 4r = 2$

Problem Solving USING EQUATIONS

▶ **Decide which equation would be used to solve each problem. Then solve the problem.**

21. Don bowled three games. His score in the second game was 14 higher than his first game score. His score in the third game was twice his second game score. His total score was 482. What was his score in the first game?

Equation A:	$n + (n + 14) + 2n = 482$
Equation B:	$n + (n - 14) + 2n = 482$
Equation C:	$n + (n + 14) + 2(n + 14) = 482$

22. Dustin bowled three games. His score in the second game was 14 lower than his first game score. His score in the third game was twice his first game score. His total score was 482. What was his score in the first game?

23. Colleen bowled three games. Her score in the second game was 14 higher than her first game score. Her score in the third game was twice her first game score. Her total score was 482. What was her score in the first game?

14-8 Problem Solving—Using Equations

OBJECTIVE: To use equations to solve coin problems.

In this lesson, you will solve coin problems by writing equations.

| EXAMPLE | Here's how to use an equation to solve a coin problem. |

Problem: Eight coins, consisting of dimes and nickels, are worth 70¢. How many dimes are there?

Step 1. Choose a variable. Use it and a table to organize the facts in the problem.

Let d = the number of dimes.

Coins	Number	Value
Dimes	d	$10d$
Nickels	$8 - d$	$5(8 - d)$

Total value: 70

Step 2. Write an equation based on the facts.

$$10d + 5(8 - d) = 70$$

These are two equal expressions for the total value of the coins.

Step 3. Solve the equation.

$$10d + 5(8 - d) = 70$$
$$10d + 40 - 5d = 70$$
$$5d + 40 = 70$$
$$5d + 40 - 40 = 70 - 40$$
$$5d = 30$$
$$\frac{5d}{5} = \frac{30}{5}$$
$$d = 6$$

There are 6 dimes.

CHECK for Understanding

1. Look at the example above.

   a. In Step 1, we let d equal the number of dimes. Then the expression $8 -$ _?_ equals the number of nickels. The expression $10d$ equals the value of the dimes, and the expression _?_ equals the value of the nickels.

   b. In Step 2, the two equal expressions for the total value of the coins are _?_ and 70.

EXERCISES

▶ **Read the facts. Then complete the steps to answer the question.**

2. *Facts:* Six coins, consisting of dimes and nickels, are worth 40¢.
   Question: How many dimes are there?

   a. Organize the facts in a table.

Coins	Number	Value
Dimes	d	$10d$
Nickels	$6 - d$	$5(6 - d)$

   If d equals the number of dimes, then $6 -$? equals the number of nickels. If $10d$ equals the value of the dimes, then ? equals the value of the nickels.

   b. To find the number of dimes, you can solve the equation:

   $$10d + 5(6 - d) = \underline{\ ?\ }$$

   c. Solve the equation in Part b. There are ? dimes.

3. *Facts:* Five coins, consisting of quarters and dimes, are worth 95¢.
   Question: How many quarters are there?

   a. Organize the facts in a table.

Coins	Number	Value
Quarters	q	$25q$
Dimes	$5 - q$	$10(5 - q)$

   If q equals the number of quarters, then $5 -$? equals the number of dimes. If $25q$ equals the value of the quarters, then ? equals the value of the dimes.

   b. To find the number of quarters, you can solve the equation:

   $$25q + \underline{\ ?\ } = 95.$$

   c. Solve the equation in Part b. There are ? quarters.

▶ **Write an equation and solve the problem.**

4. Nine coins, consisting of dimes and nickels, are worth 70¢. How many dimes are there?

Coins	Number	Value
Dimes	d	$10d$
Nickels	$9 - d$	$5(9 - d)$

5. Six coins, consisting of quarters and dimes, are worth 90¢. How many quarters are there?

Coins	Number	Value
Quarters	q	$25q$
Dimes	$6 - q$	$10(6 - q)$

6. Seven coins, consisting of quarters and nickels, are worth 75¢. How many quarters are there?

Coins	Number	Value
Quarters	q	$25q$
Nickels	$7 - q$	$5(7 - q)$

7. Twelve coins, consisting of dimes and nickels, are worth 90¢. How many dimes are there?

Coins	Number	Value
Dimes	d	$10d$
Nickels	$12 - d$	$5(12 - d)$

8. Eight coins, consisting of quarters and dimes, are worth 140¢. How many quarters are there?

9. Sixteen coins, consisting of dimes and nickels, are worth 130¢. How many dimes are there?

OBJECTIVE: To use a calculator for solving problems.

When the wind is blowing, the temperature feels colder than it actually is. The windchill chart shows, for example, that when the thermometer reading is 10° Fahrenheit (F) and the wind speed is 20 miles per hour (mph), the windchill is ⁻24°F. In other words, the wind makes 10°F feel like ⁻24°F.

▶ **Use the windchill chart to complete each statement.**

1. A 20°F temperature with a 15-mph wind feels the same as _?_°F with no wind.

2. A ⁻10°F temperature with a 20-mph wind feels the same as _?_°F with no wind.

3. A 30°F temperature with a _?_-mph wind feels the same as ⁻2°F with no wind.

4. A ⁻30°F temperature with a _?_-mph wind feels the same as ⁻81°F with no wind.

5. A _?_°F temperature with a 25-mph wind feels the same as ⁻29°F with no wind.

Windchill Chart							
Wind speed (mph)	Thermometer reading in degrees Fahrenheit						
	30°	20°	10°	0°	⁻10°	⁻20°	⁻30°
5	27°	19°	7°	⁻5°	⁻15°	⁻26°	⁻36°
10	16°	3°	⁻9°	⁻22°	⁻34°	⁻46°	⁻58°
15	9°	⁻5°	⁻18°	⁻31°	⁻45°	⁻58°	⁻72°
20	4°	⁻10°	⁻24°	⁻39°	⁻53°	⁻67°	⁻81°
25	1°	⁻15°	⁻29°	⁻44°	⁻59°	⁻74°	⁻88°
30	⁻2°	⁻18°	⁻33°	⁻49°	⁻64°	⁻79°	⁻93°
35	⁻4°	⁻20°	⁻35°	⁻52°	⁻67°	⁻82°	⁻97°

▶ **Use the formula. Fill in the facts that fit the calculator sequence.**

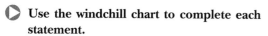
1.8 ✕ 41 ⁺⁄₋ ⊞ 32 ⊟ (⁻41.8)

6. The lowest temperature recorded in Nova Scotia, Canada, is __a.__ °C or __b.__ °F.

Formula

$$F = 1.8C + 32$$

degrees Fahrenheit   degrees Celsius

▶ **Use the formula. Decide when a calculator would be helpful.**

7. The lowest temperature recorded in the Yukon Territory is ⁻63°C. How many degrees Fahrenheit is that?

8. The highest temperature recorded in Alberta, Canada, is 42°C. How many degrees Fahrenheit is that?

9. When the temperature in Edmonton is 25°F and the wind speed is 20 mph, is the windchill above or below ⁻10°?

10. **Write a problem** that can be solved using the formula. Then solve the problem.

▶ **Solve and check.** *(page 226)*

1. $5y + 2y = 35$ **2.** $7c - c = 21$ **3.** $4j - 3j = {}^-23$

4. $6n + n - 3 = 18$ **5.** $6d + 3 - 5d = {}^-23$ **6.** $8 + m - 3m = 15$

7. $12z - 2z + 6 = {}^-34$ **8.** $9t - 3 + 3t = {}^-15$ **9.** $6 - 4z + 10z = 30$

10. ${}^-5r - 12r - 15 = {}^-18$ **11.** ${}^-4a - 10 - 11a = 0$ **12.** ${}^-9 + 3b - 17b = {}^-9$

▶ **Solve and check.** *(page 228)*

13. $2d = d + 5$ **14.** ${}^-3k = 10 - k$ **15.** ${}^-4c = {}^-8 + 2c$

16. $j + 3 = 3j - 9$ **17.** $4x - 2 = x + 7$ **18.** $5w - 3 = 2w + 15$

19. ${}^-2n + 3 = 4n - 21$ **20.** $5z - 18 = z + 2$ **21.** ${}^-3k + 6 = {}^-6k - 10$

22. $15 - 3y = y + 8$ **23.** ${}^-12 + 4j = j - 9$ **24.** $20 - 5z = {}^-2z + 16$

25. ${}^-8 + 9k = 2k + 6$ **26.** ${}^-6 - 3r = r - 8$ **27.** $18 + 9j = j + 16$

▶ **Solve each proportion.** *(page 240)*

28. $\dfrac{n}{8} = \dfrac{4}{7}$ **29.** $\dfrac{3}{n} = \dfrac{5}{11}$ **30.** $\dfrac{20}{16} = \dfrac{n}{8}$ **31.** $\dfrac{16}{7} = \dfrac{9}{n}$ **32.** $\dfrac{n}{11} = \dfrac{4}{5}$

33. $\dfrac{9}{8} = \dfrac{n}{12}$ **34.** $\dfrac{11}{20} = \dfrac{33}{n}$ **35.** $\dfrac{n}{15} = \dfrac{4}{10}$ **36.** $\dfrac{20}{n} = \dfrac{4}{5}$ **37.** $\dfrac{15}{25} = \dfrac{n}{10}$

38. $\dfrac{3}{1\frac{1}{2}} = \dfrac{17}{n}$ **39.** $\dfrac{n}{2\frac{1}{4}} = \dfrac{8}{3}$ **40.** $\dfrac{9}{n} = \dfrac{4}{5\frac{2}{3}}$ **41.** $\dfrac{6}{2\frac{3}{4}} = \dfrac{n}{11}$ **42.** $\dfrac{3\frac{5}{6}}{9} = \dfrac{4}{n}$

▶ **Solve.** *(page 252)*

43. 10% of $96 = n$ **44.** 25% of $44 = n$ **45.** $33\frac{1}{3}\%$ of $66 = n$

46. 35% of $50 = n$ **47.** 65% of $84 = n$ **48.** 112% of $216 = n$

49. 8.6% of $150 = n$ **50.** 9.9% of $90 = n$ **51.** 23.5% of $156 = n$

▶ **Write and solve an equation to find the measure of each angle.** *(page 270)*

52.
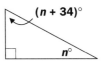

53.

54.
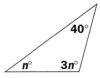

Here are scrambled answers for the review exercises:

⁻8	2	18	distributive	opposite	term
⁻5	4	binomial	grouping	polynomial	trinomial
⁻1	5	combine	monomials	product	

1. Expressions with one term such as n^2, n, and 1 are called __?__ .
   A monomial is a variable, a number, or a __?__ of variables and numbers.
   A __?__ is the sum or difference of monomials.
   A polynomial with two terms is called a __?__ .
   A polynomial with three terms is called a __?__ . *(page 400)*

2. To add polynomials, first write the sum without grouping symbols and then __?__
   like terms.

   Example: $(4x^2 + 3) + (^-3x^2 + 2) = 4x^2 + 3 + ^-3x^2 + 2$
   $$= x^2 + \underline{\ ?\ } \ (page\ 402)$$

3. To subtract a polynomial, add the __?__ of the polynomial. To find the opposite
   of a polynomial, you can multiply the polynomial by __?__ .

   Example: $(6x + 8) - (5x + 13) = (6x + 8) + (^-1)(5x + 13)$
   $$= 6x + 8 + ^-5x + ^-13$$
   $$= x + \underline{\ ?\ } \ (page\ 404)$$

4. You can use the __?__ property to multiply a polynomial by a monomial.

   Example: $2x(^-4x + y) = 2x \cdot \ ^-4x + 2x \cdot y$
   $$= \underline{\ ?\ } x^2 + 2xy \ (page\ 410)$$

5. To divide a polynomial by a monomial, divide each __?__ of the polynomial by
   the monomial.

   Example: $\dfrac{12t^2 + 6t}{3t} = \dfrac{12t^2}{3t} + \dfrac{6t}{3t}$
   $$= \dfrac{12 \cdot t \cdot t}{3 \cdot t} + \dfrac{6 \cdot t}{3 \cdot t}$$
   $$= 4t + \underline{\ ?\ } \ (page\ 412)$$

6. To solve the equation $y + 3(y + 8) - 6 = 26$, you would first simplify to remove
   the __?__ symbols and then combine like terms. Finally you would subtract __?__
   from both sides and then divide both sides by __?__ . *(page 414)*

▷ **Tell whether each polynomial is a monomial, binomial, or trinomial.** *(page 400)*

1. $4x$
2. $3r + 2s + t$
3. $5y + 8x$
4. rs

5. $y^2 + 2y + 1$
6. n^2
7. $ab + 30$
8. $x^2y + xy + {}^-11$

▷ **Give each sum.** *(page 402)*

9. $(5r + s) + ({}^-2r + 4s)$
10. $(7j^2 + 2j) + (6j^2 + {}^-5j)$

11. $(9t^2 + 7t + 3) + (2t^2 + {}^-5t)$
12. $(3x + 7y + {}^-2z) + (5x + {}^-4y + 3z)$

▷ **Give each difference.** *(page 404)*

13. $(5x + 2y) - (2x + 4y)$
14. $(4x^2 + 3y) - (3x^2 + 2y)$

15. $(4a + 3b + 3c) - (3a + b + 2c)$
16. $(10y^2 + 8y + 6) - (7y^2 + 3)$

▷ **Multiply.** *(page 410)*

17. $7(2x + {}^-8)$
18. $5j(2j + 3)$
19. ${}^-3(4c^2 + 2d + 1)$

20. ${}^-2y(y + 4)$
21. $r(3r + {}^-2t + 4)$
22. $4a({}^-2b + c + 1)$

▷ **Divide.** *(page 412)*

23. $\dfrac{12n + 8}{4}$
24. $\dfrac{n^2 + n}{n}$
25. $\dfrac{{}^-12x^2 + 4x}{2x}$

26. $\dfrac{10c^2 + 8c + 6}{2}$
27. $\dfrac{16x^3 + 12x^2 + 20x}{4x}$
28. $\dfrac{18x^2 + 15xy + {}^-21x}{{}^-3x}$

▷ **Solve and check.** *(page 414)*

29. $4c + 2(c + 1) + 3 = 35$
30. $2(x + 1) + 3(x + 4) = 9$

31. $2t + 4(t + 7) + 3(t - 1) = {}^-2$
32. ${}^-5a + 8(a + 1) + 7(a - 1) = 31$

33. $3(y + 2) + 4(y - 1) + y = 2$
34. ${}^-8r + 3(r - 2) + 4r = 0$

▷ **Write an equation and solve the problem.** *(page 416)*

35. Eight coins, consisting of dimes and nickels, are worth 55¢. How many dimes are there?

36. Five coins, consisting of quarters and dimes, are worth 80¢. How many quarters are there?

Coins	Number	Value
Dimes	d	$10d$
Nickels	$8 - d$	$5(8 - d)$

Coins	Number	Value
Quarters	q	$25q$
Dimes	$5 - q$	$10(5 - q)$

▶ **Choose the correct letter.**

---

1. ⁻1.375 written in simplest fractional form is

A. $^-1\frac{3}{8}$ **B.** $^-1\frac{5}{8}$

C. $1\frac{3}{8}$ **D.** none of these

---

2. Give the difference.

$$\frac{^-3}{4} - \frac{5}{6}$$

A. $\frac{^-1}{12}$ **B.** $\frac{1}{12}$

C. $^-1\frac{7}{12}$ **D.** none of these

---

3. Give the quotient.

$$\frac{^-5}{8} \div \frac{15}{32}$$

A. $1\frac{1}{3}$ **B.** $\frac{3}{4}$

C. $\frac{^-3}{4}$ **D.** none of these

---

4. Solve.

$$^-20 - 12y = ^-35$$

A. $4\frac{7}{12}$ **B.** $^-1\frac{1}{4}$

C. $1\frac{1}{4}$ **D.** none of these

---

5. Simplify

$$19t + 8 - t - 20$$

by combining like terms.

A. $20t - 12$

B. $18t + 20$

C. $18t - 12$

D. none of these

---

6. Solve.

$$18 - 3t + 8t = ^-30$$

A. $^-4\frac{4}{11}$ **B.** $^-2\frac{2}{5}$

C. $9\frac{3}{5}$ **D.** none of these

---

7. Solve.

$$^-18 + 3x = x + 6$$

A. $^-6$

B. $^-3$

C. 6

D. none of these

---

8. Solve.

$$\frac{n}{2\frac{1}{3}} = \frac{6}{11}$$

A. $4\frac{5}{18}$ **B.** $1\frac{3}{11}$

C. $2\frac{4}{7}$ **D.** none of these

---

9. Solve.

5.4% of 33 = n

A. 1.782

B. 17.82

C. 6.11

D. none of these

---

10. The percent of increase from 48 to 60 is

A. 25%

B. 20%

C. 12%

D. none of these

---

11. Find the area of this circle. Use 3.14 for π.

6 cm

A. 18.84 cm^2

B. 28.26 cm^2

C. 113.04 cm^2

D. none of these

---

12. Choose the equation.

Ten more than the sum of 5 times a number and the number itself is ⁻60.

A. $5n + n + 10 = ^-60$

B. $5n - n + 10 = ^-60$

C. $5n + n - 10 = ^-60$

D. $5n - n - 10 = ^-60$

Similar and Right Triangles

$$a^2 + b^2 = c^2$$

In a right triangle, the sum of the squares of the lengths of the legs is equal to the square of the length of the hypotenuse. (page 442)

The Pythagorean Theorem, stated above, can be used to find the height to the main deck if you know the length of the gangplank and the horizontal distance from the foot of the gangplank to the ship.

423

OBJECTIVE: To use a pair of similar figures to find the length of one of their sides.

The sails in the photographs are the same shape. Two figures that have the same shape are called **similar figures.**

Look at these two similar triangles.

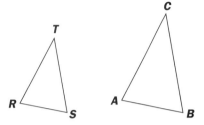

Side *AB* corresponds to side *RS*.

1. What side corresponds to side *BC*? To side *AC*?

∠*A* corresponds to ∠*R*.

2. What angle corresponds to ∠*B*? To ∠*C*?

In similar figures, the corresponding angles are **congruent** (have the same measure), and the ratios of the lengths of corresponding sides are equal.

EXAMPLE | **Here's how to use a proportion to solve a similar-figure problem.**

Problem: The two flags are similar. What is the length *n*?

3 cm

1.5 cm

Step 1. Write a proportion.

$$\begin{array}{l} \text{Small flag} \rightarrow \dfrac{\textbf{1.5}}{\textbf{2.5}} = \dfrac{\textbf{3}}{\textbf{n}} \leftarrow \text{Small flag} \\ \text{Large flag} \rightarrow \phantom{\dfrac{1.5}{2.5}} \leftarrow \text{Large flag} \end{array}$$

n

Step 2. Solve the proportion.

$$1.5n = 3 \cdot 2.5$$
$$1.5n = 7.5 \quad 7.5 \div 1.5 = \boxed{5}$$
$$n = 5 \leftarrow$$

2.5 cm

The length of the side is 5 centimeters.

CHECK for Understanding

3. Look at the example. What is the length *n*?

EXERCISES

▶ **The two figures are similar. Solve a proportion to find the length *n*.**

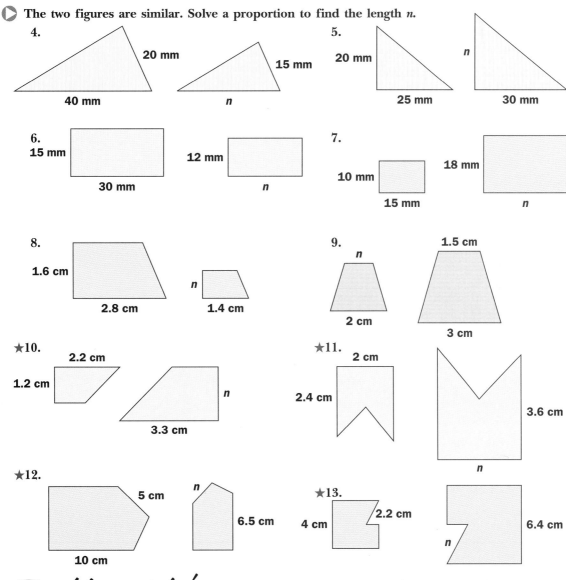

4. 20 mm, 40 mm, 15 mm, *n*

5. 20 mm, 25 mm, *n*, 30 mm

6. 15 mm, 30 mm, 12 mm, *n*

7. 10 mm, 15 mm, 18 mm, *n*

8. 1.6 cm, 2.8 cm, *n*, 1.4 cm

9. *n*, 2 cm, 1.5 cm, 3 cm

★**10.** 2.2 cm, 1.2 cm, 3.3 cm, *n*

★**11.** 2 cm, 2.4 cm, *n*, 3.6 cm

★**12.** 5 cm, 10 cm, *n*, 6.5 cm

★**13.** 2.2 cm, 4 cm, *n*, 6.4 cm

Problem Solving USING PROPORTIONS

▶ **Solve by solving a proportion. Refer to the triangles on page 424.**

14. Suppose that the length of side *AB* is 8 meters and the length of side *BC* is 12 meters. If the length of side *ST* is 7.5 meters, what is the length of side *RS*?

15. Use the information in exercise 14 and the fact that the length of side *AC* is 13 meters to find the length of side *RT*.

15-2 Similar Right Triangles

OBJECTIVE: To use pairs of similar right triangles to find various heights.

A **right triangle** is a triangle that has a right angle.

The right triangle made by the stop sign and its shadow is similar to the right triangle made by the flagpole and its shadow.

1. What is the height of the stop sign?

2. What is the length of the shadow of the stop sign?

3. What is the length of the shadow of the flagpole?

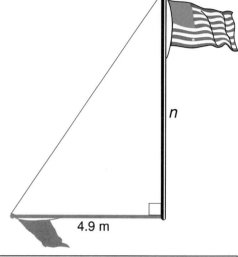

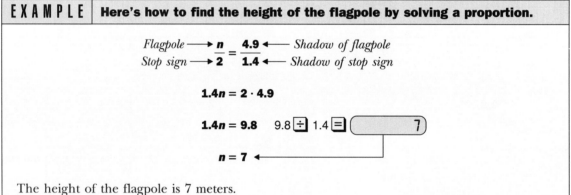

2.0 m

1.4 m

4.9 m

n

| EXAMPLE | Here's how to find the height of the flagpole by solving a proportion. |

Flagpole ⟶ $\dfrac{n}{2} = \dfrac{4.9}{1.4}$ ⟵ Shadow of flagpole
Stop sign ⟶ ⟵ Shadow of stop sign

$$1.4n = 2 \cdot 4.9$$

$$1.4n = 9.8 \quad 9.8 \boxed{\div} 1.4 \boxed{=} \boxed{7}$$

$$n = 7$$

The height of the flagpole is 7 meters.

CHECK for Understanding

4. Look at the example. What is the height of the flagpole?

EXERCISES

▶ **The two right triangles are similar. Solve a proportion to find the length *n*.**

5.

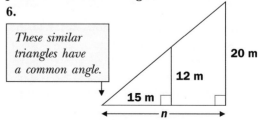

n

10 m

6 m

9 m

6.

These similar triangles have a common angle.

20 m

12 m

15 m

n

The triangles in each exercise are similar. Find the length *n* rounded to the nearest tenth of a meter.

7.
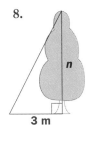

3 m

2 m   4 m

8.
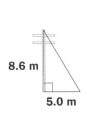

2 m

3 m   1 m

9.
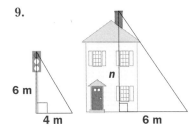

6 m

4 m   6 m

10.

n

0.6 m

8.2 m   0.5 m

11.

8.6 m

n

5.0 m   8.2 m

12.
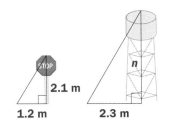

2.1 m

1.2 m   2.3 m

Problem Solving **USING A DIAGRAM**

First draw a diagram. Then solve the problem. Round answers to the nearest hundredth of a meter.

13. When an elephant casts a 2.5-meter shadow, a man 2 meters tall casts a 1.5-meter shadow. How tall is the elephant?

14. When a Ferris wheel casts a 20-meter shadow, a man 1.9 meters tall casts a 2.4-meter shadow. How tall is the Ferris wheel?

15. When a high-diving pole casts a 25-meter shadow, a man 2 meters tall casts a 1-meter shadow. How tall is the high-diving pole?

16. When a woman 1.6 meters tall casts a 0.75-meter shadow, a diving tank casts a 1.2-meter shadow. How high is the diving tank?

17. When an animal trainer 1.8 meters tall casts a 2.2-meter shadow, a black bear casts a 3.3-meter shadow. How tall is the black bear?

18. When a 2-meter sign casts a 3.4-meter shadow, a flagpole casts a 45-meter shadow. How tall is the flagpole?

19. A TV tower is 31.6 meters high and casts a 10-meter shadow. How tall is a nearby tree that casts a 6-meter shadow at the same time?

20. A fence around a water tower is 2.5 meters high and casts a 4-meter shadow. How tall is the water tower that casts a 46.4-meter shadow at the same time?

15-3 Tangent of an Angle

OBJECTIVE: To find the tangent of an angle.

Pictured at the right are three similar right triangles.

Look at △ABC. ← | *Read as "triangle ABC."* |

The two sides that form the right angle (side *AC* and side *BC*) are called the **legs** of the right triangle.

1. Which two sides are the legs of △*RST*? Of △*WXY*?

The length of the leg **opposite** ∠*A* is 3 units.

2. What is the length of the leg opposite ∠*R*? ∠*W*?

3. What is the length of the leg opposite ∠*B*? ∠*S*? ∠*X*?

The length of the leg **adjacent** to ∠*A* is 4 units.

4. What is the length of the leg adjacent to ∠*R*? ∠*W*?

5. What is the length of the leg adjacent to ∠*B*? ∠*S*? ∠*X*?

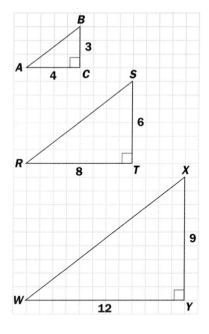

In a **right** triangle, the ratio of the length of the leg opposite an acute angle to the length of the leg adjacent to the acute angle is called the **tangent** of the angle.

E X A M P L E	**Here's how to find the tangent of angle A.**

Read as "tangent of ∠A."	→ $\tan \angle A = \dfrac{\text{length of leg opposite } \angle A}{\text{length of leg adjacent to } \angle A}$
	$= \dfrac{3}{4}$
	$= 0.75$

CHECK for Understanding

6. Look at the example. What is the tan ∠*A*?

7. Complete these examples.

    a. $\tan \angle R = \dfrac{6}{?}$ **b.** $\tan \angle W = \dfrac{9}{?}$ **c.** $\tan \angle B = \dfrac{4}{3}$ **d.** $\tan \angle X = \dfrac{12}{?}$

         $= 0.75$ $= \underline{?}$ $= \underline{?}$ $= \underline{?}$

8. Remember that △*RST* and △*WXY* are similar triangles. Look at your answers to 7a and 7b. Are they equal? Does the tangent of an angle depend on the measure of the angle or the size of the triangle?

EXERCISES

▶ Give the tan ∠X as a decimal rounded to the nearest hundredth.

9.

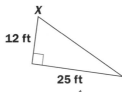
5 ft
X
12 ft

10.

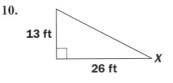
13 ft
26 ft
X

11.

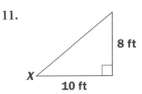
8 ft
X
10 ft

12.

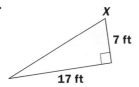
X
7 ft
17 ft

13.

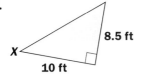
X
12 ft
25 ft

14.

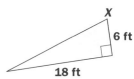
X
6 ft
18 ft

15.

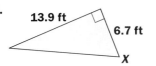
8.5 ft
X
10 ft

16.

6.4 ft
X
16.2 ft

17.

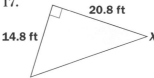
20.8 ft
14.8 ft
X

18.

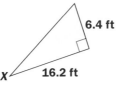
13.9 ft
6.7 ft
X

19.

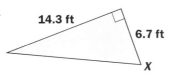
14.8 ft
9.4 ft
X

20.

X
16.5 ft
34.0 ft

21.

X
11.6 ft
5.4 ft

22.

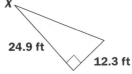
14.3 ft
6.7 ft
X

23.

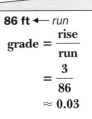
X
24.9 ft
12.3 ft

Challenge! CALCULATOR APPLICATION

The slope, or grade, of a road is the ratio of the "rise" to the "run."

Example:

3 ft ← rise

86 ft ← run

$$grade = \frac{rise}{run}$$

$$= \frac{3}{86}$$

$$\approx 0.03$$

24. Copy and complete this table. Round answers to the nearest hundredth.

	Rise (ft)	Run (ft)	Grade
a.	3	74	?
b.	9	318	?
c.	7	109	?

OBJECTIVE: To find the sine of an angle. To find the cosine of an angle.

Remember that the sides that form the right angle in a right triangle are called the legs of the right triangle. The side opposite the right angle of a right triangle is called the **hypotenuse.**

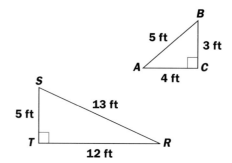

1. Which side of $\triangle ABC$ is the hypotenuse?

2. Which side of $\triangle RST$ is the hypotenuse?

3. What is the length of the hypotenuse of $\triangle ABC$? Of $\triangle RST$?

In a right triangle, the ratio of the length of the leg opposite an acute angle to the length of the hypotenuse is called the **sine** of the angle.

In a right triangle, the ratio of the length of the leg adjacent to an acute angle to the length of the hypotenuse is called the **cosine** of the angle.

EXAMPLES | **Here's how to find the sine and cosine of angle A.**

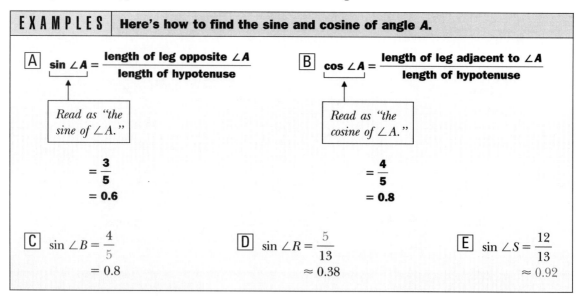

A $\sin \angle A = \dfrac{\text{length of leg opposite } \angle A}{\text{length of hypotenuse}}$

Read as "the sine of $\angle A$."

$= \dfrac{3}{5}$

$= 0.6$

B $\cos \angle A = \dfrac{\text{length of leg adjacent to } \angle A}{\text{length of hypotenuse}}$

Read as "the cosine of $\angle A$."

$= \dfrac{4}{5}$

$= 0.8$

C $\sin \angle B = \dfrac{4}{5}$

$= 0.8$

D $\sin \angle R = \dfrac{5}{13}$

≈ 0.38

E $\sin \angle S = \dfrac{12}{13}$

≈ 0.92

CHECK for Understanding

4. Look at Examples A and B. What is the $\sin \angle A$? What is the $\cos \angle A$?

5. Explain how to get the numbers that are printed in red in Examples C, D, and E.

6. Complete these examples. Give answers as decimals rounded to the nearest hundredth.

   a. $\cos \angle B = \dfrac{3}{5}$

   $= \underline{\ ?\ }$

   b. $\cos \angle R = \dfrac{12}{?}$

   $\approx \underline{\ ?\ }$

   c. $\cos \angle S = \dfrac{?}{13}$

   $\approx \underline{\ ?\ }$

EXERCISES

▶ **Give each ratio as a decimal rounded to the nearest hundredth.**

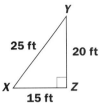

7. tan ∠X

8. tan ∠Y

9. sin ∠X

10. sin ∠Y

11. cos ∠X

12. cos ∠Y

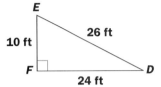

13. tan ∠D

14. tan ∠E

15. sin ∠D

16. sin ∠E

17. cos ∠D

18. cos ∠E

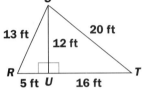

19. tan ∠URS

20. sin ∠URS

21. cos ∠URS

22. tan ∠RSU

23. sin ∠RSU

24. cos ∠RSU

25. tan ∠UTS

26. sin ∠UTS

27. cos ∠UTS

28. tan ∠UST

29. sin ∠UST

30. cos ∠UST

31. tan ∠GHI

32. sin ∠GHI

33. cos ∠GHI

★34. tan ∠GJK

★35. sin ∠GJK

★36. cos ∠GJK

37. tan ∠GIH

38. sin ∠GIH

39. cos ∠GIH

★40. tan ∠GKJ

★41. sin ∠GKJ

★42. cos ∠GKJ

Challenge! USING A DIAGRAM

▶ **First draw a diagram. Then answer the question.**

43. If $\sin \angle A = \frac{3}{5}$ and $\cos \angle A = \frac{4}{5}$, what is the tan ∠A?

44. If $\sin \angle A = \frac{5}{13}$ and $\tan \angle A = \frac{5}{12}$, what is the cos ∠A?

45. If the tangent of one acute angle of a right triangle is $\frac{12}{19}$, what is the tangent of the other acute angle?

46. If the sine of one acute angle of a right triangle is $\frac{9}{16}$, what is the cosine of the other acute angle?

47. If the cosine of one acute angle of a right triangle is $\frac{7}{12}$, what is the sine of the other acute angle?

48. If an acute angle of a right triangle has a sine of $\frac{12}{13}$ and a cosine of $\frac{5}{13}$, what is the tangent of the acute angle?

Similar and Right Triangles | 431

You will use a variety of problem-solving strategies on these two pages. For some problems, you may wish to use a calculator.

1 TRAVEL TIME

Robin and Brad are returning to Key West from Tallahassee, Florida, a distance of 650 miles. They averaged 50 miles per hour going to Tallahassee. How much would they have to increase their average speed to decrease their travel time by 1 hour?

2 PERFORMING-ARTS SURVEY

Here are the results of a performing-arts survey taken at Sonoma High School:

61 students are in the band.
40 students belong to the drama club.
26 students are in the chorus.
17 students are in the band and the chorus.
19 students are in the band and the drama club.
13 students are in the chorus and the drama club.
8 students belong to all three groups.
6 students do not belong to any group.

How many students took part in the survey?

3 CASH-OUT TIME

The cashier at Video Hut made these notes when she closed the cash register.

• just 4 kinds of coins
• 3 times as many dimes as nickels
• 8 fewer pennies than nickels
• 11 more quarters than nickels
• total amount in coins: $11.21

How many of each coin were in the register? *Hint: Make a table.*

4 ORDERLY ORANGES

Mary asked Fred, who had 15 oranges, and Tim, who had 9 oranges, to share their oranges equally with her. Mary offered to pay them with 24 dimes. Fred thought he should get 15 of the coins and Tim should get 9. Tim thought he and Fred should each get 12 coins since they would share equally. Mary said that Fred should get 21 coins and Tim only 3.

What reasoning did Mary use?

5 GEOMETRY—VISUAL THINKING

Can you move two of the tees to form a square with twice the area of the one shown?

Hint: Make a drawing.

6 GEOMETRY—USING DIAGRAMS

The square is cut into four pieces. The pieces are rearranged into the rectangle. Find the area of the square and the area of the rectangle. Explain your findings.

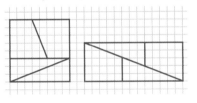

7 PATTERNS

Look for the pattern. Find the missing number.

$$1^2 = 1$$
$$11^2 = 121$$
$$111^2 = 12,321$$
$$1,111^2 = 1,234,321$$
•
•
•
$$111,111,111^2 = \underline{\ ?\ }$$

TRIGONOMETRY—APPLICATIONS

A surveyor was watching a sailboat from a cliff, 50 meters above the water. She used a transit (a surveying instrument) to measure the angles and drew this sketch.

She used her calculator to find these approximate tangent ratios:
tan 20° ≈ 0.36 and tan 40° ≈ 0.84.

8 How far is the buoy from the base of the cliff?

Hint: $\tan \angle A = \dfrac{\text{length of leg opposite } \angle A}{\text{length of leg adjacent to } \angle A}$

$$0.36 = \frac{b}{50 \text{ m}}$$

9 How far is the sailboat from the buoy?

10 What is the tangent ratio for $\angle A$ when the sailboat is 200 meters from the base of the cliff?

Rate Your Problem-Solving Power: 6–7 correct = Good;   8–9 correct = Excellent;   10 correct = Exceptional

OBJECTIVE: To use a grocery ad to solve problems.

▶ **Use the grocery ad to solve each problem. Round answers to the nearest cent.**

1. How much would 2.2 pounds of cherries cost?

   Hint: $\dfrac{1.39}{1} = \dfrac{n}{2.2}$

2. How much would 5 oranges cost?

   Hint: $\dfrac{1.19}{6} = \dfrac{n}{5}$

3. What is the cost of 3.5 pounds of grapes?

4. What is the cost of $\dfrac{1}{2}$ dozen apples?

5. How much would 5 apples cost?

6. How much would 18 apples cost?

7. If you gave the cashier $5 for 2.4 pounds of grapes, how much change should you receive?

8. If you gave the cashier $5 for some cantaloupes and received $1.20 in change, how many cantaloupes did you buy?

▶ **Write an equation and solve the problem.**

9. Henry bought a cantaloupe and some bananas. He spent a total of $2.42. How many pounds of bananas did he buy? (Let h = the number of pounds of bananas Henry bought.)

10. Edna paid $5.45 for 3 pounds of grapes and some cherries. How many pounds of cherries did she buy? (Let e = the number of pounds of cherries Edna bought.)

11. Eliza bought 3 times as many cantaloupes as Gerald did. Together they bought 12 cantaloupes. How many cantaloupes did Gerald buy? (Let g = the number of cantaloupes Gerald bought.)

12. Lucas bought twice as many pounds of grapes as Nick did. Together they bought 7.5 pounds of grapes. How many pounds of grapes did Nick buy? (Let n = the number of pounds of grapes Nick bought.)

▶ **Solve and check.** *(page 112)*

1. $5x + 8 = {}^-22$ **2.** $3y - 2 = {}^-23$ **3.** $\dfrac{w}{2} + 4 = {}^-1$ **4.** $\dfrac{z}{3} - 5 = {}^-9$

5. ${}^-8w + 17 = 1$ **6.** ${}^-3z - 9 = {}^-12$ **7.** $\dfrac{k}{2} + 10 = {}^-6$ **8.** $\dfrac{h}{{}^-3} - 9 = {}^-14$

9. $\dfrac{c}{8} - 35 = {}^-32$ **10.** ${}^-5r - 9 = 26$ **11.** ${}^-8t + 7 = 63$ **12.** $\dfrac{y}{{}^-10} - 16 = {}^-21$

▶ **Solve. Give answers in simplest form.** *(page 196)*

13. $\dfrac{1}{2}$ of $42 = n$ **14.** $\dfrac{1}{4}$ of $44 = n$ **15.** $\dfrac{3}{4}$ of $36 = n$ **16.** $\dfrac{2}{3}$ of $51 = n$

17. $\dfrac{7}{8}$ of $32 = n$ **18.** $\dfrac{5}{2}$ of $50 = n$ **19.** $\dfrac{2}{5}$ of $21 = n$ **20.** $\dfrac{5}{3}$ of $35 = n$

21. $\dfrac{9}{4}$ of $41 = n$ **22.** $\dfrac{3}{5}$ of $43 = n$ **23.** $\dfrac{4}{9}$ of $35 = n$ **24.** $\dfrac{7}{4}$ of $110 = n$

▶ **Complete.** *(page 198)*

25. $1\dfrac{1}{2}$ ft = __?__ in. **26.** $2\dfrac{2}{3}$ ft = __?__ in. **27.** $3\dfrac{2}{3}$ yd = __?__ ft **28.** $1\dfrac{1}{4}$ days = __?__ h

29. $2\dfrac{1}{2}$ days = __?__ h **30.** $2\dfrac{3}{4}$ h = __?__ min **31.** $2\dfrac{2}{3}$ h = __?__ min **32.** $1\dfrac{1}{5}$ min = __?__ s

33. $2\dfrac{5}{6}$ min = __?__ s **34.** $1\dfrac{1}{4}$ gal = __?__ qt **35.** $3\dfrac{1}{2}$ qt = __?__ pt **36.** $1\dfrac{3}{4}$ gal = __?__ pt

▶ **Solve. Give answers in simplest form.** *(page 200)*

37. $\dfrac{1}{2} n = 17$ **38.** $\dfrac{1}{4} n = 37$ **39.** $\dfrac{2}{3} n = 48$ **40.** $\dfrac{3}{4} n = 51$

41. $\dfrac{5}{2} n = 75$ **42.** $\dfrac{7}{3} n = 63$ **43.** $\dfrac{7}{12} n = 19$ **44.** $\dfrac{5}{6} n = 28$

45. $\dfrac{6}{5} n = 35$ **46.** $\dfrac{4}{5} n = 43$ **47.** $\dfrac{5}{4} n = 51$ **48.** $\dfrac{7}{8} n = 47$

▶ **Simplify by combining like terms.** *(page 224)*

49. $8x + x$ **50.** $7z - z$ **51.** $5y - 5y$

52. ${}^-3w + 5w - 3$ **53.** ${}^-3w + 5 - 3w$ **54.** ${}^-3 + 5w - 3w$

55. ${}^-9n - n - 9$ **56.** ${}^-9n + n - 8$ **57.** ${}^-9n + 1 - 8$

58. $12y + 10 + 3y - 5$ **59.** $12 + 10y + 3y - 5$ **60.** $12 + 10y + 3 - 5y$

OBJECTIVE: To use trigonometric ratios to find the length of a side of a right triangle.

In this lesson you will use sine, cosine, and tangent ratios to find the length of a side of a right triangle. These ratios are easily obtained on a scientific calculator using the keys labeled *sin, cos,* and *tan.* Some of the ratios rounded to the nearest ten thousandth are shown in the table at the right. A more complete table of trigonometric ratios is on page 508.

Table of Trigonometric Ratios			
Angle	**Sine**	**Cosine**	**Tangent**
35°	0.5736	0.8192	0.7002
40°	0.6428	0.7660	0.8391
45°	0.7071	0.7071	1.0000
50°	0.7660	0.6428	1.1918
55°	0.8192	0.5736	1.4281
60°	0.8660	0.5000	1.7321
65°	0.9063	0.4226	2.1445
70°	0.9397	0.3420	2.7475

1. What is the sin 40°?

2. What is the cos 50°?

3. What is the tan 45°?

EXAMPLE | **Here's how to use the table of trigonometric ratios to find the length of a side of a right triangle.**

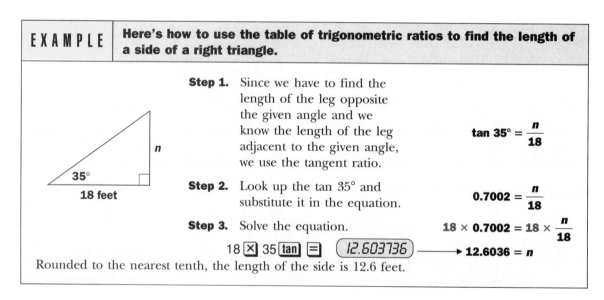

Step 1. Since we have to find the length of the leg opposite the given angle and we know the length of the leg adjacent to the given angle, we use the tangent ratio.

$$\tan 35° = \frac{n}{18}$$

Step 2. Look up the tan 35° and substitute it in the equation.

$$0.7002 = \frac{n}{18}$$

Step 3. Solve the equation.

$$18 \times 0.7002 = 18 \times \frac{n}{18}$$

18 ☒ 35 tan ⊟ ⟮ *12.603736* ⟯ ⟶ 12.6036 = *n*

Rounded to the nearest tenth, the length of the side is 12.6 feet.

CHECK for Understanding

4. Look at the example. What value was substituted for tan 35°? What was *n* rounded to the nearest tenth?

5. Complete these examples.

a. $\sin 60° = \dfrac{n}{20}$

$\dfrac{?}{} = \dfrac{n}{20}$

$20 \times 0.8660 = n$

$\underline{\;?\;} = n$

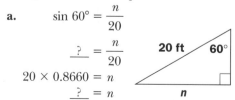

b. $\cos 45° = \dfrac{n}{27}$

$\dfrac{?}{} = \dfrac{n}{27}$

$27 \times 0.7071 = n$

$\underline{\;?\;} = n$

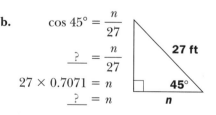

EXERCISES

▶ **Find *n* rounded to the nearest tenth of a foot.**

Use a calculator or the tangent ratios in the table on page 436.

6.

7.
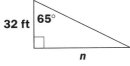

8.

Use a calculator or the sine ratios in the table on page 436.

9.

10.
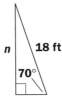

11.
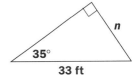

Use a calculator or the cosine ratios in the table on page 436.

12.

13.
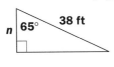

14.
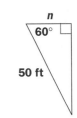

▶ **Find *n* rounded to the nearest tenth of a foot. Use a calculator or the table on page 508.**

15.
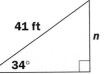

16.

17.
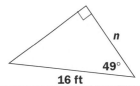

Problem Solving USING A DIAGRAM

▶ **First draw a diagram. Then solve the problem. Round answers to the nearest tenth. Use a calculator or the table on page 508.**

18. A hot-air balloon is directly over a small town. A man 10 miles away can look directly at the balloon by sighting 32° above horizontal. How high is the balloon?

19. How tall is a television tower if someone standing 800 feet away can look directly at the top by sighting 23° above horizontal?

20. When a kite flier had let out 240 feet of string, she noticed that the kite was 58° above horizontal. How high was the kite? Disrgard any sag in the kite string.

21. When a 500-foot cable is attached to a sunken treasure, the cable forms a 37° angle with the surface of the ocean. How deep is the treasure? Disregard any sag in the cable.

OBJECTIVE: To use trigonometric ratios to find the measure of an angle of a right triangle.

Use the table to answer the following questions.

1. What is the measure of an angle that has
   a. a sine of 0.5446?
   b. a cosine of 0.7771?
   c. a tangent of 0.7265?

An angle that has a sine of 0.5162 has a measure between 31° and 32°. Since 0.5162 is closer to 0.5150 (sin 31°) than to 0.5299 (sin 32°), the measure of the angle to the nearest degree is 31°.

2. What is the measure of an angle (to the nearest degree) that has a cosine of 0.8000?

Table of Trigonometric Ratios			
Angle	Sine	Cosine	Tangent
31°	0.5150	0.8572	0.6009
32°	0.5299	0.8480	0.6249
33°	0.5446	0.8387	0.6494
34°	0.5592	0.8290	0.6745
35°	0.5736	0.8192	0.7002
36°	0.5878	0.8090	0.7265
37°	0.6018	0.7986	0.7536
38°	0.6157	0.7880	0.7813
39°	0.6293	0.7771	0.8098
40°	0.6428	0.7660	0.8391

EXAMPLE	Here's how to use the table of trigonometric ratios to approximate an angle of a right triangle.

Step 1. Write a ratio.

$$\sin R = \frac{5}{9}$$

Step 2. Write the ratio as a 4-place decimal.

$$\approx 0.5556 \leftarrow$$

The measure of $\angle R$ is about 34°.

Step 3. Use the table above to find the measure of $\angle R$ to the nearest degree.

9 ft   5 ft

R   ?

The decimal is nearer 0.5592 (sin 34°) than 0.5446 (sin 33°).

CHECK for Understanding

3. Look at the example. Between what two values in the table is 0.5556? Which of the two values is it closer to?

4. What is the approximate measure of $\angle R$?

5. Complete these examples. Use the table above.

   a. $\cos \angle X = \dfrac{30}{38}$
      ≈ 0.7895
      What is the measure of $\angle X$ to the nearest degree?

      38 ft

      X   30 ft

   b. $\tan \angle Y = \dfrac{21}{30}$
      $= 0.7000$
      What is the measure of $\angle Y$ to the nearest degree?

      30 ft   21 ft

      Y

EXERCISES

▶ **Find the measure of the labeled angle to the nearest degree. Use the trigono-metric table on page 508. You may wish to use a calculator to express each ratio as a 4-place decimal.**

6.

A triangle with legs labeled 9 ft (right side) and 10 ft (bottom), vertex A at lower left, right angle at lower right.

7.

A triangle with side 22 ft (left) and 7 ft (bottom), right angle at B (bottom).

8.

A triangle with 30 ft (top) and 10 ft (right), vertex C at left, right angle at lower right.

9.

A triangle with 28 ft (top right) and 25 ft (bottom), vertex D at right, right angle at lower left.

10.

A triangle with 20 ft (left) and 48 ft (bottom), right angle at upper area, vertex E at right.

11.

A triangle with 26 ft (left) and 29 ft (right), right angle at bottom, vertex F at lower right.

12.

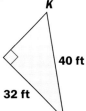 A triangle with 16 ft (top), 24 ft (left), vertex G at top right, right angle at top left.

13.

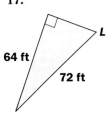 A triangle with 40 ft (top) and 31 ft (right), vertex H at left, right angle at lower right.

14.

A triangle with 35 ft (left) and 30 ft (bottom), vertex I at top, right angle at right.

15.

A triangle with 22 ft (top) and 31 ft (bottom), vertex J at top, right angle at middle right.

16.

A triangle with 40 ft (right) and 32 ft (bottom), vertex K at top, right angle at left.

17.

A triangle with 64 ft (left) and 72 ft (right), vertex L at right, right angle at top.

Problem Solving USING A DIAGRAM

▶ **First draw a diagram. Then solve the problem. Give angles to the nearest de-gree and lengths to the nearest tenth of a foot. Use the table on page 508.**

18. When a submarine was ordered to dive, it traveled 240 feet to get to 100 feet below the surface of the water. What angle did the submarine's course make with the surface of the water?

★19. From the top of a 185-foot radio tower, a construction worker observed a stop sign. If his line of sight made a 60° angle with the tower, how far from the tower was the stop sign?

20. A 40-foot inclined ramp is used to load freight onto an airplane. The bottom of the cargo door is 18 feet above the ground. What angle does the ramp make with the runway?

★21. A 60-foot pine tree was struck by light-ning. At a point 20 feet above the ground, the tree bent over, allowing the top of the tree to touch the ground. What angle did the upper part of the tree make with the ground?

OBJECTIVE: To find the square root of a number by using a table or the divide-and-average method.

Ancient mathematicians often thought about numbers in geometric terms. For example, if they multiplied a number by itself, they thought about finding the area of a square. That is the reason we still talk about squaring numbers.

From the figure, we have 16 is the square of 4:

$$4^2 = 16$$

Read as "4 squared is 16."

Table of Square Roots			
n	$\sqrt{n}$	n	$\sqrt{n}$
1	1	11	3.317
2	1.414	12	3.464
3	1.732	13	3.606
4	2	14	3.742
5	2.236	15	3.873
6	2.449	16	4
7	2.646	17	4.123
8	2.828	18	4.243
9	3	19	4.359
10	3.162	20	4.472

To find the **square root** of a number, we can first think about a square having the number as its area. Then we find the square root by finding the length of a side.

From the same figure, we also have

$$\sqrt{16} = 4$$

Read as "the square root of 16 is 4."

You will generally be working with a decimal approximation of the square root of a number. From the table above, we have

$$\sqrt{12} \approx 3.464$$

1. Multiply 3.464 by itself. Is the product near 12?

Sometimes you may need to find the square root of a number when you do not have a table of square roots or a calculator with a $\sqrt{}$ key.

EXAMPLE	Here's how to find the square root of a number by using the divide-and-average method.

Find $\sqrt{46}$ to the nearest tenth.

Step 1.

Estimate $\sqrt{46}$.

$\sqrt{36} = 6$
$\sqrt{46} = ?$
$\sqrt{49} = 7$

$\sqrt{46}$ is between 6 and 7. We will try 6.5.

Step 2.

Divide 46 by 6.5.

$7.07 \approx 7.1$
$6.5\overline{)46.0\,00}$

$\sqrt{46}$ is between 6.5 and 7.1

Step 3.

Average 6.5 and 7.1.

$$\frac{6.5 + 7.1}{2} = 6.8$$

Step 4.

Divide 46 by 6.8.

$6.76 \approx 6.8$
$6.8\overline{)46.0\,00}$

$\sqrt{46}$ is between 6.8 and 6.8. When rounded to the nearest tenth, $\sqrt{46} = 6.8$.

CHECK for Understanding

2. Look at the example in the box on page 440. We first found that $\sqrt{46}$ was between what two whole numbers?

3. In Step 2, we found that $\sqrt{46}$ was between what two numbers?

4. In Step 4, we found that $\sqrt{46}$ was between what two numbers?

5. When rounded to the nearest tenth, $\sqrt{46} = \underline{\ ?\ }$.

EXERCISES

▶ **Simplify.**
Here are scrambled answers for the next row of exercises: 5 3 1 4 2 8

6. $\sqrt{4}$ **7.** $\sqrt{1}$ **8.** $\sqrt{16}$ **9.** $\sqrt{25}$ **10.** $\sqrt{9}$ **11.** $\sqrt{64}$

12. $\sqrt{144}$ **13.** $\sqrt{256}$ **14.** $\sqrt{121}$ **15.** $\sqrt{1.44}$ **16.** $\sqrt{2.56}$ **17.** $\sqrt{1.21}$

▶ **Give the approximate square root found in the tables on pages 506 and 507.**
18. $\sqrt{29}$ **19.** $\sqrt{47}$ **20.** $\sqrt{71}$ **21.** $\sqrt{87}$ **22.** $\sqrt{98}$ **23.** $\sqrt{127}$

▶ **True or false?**
24. $\sqrt{31}$ is between 5 and 6. **25.** $\sqrt{150}$ is between 13 and 14. **26.** $\sqrt{200}$ is between 14 and 15.

▶ **Use the divide-and-average method to give an approximation to the nearest tenth.**
27. $\sqrt{300}$ **28.** $\sqrt{421}$ **29.** $\sqrt{506}$ **30.** $\sqrt{766}$ **31.** $\sqrt{919}$

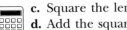 AN INVESTIGATION *Does It Measure Up?*

▶ **Work in a small group.**

32. Each group member works through the following steps.

a. Use a protractor and carefully draw a right triangle.

b. Measure each side to the nearest millimeter.

c. Square the length of each leg.

d. Add the squares of the lengths of the legs.

e. Square the length of the hypotenuse.

f. Compare the sum of the squares of the lengths of the legs (Part d) and the square of the length of the hypotenuse (Part e). Are they the same?

33. Compare your answer for Part f to the answers of other group members.

34. Repeat Exercises 32 and 33 with a different right triangle.

OBJECTIVE: To use the Pythagorean theorem to find the length of a side of a right triangle.

Look at the figure at the right.

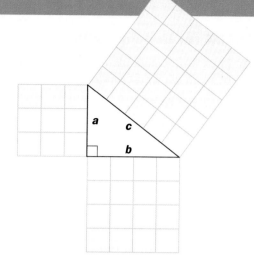

1. The area of the square on leg *a* is 9 square units. What is the area of the square on leg *b*? What is the sum of the two areas?

2. What is the area of the square on the hypotenuse?

3. Is the sum of the areas of the squares on the legs the same as the area of the square on the hypotenuse?

The Pythagorean Theorem

In a right triangle, the sum of the squares of the lengths of the legs is equal to the square of the length of the hypotenuse.

The Pythagorean theorem can be written as $a^2 + b^2 = c^2$, where *a* and *b* are the lengths of the legs and *c* is the length of the hypotenuse of a right triangle.

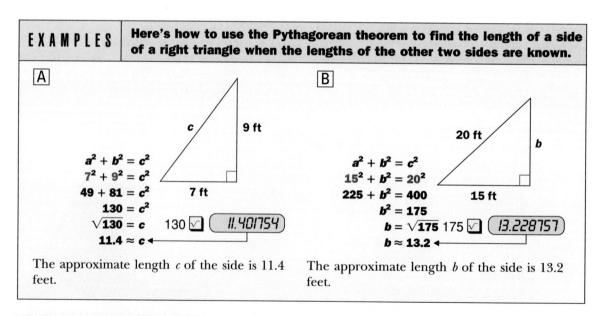

EXAMPLES Here's how to use the Pythagorean theorem to find the length of a side of a right triangle when the lengths of the other two sides are known.

A

$$a^2 + b^2 = c^2$$
$$7^2 + 9^2 = c^2$$
$$49 + 81 = c^2$$
$$130 = c^2$$
$$\sqrt{130} = c \quad 130 \;☑\; \boxed{11.401754}$$
$$11.4 \approx c$$

The approximate length *c* of the side is 11.4 feet.

B

$$a^2 + b^2 = c^2$$
$$15^2 + b^2 = 20^2$$
$$225 + b^2 = 400$$
$$b^2 = 175$$
$$b = \sqrt{175} \quad 175 \;☑\; \boxed{13.228757}$$
$$b \approx 13.2$$

The approximate length *b* of the side is 13.2 feet.

CHECK for Understanding

4. Look at the examples. In which example was the length of the hypotenuse found? The length of a leg found?

5. What was the approximate length *c* of the side in Example A?

6. What was the approximate length *b* of the side in Example B?

EXERCISES

▶ Use the Pythagorean theorem to find the length *n* rounded to the nearest tenth of a foot. Use a calculator or the square-root tables on pages 506 and 507.

7.

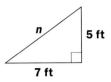
n 5 ft 7 ft

8.

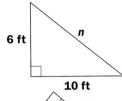
6 ft *n* 10 ft

9.

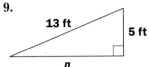
13 ft 5 ft *n*

10.

4 ft *n* 4 ft

11.

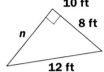
8 ft *n* 12 ft

12.

7 ft *n* 11 ft

13.

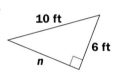
10 ft 6 ft *n*

14.

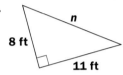
8 ft *n* 11 ft

15.

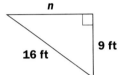
n 9 ft 16 ft

Problem Solving USING THE PYTHAGOREAN THEOREM

▶ Use the Pythagorean theorem to solve each problem. Round answers to the nearest tenth. Use a calculator or the square-root tables on pages 506 and 507.

16. Find the length of the diagonal of the rectangle.

? 4 ft 9 ft

17. Find the width of the rectangle.

9 ft ? 7 ft

18. Find the diameter of the circle.

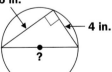
6 in. 4 in. ?

19. Find the height of the cone.

10 in. ? 5 in.

Challenge! USING THE PYTHAGOREAN THEOREM

▶ Is the triangle with the given lengths a right triangle? Write *Yes* or *No*.

20. 5, 12, 13 **21.** 1, 2, 3 **22.** 8, 15, 16 **23.** 4, 4, 6

24. 20, 21, 29 **25.** 16, 30, 34 **26.** 15, 20, 25 **27.** 8, 9, 15

OBJECTIVE: To use the Pythagorean theorem to solve problems.

In this lesson you will solve problems by using the Pythagorean theorem. In solving these problems, it will be helpful to use diagrams that show the facts in the problem.

EXAMPLE	**Here's how to use a diagram and the Pythagorean theorem to solve a problem.**

Problem: A 17-foot ladder is placed against a wall. The bottom of the ladder is 8 feet from the base of the wall. How high up the wall does the ladder reach?

Step 1. Draw a diagram to show the facts in the problem.

17 ft

8 ft

Step 2. Write the Pythagorean theorem. Substitute 17 for c and 8 for b. Simplify and solve for a.

$$a^2 + b^2 = c^2$$
$$a^2 + 8^2 = 17^2$$
$$a^2 + 64 = 289$$
$$a^2 = 225$$
$$a = \sqrt{225}$$
$$= 15$$

The ladder reaches 15 feet up the wall.

CHECK for Understanding

1. Look at the example.
   a. In Step 1, the ladder forms a right triangle with the wall and the ground. The hypotenuse of the right triangle is __?__ feet. The shorter leg is __?__ feet.
   b. In Step 2, we substituted __?__ for c and 8 for __?__ .
   c. An equation for the problem is $a^2 + 64 =$ __?__ .

2. To check the solution, ask yourself if the answer fits the facts in the problem. Does $8^2 + 15^2 = 17^2$?

EXERCISES

▶ Decide which diagram would be used to show the facts in each problem. Then use the diagram and the Pythagorean theorem to solve the problem. Round your answers to the nearest tenth. Use a calculator or the square-root tables on pages 506 and 507.

Diagram A

Diagram B

3. An airplane is 6 miles directly above a control tower. An observer on the ground is 11 miles from the control tower. How far is the observer from the airplane?

4. An 11-foot rope is fastened to the top of a flagpole. The rope reaches a point on the ground 6 feet from the base of the flagpole. What is the height of the flagpole?

5. An 11-foot ladder is placed so that it reaches the top of a wall. The bottom of the ladder is 6 feet from the wall. How high is the wall?

6. Keith and Hillary biked 11 miles east and then 6 miles north. How far are they from their starting point?

▶ First draw a diagram. Then solve the problem. Round your answers to the nearest tenth.

7. A boat traveled 15 miles west and then 10 miles south. How far is the boat from its starting point?

8. A 27-foot rope is attached to the top of a 17-foot pole. If the rope is stretched to the ground and fastened, how far from the base of the pole is it fastened?

9. The bases on a baseball diamond are 90 feet apart. How long a throw is it from second base to home plate?

10. A square empty lot that is often used as a shortcut is 50 feet on a side. How much shorter is the diagonal across the lot than the walk along two of its sides?

11. A 100-foot cable is fastened to the top of a television tower and to a stake that is 25 feet from the base of the tower. How tall is the tower?

12. A cable is fastened to the top of a 100-foot television tower and to a stake that is 25 feet from the base of the tower. How long is the cable?

Challenge! **FINDING AREAS OF SQUARES**

13. A side of a square is $\sqrt{20}$ feet long. What is the area of the square?

14. A diagonal of a square is $\sqrt{20}$ feet long. What is the area of the square?

OBJECTIVE: To use a calculator for solving problems.

The higher you go, the farther you see. The chart shows how far you can see on a clear day at different heights.

Height (miles)	Distance (miles)
0.1	28.3
0.2	40.0
0.3	48.9
0.4	56.5
0.5	63.2
0.6	69.2

1. On a clear day, how many miles can you see from a hang glider at a height of 0.1 mile?

2. Can you see twice as far at a height of 0.4 mile as you can see at a height of 0.2 mile?

3. Are you above or below 500 feet if you can see a distance of 30 miles?
   Hint: 5280 feet = 1 mile.

▶ **Use the formula. Fill in the facts that fit the calculator sequence.**

0.7 ☑ ☒ 89.4 ☰ (74.797406)

4. At a height of ___**a.**___ miles, you can see
   (number)
   about 74.8 ___**b.**___ .
   (unit of length)

$$d = 89.4\sqrt{h}$$

distance seen   height
in miles   in miles

▶ **Use the formula. Decide when a calculator would be helpful.**

5. How far can you see at a height of 1 mile?

6. Can you see a distance of 150 miles at a height of 2 miles?

7. To the nearest mile, how much farther can you see at a height of 2 miles than you can see at a height of 1 mile?

8. **Write a problem** that can be solved using the formula. Then solve the problem.

Cumulative Algebra Practice

▶ **Solve and check.** *(page 226)*

1. $3x + x = 32$

2. $7n - n = 15$

3. $5e + 2e = 18$

4. $5y + y - 2 = 17$

5. $5y + 1 - 2y = 17$

6. $5 + y - 2y = 17$

7. $8z - 2z + 7 = {}^-11$

8. $8z - 2 + 7z = {}^-11$

9. $8 - 2z + 7z = {}^-11$

10. $^-6y - 10y - 13 = {}^-19$

11. $^-6y - 10 - 13y = {}^-19$

12. $^-6 - 10y - 13y = {}^-19$

▶ **Solve and check.** *(page 228)*

13. $5x = 8 - x$

14. $^-4j = 8 + 2j$

15. $6y = 9 - 2y$

16. $4n + 6 = 2n - 14$

17. $8w - 3 = 10w - 11$

18. $^-4c - 9 = 2c + 6$

19. $13 + 3z = z + 10$

20. $12 - 7a = a + 4$

21. $10 - 7t = 2t + 6$

22. $9d + 7 = 5d$

23. $e - 12 = {}^-2e$

24. $^-4f + 1 = f$

▶ **Solve each proportion.** *(page 240)*

25. $\dfrac{8}{n} = \dfrac{24}{30}$

26. $\dfrac{3}{5} = \dfrac{n}{8}$

27. $\dfrac{16}{11} = \dfrac{8}{n}$

28. $\dfrac{n}{12} = \dfrac{13}{7}$

29. $\dfrac{19}{n} = \dfrac{15}{8}$

30. $\dfrac{6}{5} = \dfrac{n}{12}$

31. $\dfrac{16}{20} = \dfrac{32}{n}$

32. $\dfrac{n}{10} = \dfrac{3}{7}$

33. $\dfrac{16}{n} = \dfrac{5}{4}$

34. $\dfrac{12}{21} = \dfrac{n}{30}$

35. $\dfrac{4}{3\frac{1}{2}} = \dfrac{8}{n}$

36. $\dfrac{n}{3\frac{1}{2}} = \dfrac{8}{5}$

37. $\dfrac{9}{n} = \dfrac{4}{2\frac{2}{3}}$

38. $\dfrac{3}{1\frac{1}{4}} = \dfrac{n}{6}$

39. $\dfrac{2\frac{3}{4}}{4} = \dfrac{9}{n}$

▶ **Solve.** *(page 256)*

40. 25% of $n = 9$

41. 60% of $n = 30$

42. 75% of $n = 45$

43. 50% of $n = 17$

44. 150% of $n = 96$

45. 10% of $n = 16$

46. 12.5% of $n = 25$

47. 62.5% of $n = 75$

48. 37.5% of $n = 120$

49. 0.5% of $n = 6$

50. 6.4% of $n = 16$

51. 18.2% of $n = 10.92$

▶ **Find the percent of increase or decrease. Use *i* to indicate increase and *d* to indicate decrease. Give answers to the nearest tenth of a percent.** *(page 258)*

52. from 20 to 30

53. from 30 to 20

54. from 18 to 36

55. from 36 to 18

56. from 68 to 92

57. from 75 to 108

58. from 92 to 67

59. from 49 to 65

60. from 118 to 83

61. from 4 to 5

62. from 4 to 9

63. from 9 to 4

Similar and Right Triangles | **447**

Here are scrambled answers for the review exercises:

0.8480	6.2	10	26	congruent	hypotenuse	sum
5	6.4	12	31	cosine	legs	tangent
6	9	13	average	equal	square	

1. In similar figures, corresponding angles are __?__, and the ratios of the lengths of corresponding sides are __?__. *(page 424)*

2. The two right triangles are similar. To find the length n of the side, you can solve the proportion. *(page 426)*

   $$\frac{n}{25} = \frac{?}{?}$$

   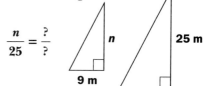

3. In a right triangle, the two sides that form the right angle are called the __?__. In a right triangle, the ratio of the length of the leg opposite an acute angle to the length of the leg adjacent to the acute angle is called the __?__ of the angle. *(page 428)*

4. The side opposite the right angle in a right triangle is called the __?__. In the triangle at the right, the sin $\angle R = \frac{?}{13}$; the cos $\angle R = \frac{?}{13}$. *(page 430)*

   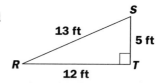

5. To find the length n, you would first write $\cos 32° = \frac{n}{?}$.

   Then you would look up the __?__ of 32° in the table. The last step would be to solve $\underline{\ ?\ } = \frac{n}{10}$. *(page 436)*

Table of Trigonometric Ratios			
Angle	Sine	Cosine	Tangent
31°	0.5150	0.8572	0.6009
32°	0.5299	0.8480	0.6249
33°	0.5446	0.8387	0.6494

6. To approximate $\angle X$, you would first write $\sin \angle X = \frac{?}{50}$

   $= 0.5200$

   Then you would use the table to approximate $\angle X$ to the nearest degree. The measure of $\angle X$ to the nearest degree is __?__°. *(page 438)*

7. You can use the divide-and-__?__ method to find the square root of a number. The $\sqrt{39}$ is between __?__ and 7. Try 6.1. From the division, you know that the $\sqrt{39}$ is between 6.1 and __?__. Next you would divide 39 by the average of 6.1 and 6.4. When rounded to the nearest tenth, $\sqrt{39} = \underline{\ ?\ }$. *(page 440)*

   $$6.1\overline{)39.00}\quad 6.39$$

8. In a right triangle, the __?__ of the squares of the lengths of the legs is equal to the __?__ of the length of the hypotenuse. *(page 442)*

▶ **The two figures are similar. Solve a proportion to find the length _n_.** *(pages 424, 426)*

1.

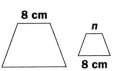
8 cm
_n_
8 cm
16 cm

2.

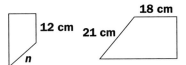
18 cm
12 cm 21 cm

3.

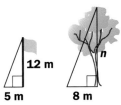
12 m
_n_
5 m   8 m

▶ **Give each ratio as a decimal rounded to the nearest hundredth.** *(pages 428, 430)*

4. tan ∠_A_   **5.** sin ∠_A_   **6.** cos ∠_A_

7. tan ∠_B_   **8.** sin ∠_B_   **9.** cos ∠_B_

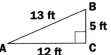
B
13 ft   5 ft
A   12 ft   C

▶ **Find the length _n_ rounded to the nearest tenth of a foot. Use the table of trigonometric ratios on page 508.** *(page 436)*

10.

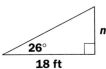
_n_
26°
18 ft

11.

15 ft
34°
_n_

12.

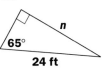
_n_
65°
24 ft

▶ **Find the measure of the labeled angle to the nearest degree. Use the table of trigonometric ratios on page 508.** *(page 438)*

13.

7 ft
25 ft   _X_

14.

18 ft   8 ft
_A_

15.

_T_
5 ft   18 ft

▶ **Use the divide-and-average method to give an approximation of the square root to the nearest tenth.** *(page 440)*

16. $\sqrt{327}$ **17.** $\sqrt{360}$ **18.** $\sqrt{425}$ **19.** $\sqrt{600}$

▶ **Find the length _n_ rounded to the nearest tenth of a foot. Use the square-root tables found on pages 506 and 507.** *(page 442)*

20.

16 ft
_n_
12 ft

21.

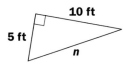
10 ft
5 ft
_n_

22.

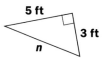
5 ft
3 ft
_n_

▶ **Solve.** *(page 444)*

23. A ladder is 10 feet long. The bottom is 6 feet from a wall. How high up the wall does the ladder reach?

24. A family traveled 6 miles north from their home and then 8 miles west. How far from their home were they then?

Similar and Right Triangles | **449**

▶ **Choose the correct letter.**

---

1. Solve.

$$^-8 = \dfrac{t}{^-2} - 11$$

  A. 6   **B.** $^-38$

  C. 38   **D.** none of these

---

2. Solve.

$$\dfrac{5}{6} \text{ of } 60 = n$$

  A. 72   **B.** 50

  C. 60   **D.** none of these

---

3. $2\dfrac{2}{3}$ yd = __?__ in.

  A. 8

  B. 32

  C. 96

  D. none of these

---

4. Solve.

$$\dfrac{9}{4}n = 45$$

  A. 20   **B.** 72

  C. $101\dfrac{1}{4}$ **D.** none of these

---

5. Simplify by combining like terms.

$$6 + 9y - 9 - y$$

  A. $8y - 3$ **B.** $10y - 3$

  C. $8y + 3$ **D.** none of these

---

6. Solve.

$$^-5x - 4 + 7x = ^-11$$

  A. $^-3\dfrac{1}{2}$ **B.** $^-7\dfrac{1}{2}$

  C. $3\dfrac{1}{2}$ **D.** none of these

---

7. Solve.

$$16 - 8c = 2c + 20$$

  A. $\dfrac{^-2}{3}$ **B.** $\dfrac{2}{5}$

  C. $^-6$ **D.** none of these

---

8. Solve.

$$\dfrac{n}{2\frac{1}{2}} = \dfrac{4}{7}$$

  A. $1\dfrac{3}{7}$ **B.** $4\dfrac{3}{8}$

  C. $11\dfrac{1}{5}$ **D.** none of these

---

9. Solve.

$$75\% \text{ of } n = 96$$

  A. 128   **B.** 72

  C. 64   **D.** none of these

---

10. Solve.

$$6.5\% \text{ of } n = 1.56$$

  A. 24

  B. 12

  C. 16

  D. none of these

---

11. The percent of decrease from 48 to 42 is

  A. 25%

  B. 20%

  C. 12.5%

  D. none of these

---

12. Choose the equation.

Twelve less than the sum of a number multiplied by 6 and the number multiplied by 3 is 51. What is the number?

  A. $6n + 3n + 12 = 51$

  B. $6n + 3n - 12 = 51$

  C. $6n - 3n + 12 = 51$

  D. $6n - 3n - 12 = 51$

SKILL TEST

Pages 452–462
This test will help you find out which skills you know well and which skills to practice more.

Prerequisite whole-number arithmetic skills are tested on page 452. These skills are not taught in this textbook, but review and practice of them is provided on pages 464–467.

Prerequisite Skill	Test Items			Skill Practice
1. Rounding whole numbers	Round to the nearest hundred.			**page 464**
	63	1250	6829	5983	
	Round to the nearest thousand.			
	4460	9672	28,095	65,817	
2. Adding whole numbers	$7549 + 4261$		$80,665 + 24,364$	**page 464**
	$496 + 3081 + 2566$		$26,245 + 6518 + 276$	
3. Subtracting whole numbers	$882 - 378$		$608 - 374$	**page 465**
	$4500 - 3492$		$7554 - 1429$	
	$81,365 - 62,471$		$53,800 - 14,529$	
4. Multiplying by multiples of 10, 100, or 1000	35×100	40×60	324×10	**page 465**
	78×1000	70×500	500×600	
	92×1000	90×6000	300×400	
5. Multiplying by a 1-digit number	527×5		984×7	**page 466**
	3168×4		5907×9	
	4190×3		6105×6	
6. Multiplying by 2- and 3-digit numbers	59×47		6035×81	**page 466**
	261×103		7448×517	
	212×112		3410×203	
7. Dividing by a 1-digit number	$474 \div 6$		$3224 \div 4$	**page 467**
	$7182 \div 3$		$78,408 \div 9$	
	$4582 \div 2$		$63,240 \div 5$	
8. Dividing by 2- and 3-digit numbers	$5355 \div 18$		$3448 \div 32$	**page 467**
	$6135 \div 24$		$7506 \div 25$	
	$8094 \div 120$		$51,921 \div 405$	

Lesson	Skill	Test Items	Skill Practice
1-2	**Evaluating expressions** *page 4*	Evaluate for $x = 12$, $y = 7$, and $z = 9$. $x + y + z$ $\qquad$ $x + z - y$ $x + z + z$ $\qquad$ $(x - z) + 10$ $(14 - z) + x$ $\qquad$ $(23 - z) - z$	**page 468**
1-3	**Rounding decimals** *page 6*	Round to the nearest tenth. 1.38 $\qquad$ 2.50 $\qquad$ 63.056 $\qquad$ 36.9521 Round to the nearest hundredth. 18.342 $\qquad$ 0.375 $\qquad$ 0.4968 $\qquad$ 8.6402	**page 468**
1-4	**Adding decimals** *page 8*	$2.34 + 1.7$ $\qquad\qquad$ $5.62 + 2.94$ $8.04 + 7 + 9.6$ $\qquad\quad$ $0.483 + 1.56 + 4.4$	**page 469**
1-5	**Subtracting decimals** *page 10*	$5 - 2.7$ $\qquad$ $7 - 2.4$ $\qquad$ $25.3 - 6$ $4.23 - 2.849$ $\qquad$ $13 - 6.7$ $\qquad$ $16.2 - 3.571$	**page 469**
1-7	**Solving addition equations** *page 16*	$c + 17 = 24$ $\qquad\qquad$ $y + 21 = 46$ $x + 16 = 57$ $\qquad\qquad$ $w + 31 = 53$ $r + 3.8 = 17.2$ $\qquad\quad$ $t + 19.5 = 25.3$	**page 470**
1-8	**Solving subtraction equations** *page 18*	$z - 19 = 28$ $\qquad\qquad$ $d - 26 = 35$ $n - 48 = 48$ $\qquad\qquad$ $r - 33 = 47$ $j - 14.3 = 38$ $\qquad\quad$ $m - 14.7 = 8.52$	**page 470**
1-9	**Solving equations with the variable on the right side** *page 20*	$93 = y + 37$ $\qquad\qquad$ $46 = w - 35$ $81 = z + 81$ $\qquad\qquad$ $106 = k + 61$ $132 = j - 56$ $\qquad\qquad$ $200 = n - 118$	**page 471**
2-2	**Evaluating expressions** *page 32*	Evaluate for $a = 24$, $b = 12$, and $c = 8$. $(bc) - a$ $\qquad$ $\dfrac{a}{c} + b$ $\qquad$ $\dfrac{b}{2} + c$ $(5b) + a$ $\qquad$ $\dfrac{120}{b} - c$ $\qquad$ $(10a) - b$	**page 471**

Lesson	Skill	Test Items			Skill Practice
2-3	**Multiplying decimals** *page 34*	5.4×0.36 6.05×0.39	6.3×1.2 2.04×1.6	34×0.88 8.25×2.06	**page 472**
2-4	**Multiplying decimals by 10, 100, or 1000** *page 36*	0.93×100 5.28×1000 3.8×10	0.3×10 4.1×1000 5.22×100	4.7×10 0.004×100 0.03×1000	**page 472**
2-5	**Dividing a decimal by a whole number** *page 38*	Round each quotient to the nearest hundredth. $12.17 \div 8$ $\qquad$ $6.34 \div 8$ $\qquad$ $27.62 \div 9$ $6.335 \div 24$ $\qquad$ $7.873 \div 25$ $\qquad$ $1.6238 \div 44$			**page 473**
2-6	**Dividing by a decimal** *page 40*	$2.04 \div 0.6$ $0.0644 \div 0.04$ $4.221 \div 0.21$	$1.64 \div 0.4$ $0.1206 \div 0.18$ $0.3233 \div 5.3$		**page 473**
2-7	**Dividing decimals by 10, 100, or 1000** *page 42*	$9.45 \div 10$ $450.5 \div 100$ $6.94 \div 1000$	$8.3 \div 10$ $53.5 \div 1000$ $0.84 \div 100$		**page 474**
2-9	**Solving multiplication equations** *page 48*	$7x = 84$ $12b = 0$ $3r = 1.95$	$9y = 99$ $25n = 275$ $2y = 4.08$	$6c = 192$ $10w = 350$ $0.5m = 150$	**page 474**
2-10	**Solving division equations** *page 50*	$\dfrac{y}{3} = 21$ $\dfrac{a}{12} = 10$ $\dfrac{w}{0.7} = 1.6$	$\dfrac{x}{7} = 17$ $\dfrac{c}{6} = 31$ $\dfrac{c}{0.8} = 7$	$\dfrac{k}{4} = 48$ $\dfrac{r}{15} = 2$ $\dfrac{s}{10} = 0.6$	**page 475**
2-11	**Solving equations with the variable on the right side** *page 52*	$62 = w + 17$ $75 = j + 37$ $15 = \dfrac{c}{7}$	$53 = m - 35$ $280 = 20f$ $12 = \dfrac{m}{4}$		**page 475**

Lesson	Skill	Test Items			Skill Practice
3-1	**Simplifying expressions** *page 62*	$12 - 8 + 4$ $24 \times (12 - 2)$ $36 + 4 \times 7 - 3$		$18 \div 6 \times 2$ $16 + 3 \times 4 \div 2$ $(36 + 4) \times 7 - 3$	**page 476**
3-4	**Solving equations using the commutative properties** *page 68*	$56 = 33 + b$ $1.6 = d(2)$ $59 = 10 + r$	$69 = 27 + j$ $7.9 = 4 + x$ $6.4 = 3 + s$	$132 = n(12)$ $2.4 = m(3)$ $4.8 = m(4)$	**page 476**
3-6	**Solving two-step equations** *page 74*	$4n + 9 = 37$ $39 = 10x + 9$ $12c - 5.1 = 18.9$		$2w - 6 = 52$ $15.8 = 5j - 4.2$ $9.6 = 7k + 6.1$	**page 477**
3-7	**Solving two-step equations** *page 76*	$\dfrac{n}{6} + 3 = 12$ $28 = 4r - 8$ $15 = \dfrac{d}{8} + 3$		$8k - 8 = 88$ $16r + 37 = 37$ $\dfrac{f}{10} - 21 = 0$	**page 477**
3-9	**Combining like terms** *page 80*	$8c + 3c$ $3n + 4n + 2n$ $5x + 3 + 4 + 2x$		$11t + t$ $11g + g + 5g$ $9y + 8 + y + 5$	**page 478**
3-10	**Solving equations by first combining like terms** *page 82*	$6a + 3a = 72$ $9n + 6n = 165$ $17.4 = 2t + t$		$5j + j = 96$ $0 = 8w + 3w$ $5n + 4n = 3.6$	**page 478**
4-1	**Comparing integers** *page 92*	$< \text{ or } > ?$ $^-3 \diamondsuit \, ^-2$ $^+12 \diamondsuit \, ^+11$		$^+9 \diamondsuit \, 0$ $^+10 \diamondsuit \, ^-13$	**page 479**
4-3	**Adding integers** *page 96*	$^+4 + \, ^+7$ $^-18 + \, ^+18$		$^+9 + \, ^-3$ $^-22 + \, ^-15$	**page 479**

Lesson	Skill	Test Items			Skill Practice
4-4	Subtracting integers *page 98*	$^-7 - {}^+3$ $^+34 - {}^+34$	$^+8 - {}^-9$ $^-46 - {}^-20$		**page 480**
4-5	Multiplying integers *page 100*	$^+6 \cdot {}^-7$ $^-5 \cdot {}^+19$	$^+8 \cdot {}^+10$ $^-23 \cdot 0$		**page 480**
4-6	Dividing integers *page 102*	$^-24 \div {}^+8$ $^-45 \div {}^-5$	$^+54 \div {}^-9$ $^+72 \div {}^-18$		**page 481**
4-8	Solving addition and subtraction equations involving integers *page 108*	$c + 18 = {}^-4$ $11 = c + 19$ $c - 11 = {}^-23$ $c + 10 = {}^-15$	$c + {}^-6 = {}^-5$ $c - 5 = 9$ $27 = c - 32$ $c - 8 = {}^-12$		**page 481**
4-9	Solving multiplication and division equations involving integers *page 110*	$^-3d = 15$ $^-42 = 14d$ $\dfrac{d}{-5} = {}^-8$	$^-12d = 0$ $\dfrac{d}{6} = {}^-4$ $15 = \dfrac{d}{-7}$		**page 482**
4-10	Solving two-step equations involving integers *page 112*	$^-12n + 15 = {}^-45$ $\dfrac{n}{-7} - 3 = {}^-9$	$\dfrac{n}{8} + 2 = 6$ $\dfrac{n}{-9} + 12 = 12$		**page 482**
5-1	Writing expressions using exponents *page 122*	Simplify, using exponents for the variables. $a \cdot b \cdot a \cdot b \cdot a$ $3 \cdot 2 \cdot b \cdot b \cdot 4 \cdot b$	$9 \cdot a \cdot a \cdot 2 \cdot a$ $7 \cdot a \cdot b \cdot 5 \cdot a \cdot b \cdot b \cdot b$		**page 483**
5-2	Multiplying and dividing expressions with like bases. *page 124*	$10^5 \cdot 10^4$ $\dfrac{5^{10}}{5^2}$	$a^4 \cdot a^2$ $\dfrac{x^9}{x^3}$	$n^{10} \cdot n^1$ $\dfrac{y^5}{y^1}$	**page 483**
5-3	Writing numbers in scientific notation *page 126*	42,000 0.084	345,000 0.0091	1,960,000 0.000675	**page 484**

Lesson	Skill	Test Items	Skill Practice
5-5	**Writing the prime or algebraic factorization** *page 130*	Give the prime factorization. 15 18 36 48 Give the algebraic factorization. $21x^2$ $30x^3$ $42x^2y$ $45x^2y^3$	page 484
5-6	**Writing the GCF and the LCM** *page 132*	Give the greatest common factor. 25, 20 15, 24 32, 48 $6, 9w$ $8w, 12w^2$ $36wz, 45w^2z$ Give the least common multiple. 3, 4 9, 6 12, 8 $8, 12u$ $5u, 7u^2$ $10u^2v, 25uv^2$	page 485
5-8	**Writing an equivalent fraction** *page 138*	$\dfrac{3}{c} = \dfrac{?}{4c}$ $\dfrac{9c}{d} = \dfrac{?}{4cd}$ $\dfrac{4c}{3d} = \dfrac{?}{12cd^2}$	page 485
5-9	**Writing fractions in lowest terms** *page 140*	$\dfrac{8}{12}$ $\dfrac{16}{40}$ $\dfrac{30}{20}$ $\dfrac{35}{50}$ $\dfrac{3}{6j}$ $\dfrac{j}{jk}$ $\dfrac{14j}{21j^2}$ $\dfrac{11jk^2}{33k}$	page 486
5-10	**Finding the least common denominator** *page 142*	Find the least common denominator. $\dfrac{4}{m}, \dfrac{1}{3m}$ $\dfrac{6}{m}, \dfrac{5}{n}$ $\dfrac{1}{mn^2}, \dfrac{3}{m^2n^2}$ $\dfrac{m}{4mn}, \dfrac{6}{3m^2n}$	page 486
5-11	**Comparing fractions** *page 144*	$<$ or $>$? $\dfrac{4}{5} \diamondsuit \dfrac{3}{5}$ $\dfrac{1}{4} \diamondsuit \dfrac{3}{8}$ $\dfrac{3}{4} \diamondsuit \dfrac{2}{3}$ $\dfrac{5}{6} \diamondsuit \dfrac{7}{8}$	page 487
6-1	**Writing whole and mixed numbers as fractions** *page 154*	Change to fourths. 2 4 6 3 Change to a fraction. $1\frac{1}{4}$ $1\frac{2}{3}$ $2\frac{3}{4}$ $3\frac{5}{6}$	page 487

Lesson	Skill	Test Items				Skill Practice
6-2	Writing fractions as whole numbers or mixed numbers *page 156*	Change to a whole number or mixed number. $\dfrac{8}{2}$ $\dfrac{4}{3}$	$\dfrac{9}{3}$ $\dfrac{5}{2}$	$\dfrac{16}{4}$ $\dfrac{13}{5}$	$\dfrac{18}{6}$ $\dfrac{11}{4}$	page 488
6-3	Writing fractions and mixed numbers in simplest form *page 158*	Write in simplest form. $\dfrac{6}{9}$ $\dfrac{2a}{6a^2}$	$3\dfrac{2}{4}$ $\dfrac{8b^3}{2b}$	$\dfrac{15}{3}$ $\dfrac{12c^2}{16c^4}$	$\dfrac{14}{6}$ $\dfrac{6ab^2}{8a^2b^2}$	page 488
6-4	Writing fractions as decimals and decimals as fractions *page 160*	Change to a decimal. $\dfrac{1}{4}$ $\qquad$ $\dfrac{2}{3}$ $\qquad$ $1\dfrac{3}{8}$ $\qquad$ $2\dfrac{5}{6}$ Change to a fraction or mixed number in simplest form. 0.8 $\qquad$ 0.06 $\qquad$ 0.625 $\qquad$ 1.75				page 489
6-6	Adding and subtracting fractions with common denominators *page 166*	Give each sum or difference in simplest form. $\dfrac{1}{8}+\dfrac{5}{8}$ $\dfrac{7}{12}-\dfrac{1}{12}$	$\dfrac{5}{6}+\dfrac{1}{6}$ $\dfrac{5}{9}-\dfrac{2}{9}$	$\dfrac{r}{s}+\dfrac{t}{s}$ $\dfrac{r}{t}-\dfrac{s}{t}$	$\dfrac{19}{t^2}+\dfrac{4}{t^2}$ $\dfrac{16}{rs}-\dfrac{9}{rs}$	page 489
6-7	Adding and subtracting fractions with different denominators *page 168*	Give each sum or difference in simplest form. $\dfrac{1}{3}+\dfrac{1}{2}$ $\dfrac{5}{6}-\dfrac{1}{2}$	$\dfrac{5}{6}+\dfrac{5}{8}$ $\dfrac{11}{12}-\dfrac{2}{3}$	$\dfrac{m}{8}+\dfrac{3}{4}$ $\dfrac{m}{3}-\dfrac{n}{9}$	$\dfrac{3m}{5}+\dfrac{4n}{7}$ $\dfrac{5m}{6}-\dfrac{2n}{5}$	page 490
6-8	Adding and subtracting fractions with algebraic denominators *page 170*	$\dfrac{1}{b}+\dfrac{a}{3}$ $\dfrac{1}{a}-\dfrac{b}{2}$	$\dfrac{4}{x}+\dfrac{2}{y}$ $\dfrac{3}{x}-\dfrac{1}{y}$	$\dfrac{r}{1}+\dfrac{3}{t}$ $\dfrac{2}{rs}-\dfrac{3}{r}$	$\dfrac{2}{mn}+\dfrac{3}{m}$ $\dfrac{5}{t^2}-\dfrac{2}{t}$	page 490
6-9	Adding and subtracting mixed numbers without regrouping *page 172*	Give each sum or difference in simplest form. $3\dfrac{1}{2}+2\dfrac{1}{4}$ $5\dfrac{2}{3}-2\dfrac{1}{2}$	$5\dfrac{2}{3}+3\dfrac{1}{8}$ $8\dfrac{5}{6}-6\dfrac{1}{4}$	$6+9\dfrac{5}{12}$ $11\dfrac{7}{8}-9\dfrac{2}{3}$		page 491

Lesson	Skill	Test Items	Skill Practice
6-10	**Adding and subtracting mixed numbers with regrouping** *page 174*	Give each sum or difference in simplest form. $6\frac{2}{3} + 4\frac{3}{4}$ $8\frac{7}{8} + 7\frac{5}{12}$ $9\frac{7}{10} + 7\frac{5}{6}$ $8\frac{1}{3} - 3\frac{1}{2}$ $12\frac{1}{4} - 9\frac{5}{6}$ $11\frac{3}{5} - 4\frac{7}{8}$	**page 491**
7-1	**Multiplying fractions** *page 184*	Give each product in simplest form. $\frac{1}{2} \cdot \frac{1}{3}$ $\frac{2}{3} \cdot \frac{3}{4}$ $3 \cdot \frac{3}{4}$ $\frac{6}{5} \cdot \frac{5}{6}$ $r \cdot \frac{5}{t}$ $\frac{5}{6} \cdot \frac{3s}{4t}$ $\frac{s^2}{14} \cdot \frac{1}{s}$ $\frac{3s}{t} \cdot \frac{t^2}{8}$	**page 492**
7-2	**Multiplying mixed numbers** *page 186*	Give each product in simplest form. $1\frac{1}{4} \cdot 1\frac{1}{2}$ $2 \cdot 2\frac{1}{2}$ $1\frac{2}{3} \cdot 1\frac{3}{4}$ $3\frac{1}{5} \cdot 2\frac{3}{8}$	**page 492**
7-3	**Dividing fractions** *page 188*	Give each quotient in simplest form. $\frac{2}{3} \div 5$ $\frac{3}{8} \div \frac{3}{2}$ $\frac{9}{4} \div \frac{5}{6}$ $\frac{0}{8} \div \frac{5}{12}$ $\frac{a}{b} \div \frac{c}{d}$ $\frac{3a}{b} \div \frac{b}{c}$ $\frac{a^2}{b} \div \frac{a}{c}$ $\frac{a}{6b} \div \frac{c}{3b}$	**page 493**
7-4	**Dividing mixed numbers** *page 190*	Give each quotient in simplest form. $3 \div 2\frac{1}{3}$ $3\frac{2}{3} \div 1\frac{1}{2}$ $4\frac{3}{4} \div 2$ $4\frac{1}{5} \div 2\frac{5}{8}$	**page 493**
7-6	**Find a fraction of a number** *page 196*	$\frac{1}{2}$ of $24 = n$ $\frac{1}{3}$ of $18 = n$ $\frac{2}{3}$ of $42 = n$ $\frac{3}{5}$ of $60 = n$ $\frac{2}{7}$ of $21 = n$ $\frac{3}{4}$ of $72 = n$	**page 494**
7-7	**Multiplying a whole number by a mixed number** *page 198*	$1\frac{2}{3}$ h = _?_ min $2\frac{1}{2}$ ft = _?_ in. $2\frac{2}{3}$ yd = _?_ ft $1\frac{3}{4}$ gal = _?_ qt	**page 494**
7-8	**Finding a number when a fraction of it is known** *page 200*	Give answers in simplest form. $\frac{1}{3}n = 8$ $\frac{5}{6}n = 55$ $\frac{3}{4}n = 51$ $\frac{5}{2}n = 42$ $\frac{7}{8}n = 20$ $\frac{1}{2}n = 34$	**page 495**

Lesson	Skill	Test Items	Skill Practice
8-1	**Comparing rational numbers** *page 210*	$< \text{ or } >$? $$\frac{2}{3} \diamond \frac{3}{4} \qquad \frac{^-5}{6} \diamond \frac{^-7}{8} \qquad ^-3\frac{1}{4} \diamond 3 \qquad ^-2\frac{3}{4} \diamond ^-2\frac{5}{6}$$	**page 495**
8-2a	**Writing a rational number in decimal form** *page 212*	Write each rational number in decimal form. $$\frac{2}{5} \qquad\qquad \frac{^-7}{4} \qquad\qquad \frac{1}{3} \qquad\qquad ^-3\frac{5}{6}$$	**page 496**
8-2b	**Writing a decimal in simplest fractional form** *page 212*	Write each decimal in simplest fractional form. $0.8 \qquad ^-0.75 \qquad 2.375 \qquad -6.125$ $0.25 \qquad 3.6 \qquad ^-1.5 \qquad 2.4$	**page 496**
8-3	**Adding and subtracting rational numbers** *page 214*	Give each sum or difference in simplest form. $$\frac{2}{3} + \frac{^-1}{4} \qquad \frac{^-5}{9} + \frac{^-1}{6} \qquad 4\frac{1}{3} + 3\frac{7}{8} \qquad ^-5\frac{3}{4} + 2\frac{1}{2}$$ $$\frac{^-3}{4} - \frac{1}{2} \qquad \frac{^-2}{3} - \frac{^-7}{12} \qquad 10\frac{1}{4} - 8\frac{5}{8} \qquad ^-6\frac{9}{10} - ^-4\frac{5}{6}$$	**page 497**
8-4	**Multiplying and dividing rational numbers** *page 216*	Give each product or quotient in simplest form. $$\frac{^-2}{3} \cdot \frac{4}{5} \qquad \frac{^-3}{8} \cdot ^-8 \qquad 2\frac{1}{2} \cdot 3\frac{1}{4} \qquad ^-4\frac{2}{3} \cdot ^-1\frac{1}{5}$$ $$\frac{^-3}{4} \div \frac{^-1}{4} \qquad \frac{5}{6} \div \frac{^-5}{8} \qquad 4\frac{1}{4} \div 2\frac{1}{8} \qquad 10\frac{3}{5} \div 3\frac{3}{10}$$	**page 497**
8-6	**Solving equations having rational solutions** *page 222*	$5x - 3 = ^-2 \qquad\qquad\qquad ^-4x + 17 = ^-12$ $^-12x - 6 = 11 \qquad\qquad\quad 18 + 15x = 19$ $^-10 - 8x = ^-30 \qquad\qquad 16 - 20x = ^-43$	**page 498**
8-7	**Simplifying expressions by combining like terms** *page 224*	Simplify. $8j + j \qquad\qquad\qquad\qquad ^-8j + 15j - 2$ $11j + 14 - 7j \qquad\qquad\quad 11j + 4 - 9 - 6j$ $11 + 4j - 9 - 6j \qquad\qquad 11 + 4j - 9j - 6$	**page 498**
8-8	**Solving equations by combining like terms** *page 226*	$2a + 9a = 55 \qquad\qquad\quad 9x - x = ^-24$ $8y - 10 - 4y = 20 \qquad\quad 5r + 2 - 14r = 0$	**page 499**

Lesson	Skill	Test Items	Skill Practice
8-9	**Solving equations with variables on both sides** *page 228*	Solve. $4x = 16 - 2x$ $4r - 5 = 6r + 10$ $6n - 2 = 3n - 5$ $12 + 2t = t - 6$ $17 - 6y = 3y + 8$ $^-4j + 12 = 10j - 30$	page 499
9-1	**Writing equal ratios** *page 238*	Complete to get an equal ratio. $\dfrac{14}{21} = \dfrac{?}{3}$ $\dfrac{3}{2} = \dfrac{?}{10}$ $\dfrac{7}{8} = \dfrac{?}{32}$ $\dfrac{12}{5} = \dfrac{?}{30}$ $\dfrac{a}{b} = \dfrac{?}{3b}$ $\dfrac{6}{a} = \dfrac{?}{5ab}$ $\dfrac{a^2}{b} = \dfrac{?}{8b^2}$ $\dfrac{7a}{4b} = \dfrac{?}{16ab^2}$	page 500
9-2	**Solving a proportion** *page 240*	$\dfrac{4}{n} = \dfrac{5}{8}$ $\dfrac{n}{15} = \dfrac{10}{9}$ $\dfrac{4}{13} = \dfrac{n}{6\frac{1}{2}}$ $\dfrac{2\frac{3}{8}}{5} = \dfrac{17}{n}$	page 500
9-4	**Changing a percent to a fraction or mixed number** *page 244*	Change to a fraction or mixed number in simplest form. 25% 150% $33\frac{1}{3}\%$ $62\frac{1}{2}\%$	page 501
9-5	**Changing a fraction to a percent** *page 246*	Change to a percent. $\dfrac{3}{5}$ $\dfrac{3}{2}$ $\dfrac{1}{6}$ $\dfrac{8}{3}$	page 501
9-7	**Finding a percent of a number** *page 252*	25% of $44 = n$ 9% of $37 = n$ 6.5% of $38 = n$ 0.5% of $60 = n$	page 502
9-9	**Finding the number when a percent is known** *page 256*	Round each answer to the nearest tenth. 20% of $n = 16$ 75% of $n = 48$ 8.5% of $n = 12.5$ 32.2% of $n = 34.6$	page 502
9-10	**Finding the percent of increase or decrease** *page 258*	Find the percent of increase or decrease. Use i to indicate increase and d to indicate decrease. Give answers to the nearest tenth of a percent. 18 to 27 60 to 80 80 to 60 56 to 40 81 to 100 125 to 95	page 503

Lesson	Skill	Test Items		Skill Practice
14-2	**Adding polynomials** *page 402*	$(5y + 3) + (2y + 6)$ $(3x + y + 2z)$ $\quad + (2x + 5y + z)$	$(5a^2 + b) + (2a^2 + {}^-3b)$ $(12t^2 + 3t + 5)$ $\quad + (4t^2 + t + {}^-8)$	**page 503**
14-3	**Subtracting polynomials** *page 404*	$(8n + 6) - (4n + 5)$ $(5r + 3s + 8t)$ $\quad - (4r + s + 6t)$	$(5a + 9b) - (3a + 4b)$ $(12y^2 + 8y + 9)$ $\quad - (7y^2 + 1)$	**page 504**
14-5	**Multiplying polynomials by monomials** *page 410*	$8(s + t)$ $3c(2c + {}^-9)$	$8x(x + 6)$ ${}^-2a(4a + {}^-3b + 1)$	**page 504**
14-6	**Dividing polynomials by monomials** *page 412*	$\dfrac{15z + 20}{5}$ $\dfrac{{}^-6t^2 + 15t}{3t}$	$\dfrac{18n^2 + 27n}{9n}$ $\dfrac{12x^3 + 24x^2 + {}^-18x}{6x}$	**page 505**
14-7	**Solving equations containing grouping symbols** *page 414*	$8a + 2(a + 4) + 3 = 31 \quad 2j + 2(j - 1) + 10 = 20$ $2y + 3(y + 2) + 4y = {}^-12$ ${}^-2x + 7(x + 1) + 8(x + 2) = 10$		**page 505**

Pages 464–505
These practice sets cover the skills tested on the Skill Test. Each set provides an example and practice for one skill.

The sets on pages 464–467 review and practice prerequisite whole number arithmetic skills that are not taught in this textbook.

The remaining sets review and practice important skills developed in the lessons. The sets are numbered to show their correlation with the lessons.

Example

Round 45,359 to the nearest hundred.

Rounding to this place
↓
45,359
↑

When the next digit to the right is 5 or greater, round up.

45,359 rounds to 45,400

Round to the nearest ten.

1. 74	**2.** 37	**3.** 42	**4.** 75
5. 183	**6.** 366	**7.** 805	**8.** 411
9. 4336	**10.** 3721	**11.** 3605	**12.** 2398

Round to the nearest hundred.

▲**13.** 276	▲**14.** 550	▲**15.** 743	▲**16.** 849
17. 3408	**18.** 3423	**19.** 6660	**20.** 8050
21. 20,305	**22.** 32,780	**23.** 42,912	**24.** 62,950

Round to the nearest thousand.

■**25.** 4841	■**26.** 6851	■**27.** 9310	■**28.** 6500
29. 35,431	**30.** 42,573	**31.** 719,527	**32.** 273,500

Round to the nearest ten thousand.

●**33.** 24,146	●**34.** 52,700	●**35.** 56,913	**36.** 49,430
37. 92,604	**38.** 28,911	**39.** 249,300	**40.** 613,812

Example

$$245 + 92 + 3916 = ?$$

Line up the digits vertically.

```
  245
   92
+3916
```

Add.

```
   11
  245
 1 92
+3916
 4253
```

Give the sum.

1. 6438 + 8310	**2.** 5832 + 694
3. 966 + 2947	**4.** 3370 + 1938
5. 34,006 + 8825	**6.** 4721 + 76,082
7. 12,500 + 38,926	**8.** 38,842 + 27,111
9. 493 + 3493 + 977	**10.** 8218 + 739 + 1005
11. 182 + 4200 + 3628	**12.** 7467 + 941 + 604
13. 593 + 444 + 1660	**14.** 2741 + 8009 + 476
15. 4850 + 1188 + 2055	**16.** 1748 + 2966 + 1826
17. 54,388 + 2112 + 599	**18.** 4368 + 829 + 12,477
19. 29,006 + 2704 + 1822	**20.** 2864 + 31,000 + 8002
▲**21.** 4589 + 3594 + 1642	▲**22.** 563 + 2516 + 9292
■**23.** 374 + 209 + 5618	■**24.** 721 + 4441 + 356
●**25.** 2843 + 78 + 197	●**26.** 5381 + 927 + 58

Example

$$8305 - 4078 = ?$$

Line up the digits vertically and subtract in columns.

$$
\begin{array}{r}
8\ \overset{2}{\cancel{3}}\overset{10}{}\ 5 \\
-4\ 0\ 7\ 8 \\
\end{array}
$$

$$
\begin{array}{r}
8\ \overset{2}{\cancel{3}}\overset{9}{\cancel{10}}\overset{15}{\cancel{0}} \\
-4\ 0\ 7\ 8 \\
\end{array}
$$

$$
\begin{array}{r}
8\ \overset{2}{\cancel{3}}\overset{9}{\cancel{10}}\overset{15}{\cancel{0}} \\
-4\ 0\ 7\ 8 \\
\hline
4\ 2\ 2\ 7 \\
\end{array}
$$

Give the difference.

1. $828 - 411$
2. $594 - 221$
3. $710 - 463$
4. $824 - 258$
5. $504 - 356$
6. $701 - 588$
7. $806 - 529$
8. $903 - 165$
9. $800 - 361$
10. $400 - 249$
11. $700 - 318$
12. $600 - 233$
13. $800 - 444$
14. $500 - 381$
15. $4916 - 2854$
16. $5874 - 2222$
▲17. $3406 - 2153$
▲18. $7112 - 4338$
■19. $2502 - 458$
■20. $3701 - 229$
●21. $5205 - 1286$
●22. $6101 - 2255$
23. $53,621 - 47,950$
24. $64,206 - 39,178$
25. $48,000 - 21,462$
26. $72,003 - 19,740$

Example

$$30 \times 200 = ?$$

Multiply 3×2 and annex 3 zeros.

$$30 \times 200 = 6000$$

Give the product.

1. 9×10
2. 6×100
3. 5×1000
4. 3×100
5. 8×20
6. 8×200
7. 12×10
8. 15×1000
9. 20×50
10. 80×300
11. 3×800
12. 20×40
13. 30×2000
14. 40×400
15. 50×100
▲16. 50×1000
▲17. 40×200
▲18. 20×3000
■19. 30×20
■20. 40×300
■21. 400×30
●22. 145×100
●23. 256×1000
●24. 228×10
25. 400×60
26. 300×300
27. 600×5000
28. 300×200
29. 800×40
30. 700×6000

Example

$$527 \times 4 = ?$$

Line up the digits vertically.

$$\begin{array}{r} 527 \\ \times\, 4 \\ \hline \end{array}$$

Multiply.

$$\begin{array}{r} {\scriptstyle 12} \\ 527 \\ \times\, 4 \\ \hline 2108 \end{array}$$

Give the product.

1. 23×3 **2.** 12×4 **3.** 44×2

4. 11×6 **5.** 56×5 **6.** 78×7

7. 39×9 **8.** 82×4 **9.** 143×3

10. 481×6 **11.** 330×2 **12.** 513×5

13. 307×8 **14.** 632×7 **15.** 812×4

▲**16.** 961×9 ▲**17.** 748×6 ▲**18.** 853×7

■**19.** 571×4 ■**20.** 660×8 ■**21.** 493×3

●**22.** 3218×3 ●**23.** 5175×5 ●**24.** 4069×8

25. 3506×6 **26.** 7153×4 **27.** 9188×5

28. 2009×9 **29.** 6021×7 **30.** 5228×3

31. 4837×6 **32.** 2740×8 **33.** 6050×7

34. 8162×5 **35.** 3908×4 **36.** 3625×8

Example

$$145 \times 23 = ?$$

Multiply by 3.

$$\begin{array}{r} 145 \\ \times\, 23 \\ \hline 435 \end{array}$$

Multiply by 20.

$$\begin{array}{r} 145 \\ \times\, 23 \\ \hline 435 \\ 2900 \end{array}$$

Add.

$$\begin{array}{r} 145 \\ \times 23 \\ \hline 435 \\ 2900 \\ \hline 3335 \end{array}$$

Give the product.

1. 34×12 **2.** 26×20 **3.** 40×41

4. 51×33 **5.** 75×18 **6.** 84×25

7. 59×36 **8.** 47×29 **9.** 43×55

10. 50×62 **11.** 78×18 **12.** 95×77

▲**13.** 125×31 ▲**14.** 236×22 ▲**15.** 304×58

■**16.** 411×70 ■**17.** 638×63 ■**18.** 905×85

●**19.** 731×108 ●**20.** 592×604 ●**21.** 521×461

22. 940×218 **23.** 3015×226 **24.** 4628×314

25. 4708×316 **26.** 8162×407 **27.** 3047×902

28. 6005×584 **29.** 3714×215 **30.** 1010×101

Example

4016 ÷ 8 = ?

Not enough thousands.
Think 40 hundreds.
Divide hundreds.

$$\begin{array}{r} 5 \\ 8\overline{)4016} \\ -40 \end{array}$$

Not enough tens.
Think 16 ones.

$$\begin{array}{r} 50 \\ 8\overline{)4016} \\ -40 \\ \hline 16 \end{array}$$

Don't forget the 0.

Divide ones.

$$\begin{array}{r} 502 \\ 8\overline{)4016} \\ -40 \\ \hline 16 \\ -16 \\ \hline 0 \end{array}$$

Give the quotient.

1. 464 ÷ 8
2. 207 ÷ 3
3. 635 ÷ 5
4. 934 ÷ 2
5. 513 ÷ 9
6. 812 ÷ 7
7. 584 ÷ 4
8. 366 ÷ 6
9. 728 ÷ 2
10. 428 ÷ 4
11. 198 ÷ 9
12. 624 ÷ 6
13. 635 ÷ 5
14. 582 ÷ 6
15. 498 ÷ 6
▲16. 808 ÷ 2
▲17. 588 ÷ 7
▲18. 729 ÷ 9
■19. 508 ÷ 4
■20. 695 ÷ 5
■21. 728 ÷ 8
●22. 2016 ÷ 4
●23. 5688 ÷ 8
●24. 9396 ÷ 9
25. 8330 ÷ 7
26. 4656 ÷ 6
27. 4000 ÷ 8
28. 2945 ÷ 5
29. 3804 ÷ 4
30. 6210 ÷ 9
31. 8091 ÷ 3
32. 1908 ÷ 6
33. 7624 ÷ 8
34. 1572 ÷ 4
35. 8113 ÷ 7
36. 7155 ÷ 9

Example

3999 ÷ 48 = ?

Think about dividing 39
by 4. So try 9.

$$\begin{array}{r} 48 \\ \times 9 \\ \hline 432 \end{array} \quad 48\overline{)3999}$$

432 is too big!

Try 8.

$$\begin{array}{r} 48 \\ \times 8 \\ \hline 384 \end{array} \quad \begin{array}{r} 8 \\ 48\overline{)3999} \\ -384 \\ \hline 15 \end{array}$$

Think about dividing 15
by 4. So try 3.

$$\begin{array}{r} 48 \\ \times 3 \\ \hline 144 \end{array} \quad \begin{array}{r} 83 \ R15 \\ 48\overline{)3999} \\ -384 \\ \hline 159 \\ -144 \\ \hline 15 \end{array}$$

Divide.

1. 2946 ÷ 12
2. 9375 ÷ 32
3. 8611 ÷ 25
4. 2589 ÷ 43
5. 8526 ÷ 50
6. 6351 ÷ 81
7. 4490 ÷ 70
8. 8555 ÷ 49
9. 5773 ÷ 64
10. 6310 ÷ 38
11. 7008 ÷ 60
12. 9362 ÷ 75
13. 8610 ÷ 35
14. 6235 ÷ 91
▲15. 6192 ÷ 32
▲16. 7815 ÷ 15
■17. 38,500 ÷ 125
■18. 67,800 ÷ 150
●19. 91,372 ÷ 212
●20. 57,988 ÷ 436
21. 81,088 ÷ 516
22. 73,849 ÷ 406
23. 47,321 ÷ 616
24. 91,156 ÷ 688
25. 183,405 ÷ 361
26. 162,261 ÷ 219

Example

Evaluate the expression

$$a + b - c$$

for $a = 9$, $b = 6$, and $c = 7$.

Substitute.

$$9 + 6 - 7$$

Simplify.

8

Evaluate each expression for $x = 10$, $y = 5$, and $z = 8$.

1. $x + 3$ **2.** $x - 7$ **3.** $x + 9$

4. $y - 5$ **5.** $y + 8$ **6.** $y - 2$

7. $z + 6$ **8.** $z - 8$ **9.** $z + 5$

10. $x + y$ **11.** $y + x$ **12.** $y + z$

13. $x + z$ **14.** $x - y$ **15.** $z - y$

▲**16.** $y + x + z$ ▲**17.** $y + z - 4$ ▲**18.** $z + x - 9$

■**19.** $y + y - 4$ ■**20.** $z + z - 9$ ■**21.** $x + x - 5$

●**22.** $x + y + z$ ●**23.** $(x - y) + 4$ ●**24.** $x + z - 5$

25. $(16 - y) + x$ **26.** $15 + y - x$ **27.** $(20 - y) - y$

28. $18 + x - y$ **29.** $24 + y + z$ **30.** $30 + z - z$

Example

Round 36.487 to the nearest tenth.

Rounding to this place

↓

36.487

↑

Since the next digit to the right is 5 or greater, round up.

36.487 rounds to 36.5.

Round to the nearest whole number.

▲ **1.** 16.6   ▲ **2.** 38.3   ▲ **3.** 92.4   ▲ **4.** 35.5

5. 51.27   **6.** 38.93   **7.** 0.025   **8.** 20.19

9. 327.04   **10.** 118.40   **11.** 0.500   **12.** 12.099

Round to the nearest tenth.

■**13.** 403.38   ■**14.** 26.10   ■**15.** 5.25   ■**16.** 3.95

17. 21.39   **18.** 24.188   **19.** 22.06   **20.** 7.472

21. 204.29   **22.** 444.484   **23.** 0.0592   **24.** 0.95

Round to the nearest hundredth.

●**25.** 22.317   ●**26.** 56.208   ●**27.** 5.531

28. 54.325   **29.** 71.594   **30.** 6.30196

31. 0.0518   **32.** 1.065   **33.** 0.0946

34. 11.269   **35.** 3.9421   **36.** 0.097

37. 42.3381   **38.** 28.095   **39.** 0.6422

Example

$3 + 2.51 + 8.6 = ?$

Line up the decimal points.

$$\begin{array}{r} 3 \\ 2.51 \\ +8.6 \end{array}$$

Add.

$$\begin{array}{r} 1 \\ 3 \\ 2.51 \\ +8.6 \\ \hline 14.11 \end{array}$$

Give the sum.

1. $4.64 + 3.08$

2. $7.564 + 3.806$

3. $6.3521 + 0.5821$

4. $721.6 + 38.4$

5. $2.35 + 4.829$

6. $5.008 + 3.62$

7. $43.6 + 27.48$

8. $10.88 + 9.3$

9. $5.6 + 3.04 + 2.7$

10. $2.64 + 5.7 + 8.8$

11. $4.20 + 9.2 + 3.65$

12. $6.1 + 2.22 + 6.83$

13. $2.641 + 0.75 + 3.58$

14. $5.34 + 0.756 + 2.84$

▲**15.** $9.3645 + 2.055 + 0.221$

▲**16.** $8.471 + 0.4911 + 3.300$

■**17.** $7.4 + 4.611 + 8.5$

■**18.** $15.966 + 8.4 + 4.8$

●**19.** $32 + 3.4 + 2.08$

●**20.** $5.7 + 41 + 6.63$

21. $5 + 3.741 + 2.68$

22. $18 + 5.77 + 6.411$

23. $3.216 + 2.84 + 0.95$

24. $0.8 + 9.142 + 5.33$

Example

$20 - 14.38 = ?$

Line up the decimal points. Write the zeros.

$$\begin{array}{r} 20.00 \\ -14.38 \end{array}$$

Subtract.

$$\begin{array}{r} 1\,9 \quad 9 \\ 2^{1}0 \,.\, {}^{1}0^{1}0 \\ -1\,4\,.\,3\,8 \\ \hline 5\,.\,6\,2 \end{array}$$

Give the difference.

1. $9 - 3.2$

2. $8 - 4.6$

3. $15 - 7.2$

4. $23 - 8.6$

5. $18.01 - 9.45$

6. $14.05 - 7.75$

7. $9.4 - 6.73$

8. $8.5 - 4.55$

9. $8.3 - 6$

10. $7.4 - 2$

11. $10.3 - 8.4$

12. $30.1 - 9.7$

13. $7 - 3.44$

14. $8 - 6.45$

▲**15.** $8.23 - 0.749$

▲**16.** $6.729 - 0.88$

■**17.** $8.5 - 3.692$

■**18.** $5.1 - 0.651$

●**19.** $42 - 8.2$

●**20.** $34 - 9.5$

21. $81.64 - 33$

22. $63.89 - 18$

23. $100 - 44.63$

24. $200 - 53.87$

25. $102 - 9.4$

26. $105 - 49.7$

Skill 1-7 (Use after page 16.)

Example

Solve.

$$x + 14 = 23$$

Subtract 14 from both sides.

$$x + 14 - 14 = 23 - 14$$

Simplify.

$$x = 9$$

$$9 + 14 \stackrel{?}{=} 23$$
$$23 = 23$$

Solve and check.

1. $w + 5 = 9$

2. $h + 8 = 15$

3. $e + 10 = 12$

4. $b + 9 = 15$

5. $t + 5 = 20$

6. $p + 7 = 18$

7. $r + 11 = 20$

8. $a + 13 = 13$

9. $k + 18 = 29$

10. $x + 17 = 34$

11. $z + 19 = 42$

12. $v + 22 = 47$

13. $c + 21 = 30$

14. $m + 23 = 53$

15. $g + 34 = 58$

▲16. $r + 42 = 75$

▲17. $y + 46 = 48$

▲18. $d + 43 = 43$

■19. $t + 36 = 55$

■20. $d + 29 = 35$

■21. $f + 37 = 60$

●22. $i + 51 = 54$

●23. $u + 42 = 56$

●24. $c + 48 = 92$

25. $s + 73 = 93$

26. $g + 65 = 65$

27. $v + 59 = 87$

28. $j + 60 = 100$

29. $n + 78 = 116$

30. $s + 83 = 133$

Skill 1-8 (Use after page 18.)

Example

Solve.

$$y - 26 = 34$$

Add 26 to both sides

$$y - 26 + 26 = 34 + 26$$

Simplify.

$$y = 60$$

✔ CHECK

$$60 - 26 \stackrel{?}{=} 34$$
$$34 = 34$$

Solve and check.

1. $v - 9 = 8$

2. $r - 7 = 10$

3. $f - 6 = 9$

4. $c - 8 = 7$

5. $a - 11 = 0$

6. $q - 15 = 12$

7. $n - 19 = 15$

8. $j - 12 = 23$

9. $a - 20 = 19$

10. $w - 24 = 21$

11. $b - 32 = 50$

12. $f - 41 = 27$

13. $b - 37 = 40$

14. $d - 48 = 16$

15. $k - 55 = 24$

▲16. $s - 43 = 46$

▲17. $c - 30 = 35$

▲18. $v - 29 = 43$

■19. $x - 62 = 17$

■20. $p - 52 = 26$

■21. $e - 43 = 20$

●22. $t - 40 = 33$

●23. $e - 56 = 37$

●24. $z - 30 = 42$

25. $g - 62 = 22$

26. $d - 65 = 18$

27. $m - 44 = 44$

28. $t - 57 = 45$

29. $h - 71 = 25$

30. $y - 66 = 34$

31. $a - 46 = 50$

32. $b - 83 = 0$

33. $c - 1 = 59$

34. $d - 0 = 87$

35. $e - 62 = 62$

36. $f - 74 = 47$

Example

Solve.

$$28 = y + 11$$

Use the symmetric property of equality.

$$y + 11 = 28$$

Subtract 11 from both sides.

$$y + 11 - 11 = 28 - 11$$

Simplify.

$$y = 17$$

 CHECK

$$28 \overset{?}{=} 17 + 11$$
$$28 = 28$$

Solve and check.

1. $19 = f + 8$ **2.** $f - 6 = 6$ **3.** $20 = f - 19$

4. $a + 9 = 15$ **5.** $45 = t + 35$ **6.** $e + 12 = 56$

7. $18 = b - 16$ **8.** $p - 17 = 35$ **9.** $60 = y + 23$

10. $q + 20 = 42$ **11.** $18 = j - 16$ **12.** $49 = g + 37$

13. $45 = h - 30$ **14.** $b - 44 = 27$ **15.** $k + 42 = 42$

▲**16.** $h - 31 = 53$ ▲**17.** $60 = w + 42$ ▲**18.** $m + 25 = 100$

■**19.** $58 = x - 23$ ■**20.** $52 = c + 52$ ■**21.** $35 = u - 25$

●**22.** $92 = a + 61$ ●**23.** $g - 17 = 60$ ●**24.** $d + 23 = 84$

25. $c - 42 = 23$ **26.** $30 = d - 56$ **27.** $101 = z + 74$

28. $58 = v - 35$ **29.** $r - 35 = 76$ **30.** $j + 53 = 112$

Example

Evaluate the expression

$$\frac{r}{t} + s$$

for $r = 12$, $s = 4$, and $t = 2$.

Substitute.

$$\frac{12}{2} + 4$$

Simplify.

$$10$$

Evaluate each expression for $x = 12$, $y = 6$, and $z = 3$.

1. $5x$ **2.** $9y$ **3.** $\dfrac{x}{4}$

4. $\dfrac{y}{2}$ **5.** $3 + z$ **6.** $15 - y$

7. $(3x) + 4$ **8.** $(8y) - 3$ **9.** $(10z) + 6$

10. $(12y) + z$ **11.** $(2x) - y$ **12.** $(3x) + z$

13. $\dfrac{x}{3} + z$ **14.** $\dfrac{y}{2} - z$ **15.** $\dfrac{x}{6} + y$

▲**16.** $(xy) - z$ ▲**17.** $(xz) + y$ ▲**18.** $(yz) - x$

■**19.** $(5x) + x$ ■**20.** $(7y) - y$ ■**21.** $(8z) - z$

●**22.** $(10x) - z$ ●**23.** $(4y) + x$ ●**24.** $(2x) + x$

25. $\dfrac{42}{y} + z$ **26.** $\dfrac{30}{z} - y$ **27.** $\dfrac{24}{x} + x$

28. $\dfrac{x}{z} - z$ **29.** $\dfrac{x}{y} + z$ **30.** $\dfrac{y}{z} + x$

Example

$3.08 × 4.2 = ?$

Multiply as whole numbers.

$$\begin{array}{r} 3.0\,8 \\ \times 4.2 \\ \hline 6\,1\,6 \\ 1\,2\,3\,2 \\ \hline 1\,2\,9\,3\,6 \end{array}$$

Count the digits to the right of the decimal points.

$$\begin{array}{r} 3.0\,8 \\ \times 4.2 \\ \hline 6\,1\,6 \\ 1\,2\,3\,2 \\ \hline 1\,2.9\,3\,6 \end{array} \quad \boxed{3}$$

Count off the same number of digits in the product.

Give the product.

1. $4.2 × 12$
2. $3.8 × 10$
3. $2.6 × 2.6$
4. $5.9 × 8.7$
5. $4.06 × 0.8$
6. $2.05 × 5.5$
7. $0.94 × 0.34$
8. $0.95 × 0.55$
9. $58 × 0.25$
10. $74 × 0.78$
11. $221 × 4.6$
12. $360 × 8.2$
13. $3.62 × 0.95$
14. $2.88 × 0.47$
▲15. $6.16 × 7.5$
▲16. $2.09 × 0.8$
■17. $5.4 × 0.06$
■18. $8.8 × 0.07$
●19. $6.25 × 0.56$
●20. $8.65 × 0.44$
21. $30.5 × 20.2$
22. $56.7 × 18.4$
23. $55.5 × 21.6$
24. $63.2 × 8.94$
25. $300 × 4.8$
26. $600 × 0.52$
27. $2.54 × 2.54$
28. $3.08 × 3.08$

Examples

$2.47 × 10 = ?$

When multiplying by 10, move the decimal point 1 place to the right.

$2.47 × 10 = 24.7$

$2.47 × 100 = ?$

When multiplying by 100, move the decimal point 2 places to the right.

$2.47 × 100 = 247$

Give the product.

1. $4.2 × 10$
2. $0.38 × 100$
3. $16 × 1000$
4. $6.5 × 100$
5. $1.25 × 100$
6. $113 × 1000$
7. $8.2 × 10$
8. $0.05 × 1000$
9. $6.8 × 100$
10. $4.7 × 100$
11. $2.95 × 10$
12. $9.44 × 1000$
13. $220 × 1000$
14. $300 × 10$
15. $6.5 × 100$
16. $9.55 × 100$
17. $8.74 × 1000$
18. $0.75 × 10$
▲19. $0.005 × 10$
▲20. $0.002 × 100$
▲21. $0.008 × 1000$
■22. $8.4 × 1000$
■23. $7.2 × 100$
■24. $9.6 × 10$
●25. $6.9 × 10$
●26. $3.74 × 100$
●27. $5.34 × 1000$
28. $3.96 × 1000$
29. $6.66 × 10$
30. $8.51 × 1000$
31. $4.798 × 100$
32. $4.798 × 10$
33. $4.798 × 1000$

Example

$$12.99 \div 23 = ?$$

Divide.

$$
\begin{array}{r}
0.56 \\
23\overline{)12.99} \\
-11\,5 \\
\hline
1\,49 \\
-1\,38 \\
\hline
11
\end{array}
$$

To round quotient to nearest hundredth, write a zero and carry out the division another place.

$$
\begin{array}{r}
0.564 \\
23\overline{)12.990} \\
-11\,5 \\
\hline
1\,49 \\
-1\,38 \\
\hline
110 \\
-92 \\
\hline
18
\end{array}
$$

Give the quotient.

1. $8.1 \div 5$

2. $25.9 \div 7$

3. $4.32 \div 8$

4. $0.938 \div 2$

5. $6.75 \div 9$

6. $0.847 \div 7$

7. $1.44 \div 12$

8. $7.13 \div 23$

9. $1.008 \div 36$

10. $0.2491 \div 47$

Divide. Round the quotient to the nearest hundredth.

11. $2.5 \div 3$

12. $0.32 \div 6$

▲13. $0.53 \div 9$

▲14. $56.92 \div 6$

■15. $7.34 \div 14$

■16. $8.91 \div 49$

●17. $3.114 \div 42$

●18. $0.8113 \div 29$

19. $89.1 \div 94$

20. $5.347 \div 85$

Example

$$0.42\overline{)0.5670}$$

Move both decimal points two places to the right.

$$0.42\overline{)0.56\,70}$$

Divide.

$$
\begin{array}{r}
1.35 \\
0.42\overline{)0.56{,}70} \\
-42 \\
\hline
14\,7 \\
-12\,6 \\
\hline
2\,10 \\
-2\,10 \\
\hline
0
\end{array}
$$

Give the quotient.

1. $38.36 \div 0.7$

2. $2.634 \div 0.6$

3. $4.584 \div 0.08$

4. $2.076 \div 0.03$

5. $1.473 \div 0.03$

6. $3.605 \div 0.005$

7. $0.2656 \div 0.004$

8. $96.3 \div 0.3$

9. $6.4 \div 0.04$

10. $350.4 \div 0.6$

11. $0.0644 \div 0.07$

12. $0.963 \div 0.9$

▲13. $8.6055 \div 0.005$

▲14. $0.0152 \div 0.08$

■15. $5.2 \div 1.3$

■16. $0.144 \div 1.2$

●17. $0.6075 \div 0.15$

●18. $28.52 \div 2.3$

19. $1.3995 \div 0.45$

20. $1.2912 \div 2.4$

21. $0.22274 \div 0.37$

22. $29.011 \div 6.7$

23. $19.292 \div 5.3$

24. $0.38442 \div 0.86$

Examples

5.2 ÷ 10 = ?

When dividing by 10, move the decimal point 1 place to the left.

5.2 ÷ 10 = 0.52

5.2 ÷ 100 = ?

When dividing by 100, move the decimal point 2 places to the left.

5.2 ÷ 100 = 0.052

Give the quotient.

1. $34.2 \div 10$
2. $34.2 \div 100$
3. $34.2 \div 1000$
4. $45.8 \div 100$
5. $45.8 \div 1000$
6. $45.8 \div 10$
7. $252.5 \div 100$
8. $252.5 \div 10$
9. $252.5 \div 1000$
10. $80 \div 10$
11. $80 \div 100$
12. $80 \div 1000$
13. $23.94 \div 10$
14. $23.94 \div 100$
15. $23.94 \div 1000$
16. $2.8 \div 10$
17. $2.8 \div 100$
18. $2.8 \div 1000$
▲19. $2.84 \div 1000$
▲20. $9.05 \div 10$
■21. $0.96 \div 100$
■22. $9.05 \div 100$
●23. $9.05 \div 1000$
●24. $90.5 \div 10$
25. $3.25 \div 100$
26. $3.25 \div 10$
27. $3.25 \div 1000$
28. $82.5 \div 1000$

Example

Solve.

4c = 36

Divide both sides by 4.

$$\frac{4c}{4} = \frac{36}{4}$$

Simplify.

$$c = 9$$

 CHECK

$4(9) \stackrel{?}{=} 36$

$36 = 36$

Solve and check.

1. $8c = 16$
2. $9x = 36$
3. $5n = 15$
4. $6y = 54$
5. $7p = 49$
6. $3v = 0$
7. $9s = 63$
8. $4a = 80$
9. $2n = 42$
10. $8m = 88$
11. $5v = 65$
12. $6h = 0$
13. $3z = 81$
14. $4b = 96$
15. $8s = 104$
▲16. $10e = 50$
▲17. $15u = 135$
▲18. $12m = 48$
■19. $15q = 0$
■20. $12d = 60$
■21. $10g = 80$
●22. $11t = 77$
●23. $18r = 36$
●24. $11q = 99$
25. $20f = 160$
26. $25p = 175$
27. $30j = 150$
28. $40n = 200$
29. $16r = 320$
30. $25t = 225$

Example

Solve.

$$\frac{y}{6} = 9$$

Multiply both sides by 6.

$$6 \times \frac{y}{6} = 6 \times 9$$

Simplify.

$$y = 54$$

✔ **CHECK**

$$\frac{54}{6} \overset{?}{=} 9$$
$$9 = 9$$

Solve and check.

1. $\dfrac{x}{3} = 5$

2. $\dfrac{u}{4} = 7$

3. $\dfrac{k}{2} = 12$

4. $\dfrac{w}{6} = 10$

5. $\dfrac{s}{7} = 13$

6. $\dfrac{t}{9} = 6$

7. $\dfrac{m}{5} = 8$

8. $\dfrac{u}{3} = 9$

9. $\dfrac{v}{8} = 8$

10. $\dfrac{c}{6} = 10$

11. $\dfrac{d}{4} = 12$

12. $\dfrac{a}{3} = 9$

13. $\dfrac{z}{4} = 15$

14. $\dfrac{j}{7} = 11$

15. $\dfrac{q}{9} = 7$

▲16. $\dfrac{p}{8} = 6$

▲17. $\dfrac{b}{2} = 15$

▲18. $\dfrac{v}{5} = 12$

■19. $\dfrac{m}{10} = 17$

■20. $\dfrac{n}{12} = 16$

■21. $\dfrac{y}{11} = 5$

●22. $\dfrac{e}{15} = 8$

●23. $\dfrac{p}{18} = 3$

●24. $\dfrac{h}{16} = 2$

Example

Solve.

$$63 = 9x$$

Use the symmetric property of equality.

$$9x = 63$$

Divide both sides by 9.

$$\frac{9x}{9} = \frac{63}{9}$$

Simplify.

$$x = 7$$

 CHECK

$$63 \overset{?}{=} 9(7)$$
$$63 = 63$$

Solve and check.

1. $8 = \dfrac{r}{2}$

2. $5p = 15$

3. $\dfrac{h}{9} = 3$

4. $16 = 2q$

5. $4 = \dfrac{j}{5}$

6. $40 = 4b$

7. $0 = \dfrac{y}{9}$

8. $\dfrac{d}{3} = 11$

9. $4c = 20$

10. $60 = 3j$

11. $85 = 5k$

12. $6 = \dfrac{t}{7}$

13. $\dfrac{e}{5} = 15$

14. $30 = 10t$

15. $81 = 9c$

▲16. $8n = 72$

▲17. $14 = \dfrac{k}{3}$

▲18. $0 = 6g$

■19. $10 = \dfrac{x}{11}$

■20. $6v = 60$

■21. $\dfrac{b}{9} = 11$

●22. $88 = 11k$

●23. $\dfrac{a}{15} = 0$

●24. $13u = 52$

25. $\dfrac{f}{12} = 10$

26. $7m = 84$

27. $12 = \dfrac{a}{8}$

Skill 3-1 (Use after page 62.)

Example

$6 + 8 \times (4 - 2) = ?$

First, work within the grouping symbols.

$6 + 8 \times 2$

Next, do the multiplication and division.

$6 + 16$

Last, do the addition and subtraction.

22

Simplify.

1. $6 \div 3 \times 2$
2. $12 - 8 + 4$
3. $5 + 2 \times 5 - 1$
4. $5 \times 2 + 10 \div 2$
5. $5 + (3 + 9) \div 6$
6. $(4 + 5) \times 2 - 8$
7. $12 \div 4 - 1$
8. $8 \times 5 - 3$
9. $24 - 4 \div 4$
10. $30 - 12 - 6$
11. $10 + 16 \div 4$
12. $18 + 6 \div 3$
▲13. $48 \div 8 \times 2$
▲14. $35 + 12 - 10$
■15. $18 - 6 + 6$
■16. $20 - 9 + 5$
●17. $(12 + 18) \div 6$
●18. $34 \times (8 - 3)$
19. $16 + 8 \div 4 + 4$
20. $(16 + 8) \div 4 + 4$
21. $16 + 8 \div (4 + 4)$
22. $20 + 12 \times 4 - 1$

Skill 3-4 (Use after page 68.)

Example

Solve.

$19 + j = 63$

Use the commutative property of addition.

$j + 19 = 63$
$j + 19 - 19 = 63 - 19$
$j = 44$

 CHECK

$19 + 44 \overset{?}{=} 63$
$63 = 63$

Solve and check.

1. $15 + g = 36$
2. $j(12) = 120$
3. $100 = y(20)$
4. $37 = 21 + b$
5. $110 = x(11)$
6. $27 + h = 91$
7. $50 = 33 + c$
8. $n(25) = 225$
9. $74 = 43 + y$
10. $8k = 128$
11. $p(6) = 144$
12. $56 = u - 19$
13. $220 = t(20)$
14. $6 = \dfrac{g}{12}$
15. $64 = w + 23$
▲16. $\dfrac{h}{5} = 13$
▲17. $56 = k(2)$
▲18. $17 = \dfrac{m}{4}$
■19. $c + 3.4 = 6$
■20. $5.1 = \dfrac{j}{10}$
■21. $d - 9.4 = 19$
●22. $18.5 = 4.6 + d$
●23. $12.3 = n - 7.7$
●24. $23.2 = 23.2 + p$
25. $20.4 = t(5)$
26. $12 = \dfrac{r}{0.1}$
27. $12.6 = u(4)$

Skill 3-6 (Use after page 74.)

Example

Solve.

$$8n + 3 = 59$$

Subtract 3 from both sides.
Then simplify.

$$8n + 3 - 3 = 59 - 3$$
$$8n = 56$$

Divide both sides by 8.
Then simplify.

$$\frac{8n}{8} = \frac{56}{8}$$
$$n = 7$$

 CHECK

$$8(7) + 3 \stackrel{?}{=} 59$$
$$59 = 59$$

Solve and check.

1. $3b + 9 = 27$ 2. $5c - 4 = 31$

3. $7f + 4 = 60$ 4. $4h - 8 = 16$

5. $9g + 10 = 100$ 6. $6d - 16 = 50$

7. $39 = 2y + 5$ 8. $0 = 4z - 16$

9. $57 = 8x + 17$ 10. $20 = 7t - 15$

11. $93 = 6u + 21$ 12. $29 = 3r - 13$

13. $12s + 8 = 116$ 14. $15n - 23 = 127$

15. $20q + 30 = 250$ 16. $18x + 17 = 17$

▲17. $12w - 50 = 82$ ▲18. $25z + 33 = 283$

■19. $2c + 4.4 = 9$ ■20. $5a - 6.7 = 18$

●21. $4b + 8 = 15.4$ ●22. $12 = 3m - 3.6$

23. $20 = 9p + 1.1$ 24. $27 = 6r - 9.6$

25. $5z - 4.3 = 10.9$ 26. $2y + 8.4 = 8.4$

Skill 3-7 (Use after page 76.)

Example

Solve.

$$\frac{n}{5} - 3 = 6$$

Add 3 to both sides. Then simplify.

$$\frac{n}{5} - 3 + 3 = 6 + 3$$

$$\frac{n}{5} = 9$$

Multiply both sides by 5. Then simplify.

$$5 \times \frac{n}{5} = 5 \times 9$$

$$n = 45$$

 CHECK

$$\frac{45}{5} - 3 \stackrel{?}{=} 6$$

$$6 = 6$$

Solve and check.

1. $\frac{n}{4} + 3 = 7$ 2. $\frac{r}{6} - 4 = 1$ 3. $12 = \frac{d}{2} + 8$

4. $6 + \frac{t}{3} = 6$ 5. $4n + 6 = 30$ 6. $3j - 5 = 43$

7. $15 = 7w - 13$ 8. $6m + 7 = 67$ 9. $\frac{r}{9} - 23 = 27$

10. $\frac{m}{6} + 8 = 9$ 11. $\frac{c}{3} - 4 = 7$ 12. $24 = \frac{f}{5} + 9$

13. $9p - 4 = 41$ 14. $68 = 5v + 13$ 15. $15h - 22 = 38$

▲16. $7y + 14 = 56$ ▲17. $\frac{n}{3} - 12 = 8$ ▲18. $126 = 10j + 16$

■19. $23 = \frac{d}{8} + 19$ ■20. $102 = 12f - 18$ ■21. $3 = \frac{g}{2} - 21$

●22. $\frac{n}{6} + 1 = 20$ ●23. $5j - 31 = 34$ ●24. $\frac{t}{16} + 37 = 37$

25. $2t - 8.4 = 0$ 26. $9.6 = 4y + 2.4$ 27. $10j - 3.9 = 7.4$

28. $\frac{f}{3} + 2.6 = 13.4$ 29. $\frac{n}{8} - 1.4 = 8.7$ 30. $19.7 = 8t + 4.5$

Skill 3-9 (Use after page 80.)

Example

Simplify.

$$4x + 3 + x + 4$$

Combine like terms.

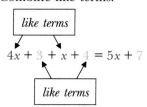

$$4x + 3 + x + 4 = 5x + 7$$

Simplify by combining like terms.

1. $5x + 3x$
2. $9y + 6y$
3. $7w + w$
4. $n + 12n$
5. $3t + 2t + 5t$
6. $6v + 3v + 8v$
7. $m + m + 3m$
8. $j + j + j$
9. $5y + 2y + 7$
10. $5y + 2 + 7y$
11. $8c + 5 + 3c$
12. $8 + 5c + 3c$
13. $6f + 3f + 16$
14. $6 + 3f + 16f$
▲15. $18d + d + 13$
▲16. $18 + d + 13d$
■17. $15x + 3x + 6 + 8$
■18. $15x + 3 + 6x + 8$
●19. $10z + 11 + 4 + 9z$
●20. $10 + 11z + 4 + 9z$
21. $22 + 11j + 9 + 15j$
22. $22j + 11 + 9j + 15$

Skill 3-10 (Use after page 82.)

Example

Solve.

$$8v + 2v = 19.6$$

Simplify by combining like terms.

$$10v = 19.6$$
$$\frac{10v}{10} = \frac{19.6}{10}$$
$$v = 1.96$$

✔ CHECK

$$8(1.96) + 2(1.96) \stackrel{?}{=} 19.6$$
$$19.6 = 19.6$$

Solve and check.

1. $3a + 2a = 55$
2. $49 = 4n + 3n$
3. $63 = 8r + r$
4. $5c + 7c = 96$
5. $114 = 11f + 8f$
6. $y + 10y = 132$
7. $m + m = 78$
8. $15 = 9f + 6f$
9. $0 = 13e + e$
10. $12c + 8c = 180$
11. $144 = n + 11n$
12. $156 = 2k + k$
13. $16j + 6j = 88$
14. $162 = 9x + 9x$
15. $13y + 3y = 0$
▲16. $21d + 4d = 275$
▲17. $300 = g + 9g$
▲18. $168 = 2n + 10n$
■19. $3k + k = 18$
■20. $37 = v + v$
■21. $6y + 2y = 118$
●22. $y + 9y = 6.8$
●23. $3y + 9y = 19.2$
●24. $c + c = 38.6$
25. $17.6 = 2n + 2n$
26. $36.12 = t + 3t$
27. $4m + 4m = 1.92$
28. $2a + 5a = 4.9$
29. $37.8 = n + 2n$
30. $27 = 2c + 4c$
31. $8 = 6b + 4b$
32. $3f + f = 0.24$
33. $5d + 3d = 8.4$

Examples

< or >?

$$0 \diamond +2$$

$$-2 \diamond -1$$

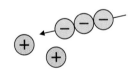

← *smaller* *larger* →

$$0 < +2$$

$$-2 < -1$$

< or >?

1. $+5 \diamond +8$	**2.** $+5 \diamond -8$	**3.** $-5 \diamond -8$
4. $-6 \diamond +1$	**5.** $+6 \diamond -1$	**6.** $+6 \diamond +1$
7. $0 \diamond -5$	**8.** $0 \diamond +5$	**9.** $-7 \diamond 0$
10. $+9 \diamond -4$	**11.** $-9 \diamond +4$	**12.** $+9 \diamond +4$
13. $-7 \diamond -2$	**14.** $-8 \diamond +3$	**15.** $0 \diamond +6$
16. $+9 \diamond +4$	**17.** $-6 \diamond +6$	**18.** $-3 \diamond -5$
19. $-11 \diamond +10$	**20.** $+16 \diamond -17$	**21.** $+15 \diamond +17$
▲**22.** $0 \diamond -12$	▲**23.** $+15 \diamond -15$	▲**24.** $-19 \diamond -13$
■**25.** $-16 \diamond +11$	■**26.** $-19 \diamond -18$	■**27.** $+14 \diamond +17$
●**28.** $-22 \diamond +22$	●**29.** $-26 \diamond -21$	●**30.** $+23 \diamond +24$
31. $+27 \diamond -20$	**32.** $-29 \diamond +23$	**33.** $-28 \diamond -26$
34. $0 \diamond +32$	**35.** $-36 \diamond 0$	**36.** $+37 \diamond -31$

Examples

To add integers, think about combining charges.

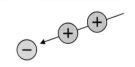

$$+2 + -3 = -1$$

$$-1 + +2 = +1$$

Give the sum.

1. $+3 + +2$	**2.** $+3 + -2$	**3.** $-3 + -2$
4. $-1 + +5$	**5.** $-1 + -5$	**6.** $+1 + -5$
7. $0 + -8$	**8.** $0 + +8$	**9.** $-6 + 0$
10. $+4 + -4$	**11.** $-4 + +4$	**12.** $+4 + +4$
13. $+7 + -3$	**14.** $-7 + +3$	**15.** $-7 + -3$
16. $-6 + +9$	**17.** $+6 + -9$	**18.** $-6 + -9$
19. $+10 + +12$	**20.** $+11 + -11$	**21.** $-17 + +14$
▲**22.** $-19 + -13$	▲**23.** $+18 + -19$	▲**24.** $-16 + +11$
■**25.** $+15 + +14$	■**26.** $-18 + -12$	■**27.** $-11 + +19$
●**28.** $+17 + -17$	●**29.** $-19 + +14$	●**30.** $+12 + +15$
31. $-20 + -20$	**32.** $-21 + +25$	**33.** $+27 + -22$
34. $-31 + +31$	**35.** $0 + +34$	**36.** $-32 + -36$

Examples

To subtract an integer, add the opposite of the integer.

$$^+5 - {}^+2 = {}^+5 + {}^-2$$
$$= {}^+3$$

$$^-5 - {}^+2 = {}^-5 + {}^-2$$
$$= {}^-7$$

$$^+5 - {}^-2 = {}^+5 + {}^+2$$
$$= {}^+7$$

$$^-5 - {}^-2 = {}^-5 + {}^+2$$
$$= {}^-3$$

Give the difference.

1. $^+4 - {}^-6$ 2. $^-4 - {}^+6$ 3. $^+4 - {}^+6$

4. $^-7 - {}^-3$ 5. $^+7 - {}^-3$ 6. $^-7 - {}^+3$

7. $^+8 - {}^+1$ 8. $^-8 - {}^+1$ 9. $^+8 - {}^-1$

10. $^-7 - 0$ 11. $0 - {}^+7$ 12. $0 - {}^-7$

13. $^+5 - {}^+9$ 14. $^+5 - {}^-9$ 15. $^-5 - {}^+9$

16. $^-5 - {}^-5$ 17. $^+5 - {}^+5$ 18. $^-5 - {}^+5$

19. $^+12 - {}^-11$ 20. $^-13 - {}^+16$ 21. $^-14 - {}^-14$

▲22. $^-16 - {}^+19$ ▲23. $^+18 - {}^+11$ ▲24. $^+19 - {}^+14$

■25. $^+17 - {}^-10$ ■26. $^-10 - {}^+17$ ■27. $^-16 - {}^-16$

●28. $^+11 - {}^+15$ ●29. $^-15 - {}^+18$ ●30. $^+17 - {}^-14$

31. $^-23 - {}^-27$ 32. $^+25 - {}^+25$ 33. $^-28 - {}^+24$

34. $0 - {}^-34$ 35. $^-36 - {}^+32$ 36. $^+33 - {}^-38$

Examples

The product of two integers with the same signs is positive.
$$^+2 \times {}^+3 = {}^+6$$
$$^-4 \times {}^-5 = {}^+20$$

The product of two integers with different signs is negative.
$$^+6 \times {}^-3 = {}^-18$$
$$^-5 \times {}^+6 = {}^-30$$

The product of any integer and 0 is 0.
$$^+7 \times 0 = 0$$
$$0 \times {}^-4 = 0$$

Give the product.

1. $^+3 \times {}^+4$ 2. $^+3 \times {}^-4$ 3. $^-3 \times {}^-4$

4. $^-6 \times {}^+5$ 5. $^+6 \times {}^+5$ 6. $^+6 \times {}^-5$

7. $^-5 \times {}^+7$ 8. $^-5 \times {}^-7$ 9. $^+5 \times {}^+7$

10. $^+2 \times {}^-6$ 11. $^-6 \times {}^+2$ 12. $^-6 \times {}^-2$

13. $^+7 \times 0$ 14. $^-7 \times 0$ 15. 0×0

16. $^-9 \times {}^-8$ 17. $^+8 \times {}^-8$ 18. $^-6 \times {}^+5$

19. $^+4 \times {}^+6$ 20. $^-5 \times {}^+5$ 21. $^+6 \times {}^-6$

▲22. $^-5 \times {}^-9$ ▲23. $^+7 \times {}^-8$ ▲24. $^-8 \times {}^+7$

■25. $^+9 \times {}^+7$ ■26. $^-7 \times {}^-7$ ■27. $0 \times {}^+8$

●28. $^+10 \times {}^-6$ ●29. $^-10 \times {}^+6$ ●30. $^+10 \times {}^+6$

31. $^-11 \times {}^-11$ 32. $^+11 \times {}^-11$ 33. $^+11 \times {}^+11$

34. $^-13 \times {}^+14$ 35. $^+18 \times {}^+15$ 36. $^-19 \times {}^-13$

Examples

The quotient of two integers with the same signs is positive.

$$^+12 \div {}^+6 = {}^+2$$
$$^-21 \div {}^-7 = {}^+3$$

The quotient of two integers with different signs is negative.

$$^+20 \div {}^-4 = {}^-5$$
$$^-32 \div {}^+8 = {}^-4$$

The quotient of 0 divided by any nonzero integer is 0.

$$0 \div {}^-6 = 0$$
$$0 \div {}^+7 = 0$$

Give the quotient.

1. $^+12 \div {}^+3$
2. $^+12 \div {}^-3$
3. $^-12 \div {}^-3$

4. $^-8 \div {}^+2$
5. $^+8 \div {}^+2$
6. $^+8 \div {}^-2$

7. $^-16 \div {}^+4$
8. $^-16 \div {}^-4$
9. $^+16 \div {}^-4$

10. $^-18 \div {}^+6$
11. $^+18 \div {}^+6$
12. $^+18 \div {}^-6$

13. $^-25 \div {}^+5$
14. $^-25 \div {}^-5$
15. $^+25 \div {}^-5$

16. $0 \div {}^+4$
17. $0 \div {}^-4$
18. $0 \div {}^-9$

19. $^+49 \div {}^-7$
20. $^-42 \div {}^+6$
21. $^-45 \div {}^-5$

▲22. $^+36 \div {}^+6$
▲23. $^-32 \div {}^+8$
▲24. $^+35 \div {}^-7$

■25. $^-36 \div {}^-9$
■26. $^+54 \div {}^+6$
■27. $^-64 \div {}^+8$

●28. $^+72 \div {}^-9$
●29. $^-81 \div {}^-9$
●30. $^+63 \div {}^-7$

31. $^-70 \div {}^+10$
32. $^+90 \div {}^+10$
33. $^-60 \div {}^-10$

34. $^-132 \div {}^+12$
35. $^+182 \div {}^-13$
36. $^+224 \div {}^+16$

Examples

Solve.

$$x + 13 = {}^-6$$
$$x + 13 - 13 = {}^-6 - 13$$
$$x = {}^-19$$

 CHECK

$$^-19 + 13 \overset{?}{=} {}^-6$$
$$^-6 = {}^-6$$

Solve.

$$y - 17 = {}^-4$$
$$y - 17 + 17 = {}^-4 + 17$$
$$y = 13$$

 CHECK

$$13 - 17 \overset{?}{=} {}^-4$$
$$^-4 = {}^-4$$

Solve and check.

1. $x + 8 = {}^-4$
2. $g - 9 = 7$
3. $d + {}^-6 = 9$

4. $e - 10 = 6$
5. $y + {}^-12 = 13$
6. $m - 18 = {}^-5$

7. $h - {}^-20 = 11$
8. $f - 11 = 18$
9. $z + {}^-5 = {}^-16$

10. $j - 13 = {}^-15$
11. $w + {}^-19 = 20$
12. $k + 23 = 23$

13. $a + {}^-16 = 15$
14. $n - 12 = {}^-12$
15. $q + {}^-26 = 35$

▲16. $c - 25 = 11$
▲17. $p - {}^-32 = 22$
▲18. $b + 42 = 50$

■19. $^-3 = r + {}^-13$
■20. $18 = y - 11$
■21. $12 = u + 9$

●22. $^-21 = x - 23$
●23. $24 = s - {}^-17$
●24. $^-26 = z + {}^-18$

25. $^-15 = w + {}^-15$
26. $11 = m - 36$
27. $^-30 = t + 33$

28. $40 = k + {}^-41$
29. $^-36 = v - 52$
30. $40 = j - {}^-49$

Examples

Solve.

$$^-6n = 42$$

$$\frac{^-6n}{^-6} = \frac{42}{^-6}$$

$$n = {}^-7$$

Solve.

$$\frac{m}{^-3} = 8$$

$$^-3 \cdot \frac{m}{^-3} = {}^-3 \cdot 8$$

$$m = {}^-24$$

Solve and check.

1. $^-4n = 24$
2. $\frac{r}{2} = 16$
3. $9b = {}^-36$
4. $\frac{s}{^-3} = {}^-5$
5. $\frac{y}{8} = 13$
6. $^-7n = 56$
7. $\frac{x}{^-6} = 11$
8. $^-3a = {}^-36$
9. $\frac{w}{5} = {}^-12$
10. $11j = 33$
11. $\frac{v}{9} = {}^-20$
12. $10m = {}^-110$
13. $^-10c = 80$
14. $\frac{z}{^-15} = 8$
15. $\frac{t}{21} = 20$
▲16. $20c = 160$
▲17. $\frac{u}{32} = 10$
▲18. $^-12d = {}^-120$
■19. $5 = \frac{a}{^-8}$
■20. $0 = 10h$
■21. $^-5 = \frac{c}{7}$
●22. $80 = 5g$
●23. $^-7 = \frac{b}{^-9}$
●24. $90 = 15j$

Examples

Solve.

$$3n + {}^-6 = 6$$

$$3n + {}^-6 - {}^-6 = 6 - {}^-6$$

$$3n = 12$$

$$\frac{3n}{3} = \frac{12}{3}$$

$$n = 4$$

Solve.

$$\frac{m}{^-2} + 8 = 3$$

$$\frac{m}{^-2} + 8 - 8 = 3 - 8$$

$$\frac{m}{^-2} = {}^-5$$

$$^-2 \cdot \frac{m}{^-2} = {}^-2 \cdot {}^-5$$

$$m = 10$$

Solve and check.

1. $^-5a + 4 = 19$
2. $\frac{j}{3} + 4 = {}^-10$
3. $\frac{p}{6} - 11 = {}^-9$
4. $7q + 15 = 36$
5. $4s - {}^-12 = 0$
6. $^-9c + 15 = {}^-12$
7. $\frac{g}{^-10} + {}^-7 = {}^-3$
8. $\frac{f}{7} - 14 = {}^-14$
9. $20d + 10 = 70$
10. $^-12t - {}^-3 = 15$
▲11. $\frac{n}{9} + {}^-8 = {}^-11$
▲12. $\frac{h}{16} - {}^-6 = {}^-7$
■13. $2 = 3y + 8$
■14. $^-4 = \frac{s}{^-3} - 2$
●15. $0 = \frac{t}{9} + {}^-7$
●16. $^-15 = {}^-5r + 10$
17. $1 = \frac{w}{^-4} + {}^-8$
18. $5 = {}^-3h - 4$
19. $0 = {}^-4j + 8$
20. $2 = \frac{m}{4} + 6$

Examples

Write using exponents.

$$4 \cdot 4 \cdot 4 = 4^3$$

Write using exponents for the variables.

$$a \cdot a \cdot a \cdot b \cdot b = a^3 b^2$$
$$2 \cdot c \cdot d \cdot 3 \cdot c \cdot d \cdot d = 6c^2 d^3$$

Write using exponents.

1. $2 \cdot 2 \cdot 2$

2. $5 \cdot 5 \cdot 5 \cdot 5$

3. $3 \cdot 3 \cdot 4 \cdot 4 \cdot 4$

4. $6 \cdot 6 \cdot 6 \cdot 6 \cdot 10 \cdot 10 \cdot 10$

5. $2 \cdot 7 \cdot 7 \cdot 7 \cdot 7 \cdot 7 \cdot 7$

6. $3 \cdot 3 \cdot 3 \cdot 3 \cdot 3 \cdot 8 \cdot 8 \cdot 8 \cdot 8$

Write using exponents for the variables.

7. $x \cdot x \cdot y$

8. $m \cdot m \cdot n \cdot n \cdot n$

9. $a \cdot a \cdot b \cdot b \cdot b$

10. $r \cdot s \cdot s \cdot s \cdot s$

11. $4 \cdot d \cdot 3 \cdot d$

12. $2 \cdot j \cdot j \cdot 7 \cdot j$

13. $9 \cdot 3 \cdot m \cdot n \cdot n$

14. $x \cdot 4 \cdot x \cdot 3 \cdot x \cdot 2$

▲15. $3 \cdot a \cdot a \cdot c \cdot c$

▲16. $3 \cdot a \cdot c \cdot c \cdot c$

■17. $8 \cdot x \cdot x \cdot 2 \cdot y$

■18. $8 \cdot x \cdot x \cdot 2 \cdot y \cdot y$

●19. $p \cdot q \cdot 3 \cdot p \cdot 7 \cdot q$

●20. $y \cdot z \cdot z \cdot 5 \cdot y \cdot 3$

21. $c \cdot d \cdot 4 \cdot d \cdot c \cdot d$

22. $m \cdot 3 \cdot n \cdot 2 \cdot m \cdot n \cdot n$

23. $9 \cdot y \cdot z \cdot 3 \cdot y \cdot z \cdot y$

24. $j \cdot k \cdot 3 \cdot k \cdot k \cdot 4 \cdot j \cdot k$

Examples

Simplify. $5^2 \cdot 5^4$

$$5^2 \cdot 5^4 = 5^{(2+4)} = 5^6$$

Simplify. $\dfrac{x^7}{x^3}$

$$\frac{x^7}{x^3} = x^{(7-3)} = x^4$$

Simplify.

▲ 1. $3^2 \cdot 3^7$

■ 2. $7^3 \cdot 7^5$

● 3. $2^4 \cdot 2^6$

▲ 4. $5^a \cdot 5^b$

■ 5. $c^3 \cdot c^7$

● 6. $d^2 \cdot d^{10}$

▲ 7. $\dfrac{9^7}{9^2}$

■ 8. $\dfrac{3^8}{3^4}$

● 9. $\dfrac{7^{10}}{7^2}$

▲10. $\dfrac{15^a}{15^b}$

■11. $\dfrac{x^6}{x^4}$

●12. $\dfrac{z^8}{z^2}$

Examples

Large number	Small number
830,000	**0.000381**

Locate the decimal point to get a number between 1 and 10.

830000   0.000381

Count the number of places the decimal point was moved to get a number between 1 and 10.

830000   0.000381

5 places to left   *4 places to right*

Write the number as a product of a number between 1 and 10 and a power of 10.

8.3×10^{5} 3.81×10^{-4}

Write in scientific notation.

1. 4100
2. 600
3. 57,000
4. 9300
5. 94,000
6. 536,000
7. 7,420,000
8. 390,000
9. 8000
10. 91,000
11. 6,800,000
12. 43,200,000
13. 596,000
14. 8,160,000
15. 16,000,000
16. 420
17. 40,200
18. 4,200,000
▲19. 0.0035
▲20. 0.057
▲21. 0.82
■22. 0.00017
■23. 0.0039
■24. 0.00006
●25. 0.0534
●26. 0.000174
●27. 0.00018
28. 0.000039
29. 0.000751
30. 0.0000048
31. 0.000023
32. 0.0135
33. 0.0000712
34. 0.000042
35. 0.42
36. 0.00402

Examples

Give the prime factorization.

$20 = 4 \cdot 5$

Not prime! Factor again.

$= 2 \cdot 2 \cdot 5$

Give the algebraic factorization.

$12a^{2}b^{3} =$
$2 \cdot 2 \cdot 3 \cdot a \cdot a \cdot b \cdot b \cdot b$

Give the prime factorization.

1. 10
2. 15
3. 9
4. 18
5. 8
6. 21
7. 30
8. 25
9. 39
10. 45
11. 48
12. 56
13. 72
14. 64
15. 100
16. 120

Give the algebraic factorization.

17. $3a^{2}$
18. $4y^{2}$
19. $11t^{3}$
20. $10r^{4}$
▲21. $16x^{2}$
▲22. $20z^{3}$
▲23. $18n^{4}$
▲24. $32t^{3}$
■25. $10ab^{2}$
■26. $27rs^{2}$
■27. $30st^{2}$
■28. $15mn^{3}$
●29. $35y^{2}z^{2}$
●30. $42x^{3}y^{2}$
●31. $65m^{4}n^{2}$
●32. $84c^{2}d^{3}$

Examples

GCF of 12, 18 = ?

$$12 = 2 \cdot 2 \cdot 3$$
$$18 = 2 \cdot 3 \cdot 3$$
$$GCF = 2 \cdot 3$$
$$= 6$$

GCF of $4r^2s^3$, $10rs^2$ = ?

$$4r^2s^3 = 2 \cdot 2 \cdot r \cdot r \cdot s \cdot s \cdot s$$
$$10rs^2 = 2 \cdot 5 \cdot r \cdot s \cdot s$$
$$GCF = 2 \cdot r \cdot s \cdot s$$
$$= 2rs^2$$

LCM of 12, 18 = ?

$$12 = 2 \cdot 2 \cdot 3$$
$$18 = 2 \cdot 3 \cdot 3$$
$$LCM = 2 \cdot 2 \cdot 3 \cdot 3$$
$$= 36$$

LCM of $4r^2s^3$, $10rs^2$ = ?

$$4r^2s^3 = 2 \cdot 2 \cdot r \cdot r \cdot s \cdot s \cdot s$$
$$10rs^2 = 2 \cdot 5 \cdot r \cdot s \cdot s$$
$$LCM = 2 \cdot 2 \cdot 5 \cdot r \cdot r \cdot s \cdot$$
$$s \cdot s$$
$$= 20r^2s^3$$

Give the greatest common factor (GCF).

1. 6, 18
2. 4, 28
3. 16, 30
4. 20, 35
5. 27, 45
6. 40, 30
7. 27, 64
8. 54, 90
9. 45, 75
▲10. $3, 6a$
▲11. $12, 9d$
▲12. $5d, 15$
■13. $j, 7j$
■14. $12e, 15e$
■15. $15w, 11w$
●16. $5x, 10x^2$
●17. $3yz, z^2$
●18. $10a^2b, 15ab^2$

Give the least common multiple (LCM).

19. 9, 12
20. 8, 12
21. 16, 24
22. 20, 30
23. 25, 20
24. 36, 24
25. 21, 15
26. 30, 24
27. 25, 30
▲28. $10, 5x$
▲29. $6k, 9$
▲30. $15, 20t$
■31. $7j, j$
■32. $m, 9m$
■33. $8r, r$
●34. $j^2, 3j$
●35. $8c, 4c^2$
●36. $5m^2n, mn$

Example

Complete to get an equivalent fraction.

$$\frac{x}{2y} = \frac{?}{6yz} \leftarrow \boxed{2y \cdot 3z}$$

Multiply the numerator by $3z$.

$$\frac{x}{2y} = \frac{3xz}{6yz}$$

Complete to get an equivalent fraction.

1. $\dfrac{3}{5} = \dfrac{?}{20}$
2. $\dfrac{5}{8} = \dfrac{?}{24}$
3. $\dfrac{3}{8} = \dfrac{?}{40}$
4. $\dfrac{4}{3} = \dfrac{?}{12}$

5. $\dfrac{7}{2} = \dfrac{?}{22}$
6. $\dfrac{5}{8} = \dfrac{?}{32}$
7. $\dfrac{7}{10} = \dfrac{?}{60}$
8. $\dfrac{6}{5} = \dfrac{?}{30}$

▲ 9. $\dfrac{2}{c} = \dfrac{?}{4c}$
▲10. $\dfrac{5}{d} = \dfrac{?}{6d}$
▲11. $\dfrac{a}{b} = \dfrac{?}{5b}$
▲12. $\dfrac{12}{g} = \dfrac{?}{4g}$

■13. $\dfrac{f}{g} = \dfrac{?}{g^2}$
■14. $\dfrac{r}{s^2} = \dfrac{?}{3s^2}$
■15. $\dfrac{2a}{3b} = \dfrac{?}{6ab}$
■16. $\dfrac{5d}{2e} = \dfrac{?}{8de}$

17. $\dfrac{11}{3m} = \dfrac{?}{3mn^2}$
18. $\dfrac{5a}{6b} = \dfrac{?}{18ab}$
19. $\dfrac{12e}{5f} = \dfrac{?}{10ef}$
20. $\dfrac{6g}{11h} = \dfrac{?}{11gh^2}$

Examples

Write in lowest terms.

$$\frac{18}{24} = \frac{\overset{1}{\cancel{2}} \cdot \overset{1}{\cancel{3}} \cdot 3}{\cancel{2} \cdot 2 \cdot 2 \cdot \cancel{3}}$$

$$= \frac{3}{4}$$

$$\frac{9x^2}{6xy} = \frac{\overset{1}{\cancel{3}} \cdot 3 \cdot \overset{1}{\cancel{x}} \cdot x}{2 \cdot \cancel{3} \cdot \cancel{x} \cdot y}$$

$$= \frac{3x}{2y}$$

Write in lowest terms.

1. $\dfrac{3}{6}$ 2. $\dfrac{3}{9}$ 3. $\dfrac{2}{8}$ 4. $\dfrac{6}{9}$

5. $\dfrac{9}{6}$ 6. $\dfrac{14}{16}$ 7. $\dfrac{9}{12}$ 8. $\dfrac{10}{12}$

9. $\dfrac{8}{10}$ 10. $\dfrac{5}{15}$ 11. $\dfrac{6}{4}$ 12. $\dfrac{7}{14}$

13. $\dfrac{8}{16a}$ 14. $\dfrac{15}{6b}$ 15. $\dfrac{2d}{10}$ 16. $\dfrac{16f}{12}$

▲17. $\dfrac{cd}{6d}$ ▲18. $\dfrac{x}{xy}$ ▲19. $\dfrac{6a}{16b}$ ▲20. $\dfrac{4m}{12n}$

■21. $\dfrac{3y}{12xy}$ ■22. $\dfrac{4m}{6n}$ ■23. $\dfrac{3r^2}{18r}$ ■24. $\dfrac{15uv}{18u^2}$

●25. $\dfrac{10r}{6s^2}$ ●26. $\dfrac{15e^2}{20de}$ ●27. $\dfrac{5yz^2}{10z}$ ●28. $\dfrac{m^2n}{15n^2}$

Example

The least common denominator (LCD) is the least common multiple of the denominators.

LCD of $\dfrac{8}{3x}, \dfrac{1}{4x} = ?$

$3x = 3 \cdot x$
$4x = 2 \cdot 2 \cdot x$
$\text{LCM} = 2 \cdot 2 \cdot 3 \cdot x$
$\qquad = 12x$

So the LCD of $\dfrac{8}{3x}$ and $\dfrac{1}{4x}$ is $12x$

Find the least common denominator.

1. $\dfrac{1}{6}, \dfrac{1}{5}$ 2. $\dfrac{3}{4}, \dfrac{1}{2}$ 3. $\dfrac{1}{5}, \dfrac{2}{9}$

4. $\dfrac{1}{10}, \dfrac{2}{5}$ 5. $\dfrac{3}{20}, \dfrac{1}{10}$ 6. $\dfrac{1}{4}, \dfrac{1}{6}$

7. $\dfrac{5}{6}, \dfrac{3}{8}$ 8. $\dfrac{1}{6}, \dfrac{1}{8}$ 9. $\dfrac{1}{8}, \dfrac{4}{3}$

▲10. $\dfrac{7}{8}, \dfrac{1}{12}$ ▲11. $\dfrac{1}{6}, \dfrac{3}{4}$ ▲12. $\dfrac{1}{5}, \dfrac{3}{7}$

■13. $\dfrac{2}{5a}, \dfrac{1}{4a}$ ■14. $\dfrac{1}{10b}, \dfrac{4}{5b^2}$ ■15. $\dfrac{5}{9c}, \dfrac{1}{2c}$

●16. $\dfrac{3}{10}, \dfrac{1}{4e}$ ●17. $\dfrac{1}{6f}, \dfrac{4}{5g}$ ●18. $\dfrac{7}{6h}, \dfrac{5}{9hi}$

19. $\dfrac{5}{6k^2}, \dfrac{1}{15k^3}$ 20. $\dfrac{1}{7m}, \dfrac{1}{3m^2}$ 21. $\dfrac{2}{5np}, \dfrac{1}{8n}$

22. $\dfrac{1}{4st}, \dfrac{1}{7st^2}$ 23. $\dfrac{1}{9u^2v}, \dfrac{2}{3v^2}$ 24. $\dfrac{1}{3w^3}, \dfrac{2}{9wx}$

Example

$< , >$, or $=$?

$$\frac{2}{3} \diamondsuit \frac{3}{4}$$

Find the LCD.

$$\frac{2}{3} \diamondsuit \frac{3}{4}$$

$$\boxed{12}$$

Write equivalent fractions and compare.

$$\frac{8}{12} \diamondsuit \frac{9}{12}$$

$$\frac{2}{3} < \frac{3}{4}$$

$< , >$, or $=$?

1. $\frac{1}{4} \diamondsuit \frac{3}{4}$ 2. $\frac{3}{5} \diamondsuit \frac{2}{5}$ 3. $\frac{3}{7} \diamondsuit \frac{4}{7}$ 4. $\frac{0}{8} \diamondsuit \frac{5}{8}$

5. $\frac{5}{4} \diamondsuit \frac{4}{4}$ 6. $\frac{7}{5} \diamondsuit \frac{9}{5}$ 7. $\frac{5}{8} \diamondsuit \frac{7}{8}$ 8. $\frac{7}{3} \diamondsuit \frac{5}{3}$

▲ 9. $\frac{2}{3} \diamondsuit \frac{4}{6}$ ▲10. $\frac{5}{4} \diamondsuit \frac{3}{2}$ ▲11. $\frac{2}{7} \diamondsuit \frac{1}{3}$ ▲12. $\frac{3}{4} \diamondsuit \frac{6}{8}$

■13. $\frac{1}{8} \diamondsuit \frac{1}{6}$ ■14. $\frac{3}{8} \diamondsuit \frac{1}{4}$ ■15. $\frac{3}{10} \diamondsuit \frac{1}{3}$ ■16. $\frac{1}{4} \diamondsuit \frac{2}{5}$

●17. $\frac{1}{4} \diamondsuit \frac{1}{3}$ ●18. $\frac{2}{9} \diamondsuit \frac{3}{4}$ ●19. $\frac{2}{3} \diamondsuit \frac{3}{4}$ ●20. $\frac{5}{8} \diamondsuit \frac{4}{7}$

21. $\frac{3}{5} \diamondsuit \frac{9}{15}$ 22. $\frac{3}{4} \diamondsuit \frac{5}{6}$ 23. $\frac{8}{9} \diamondsuit \frac{7}{8}$ 24. $\frac{1}{3} \diamondsuit \frac{5}{12}$

25. $\frac{7}{8} \diamondsuit \frac{5}{6}$ 26. $\frac{2}{3} \diamondsuit \frac{7}{10}$ 27. $\frac{8}{12} \diamondsuit \frac{2}{3}$ 28. $\frac{9}{2} \diamondsuit \frac{9}{4}$

Examples

$$4 = \frac{?}{3}$$

Write the whole number over 1 and multiply both numerator and denominator by 3.

$$\frac{4}{1} = \frac{12}{3}$$

$$2\frac{3}{4} = ?$$

To change a mixed number to a fraction, multiply the denominator by the whole number and add the numerator.

$$2\frac{3}{4} = \frac{11}{4}$$

Change to thirds.

1. 2   2. 1   3. 4   4. 5

▲ 5. 3   ▲ 6. 8   ▲ 7. 10   ▲ 8. 6

Change to fourths.

■ 9. 3   ■10. 1   ■11. 4   ■12. 2

●13. 7   ●14. 9   ●15. 10   ●16. 8

Change to a fraction.

17. $1\frac{1}{3}$ 18. $2\frac{1}{4}$ 19. $2\frac{1}{2}$ 20. $1\frac{1}{5}$

▲21. $1\frac{2}{3}$ ▲22. $2\frac{1}{3}$ ▲23. $1\frac{1}{4}$ ▲24. $3\frac{1}{4}$

■25. $1\frac{1}{2}$ ■26. $2\frac{7}{8}$ ■27. $3\frac{3}{8}$ ■28. $2\frac{5}{6}$

●29. $5\frac{1}{4}$ ●30. $2\frac{2}{3}$ ●31. $3\frac{1}{2}$ ●32. $4\frac{5}{8}$

33. $4\frac{4}{5}$ 34. $3\frac{1}{3}$ 35. $5\frac{3}{10}$ 36. $4\frac{1}{5}$

Examples

To change a fraction to a whole number or mixed number, divide the numerator by the denominator.

$$\frac{18}{3} = 6$$

$$\frac{19}{4} = 4\frac{3}{4}$$

$$\begin{array}{r} 4\overline{)19} \\ -16 \\ \hline 3 \end{array}$$

There are 3 fourths left over.

Change to a whole number.

▲ 1. $\frac{4}{2}$ ▲ 2. $\frac{40}{5}$ ▲ 3. $\frac{10}{2}$ ▲ 4. $\frac{9}{3}$

■ 5. $\frac{25}{5}$ ■ 6. $\frac{8}{2}$ ■ 7. $\frac{32}{4}$ ■ 8. $\frac{30}{3}$

● 9. $\frac{18}{3}$ ●10. $\frac{35}{5}$ ●11. $\frac{16}{4}$ ●12. $\frac{50}{5}$

Change to a mixed number.

▲13. $\frac{3}{2}$ ▲14. $\frac{5}{4}$ ▲15. $\frac{5}{3}$ ▲16. $\frac{9}{2}$

■17. $\frac{11}{4}$ ■18. $\frac{4}{3}$ ■19. $\frac{13}{5}$ ■20. $\frac{7}{6}$

●21. $\frac{16}{3}$ ●22. $\frac{13}{10}$ ●23. $\frac{5}{2}$ ●24. $\frac{8}{3}$

25. $\frac{23}{10}$ 26. $\frac{9}{5}$ 27. $\frac{13}{12}$ 28. $\frac{19}{6}$

Examples

Write in simplest form.

$$\frac{6}{9} = \frac{2}{3}$$

$$3\frac{4}{6} = 3\frac{2}{3}$$

$$\frac{18}{3} = 6$$

$$\frac{14}{4} = 3\frac{2}{4} = 3\frac{1}{2}$$

$$\frac{5a^2}{15ab} = \frac{a}{3b}$$

$$\frac{12cd}{8c^2d} = \frac{3}{2c}$$

Write in simplest form.

1. $\frac{10}{12}$ 2. $\frac{6}{15}$ 3. $\frac{4}{3}$ 4. $\frac{15}{5}$

▲ 5. $\frac{9}{2}$ ▲ 6. $1\frac{8}{10}$ ▲ 7. $1\frac{3}{6}$ ▲ 8. $2\frac{4}{6}$

■ 9. $1\frac{2}{4}$ ■10. $\frac{10}{2}$ ■11. $\frac{11}{3}$ ■12. $3\frac{6}{16}$

●13. $\frac{23}{4}$ ●14. $4\frac{4}{10}$ ●15. $5\frac{9}{24}$ ●16. $\frac{20}{24}$

▲17. $\frac{36x}{6}$ ▲18. $\frac{17y^2}{2y}$ ▲19. $\frac{25z}{5z^2}$ ▲20. $\frac{13r}{3r^2}$

■21. $\frac{18m^2}{5m}$ ■22. $\frac{24r^2}{3rs}$ ■23. $\frac{12y^2}{15y^2}$ ■24. $\frac{9ab}{12b}$

●25. $\frac{10st}{12t^2}$ ●26. $\frac{19c^2d}{4cd}$ ●27. $\frac{12a^2b}{16b^2}$ ●28. $\frac{17jk}{5j^2k^2}$

Examples

Write $\dfrac{2}{3}$ as a decimal.

To change a fraction to a decimal, divide the numerator by the denominator.

$$\begin{array}{r} 0.66\frac{2}{3} \\ 3\overline{)2.00} \\ -18 \\ \hline 20 \\ -18 \\ \hline 2 \end{array}$$

Write 2.8 as a fraction in simplest form.

Read as a decimal.   2 and 8 tenths

Write as a mixed number. $2\dfrac{8}{10}$

Write in simplest form. $2\dfrac{8}{10} = 2\dfrac{4}{5}$

Change to a decimal.

1. $\dfrac{1}{4}$ **2.** $\dfrac{1}{2}$ **3.** $\dfrac{3}{4}$ **4.** $\dfrac{1}{8}$

▲ **5.** $\dfrac{5}{8}$ ▲ **6.** $\dfrac{2}{5}$ ▲ **7.** $\dfrac{1}{3}$ ▲ **8.** $\dfrac{5}{6}$

■ **9.** $1\dfrac{1}{2}$ ■**10.** $1\dfrac{3}{4}$ ■**11.** $2\dfrac{3}{8}$ ■**12.** $2\dfrac{4}{5}$

●**13.** $1\dfrac{2}{3}$ ●**14.** $2\dfrac{1}{6}$ ●**15.** $2\dfrac{5}{12}$ ●**16.** $3\dfrac{5}{16}$

Change to a fraction in simplest form.

17. 0.6   **18.** 0.25   **19.** 0.32   **20.** 0.5

▲**21.** 0.75   ▲**22.** 0.08   ▲**23.** 0.05   ▲**24.** 0.35

■**25.** 0.375   ■**26.** 0.625   ■**27.** 1.4   ■**28.** 1.05

●**29.** 1.875   ●**30.** 2.75   ●**31.** 3.5   ●**32.** 2.24

Examples

$$\dfrac{9}{c} + \dfrac{4}{c} = \dfrac{9+4}{c}$$
$$= \dfrac{13}{c}$$

$$\dfrac{11}{c} - \dfrac{3}{c} = \dfrac{11-3}{c}$$
$$= \dfrac{8}{c}$$

Give the sum in simplest form.

▲ **1.** $\dfrac{1}{9} + \dfrac{1}{9}$ ▲ **2.** $\dfrac{2}{7} + \dfrac{2}{7}$ ▲ **3.** $\dfrac{2}{9} + \dfrac{1}{9}$ ▲ **4.** $\dfrac{1}{8} + \dfrac{3}{8}$

■ **5.** $\dfrac{r}{t} + \dfrac{s}{t}$ ■ **6.** $\dfrac{a}{c} + \dfrac{b}{c}$ ■ **7.** $\dfrac{3x}{y} + \dfrac{3}{y}$ ■ **8.** $\dfrac{4g}{h} + \dfrac{5}{h}$

Give the difference in simplest form.

▲ **9.** $\dfrac{4}{5} - \dfrac{1}{5}$ ▲**10.** $\dfrac{5}{6} - \dfrac{1}{6}$ ▲**11.** $\dfrac{5}{4} - \dfrac{3}{4}$ ▲**12.** $\dfrac{1}{6} - \dfrac{1}{6}$

■**13.** $\dfrac{p}{r} - \dfrac{q}{r}$ ■**14.** $\dfrac{w}{y} - \dfrac{x}{y}$ ■**15.** $\dfrac{4a}{b} - \dfrac{3}{b}$ ■**16.** $\dfrac{7c}{d} - \dfrac{3}{d}$

Skill 6-7 (Use after page 168.)

Examples

$$\frac{3}{8} + \frac{1}{6} = \frac{9}{24} + \frac{4}{24}$$
$$= \frac{13}{24}$$

$$\frac{x}{3} + \frac{y}{4} = \frac{4x}{12} + \frac{3y}{12}$$
$$= \frac{4x + 3y}{12}$$

$$\frac{3}{8} - \frac{1}{3} = \frac{9}{24} - \frac{8}{24}$$
$$= \frac{1}{24}$$

$$\frac{a}{6} - \frac{b}{9} = \frac{3a}{18} - \frac{2b}{18}$$
$$= \frac{3a - 2b}{18}$$

Give the sum in simplest form.

1. $\frac{1}{2} + \frac{1}{4}$ 2. $\frac{1}{6} + \frac{2}{3}$ 3. $\frac{3}{8} + \frac{1}{4}$ 4. $\frac{1}{2} + \frac{5}{8}$

▲ 5. $\frac{1}{3} + \frac{1}{6}$ ▲ 6. $\frac{4}{5} + \frac{3}{10}$ ▲ 7. $\frac{1}{5} + \frac{3}{10}$ ▲ 8. $\frac{2}{3} + \frac{3}{4}$

■ 9. $\frac{d}{3} + \frac{3}{7}$ ■10. $\frac{a}{9} + \frac{1}{6}$ ■11. $\frac{3k}{8} + \frac{3}{4}$ ■12. $\frac{s}{5} + \frac{t}{4}$

●13. $\frac{2a}{3} + \frac{b}{9}$ ●14. $\frac{4c}{3} + \frac{d}{2}$ ●15. $\frac{2r}{3} + \frac{3s}{8}$ ●16. $\frac{5j}{2} + \frac{7k}{6}$

Give the difference in simplest form.

17. $\frac{1}{3} - \frac{1}{4}$ 18. $\frac{1}{2} - \frac{1}{4}$ 19. $\frac{3}{4} - \frac{1}{2}$ 20. $\frac{5}{8} - \frac{1}{4}$

▲21. $\frac{3}{4} - \frac{0}{2}$ ▲22. $\frac{2}{3} - \frac{1}{2}$ ▲23. $\frac{1}{4} - \frac{1}{8}$ ▲24. $\frac{1}{3} - \frac{1}{8}$

■25. $\frac{t}{4} - \frac{3}{8}$ ■26. $\frac{a}{8} - \frac{2}{3}$ ■27. $\frac{3k}{2} - \frac{1}{3}$ ■28. $\frac{s}{3} - \frac{t}{8}$

●29. $\frac{5c}{9} - \frac{d}{6}$ ●30. $\frac{2a}{3} - \frac{b}{6}$ ●31. $\frac{5j}{8} - \frac{2k}{5}$ ●32. $\frac{7m}{10} - \frac{2n}{5}$

Skill 6-8 (Use after page 170.)

Examples

Add. $\frac{2}{5} + \frac{3}{x}$

$$\frac{2 \cdot x}{5 \cdot x} + \frac{5 \cdot 3}{5 \cdot x} = \frac{2x + 15}{5x}$$

Subtract. $\frac{b}{a} - \frac{2}{ab}$

$$\frac{b \cdot b}{a \cdot b} - \frac{2}{ab} = \frac{b^2 - 2}{ab}$$

Add.

▲ 1. $\frac{2}{a} + \frac{a}{3}$ ▲ 2. $\frac{3}{b} + \frac{4}{c}$

■ 3. $\frac{c}{d^2} + \frac{1}{cd}$ ■ 4. $\frac{z}{2} + \frac{3}{x}$

● 5. $\frac{2}{ab} + \frac{3}{a}$ ● 6. $\frac{2}{k^2} + \frac{5}{k}$

Subtract.

▲ 7. $\frac{a}{b} - \frac{2}{5}$ ▲ 8. $\frac{b}{a} - \frac{b}{c}$

■ 9. $\frac{3}{bc} - \frac{1}{c^2}$ ■10. $\frac{2}{x} - \frac{x}{c}$

●11. $\frac{4}{m} - \frac{1}{mn}$ ●12. $\frac{3}{r} - \frac{2}{r^3}$

Examples

$$3\frac{1}{3} + 2\frac{1}{4} = ?$$

$$3\frac{1}{3} = \quad 3\frac{4}{12}$$

$$+2\frac{1}{4} = +2\frac{3}{12}$$

$$\overline{\qquad\qquad 5\frac{7}{12}}$$

$$8\frac{5}{6} - 2\frac{1}{12} = ?$$

$$8\frac{5}{6} = \quad 8\frac{10}{12}$$

$$-2\frac{1}{12} = -2\frac{1}{12}$$

$$\overline{\qquad\qquad 6\frac{9}{12} = 6\frac{3}{4}}$$

Give each sum in simplest form.

1. $2\frac{1}{4} + 3\frac{1}{2}$
2. $3\frac{1}{8} + 3\frac{3}{4}$
3. $1\frac{3}{8} + 4\frac{1}{4}$

▲ 4. $5\frac{1}{3} + 2\frac{1}{4}$
▲ 5. $8\frac{5}{9} + 2\frac{1}{6}$
▲ 6. $9\frac{1}{4} + 3\frac{2}{5}$

■ 7. $6\frac{1}{9} + 8\frac{2}{3}$
■ 8. $4\frac{1}{6} + 5\frac{1}{2}$
■ 9. $7\frac{3}{8} + 4\frac{1}{6}$

●10. $8\frac{1}{6} + 5\frac{2}{3}$
●11. $7\frac{1}{8} + 8\frac{5}{16}$
●12. $8\frac{1}{4} + 6\frac{5}{12}$

Give each difference in simplest form.

13. $3\frac{5}{8} - 2\frac{1}{4}$
14. $6\frac{3}{4} - 4\frac{3}{8}$
15. $4\frac{1}{2} - 2\frac{1}{4}$

▲16. $5\frac{5}{9} - 3\frac{1}{6}$
▲17. $8\frac{1}{3} - 4\frac{1}{6}$
▲18. $5\frac{3}{4} - 1\frac{1}{2}$

■19. $9\frac{5}{8} - 3\frac{2}{5}$
■20. $5\frac{7}{8} - 3\frac{2}{3}$
■21. $7\frac{2}{3} - 6\frac{5}{8}$

●22. $5\frac{7}{10} - 1\frac{2}{5}$
●23. $7\frac{1}{2} - 2\frac{3}{8}$
●24. $3\frac{9}{10} - 1\frac{2}{3}$

Examples

$$2\frac{5}{8} + 1\frac{2}{3} = ?$$

$$2\frac{5}{8} = \quad 2\frac{15}{24}$$

$$+1\frac{2}{3} = +1\frac{16}{24}$$

$$\overline{\qquad\qquad 3\frac{31}{24} = 4\frac{7}{24}}$$

$$6\frac{1}{4} - 2\frac{2}{3} = ?$$

$$6\frac{1}{4} = \quad 6\frac{\overset{5}{\cancel{3}}\overset{15}{12}}{}$$

$$-2\frac{2}{3} = -2\frac{8}{12}$$

$$\overline{\qquad\qquad 3\frac{7}{12}}$$

Give each sum in simplest form.

1. $4\frac{3}{4} + 3\frac{1}{2}$
2. $5\frac{3}{8} + 1\frac{3}{4}$
3. $8\frac{2}{3} + 3\frac{5}{6}$

▲ 4. $6\frac{7}{8} + 6\frac{1}{4}$
▲ 5. $9\frac{2}{5} + 4\frac{1}{2}$
▲ 6. $7\frac{5}{8} + 2\frac{2}{3}$

■ 7. $8\frac{3}{4} + 7\frac{2}{3}$
■ 8. $6\frac{1}{8} + 9\frac{2}{5}$
■ 9. $5\frac{2}{3} + 5\frac{3}{8}$

●10. $4\frac{5}{6} + 8\frac{5}{8}$
●11. $3\frac{3}{4} + 7\frac{4}{5}$
●12. $9\frac{2}{3} + 8\frac{7}{10}$

Give each difference in simplest form.

13. $4\frac{1}{2} - 3\frac{3}{4}$
14. $5\frac{5}{8} - 1\frac{3}{4}$
15. $7\frac{3}{4} - 4\frac{1}{2}$

▲16. $8\frac{2}{5} - 2\frac{1}{2}$
▲17. $9\frac{1}{3} - 6\frac{3}{4}$
▲18. $6\frac{1}{4} - 1\frac{3}{8}$

■19. $8\frac{1}{8} - 3\frac{1}{3}$
■20. $5\frac{2}{3} - 4\frac{7}{8}$
■21. $4\frac{1}{5} - 3\frac{2}{3}$

●22. $9\frac{5}{9} - 3\frac{5}{6}$
●23. $6\frac{2}{3} - 5\frac{3}{4}$
●24. $7\frac{5}{16} - 2\frac{7}{8}$

Example

$$\frac{a}{b} \cdot \frac{b}{c} = ?$$

Multiply numerators and denominators. Then simplify.

$$\frac{a}{b} \cdot \frac{b}{c} = ?$$

$$\frac{a}{b} \cdot \frac{b}{c} = \frac{a \cdot b}{b \cdot c}$$

$$= \frac{a}{c}$$

Or use canceling.

$$\frac{a}{b} \cdot \frac{\overset{1}{b}}{c} = \frac{a}{c}$$

Give the product in simplest form.

1. $\dfrac{4}{3} \cdot \dfrac{3}{2}$ 2. $\dfrac{3}{4} \cdot \dfrac{16}{3}$ 3. $\dfrac{1}{3} \cdot \dfrac{3}{8}$

4. $\dfrac{7}{4} \cdot \dfrac{4}{3}$ 5. $\dfrac{1}{3} \cdot \dfrac{4}{5}$ 6. $\dfrac{1}{2} \cdot \dfrac{4}{9}$

7. $\dfrac{5}{8} \cdot \dfrac{4}{5}$ 8. $\dfrac{1}{4} \cdot \dfrac{8}{5}$ 9. $\dfrac{3}{2} \cdot \dfrac{2}{3}$

10. $\dfrac{m}{n} \cdot \dfrac{3}{m}$ 11. $r \cdot \dfrac{s}{t}$ 12. $\dfrac{a}{b} \cdot \dfrac{c}{d}$

▲13. $\dfrac{p}{q} \cdot r$ ▲14. $\dfrac{m}{n} \cdot \dfrac{5}{n}$ ▲15. $\dfrac{y}{3} \cdot \dfrac{1}{y}$

■16. $\dfrac{7}{p} \cdot \dfrac{q}{14}$ ■17. $\dfrac{8}{c} \cdot \dfrac{b}{8}$ ■18. $\dfrac{c}{3} \cdot \dfrac{12}{d}$

●19. $\dfrac{r}{s} \cdot \dfrac{s}{r}$ ●20. $\dfrac{4}{9} \cdot \dfrac{3}{2d}$ ●21. $\dfrac{8v}{3} \cdot \dfrac{1}{8}$

22. $\dfrac{r}{t} \cdot \dfrac{s}{r^2}$ 23. $\dfrac{p^2}{q} \cdot \dfrac{q}{p}$ 24. $\dfrac{m}{n} \cdot \dfrac{n^2}{5}$

Example

$$2\frac{1}{2} \cdot 1\frac{2}{3} = ?$$

Change to fractions.

$$2\frac{1}{2} \cdot 1\frac{2}{3} = \frac{5}{2} \cdot \frac{5}{3}$$

Multiply.

$$2\frac{1}{2} \cdot 1\frac{2}{3} = \frac{5}{2} \cdot \frac{5}{3}$$

$$= \frac{25}{6}$$

$$= 4\frac{1}{6}$$

Give the product in simplest form.

1. $2 \cdot 1\frac{1}{2}$ 2. $1\frac{1}{2} \cdot 1\frac{1}{3}$ 3. $2\frac{2}{3} \cdot 1\frac{1}{4}$

4. $1\frac{3}{4} \cdot 1\frac{3}{4}$ 5. $3 \cdot 2\frac{1}{3}$ 6. $2\frac{1}{3} \cdot 2$

7. $2\frac{2}{5} \cdot 3$ 8. $1\frac{5}{6} \cdot 2\frac{1}{3}$ 9. $3\frac{1}{4} \cdot 3\frac{1}{4}$

▲10. $4\frac{1}{6} \cdot 2\frac{1}{3}$ ▲11. $2\frac{2}{3} \cdot 2\frac{1}{2}$ ▲12. $3 \cdot 4\frac{1}{2}$

■13. $2 \cdot 1\frac{2}{3}$ ■14. $1\frac{1}{2} \cdot 2\frac{1}{2}$ ■15. $3\frac{3}{4} \cdot 2$

●16. $1\frac{3}{8} \cdot 2\frac{1}{2}$ ●17. $3\frac{3}{4} \cdot 3\frac{1}{8}$ ●18. $1\frac{5}{8} \cdot 1\frac{5}{8}$

Examples

To divide by a fraction, multiply by its reciprocal.

$$\frac{3}{4} \div \frac{9}{2} = ?$$

$$\frac{3}{4} \div \frac{9}{2} = \frac{3}{4} \cdot \frac{2}{9}$$

Cancel.

$$= \frac{\overset{1}{3}}{4} \cdot \frac{\overset{1}{2}}{9}$$
$$ \overset{2}{} \quad \overset{3}{}$$

$$= \frac{1}{6}$$

$$\frac{5a}{b} \div \frac{3}{b^2} = ?$$

$$\frac{5a}{b} \div \frac{3}{b^2} = \frac{5a}{b} \cdot \frac{b^2}{3}$$

Cancel.

$$= \frac{5a}{b} \cdot \frac{\overset{b}{b^2}}{3}$$
$$ \underset{1}{}$$

$$= \frac{5ab}{3}$$

Give the quotient in lowest terms.

1. $\dfrac{3}{4} \div \dfrac{1}{4}$

2. $\dfrac{2}{3} \div \dfrac{1}{3}$

3. $\dfrac{1}{2} \div \dfrac{1}{3}$

4. $\dfrac{3}{5} \div \dfrac{1}{5}$

5. $\dfrac{4}{5} \div 3$

6. $\dfrac{7}{8} \div \dfrac{7}{8}$

7. $\dfrac{5}{6} \div \dfrac{2}{3}$

8. $\dfrac{2}{3} \div \dfrac{1}{2}$

9. $\dfrac{3}{10} \div \dfrac{4}{5}$

10. $\dfrac{3}{4} \div \dfrac{3}{2}$

11. $6 \div \dfrac{3}{4}$

12. $\dfrac{5}{8} \div 3$

13. $\dfrac{5}{6} \div \dfrac{5}{8}$

14. $\dfrac{5}{8} \div \dfrac{2}{3}$

15. $\dfrac{2}{3} \div \dfrac{4}{5}$

▲16. $\dfrac{r}{s} \div \dfrac{t}{u}$

▲17. $\dfrac{a}{b} \div c$

▲18. $\dfrac{1}{j} \div \dfrac{3}{k}$

■19. $\dfrac{m}{n} \div p$

■20. $\dfrac{a}{c} \div \dfrac{d}{e}$

■21. $\dfrac{2a}{b} \div \dfrac{3}{c}$

●22. $\dfrac{r}{s} \div \dfrac{2t}{3u}$

●23. $\dfrac{6a}{b} \div \dfrac{3c}{5d}$

●24. $\dfrac{4}{m} \div \dfrac{6}{n}$

25. $\dfrac{5r}{3} \div \dfrac{7s}{12}$

26. $\dfrac{x}{y} \div \dfrac{z}{y}$

27. $\dfrac{m^2}{n} \div \dfrac{m}{p}$

28. $\dfrac{3r}{t} \div \dfrac{6r}{5}$

29. $\dfrac{y^2}{z} \div \dfrac{y}{z}$

30. $\dfrac{a}{6b} \div \dfrac{c}{3b}$

Example

$$2\frac{1}{4} \div 4\frac{1}{2} = ?$$

Change to fractions.

$$2\frac{1}{4} \div 4\frac{1}{2} = \frac{9}{4} \div \frac{9}{2}$$

Divide. Write in simplest form.

$$2\frac{1}{4} \div 4\frac{1}{2} = \frac{9}{4} \div \frac{9}{2}$$

$$= \frac{\overset{1}{9}}{4} \cdot \frac{\overset{1}{2}}{9}$$
$$ \underset{2}{} \quad \underset{1}{}$$

$$= \frac{1}{2}$$

Give the quotient in simplest form.

1. $5 \div 2\frac{1}{2}$

2. $2\frac{1}{2} \div 1\frac{1}{4}$

3. $5 \div 1\frac{1}{4}$

4. $3\frac{1}{2} \div 2$

5. $10 \div 3\frac{1}{3}$

6. $5\frac{1}{4} \div 3$

7. $4\frac{1}{6} \div 5$

8. $4\frac{3}{4} \div 2$

9. $2\frac{1}{3} \div 1\frac{1}{4}$

10. $2\frac{1}{2} \div 2\frac{1}{2}$

11. $7\frac{1}{2} \div 2\frac{1}{2}$

12. $6\frac{3}{4} \div 3\frac{1}{2}$

▲13. $3\frac{1}{2} \div 1\frac{3}{4}$

▲14. $2\frac{7}{8} \div 3\frac{1}{4}$

▲15. $4\frac{5}{8} \div 2\frac{2}{3}$

■16. $3\frac{5}{6} \div 2\frac{1}{3}$

■17. $5 \div 1\frac{1}{4}$

■18. $5\frac{3}{4} \div 2\frac{2}{3}$

●19. $6\frac{2}{3} \div 5\frac{1}{3}$

●20. $4\frac{7}{8} \div 6\frac{1}{4}$

●21. $2\frac{3}{4} \div 5\frac{2}{3}$

Example

Solve.

$$\frac{5}{6} \text{ of } 26 = n$$

$$\frac{5}{6} \cdot 26 = n$$

$$\frac{5}{\overset{}{\cancel{6}}} \cdot \overset{13}{\cancel{26}} = n$$

$$\frac{65}{3} = n$$

$$21\frac{2}{3} = n$$

Solve. Give answers in simplest form.

1. $\frac{1}{4}$ of $12 = n$ 2. $\frac{1}{2}$ of $18 = n$ 3. $\frac{1}{3}$ of $24 = n$

4. $\frac{1}{5}$ of $30 = n$ 5. $\frac{1}{8}$ of $72 = n$ 6. $\frac{1}{6}$ of $42 = n$

7. $\frac{2}{3}$ of $18 = n$ 8. $\frac{3}{4}$ of $20 = n$ 9. $\frac{2}{5}$ of $10 = n$

10. $\frac{3}{8}$ of $32 = n$ 11. $\frac{2}{3}$ of $24 = n$ 12. $\frac{5}{6}$ of $30 = n$

▲13. $\frac{3}{4}$ of $32 = n$ ▲14. $\frac{5}{8}$ of $40 = n$ ▲15. $\frac{2}{5}$ of $40 = n$

■16. $\frac{3}{5}$ of $22 = n$ ■17. $\frac{5}{6}$ of $40 = n$ ■18. $\frac{2}{3}$ of $38 = n$

●19. $\frac{9}{10}$ of $96 = n$ ●20. $\frac{2}{3}$ of $28 = n$ ●21. $\frac{4}{5}$ of $46 = n$

22. $\frac{3}{4}$ of $38 = n$ 23. $\frac{7}{10}$ of $45 = n$ 24. $\frac{7}{8}$ of $58 = n$

Example

$$2\frac{3}{4} \text{ days} = \underline{\ ?\ } \text{ h}$$

Find the hours in 2 days and $\frac{3}{4}$ of a day.

$$2\frac{3}{4} \text{ days} = 48\text{ h} + 18\text{ h}$$

Add.

$$2\frac{3}{4} \text{ days} = 48\text{ h} + 18\text{ h}$$
$$= 66\text{ h}$$

Complete.

1. $1\frac{1}{2}$ days $= \underline{\ ?\ }$ h   2. $1\frac{1}{4}$ h $= \underline{\ ?\ }$ min

3. $1\frac{3}{4}$ h $= \underline{\ ?\ }$ min   4. $2\frac{1}{2}$ min $= \underline{\ ?\ }$ s

5. $2\frac{1}{4}$ min $= \underline{\ ?\ }$ s   6. $2\frac{1}{3}$ days $= \underline{\ ?\ }$ h

7. $1\frac{2}{3}$ yd $= \underline{\ ?\ }$ ft   8. $1\frac{3}{4}$ ft $= \underline{\ ?\ }$ in.

▲ 9. $1\frac{1}{2}$ yd $= \underline{\ ?\ }$ in.   ▲10. $2\frac{2}{3}$ ft $= \underline{\ ?\ }$ in.

■11. $4\frac{1}{3}$ yd $= \underline{\ ?\ }$ ft   ■12. $1\frac{3}{4}$ yd $= \underline{\ ?\ }$ in.

●13. $1\frac{1}{4}$ gal $= \underline{\ ?\ }$ qt   ●14. $2\frac{1}{2}$ qt $= \underline{\ ?\ }$ pt

15. $3\frac{1}{2}$ pt $= \underline{\ ?\ }$ c   16. $3\frac{1}{2}$ qt $= \underline{\ ?\ }$ pt

Example

Solve.

$$\frac{3}{8} \text{ of } n = 20$$

$$\frac{3}{8}n = 20$$

$$\frac{\frac{3}{8}n}{\frac{3}{8}} = \frac{20}{\frac{3}{8}}$$

$$n = 20 \cdot \frac{8}{3}$$

$$n = \frac{160}{3}$$

$$n = 53\frac{1}{3}$$

Solve. Give answers in simplest form.

1. $\frac{1}{3}n = 7$

2. $\frac{2}{3}n = 12$

3. $\frac{3}{4}n = 27$

4. $\frac{3}{8}n = 18$

5. $\frac{2}{5}n = 30$

6. $\frac{4}{3}n = 36$

7. $\frac{1}{4}n = 17$

8. $\frac{3}{5}n = 42$

9. $\frac{5}{2}n = 10$

10. $\frac{6}{5}n = 42$

11. $\frac{7}{8}n = 35$

12. $\frac{3}{10}n = 36$

▲13. $\frac{3}{2}n = 10$

▲14. $\frac{4}{9}n = 13$

▲15. $\frac{2}{5}n = 21$

■16. $\frac{3}{4}n = 25$

■17. $\frac{5}{2}n = 28$

■18. $\frac{7}{8}n = 30$

●19. $\frac{3}{5}n = 41$

●20. $\frac{5}{12}n = 38$

●21. $\frac{9}{10}n = 46$

22. $\frac{5}{6}n = 53$

23. $\frac{6}{5}n = 49$

24. $\frac{5}{16}n = 58$

Examples

$<, =, \text{or} >?$

$$0 \Diamond \frac{1}{2}$$

$$\frac{^-2}{3} \Diamond {}^-1$$

← *smaller*   *larger* →

$$0 < \frac{1}{2}$$

$$\frac{^-2}{3} > {}^-1$$

$<, =, \text{or} >?$

1. $^-7 \Diamond {}^-5$

2. $19 \Diamond 11$

3. $^-19 \Diamond {}^-11$

4. $\frac{^-3}{4} \Diamond \frac{^-2}{3}$

5. $\frac{3}{4} \Diamond \frac{2}{3}$

6. $\frac{5}{6} \Diamond \frac{10}{12}$

7. $\frac{^-5}{6} \Diamond \frac{^-7}{8}$

8. $^-3 \Diamond {}^-3\frac{3}{8}$

9. $3 \Diamond 3\frac{3}{8}$

10. $5\frac{3}{4} \Diamond 6$

11. $\frac{0}{4} \Diamond 0$

12. $0 \Diamond \frac{^-2}{5}$

▲13. $^-4\frac{5}{8} \Diamond {}^-4\frac{3}{4}$

▲14. $4\frac{5}{8} \Diamond 4\frac{3}{4}$

▲15. $^-9 \Diamond {}^-8\frac{5}{8}$

■16. $8\frac{1}{2} \Diamond 8\frac{3}{6}$

■17. $^-1\frac{1}{2} \Diamond \frac{7}{8}$

■18. $2\frac{3}{5} \Diamond \frac{^-1}{2}$

●19. $^-7 \Diamond {}^-7\frac{1}{2}$

●20. $8\frac{5}{9} \Diamond 8$

●21. $^-3\frac{1}{4} \Diamond {}^-3\frac{1}{3}$

22. $6\frac{3}{4} \Diamond 6\frac{5}{6}$

23. $^-3\frac{1}{2} \Diamond 3\frac{1}{4}$

24. $^-6\frac{7}{8} \Diamond {}^-6\frac{5}{6}$

Example

$$\frac{5}{6} = ?$$

To write a rational number in decimal form, divide the numerator by the denominator.

$$\begin{array}{r} 0.833 \\ 6\overline{)5.000} \\ -4\,8 \\ \hline 20 \\ -18 \\ \hline 20 \\ -18 \\ \hline 2 \end{array}$$

$$\frac{5}{6} = 0.8\overline{3}$$

Write each rational number in decimal form.

1. $\dfrac{1}{4}$ 2. $\dfrac{3}{4}$ 3. $\dfrac{^-1}{5}$ 4. $\dfrac{^-2}{3}$

5. $\dfrac{9}{10}$ 6. $\dfrac{^-2}{5}$ 7. $\dfrac{7}{10}$ 8. $\dfrac{^-4}{5}$

9. $\dfrac{^-1}{8}$ 10. $\dfrac{3}{10}$ 11. $\dfrac{^-7}{4}$ 12. $\dfrac{9}{8}$

13. $\dfrac{9}{2}$ 14. $\dfrac{^-3}{8}$ 15. $\dfrac{3}{5}$ 16. $\dfrac{^-7}{8}$

17. $\dfrac{^-9}{4}$ 18. $\dfrac{11}{3}$ 19. $\dfrac{^-1}{16}$ 20. $\dfrac{11}{8}$

▲21. $\dfrac{5}{16}$ ▲22. $\dfrac{^-3}{2}$ ▲23. $\dfrac{8}{5}$ ▲24. $\dfrac{^-13}{5}$

■25. $^-2\frac{1}{2}$ ■26. $3\frac{3}{4}$ ■27. $^-3\frac{4}{5}$ ■28. $2\frac{7}{8}$

●29. $1\frac{2}{5}$ ●30. $^-4\frac{3}{8}$ ●31. $6\frac{4}{5}$ ●32. $^-3\frac{1}{3}$

33. $^-18\frac{7}{8}$ 34. $10\frac{1}{16}$ 35. $^-12\frac{5}{16}$ 36. $24\frac{4}{5}$

Examples

Write in simplest fractional form.

$$^-0.75 = \frac{^-75}{100}$$

$$= \frac{^-3}{4}$$

$$3.8 = 3\frac{8}{10}$$

$$= 3\frac{4}{5}$$

Write in simplest fractional form.

1. 0.4   2. $^-0.8$ 3. 0.6

4. $^-0.1$ 5. $^-0.9$ 6. 0.5

7. $^-0.3$ 8. 0.2   9. 0.25

10. $^-0.75$ 11. 0.15   12. $^-0.45$

13. $^-0.375$ 14. 0.625   15. $^-0.875$

16. 0.125   17. 1.2   18. $^-2.3$

▲19. 1.6   ▲20. $^-3.8$ ▲21. $^-2.4$

■22. 7.7   ■23. $^-4.5$ ■24. 5.9

●25. 3.25   ●26. $^-1.75$ ●27. 2.12

28. $^-4.48$ 29. $^-4.625$ 30. 6.875

31. $^-2.125$ 32. 5.375   33. $^-8.625$

Examples

Give the sum in simplest form.

$$\frac{^-3}{8} + \frac{1}{4} = \frac{^-3}{8} + \frac{2}{8}$$

$$= \frac{^-1}{8}$$

Give the difference in simplest form.

$$^-2\frac{1}{3} - 3\frac{1}{4} = \frac{^-7}{3} - \frac{13}{4}$$

$$= \frac{^-7}{3} + \frac{^-13}{4}$$

$$= \frac{^-28}{12} + \frac{^-39}{12}$$

$$= \frac{^-67}{12}$$

$$= {}^-5\frac{7}{12}$$

Give each sum in simplest form.

1. $\frac{3}{8} + \frac{5}{8}$
2. $\frac{^-2}{9} + \frac{^-4}{9}$
3. $\frac{7}{3} + \frac{^-5}{3}$

▲ 4. $\frac{5}{12} + \frac{^-7}{8}$
▲ 5. $\frac{^-4}{5} + \frac{3}{4}$
▲ 6. $\frac{5}{6} + \frac{0}{8}$

■ 7. $\frac{^-3}{5} + \frac{^-5}{8}$
■ 8. $\frac{5}{6} + \frac{7}{8}$
■ 9. $\frac{7}{12} + \frac{^-2}{3}$

●10. $3\frac{1}{2} + {}^-2\frac{1}{4}$
●11. $^-1\frac{2}{3} + 2\frac{3}{4}$
●12. $^-4\frac{3}{8} + {}^-2\frac{5}{6}$

Give each difference in simplest form.

13. $\frac{5}{6} - \frac{1}{6}$
14. $\frac{^-7}{8} - \frac{3}{8}$
15. $\frac{^-5}{9} - \frac{^-2}{9}$

▲16. $\frac{3}{5} - \frac{^-9}{10}$
▲17. $\frac{^-5}{12} - \frac{3}{4}$
▲18. $\frac{^-7}{8} - \frac{^-5}{6}$

■19. $\frac{9}{16} - \frac{4}{5}$
■20. $\frac{^-3}{8} - \frac{^-4}{5}$
■21. $\frac{^-5}{12} - \frac{0}{4}$

●22. $2\frac{1}{2} - 4\frac{1}{3}$
●23. $^-2\frac{3}{4} - 1\frac{5}{6}$
●24. $^-4\frac{2}{5} - {}^-2\frac{3}{10}$

Examples

Give the product in simplest form.

$$\frac{^-12}{5} \cdot \frac{10}{9} =$$

$$\frac{^-\overset{4}{\cancel{12}}}{\underset{1}{\cancel{5}}} \cdot \frac{\overset{2}{\cancel{10}}}{\underset{3}{\cancel{9}}} = \frac{^-8}{3} = {}^-2\frac{2}{3}$$

Give the quotient in simplest form.

$$^-3\frac{1}{4} \div 1\frac{1}{3} = \frac{^-13}{4} \div \frac{4}{3}$$

$$= \frac{^-13}{4} \cdot \frac{3}{4}$$

$$= \frac{^-39}{16} = {}^-2\frac{7}{16}$$

Give each product in simplest form.

▲ 1. $\frac{6}{5} \cdot \frac{5}{6}$
▲ 2. $\frac{^-2}{3} \cdot \frac{^-2}{3}$
▲ 3. $\frac{^-4}{5} \cdot \frac{3}{4}$

■ 4. $\frac{7}{8} \cdot \frac{0}{3}$
■ 5. $\frac{^-5}{6} \cdot \frac{3}{10}$
■ 6. $\frac{^-7}{8} \cdot \frac{^-8}{7}$

● 7. $2\frac{5}{6} \cdot {}^-1\frac{1}{3}$
● 8. $3\frac{1}{10} \cdot 2\frac{2}{5}$
● 9. $^-4\frac{1}{8} \cdot {}^-2\frac{3}{4}$

Give each quotient in simplest form.

▲10. $\frac{^-3}{4} \div \frac{^-2}{3}$
▲11. $\frac{^-7}{8} \div \frac{5}{4}$
▲12. $\frac{^-9}{10} \div \frac{3}{5}$

■13. $\frac{2}{3} \div {}^-6$
■14. $\frac{5}{12} \div \frac{^-10}{3}$
■15. $\frac{^-7}{8} \div \frac{^-21}{2}$

●16. $^-2\frac{1}{2} \div 1\frac{1}{4}$
●17. $3\frac{1}{5} \div 1\frac{3}{10}$
●18. $4\frac{7}{8} \div {}^-2\frac{3}{4}$

Examples

Solve.

$$^-4x + 7 = {}^-18$$

$$^-4x + 7 - 7 = {}^-18 - 7$$
$$^-4x = {}^-25$$
$$\frac{^-4x}{^-4} = \frac{^-25}{^-4}$$
$$x = 6\frac{1}{4}$$

Solve.

$$6 - 3y = {}^-8$$

$$6 + {}^-3y = {}^-8$$
$$^-3y + 6 = {}^-8$$
$$^-3y + 6 - 6 = {}^-8 - 6$$
$$^-3y = {}^-14$$
$$\frac{^-3y}{^-3} = \frac{^-14}{^-3}$$
$$y = 4\frac{2}{3}$$

Solve and check.

1. $4m + 6 = 17$
2. $7b - 6 = {}^-4$
3. $^-3 - 2n = 9$
4. $5 + 3y = {}^-4$
5. $^-5c - 10 = 9$
6. $^-12 + 5c = 9$
7. $^-3a + 12 = 13$
8. $^-15 - 4t = 16$
9. $6k + 9 = {}^-17$
10. $5 - 6p = 0$
11. $12q - 4 = 2$
12. $^-16 + 10r = {}^-7$
13. $16g - 8 = 11$
14. $^-13 - 9q = {}^-16$
15. $14 + 12d = 22$
16. $18j + 9 = {}^-11$
17. $^-8z - 14 = 0$
18. $^-12t + 30 = 6$
▲19. $^-10 + 6f = {}^-6$
▲20. $20v + 13 = {}^-7$
■21. $19 + 24g = {}^-13$
■22. $11j - 10 = {}^-10$
●23. $24 - 2t = {}^-18$
●24. $^-30 + 10j = 30$
25. $^-15n - 12 = 24$
26. $20z + 34 = 4$
27. $9c - 16 = 30$
28. $^-30n + 18 = 3$

Example

Simplify.

$$3y - 9 - y + 8$$

Combine like terms.

$$3y - 9 - y + 8 =$$

like terms

$$3y + {}^-9 + {}^-1y + 8$$

like terms

$$= 2y + {}^-1$$
$$= 2y - 1$$

Simplify by combining like terms.

1. $5j + 3j$
2. $6n + n$
3. $8y - 5y$
4. $12k - k$
5. $4x + 3x - 2$
6. $7m - 2m - 3$
7. $12c - 5 - 9c$
8. $12c - 5c - 9$
9. $d - 13d + 9$
10. $d - 13 + 9d$
11. $11n - 9n - 15$
12. $11n - 9 - 15n$
▲13. $14j - 13j + 6 - 11$
▲14. $14j - 13 + 6j - 11$
■15. $16a - 20 + 16 - a$
■16. $16 - 20a + 16 - a$
●17. $21 + 18z - 11z + 12$
●18. $21 + 18 - 11z + 12z$
19. $17 + 15f - 8 + 23f$
20. $17 + 15f - 8f + 23$
21. $20t - 19 - 25t - 6$
22. $20 - 19t - 25t - 6$
23. $25 - x - 12x + 8$
24. $25 - x - 12 + 8x$

Example

Solve.

$$6x + 5 - 2x = 7$$

Combine like terms.

$$4x + 5 = 7$$
$$4x + 5 - 5 = 7 - 5$$
$$4x = 2$$
$$\frac{4x}{4} = \frac{2}{4}$$
$$x = \frac{1}{2}$$

Solve.

▲ **1.** $2a + 8a - 4 = 6$

▲ **2.** $3b + 5 - b = 8$

3. $5c + 3 - 2c = 5$

4. $6d - 4 + d = 10$

■ **5.** $7h - 3h + 2 = 14$

■ **6.** $5k - 4 - 4k = 9$

7. $8m - 2 + 4m = 9$

8. $6n + 3 - 4n = 4$

● **9.** $4y - 3y + 2 + y = 3$

●**10.** $5x - 2 + 3x + 1 = 6$

Example

Solve.

$$8y + 7 = 2y - 5$$

$$8y - 2y + 7 = 2y - 2y - 5$$
$$6y + 7 = {}^-5$$
$$6y + 7 - 7 = {}^-5 - 7$$
$$6y = {}^-12$$
$$\frac{6y}{6} = \frac{{}^-12}{6}$$
$$y = {}^-2$$

✔ CHECK

$$8 \cdot {}^-2 + 7 = 2 \cdot {}^-2 - 5$$
$${}^-16 + 7 = {}^-4 - 5$$
$${}^-9 = {}^-9$$

Solve and check.

1. $4y = 6 - 2y$

2. $^-3n = 4 + {}^-5n$

3. $^-6x = 3 - 2x$

4. $8a + 2 = 6a + 10$

5. $4n - 3 = 6n + 12$

6. $12 - 4j = j + 4$

7. $5w - 2 = 2w - 5$

8. $10 + 2d = d - 8$

9. $^-5t = 3t + 18$

10. $^-2k + 6 = 5k - 15$

11. $4r = {}^-6r - 18$

12. $13 - 3c = 5 - 2c$

▲**13.** $4q - 1 = 8q + 13$

▲**14.** $6 + 4h = 3 - h$

■**15.** $2r + 4 = 4r - 3$

■**16.** $^-7n = 8 + 3n$

●**17.** $15 - 6f = 3f + 6$

●**18.** $9g - 6 = 3g - 9$

19. $5x = 8 - 2x$

20. $11m - 3 = 2m + 17$

21. $13 - 8k = 2k + 10$

22. $6c = 11 - 4c$

23. $5y + 4 = 2y - 18$

24. $7x - 3 = 5x - 11$

Examples

$$\frac{5}{3} = \frac{?}{12}$$

To get an equal ratio, multiply both terms by the same number (not 0) or expression.

$$\frac{5}{3} = \frac{5 \cdot 4}{3 \cdot 4}$$

$$= \frac{20}{12}$$

$$\frac{3}{2x} = \frac{?}{4xy}$$

$$\frac{3}{2x} = \frac{3 \cdot 2y}{2x \cdot 2y}$$

$$= \frac{6y}{4xy}$$

Complete to get an equal ratio.

1. $\dfrac{2}{3} = \dfrac{?}{12}$ 2. $\dfrac{1}{5} = \dfrac{?}{15}$ 3. $\dfrac{3}{8} = \dfrac{?}{16}$ 4. $\dfrac{9}{2} = \dfrac{?}{20}$

5. $\dfrac{4}{5} = \dfrac{?}{25}$ 6. $\dfrac{7}{8} = \dfrac{?}{48}$ 7. $\dfrac{7}{4} = \dfrac{?}{36}$ 8. $\dfrac{5}{6} = \dfrac{?}{66}$

9. $\dfrac{4}{3} = \dfrac{?}{36}$ 10. $\dfrac{2}{5} = \dfrac{?}{45}$ 11. $\dfrac{4}{5} = \dfrac{?}{100}$ 12. $\dfrac{9}{4} = \dfrac{?}{80}$

13. $\dfrac{5}{12} = \dfrac{?}{72}$ 14. $\dfrac{7}{2} = \dfrac{?}{30}$ 15. $\dfrac{3}{10} = \dfrac{?}{80}$ 16. $\dfrac{5}{16} = \dfrac{?}{48}$

▲17. $\dfrac{5}{c} = \dfrac{?}{3c}$ ▲18. $\dfrac{3}{r} = \dfrac{?}{rs}$ ▲19. $\dfrac{n}{6} = \dfrac{?}{6m}$ ▲20. $\dfrac{11}{d} = \dfrac{?}{d^2}$

■21. $\dfrac{u}{v} = \dfrac{?}{2v^2}$ ■22. $\dfrac{a}{b} = \dfrac{?}{3b^2}$ ■23. $\dfrac{t^2}{v} = \dfrac{?}{3v^2}$ ■24. $\dfrac{4c}{3d} = \dfrac{?}{18cd}$

●25. $\dfrac{3y}{2z} = \dfrac{?}{10z^2}$ ●26. $\dfrac{5j}{6k} = \dfrac{?}{6jk}$ ●27. $\dfrac{4s^2t}{3s} = \dfrac{?}{18s}$ ●28. $\dfrac{3ef}{e^2} = \dfrac{?}{e^2f}$

29. $\dfrac{5ab^2}{4ab} = \dfrac{?}{8a^2b}$ 30. $\dfrac{10}{3a} = \dfrac{?}{9a^2b}$ 31. $\dfrac{5c}{6d} = \dfrac{?}{24cd^2}$ 32. $\dfrac{9y^2z}{8} = \dfrac{?}{24z}$

Example

Solve.

$$\frac{n}{6} = \frac{5}{9}$$

Cross-multiply.

$$\frac{n}{6} \diagup\!\!\!\!\diagdown \frac{5}{9}$$

$$9n = 6 \cdot 5$$

$$9n = 30$$

$$\frac{9n}{9} = \frac{30}{9}$$

$$n = 3\frac{1}{3}$$

Solve each proportion.

1. $\dfrac{n}{6} = \dfrac{11}{4}$ 2. $\dfrac{8}{n} = \dfrac{2}{9}$ 3. $\dfrac{9}{8} = \dfrac{n}{7}$ 4. $\dfrac{10}{7} = \dfrac{4}{n}$

5. $\dfrac{5}{n} = \dfrac{2}{9}$ 6. $\dfrac{n}{13} = \dfrac{6}{5}$ 7. $\dfrac{7}{4} = \dfrac{n}{6}$ 8. $\dfrac{5}{8} = \dfrac{3}{n}$

9. $\dfrac{3}{7} = \dfrac{9}{n}$ 10. $\dfrac{18}{n} = \dfrac{6}{5}$ 11. $\dfrac{n}{21} = \dfrac{6}{7}$ 12. $\dfrac{n}{6} = \dfrac{15}{8}$

13. $\dfrac{9}{n} = \dfrac{3}{13}$ 14. $\dfrac{18}{n} = \dfrac{9}{16}$ 15. $\dfrac{6}{11} = \dfrac{24}{n}$ 16. $\dfrac{5}{7} = \dfrac{9}{n}$

▲17. $\dfrac{n}{7} = \dfrac{11}{4}$ ▲18. $\dfrac{6}{n} = \dfrac{8}{3}$ ▲19. $\dfrac{19}{2} = \dfrac{n}{5}$ ▲20. $\dfrac{10}{13} = \dfrac{16}{n}$

■21. $\dfrac{9}{n} = \dfrac{2}{15}$ ■22. $\dfrac{n}{9} = \dfrac{4}{3}$ ■23. $\dfrac{6}{15} = \dfrac{30}{n}$ ■24. $\dfrac{11}{4} = \dfrac{n}{8}$

●25. $\dfrac{n}{4} = \dfrac{6}{8}$ ●26. $\dfrac{1}{n} = \dfrac{7}{21}$ ●27. $\dfrac{6}{9} = \dfrac{n}{3}$ ●28. $\dfrac{6}{10} = \dfrac{18}{n}$

29. $\dfrac{1\frac{1}{2}}{15} = \dfrac{3}{n}$ 30. $\dfrac{10}{3} = \dfrac{n}{2\frac{1}{4}}$ 31. $\dfrac{7}{n} = \dfrac{2\frac{2}{3}}{9}$ 32. $\dfrac{n}{8} = \dfrac{1\frac{3}{4}}{4}$

Examples

Write 25% as a fraction.

Write the percent as a fraction with a denominator of 100. Write in simplest form.

$$25\% = \frac{25}{100} = \frac{1}{4}$$

Write $33\frac{1}{3}\%$ as a fraction.

$$33\frac{1}{3}\% = \frac{33\frac{1}{3}}{100}$$

$$= 33\frac{1}{3} \div 100$$

$$= \frac{100}{3} \times \frac{1}{100}$$

$$= \frac{100}{300} = \frac{1}{3}$$

Change to a fraction, whole number, or mixed number. Give each answer in simplest form.

1. 10%
2. 15%
3. 40%
4. 50%
5. 90%
6. 60%
7. 25%
8. 75%
9. 20%
10. 30%
11. 45%
12. 85%
13. 80%
14. 40%
15. 48%
16. 96%
17. 18%
18. 24%
19. 72%
20. 84%
21. 150%
22. 125%
23. 175%
24. 120%
25. 180%
26. 250%
27. 275%
28. 225%
▲29. 160%
▲30. 200%
▲31. 300%
▲32. 320%
■33. $33\frac{1}{3}\%$
■34. $66\frac{2}{3}\%$
■35. $37\frac{1}{2}\%$
■36. $16\frac{2}{3}\%$
●37. $87\frac{1}{2}\%$
●38. $81\frac{1}{4}\%$
●39. $162\frac{1}{2}\%$
●40. $118\frac{3}{4}\%$

Examples

Write $\frac{1}{2}$ as a percent.

Change to an equivalent fraction with a denominator of 100. Write as a percent.

$$\frac{1}{2} = \frac{50}{100} = 50\%$$

Write $\frac{1}{6}$ as a percent.

If necessary, solve a proportion.

$$\frac{1}{6} = \frac{n}{100}$$
$$6n = 100$$
$$n = 16\frac{2}{3}$$

So $\frac{1}{6} = \frac{16\frac{2}{3}}{100} = 16\frac{2}{3}\%$

Change to a percent.

1. $\frac{1}{5}$
2. $\frac{1}{4}$
3. $\frac{1}{2}$
4. $\frac{4}{5}$
5. $\frac{3}{4}$
6. $\frac{1}{3}$
7. $\frac{2}{5}$
8. $\frac{2}{3}$
9. $\frac{1}{8}$
10. $\frac{3}{5}$
11. $\frac{3}{8}$
12. $\frac{3}{2}$
13. $\frac{6}{5}$
14. $\frac{1}{6}$
15. $\frac{5}{4}$
16. $\frac{7}{5}$
17. $\frac{1}{10}$
18. $\frac{8}{5}$
19. $\frac{3}{10}$
20. $\frac{1}{12}$
▲21. $\frac{5}{2}$
▲22. 3
▲23. $\frac{7}{4}$
▲24. $\frac{1}{20}$
■25. $\frac{5}{8}$
■26. $\frac{7}{10}$
■27. $\frac{11}{8}$
■28. $\frac{7}{3}$
●29. $\frac{5}{12}$
●30. $\frac{5}{3}$
●31. 2
●32. $\frac{7}{12}$

Examples

75% of 36 = n

$$75\% \text{ of } 36 = \frac{3}{4} \times 36$$
$$= 27$$

6.5% of 9 = n

$$6.5\% \text{ of } 9 = 0.065 \times 9$$
$$= 0.585$$

Solve.

1. 50% of 28 = n
2. 25% of 24 = n
3. 40% of 60 = n
4. 20% of 45 = n
5. 80% of 40 = n
6. 10% of 50 = n
7. 125% of 40 = n
8. 250% of 72 = n
▲ 9. 14% of 26 = n
▲10. 23% of 75 = n
■11. 56% of 29 = n
■12. 41% of 83 = n
●13. 5.4% of 60 = n
●14. 6.5% of 47 = n

Examples

60% of n = 50

First change the percent to a fraction. Then solve the equation.

$$\frac{3}{5}n = 50$$

$$\frac{\frac{3}{5}n}{\frac{3}{5}} = \frac{50}{\frac{3}{5}}$$

$$n = 50 \cdot \frac{5}{3}$$

$$= \frac{250}{3}$$

$$= 83\frac{1}{3}$$

6.5% of n = 10

First change the percent to a decimal. Then solve the equation.

$$0.065n = 10$$

$$\frac{0.065n}{0.065} = \frac{10}{0.065}$$

$$n \approx 153.8$$

Solve.

1. 25% of n = 14
2. 50% of n = 19
3. 75% of n = 33
4. 60% of n = 48
▲ 5. 80% of n = 64
▲ 6. 40% of n = 46
■ 7. 10% of n = 16
■ 8. 5% of n = 120
● 9. 30% of n = 57
●10. 20% of n = 65
11. 45% of n = 27
12. 90% of n = 216
13. 8% of n = 6
14. 1% of n = 3

Solve. Round each answer to the nearest tenth.

15. 8.5% of n = 6
16. 7.5% of n = 11
▲17. 20.5% of n = 16.2
▲18. 14.2% of n = 37
■19. 34.6% of n = 18.5
■20. 42.8% of n = 10.3
●21. 1.8% of n = 2.4
●22. 4.7% of n = 2.9
23. 125% of n = 17.6
24. 250% of n = 31.3
25. 45% of n = 70.8
26. 90% of n = 145
27. 9% of n = 35
28. 3% of n = 64.7

Examples

Find the percent of increase from 40 to 50.

$increase \longrightarrow \dfrac{10}{40} = \dfrac{x}{100}$
$number \longrightarrow$
$before$
$increase$

$$40x = 1000$$

Increase of 25%

Find the percent of decrease from 50 to 40.

$decrease \longrightarrow \dfrac{10}{50} = \dfrac{x}{100}$
$number \longrightarrow$
$before$
$decrease$

$$50x = 1000$$
$$x = 20$$

Decrease of 20%

Find the percent of increase or decrease. Use *i* to indicate increase and *d* to indicate decrease. Give answers to the nearest tenth of a percent.

1. From 10 to 20

2. From 20 to 10

3. From 16 to 20

4. From 20 to 16

5. From 36 to 48

6. From 48 to 36

7. From 60 to 72

8. From 72 to 60

▲ **9.** From 75 to 100

▲**10.** From 100 to 75

■**11.** From 120 to 150

■**12.** From 150 to 120

●**13.** From 160 to 200

●**14.** From 200 to 160

Example

$(2x^2 + 3x) + (x^2 + {}^{-}5x) = ?$

First write the sum without grouping symbols and then combine like terms.

$2x^2 + 3x + x^2 + {}^{-}5x$
$= 3x^2 + {}^{-}2x$

Give each sum.

1. $(2x + 8) + (4x + 9)$

2. $(5y^2 + 2y) + (6y^2 + 3y)$

▲ **3.** $(3a + 2b) + (4a + b)$

▲ **4.** $(6x + {}^{-}8y) + (7x + 9y)$

5. $(5m + {}^{-}6n) + (10m + {}^{-}2n)$

6. $(6d + 4) + ({}^{-}7d + 7)$

■ **7.** $(7y + 4z) + ({}^{-}6y + 3z)$

■ **8.** $(8c + 11d) + ({}^{-}3c + {}^{-}5d)$

● **9.** $(5x + {}^{-}7y) + ({}^{-}3x + 4y)$

●**10.** $(5a + 2b + 3c) + (2a + b + 3c)$

11. $(6x + 4y + 7z) + (3y + z)$

12. $(4x^2 + 3x + 2) + (3x^2 + {}^{-}3x)$

Example

$(5y^2 + 8y) - (y^2 + 4y) = ?$

To subtract a polynomial, add the opposite of the polynomial.

$(5y^2 + 8y) - (y^2 + 4y)$
$= 5y^2 + 8y + (^-1)(y^2 + 4y)$
$= 5y^2 + 8y + {}^-1y^2 + {}^-4y$
$= 4y^2 + 4y$

Give each difference.

▲ **1.** $(5a + 3b) - (6a + 9b)$

▲ **2.** $(7a + 3b) - (4a + 2b)$

▲ **3.** $(7x^2 + 3y) - (2x^2 + y)$

4. $(4j + 6) - (2j + 3)$

■ **5.** $(5m + {}^-3n) - (4m + 2n)$

■ **6.** $(11cd + 3c) - (8cd + c)$

■ **7.** $(4r + 5s) - (3r + {}^-3s)$

8. $(8y^2 + 3y) - ({}^-3y^2 + 3y)$

9. $(n^2 + 6) - (3n^2 + 4)$

●**10.** $(5r + 3s + 2t) - (2r + s + 3t)$

●**11.** $(7a + 9b + 8c) - (6a + {}^-9b + 5c)$

●**12.** $(9f^2 + 8f + 11) - (3f^2 + 7f + 6)$

Example

$3x(2x + 4) = ?$

Use the distributive property to multiply a polynomial by a monomial.

$3x(2x + 4)$
$= 3x \cdot 2x + 3x \cdot 4$
$= 6x^2 + 12x$

Multiply.

▲ **1.** $3(n + 2)$ ▲ **2.** $5(2n + 6)$

▲ **3.** $4x(x + 2)$ ▲ **4.** $4y(3y + 1)$

5. $5w(w + 6)$ **6.** $3z(z + 10)$

▲ **7.** $8j(j^2 + 1)$ **8.** $t(3t + 6)$

■ **9.** $4r(r^2 + 9)$ ■**10.** $9(y + {}^-3)$

■**11.** $w(w + {}^-6)$ ■**12.** $4d(d + 7)$

13. $5b(b^2 + {}^-3)$ **14.** $2x^2(x + 6)$

15. $6y(y^2 + {}^-9)$ **16.** $d^2(3d + e)$

●**17.** $2({}^-7a + b)$ ●**18.** $^-2(6xy + {}^-3)$

●**19.** $8(w^2 + 2w + 1)$ ●**20.** $3(2r^2 + 7r + {}^-5)$

21. $6(3n^2 + {}^-2n + {}^-7)$ **22.** $3x(2x^2 + 4x + 6)$

23. $^-2r(4r^2 + r + {}^-9)$ **24.** $4z({}^-3z^2 + {}^-3z + {}^-2)$

Example

$$\frac{8x^3 + 12x}{4x} = ?$$

To divide a polynomial by a monomial, divide each term of the polynomial by the monomial.

$$\frac{8x^3 + 12x}{4x}$$

$$= \frac{8x^3}{4x} + \frac{12x}{4x}$$

$$= \frac{2 \cdot 4 \cdot x \cdot x \cdot x}{4 \cdot x} + \frac{3 \cdot 4 \cdot x}{4 \cdot x}$$

$$= 2x^2 + 3$$

Divide.

▲ 1. $\dfrac{12x + 6}{6}$

▲ 2. $\dfrac{2n^3 + 2n^2}{2n}$

▲ 3. $\dfrac{c^2 + c}{c}$

▲ 4. $\dfrac{^-15r^2 + 20}{5}$

5. $\dfrac{d^3 + d}{d}$

6. $\dfrac{r^3 + r^2}{r}$

■ 7. $\dfrac{15x^3 + 12x^2}{3x}$

■ 8. $\dfrac{20t^3 + 12t^2}{4t^2}$

■ 9. $\dfrac{3a^2b + 6ab^2}{3ab}$

■10. $\dfrac{4rs^2 + 8rs}{2rs}$

11. $\dfrac{9xy^2 + 3y}{3y}$

12. $\dfrac{16x^2y + {}^-12y}{4y}$

●13. $\dfrac{10a^2b^2 + 8ab}{2ab}$

●14. $\dfrac{6n^2 + 8n + 12}{2}$

●15. $\dfrac{16c^2 + 12c + {}^-8}{4}$

●16. $\dfrac{14x^2 + {}^-7x + {}^-35}{7}$

Example

Solve.

$$x + (x + 3) + 2(x - 5) = 17$$

Use the distributive property and remove parentheses.

$$\begin{aligned} x + x + 3 + 2x - 10 &= 17 \\ 4x - 7 &= 17 \\ 4x - 7 + 7 &= 17 + 7 \\ 4x &= 24 \\ \frac{4x}{4} &= \frac{24}{4} \\ x &= 6 \end{aligned}$$

✔ **CHECK**

$$6 + (6 + 3) + 2(6 - 5) \stackrel{?}{=} 17$$
$$6 + 6 + 3 + 2(1) \stackrel{?}{=} 17$$
$$17 = 17$$

Solve and check.

▲ 1. $5n + 2(n + 3) = 34$

▲ 2. $3m + 2(m - 1) = {}^-12$

▲ 3. $^-8y + 5(y - 6) = 42$

4. $d + 3(d + 2) + 8 = 18$

5. $j + 2(j + 1) + 3(j - 4) = 20$

■ 6. $2(t - 2) + 3(t + 2) + 4t = 65$

■ 7. $6(x + 2) + {}^-3x + 4(x + 1) = {}^-5$

■ 8. $8(d - 2) + 3(d - 6) + 5 = {}^-7$

9. $7(b - 1) + 2(b + 3) + 6 = {}^-4$

10. $^-8r + 2(r + 3) + 5r = 6$

●11. $8(t + 3) + {}^-6t + 18 = {}^-20$

●12. $7n + 6(n + 8) + 3(n - 5) = {}^-15$

●13. $^-8(x + 3) + 2(x - 4) + 4x = {}^-8$

14. $3j + 4(j + 5) + j = 4$

Table of Square Roots of Integers from 1 to 320

Number	Positive Square Root		Number	Positive Square Root		Number	Positive Square Root		Number	Positive Square Root
N	$\sqrt{N}$		N	$\sqrt{N}$		N	$\sqrt{N}$		N	$\sqrt{N}$
1	1		41	6.403		81	9		121	11
2	1.414		42	6.481		82	9.055		122	11.045
3	1.732		43	6.557		83	9.110		123	11.091
4	2		44	6.633		84	9.165		124	11.136
5	2.236		45	6.708		85	9.220		125	11.180
6	2.449		46	6.782		86	9.274		126	11.225
7	2.646		47	6.856		87	9.327		127	11.269
8	2.828		48	6.928		88	9.381		128	11.314
9	3		49	7		89	9.434		129	11.358
10	3.162		50	7.071		90	9.487		130	11.402
11	3.317		51	7.141		91	9.539		131	11.446
12	3.464		52	7.211		92	9.592		132	11.489
13	3.606		53	7.280		93	9.644		133	11.533
14	3.742		54	7.348		94	9.695		134	11.576
15	3.873		55	7.416		95	9.747		135	11.619
16	4		56	7.483		96	9.798		136	11.662
17	4.123		57	7.550		97	9.849		137	11.705
18	4.243		58	7.616		98	9.899		138	11.747
19	4.359		59	7.681		99	9.950		139	11.790
20	4.472		60	7.746		100	10		140	11.832
21	4.583		61	7.810		101	10.050		141	11.874
22	4.690		62	7.874		102	10.100		142	11.916
23	4.796		63	7.937		103	10.149		143	11.958
24	4.899		64	8		104	10.198		144	12
25	5		65	8.062		105	10.247		145	12.042
26	5.099		66	8.124		106	10.296		146	12.083
27	5.196		67	8.185		107	10.344		147	12.124
28	5.292		68	8.246		108	10.392		148	12.166
29	5.385		69	8.307		109	10.440		149	12.207
30	5.477		70	8.367		110	10.488		150	12.247
31	5.568		71	8.426		111	10.536		151	12.288
32	5.657		72	8.485		112	10.583		152	12.329
33	5.745		73	8.544		113	10.630		153	12.369
34	5.831		74	8.602		114	10.677		154	12.410
35	5.916		75	8.660		115	10.724		155	12.450
36	6		76	8.718		116	10.770		156	12.490
37	6.083		77	8.775		117	10.817		157	12.530
38	6.164		78	8.832		118	10.863		158	12.570
39	6.245		79	8.888		119	10.909		159	12.610
40	6.325		80	8.944		120	10.954		160	12.649

Exact square roots are shown in red. For the others, rational approximations are given correct to three decimal places.

Number	Positive Square Root		Number	Positive Square Root		Number	Positive Square Root		Number	Positive Square Root
N	$\sqrt{N}$		N	$\sqrt{N}$		N	$\sqrt{N}$		N	$\sqrt{N}$
161	12.689		201	14.177		241	15.524		281	16.763
162	12.728		202	14.213		242	15.556		282	16.793
163	12.767		203	14.248		243	15.588		283	16.823
164	12.806		204	14.283		244	15.620		284	16.852
165	12.845		205	14.318		245	15.652		285	16.882
166	12.884		206	14.353		246	15.684		286	16.912
167	12.923		207	14.387		247	15.716		287	16.941
168	12.961		208	14.422		248	15.748		288	16.971
169	13		209	14.457		249	15.780		289	17
170	13.038		210	14.491		250	15.811		290	17.029
171	13.077		211	14.526		251	15.843		291	17.059
172	13.115		212	14.560		252	15.875		292	17.088
173	13.153		213	14.595		253	15.906		293	17.117
174	13.191		214	14.629		254	15.937		294	17.146
175	13.229		215	14.663		255	15.969		295	17.176
176	13.266		216	14.697		256	16		296	17.205
177	13.304		217	14.731		257	16.031		297	17.234
178	13.342		218	14.765		258	16.062		298	17.263
179	13.379		219	14.799		259	16.093		299	17.292
180	13.416		220	14.832		260	16.125		300	17.321
181	13.454		221	14.866		261	16.155		301	17.349
182	13.491		222	14.900		262	16.186		302	17.378
183	13.528		223	14.933		263	16.217		303	17.407
184	13.565		224	14.967		264	16.248		304	17.436
185	13.601		225	15		265	16.279		305	17.464
186	13.638		226	15.033		266	16.310		306	17.493
187	13.675		227	15.067		267	16.340		307	17.521
188	13.711		228	15.100		268	16.371		308	17.550
189	13.748		229	15.133		269	16.401		309	17.578
190	13.784		230	15.166		270	16.432		310	17.607
191	13.820		231	15.199		271	16.462		311	17.635
192	13.856		232	15.232		272	16.492		312	17.664
193	13.892		233	15.264		273	16.523		313	17.692
194	13.928		234	15.297		274	16.553		314	17.720
195	13.964		235	15.330		275	16.583		315	17.748
196	14		236	15.362		276	16.613		316	17.776
197	14.036		237	15.395		277	16.643		317	17.804
198	14.071		238	15.427		278	16.673		318	17.833
199	14.107		239	15.460		279	16.703		319	17.861
200	14.142		240	15.492		280	16.733		320	17.889

Table of Trigonometric Ratios

Angle	Sine	Cosine	Tangent	Angle	Sine	Cosine	Tangent
1°	0.0175	0.9998	0.0175	46°	0.7193	0.6947	1.0355
2°	0.0349	0.9994	0.0349	47°	0.7314	0.6820	1.0724
3°	0.0523	0.9986	0.0524	48°	0.7431	0.6691	1.1106
4°	0.0698	0.9976	0.0699	49°	0.7547	0.6561	1.1504
5°	0.0872	0.9962	0.0875	50°	0.7660	0.6428	1.1918
6°	0.1045	0.9945	0.1051	51°	0.7771	0.6293	1.2349
7°	0.1219	0.9925	0.1228	52°	0.7880	0.6157	1.2799
8°	0.1392	0.9903	0.1405	53°	0.7986	0.6018	1.3270
9°	0.1564	0.9877	0.1584	54°	0.8090	0.5878	1.3764
10°	0.1736	0.9848	0.1763	55°	0.8192	0.5736	1.4281
11°	0.1908	0.9816	0.1944	56°	0.8290	0.5592	1.4826
12°	0.2079	0.9781	0.2126	57°	0.8387	0.5446	1.5399
13°	0.2250	0.9744	0.2309	58°	0.8480	0.5299	1.6003
14°	0.2419	0.9703	0.2493	59°	0.8572	0.5150	1.6643
15°	0.2588	0.9659	0.2679	60°	0.8660	0.5000	1.7321
16°	0.2756	0.9613	0.2867	61°	0.8746	0.4848	1.8040
17°	0.2924	0.9563	0.3057	62°	0.8829	0.4695	1.8807
18°	0.3090	0.9511	0.3249	63°	0.8910	0.4540	1.9626
19°	0.3256	0.9455	0.3443	64°	0.8988	0.4384	2.0503
20°	0.3420	0.9397	0.3640	65°	0.9063	0.4226	2.1445
21°	0.3584	0.9336	0.3839	66°	0.9135	0.4067	2.2460
22°	0.3746	0.9272	0.4040	67°	0.9205	0.3907	2.3559
23°	0.3907	0.9205	0.4245	68°	0.9272	0.3746	2.4751
24°	0.4067	0.9135	0.4452	69°	0.9336	0.3584	2.6051
25°	0.4226	0.9063	0.4663	70°	0.9397	0.3420	2.7475
26°	0.4384	0.8988	0.4877	71°	0.9455	0.3256	2.9042
27°	0.4540	0.8910	0.5095	72°	0.9511	0.3090	3.0777
28°	0.4695	0.8829	0.5317	73°	0.9563	0.2924	3.2709
29°	0.4848	0.8746	0.5543	74°	0.9613	0.2756	3.4874
30°	0.5000	0.8660	0.5774	75°	0.9659	0.2588	3.7321
31°	0.5150	0.8572	0.6009	76°	0.9703	0.2419	4.0108
32°	0.5299	0.8480	0.6249	77°	0.9744	0.2250	4.3315
33°	0.5446	0.8387	0.6494	78°	0.9781	0.2079	4.7046
34°	0.5592	0.8290	0.6745	79°	0.9816	0.1908	5.1446
35°	0.5736	0.8192	0.7002	80°	0.9848	0.1736	5.6713
36°	0.5878	0.8090	0.7265	81°	0.9877	0.1564	6.3138
37°	0.6018	0.7986	0.7536	82°	0.9903	0.1392	7.1154
38°	0.6157	0.7880	0.7813	83°	0.9925	0.1219	8.1443
39°	0.6293	0.7771	0.8098	84°	0.9945	0.1045	9.5144
40°	0.6428	0.7660	0.8391	85°	0.9962	0.0872	11.4301
41°	0.6561	0.7547	0.8693	86°	0.9976	0.0698	14.3007
42°	0.6691	0.7431	0.9004	87°	0.9986	0.0523	19.0811
43°	0.6820	0.7314	0.9325	88°	0.9994	0.0349	28.6363
44°	0.6947	0.7193	0.9657	89°	0.9998	0.0175	57.2900
45°	0.7071	0.7071	1.0000	90°	1.0000	0.0000	.....

absolute value (p. 92) The absolute value of a number is its distance from 0 on a number line.

acute angle (p. 268) An angle that measures between 0° and 90°.

acute triangle (p. 270) A triangle with three acute angles.

adding 0 property (p. 64) The sum of any number and 0 is the original number.
$$a + 0 = a$$

addition property of equality (p. 18) Adding the same number to each side of an equation does not affect the solution.
If $a = b$, then $a + c = b + c$.

algebraic denominator (p. 170) A denominator that includes a variable.

algebraic expression (p. 2) A combination of numbers, variables, and operation signs.

algebraic factorization (p. 130) An algebraic expression written as the product of its factors.

algebraic fraction (p. 158) A fraction that includes a variable.

angle (p. 268) A figure formed by two rays with the same endpoint.

associative property of addition (p. 64) Changing the grouping of the addends does not change the sum.
$$(a + b) + c = a + (b + c)$$

associative property of multiplication (p. 66) Changing the grouping of the factors does not change the product.
$$(ab)c = a(bc)$$

average (p. 382) The sum of the numbers in a set divided by the *number* of numbers in the set.

axes (p. 344) Two perpendicular lines used as a reference for graphing ordered pairs.

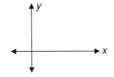

base (of an exponent) (p. 122) The number or variable that is raised to a power, for example, x in x^2.

basic counting principle (p. 364) If a first event has m outcomes and a second event has n outcomes, then the first event followed by the second event has $m \times n$ outcomes.

binomial (p. 400) A polynomial consisting of two terms.

bisector of an angle (p. 272) The ray that divides the angle into two angles that have the same measure.

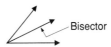

Bisector

Glossary | **509**

bisector of a segment (p. 272) A line that divides the segment into two segments that have the same length.

canceling (p. 184) Dividing a numerator and a denominator by a common factor to write a fraction in lowest terms or before multiplying fractions.

centimeter (p. 278) A metric unit of length.
1 centimeter = 0.01 meter

circumference (p. 288) The distance around a circle.

common factor (p. 132) 2 is a common factor of 4 and 6 because 2 is a factor of both 4 and 6.

common multiple (p. 132) 30 is a common multiple of 5 and 6 because it is a multiple of both 5 and 6.

commutative property of addition (p. 64) Changing the order of the addends does not change the sum.
$a + b = b + a$

commutative property of multiplication (p. 66) Changing the order of the factors does not change the product.
$ab = ba$

complementary angles (p. 268) Two angles whose measures have a sum of 90°.

composite number (p. 130) A whole number other than 0 that has more than two factors.

congruent angles (p. 268) Angles having the same measure.

congruent triangles (p. 275) Triangles whose corresponding parts are congruent.

coordinate plane (p. 344) A grid used for locating a point whose coordinates are known.

coordinates (p. 344) An ordered pair of numbers that locate a point on a grid.

cosine of an angle (p. 430) In a right triangle, the ratio of the length of the leg adjacent to an acute angle to the length of the hypotenuse.

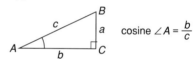

cube (p. 306) A rectangular prism whose six faces are squares.

cylinder (p. 306) A solid figure with a circular base.

denominator (p. 138) In the fraction $\frac{2}{3}$, the denominator is 3.

dependent events (p. 374) Two events such that the outcome of the first affects the outcome of the second—for instance, drawing a first card and then a second card without replacing the first.

diameter (p. 288) The distance across a circle through its center. The length of the diameter is twice the length of the radius.

Diameter

digits (p. 6) The basic symbols used to write numerals. In our system, the digits are 0, 1, 2, 3, 4, 5, 6, 7, 8, and 9.

direct variation (p. 358) A relationship between two quantities in which one is always a fixed multiple of the other.

discount (p. 252) The amount a store subtracts from the regular price during a sale.

distributive property of multiplication (p. 78) A product of a sum can be written as the sum of two products.
$$a(b + c) = ab + ac$$

divisible (p. 128) One number is divisible by another if there is no remainder after division. 84 is divisible by 2, since $84 \div 2$ leaves no remainder.

division property of equality (p. 48) Dividing both sides of an equation by the same (nonzero) number does not affect the solution.

If $a = b$, then $\dfrac{a}{c} = \dfrac{b}{c}$.

equation (p. 16) A sentence with an equal sign such as $3 \times 9 = 27$ or $8 + x = 10$.

equilateral triangle (p. 270) A triangle with three congruent sides.

equivalent fractions (p. 138) Fractions that name the same number. $\dfrac{1}{2}, \dfrac{2}{4}, \dfrac{3}{6}$ are equivalent fractions.

estimate (p. 8) To use rounded numbers to check whether an answer is reasonable. To estimate $47 + 32$, you would add $50 + 30$. The sum should be about 80.

evaluate an expression (p. 32) To replace the variables in an expression with numbers and then simplify the expression.

even number (p. 128) A whole number that is divisible by 2.

event (p. 368) A set of one or more outcomes.

expectation (p. 380) The probability of winning multiplied by the value of the prize.

exponent (p. 122) An exponent tells how many times a number is used as a factor.

$$\overset{\text{exponent}}{\underset{\underset{\text{3 factors}}{\uparrow}}{x^3 = \underbrace{x \cdot x \cdot x}}}$$

factors (p. 132) Numbers or algebraic expressions that are multiplied.

$$\underset{\text{factors}}{\overset{\uparrow\uparrow}{8b}}$$

formula (p. 21) A general way of expressing a relationship using variables.

frequency table (p. 384) A table showing the number of times different events or responses occur.

gram (p. 322) A metric unit of mass.
1 gram = 0.001 kilogram

greatest common factor (GCF) (p. 132) The greatest common factor of two numbers or expressions is the largest of their common factors. $4a$ is the GCF of $20ab^2$ and $24a^2$.

hexagon (p. 276) A polygon with six sides.

hypotenuse (p. 430) The side of a right triangle that is opposite the right angle. It is the longest side of a right triangle.

independent events (p. 372) Events such that the outcome of the first does not affect the outcome of the second—for example, tossing *heads* and rolling a *6.*

indirect variation (See inverse variation.)

inequality (p. 334) A mathematical sentence that contains a symbol such as $\neq$, $>$, $<$, $\geq$, or $\leq$.

integers (p. 92) The numbers . . . $^-3$, $^-2$, $^-1$, 0, 1, 2, 3, . . .

interest (p. 260) Money paid to a lender or depositor for the use of money (the principal).

inverse operations (p. 16) Operations that undo each other. Addition and subtraction are inverse operations.

inverse variation (p. 358) A relationship between two quantities in which the product of the two is always the same.

isosceles triangle (p. 270) A triangle with at least two congruent sides.

kilogram (p. 322) A metric unit of mass.
   1 kilogram = 1000 grams

kilometer (p. 278) A metric unit of length.
   1 kilometer = 1000 meters

least common denominator (LCD) (p. 142) The least common denominator of two fractions is the least common multiple of the denominators.

least common multiple (LCM) (p. 132) The least common multiple of two numbers or expressions is the smallest of their common multiples. $6a^2b$ is the LCM of $2ab$ and $3a^2$.

leg of a right triangle (p. 430) Either of the two shorter sides of a right triangle.

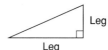

like terms (p. 80) Monomials with identical variables and identical exponents on those variables.

linear equation (p. 348) An equation for which the graph is a straight line. A linear equation has the form
   $y = mx + b$
where m is the slope of the line and b is the y-intercept.

linear inequality (p. 352) An inequality for which the graph is a region above or below a straight line.

line segment (p. 268) Part of a line that has two endpoints.

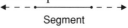

liter (p. 322) A metric unit of liquid volume.

lowest terms (p. 140) A fraction is in lowest terms if the greatest common factor of the numerator and denominator is 1.

markup (p. 178) The difference between the retail price and the wholesale price of an item.

mean (p. 382) The average of all the numbers in a set.

median (p. 382) If an odd number of numbers is ranked from least to greatest, the median is the middle number. For an even number of numbers, the median is the average of the two middle numbers.

meter (p. 278) A metric unit of length.
    1 meter = 100 centimeters

milliliter (p. 322) A metric unit of liquid volume.
    1 milliliter = 0.001 liter

millimeter (p. 278) A metric unit of length.
    1 millimeter = 0.001 meter

mixed number (p. 154) A number that has a whole-number part and a fraction part.
    $2\frac{3}{4}$ is a mixed number.

mode (p. 382) The number that occurs most often in a given data set.

monomial (p. 400) A variable, a number, or a product of variables and numbers.

multiplication property of equality (p. 50) Multiplying both sides of an equation by the same number does not affect the solution.
    If $a = b$, then $ac = bc$.

multiplying by 1 property (p. 66) The product of any number and 1 is the original number.
    $x \times 1 = x$

negative number (p. 92) A number that is less than 0.

numerator (p. 138) In the fraction $\frac{2}{3}$, the numerator is 2.

obtuse angle (p. 268) An angle that measures between 90° and 180°.

obtuse triangle (p. 270) A triangle with an obtuse angle.

odd number (p. 128) A whole number that is not divisible by 2. The numbers 1, 3, 5, 7, and so on, are odd.

operation (p. 62) Addition, subtraction, multiplication, and division are examples of operations.

opposites (p. 96) Two numbers are opposites if their sum is 0.
    $^-3 + {}^+3 = 0$
    opposites

ordered pair (p. 344) A pair of numbers or coordinates giving the location of a point on a grid.

origin (p. 344) The point of intersection of the axes of a coordinate grid.

outcome (p. 368) The possible result of an experiment, for example, tossing a coin.

parallel lines (p. 272) Lines in a plane that do not intersect.

parallelogram (p. 276) A quadrilateral with two pairs of parallel sides.

pentagon (p. 276) A polygon with five sides.

percent (%) (p. 244) *Percent* means "per hundred."

$$5\% \text{ (5 percent) equals } \frac{5}{100}$$

perimeter (p. 286) The distance around a figure; the sum of the lengths of the sides.

permutation (p. 366) An ordered arrangement of a set of objects.

perpendicular lines (p. 272) Two lines that form four right angles.

pi (p. 288) The number that is the ratio of the circumference of a circle to its diameter. It is represented by the Greek letter π and is approximately equal to 3.14.

plane (p. 276) A flat surface that extends endlessly in all directions.

polygon (p. 276) A closed plane figure made up of segments.

polynomial (p. 400) A monomial or the sum or difference of monomials.

positive number (p. 92) A number greater than 0.

prime factorization (p. 130) Expression of a composite number as a product of prime numbers.

The prime factorization of 18 is $2 \cdot 3 \cdot 3$.

prime number (p. 130) A whole number that has exactly two factors. 2, 3, 5, 7, 11, 13, and so on, are prime numbers.

principal (p. 260) The amount of money loaned to a borrower.

prism (p. 306) A space figure that has two bases that are the same size and shape and are in parallel planes. The other faces are rectangles.

probability (p. 368) The ratio of the number of favorable outcomes to the total number of outcomes.

proportion (p. 240) An equation stating that two ratios are equal.

$$\frac{5}{8} = \frac{30}{48}$$

Pythagorean theorem (p. 442) In a right triangle, the sum of the squares of the lengths of the two legs is equal to the square of the length of the hypotenuse.

$$a^2 + b^2 = c^2$$

quadrilateral (p. 276) A polygon with four sides.

quartiles (of a data set) (p. 388) Those two values that with the median subdivide the data set into four equal groups.

quotient (p. 38) The answer to a division problem.

radius (p. 288) The distance from the center of a circle to the circle. The radius is equal to one half the diameter.

range (p. 382) The difference between the least and greatest numbers in a data set.

rate (p. 242) A comparison by division of two unlike quantities.

$$\frac{87 \text{ kilometers}}{2 \text{ hours}}$$

ratio (p. 238) A comparison of two numbers by division.

rational number (p. 210) A number that can be written as the quotient of two integers (denominator not 0).

rational solutions (p. 222) Solutions that are rational numbers.

ray (p. 268) Part of a line that has just one endpoint.

Ray

reciprocal (p. 188) Two numbers are reciprocals when their product is 1.

$$\frac{3}{4} \times \frac{4}{3} = 1$$
reciprocals

rectangle (p. 276) A quadrilateral with four right angles.

rectangular prism (p. 306) A prism whose bases are rectangles.

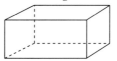

repeating decimal (p. 212) A decimal in which a digit or group of digits repeats forever.
0.3333 . . .   1.47474747 . . .

retail price (p. 178) The price a store charges for an item it sells.

right angle (p. 268) An angle whose measure is 90°.

right triangle (p. 270) A triangle with a right angle.

round a number (p. 6) To replace a number by another one of approximately the same value that is easier to use.

sample (p. 392) A small group, chosen from a larger group, that is examined carefully in order to make predictions about the larger group.

sample space (p. 370) The set of all possible outcomes of an event.

scale drawing (p. 250) A drawing of an object such that the ratio of a unit of length on the drawing to a unit of length on the object is fixed.

scalene triangle (p. 270) A triangle with no congruent sides.

scientific notation (p. 126) A notation for writing a number as the product of a number between 1 and 10 and a power of 10.
$$186.3 = 1.863 \cdot 10^2$$

segment (See line segment.)

similar figures (p. 424) Two figures that have the same shape.

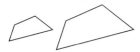

simplest form (p. 158) A fraction or mixed number is in simplest form if the fraction or fraction-part of the mixed number is less than 1 and in lowest terms.

sine of an angle (p. 430) In a right triangle, the ratio of the length of the leg opposite an acute angle to the length of the hypotenuse.

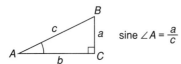

slope (p. 350) The ratio of the difference in y-coordinates to the difference in x-coordinates for any two points on the graph of a linear equation. Also defined as the rise over the run.

solve an equation (p. 16) To find all the values of the variable that make the equation true.

space figure (p. 306) A three-dimensional object.

square (p. 276) A quadrilateral with four sides the same length and four right angles.

square root of a number (p. 440) The number which when squared will produce the given number. 7 is the square root of 49, because $7^2 = 49$.

statistics (p. 382) A branch of mathematics that studies numerical facts as a basis for drawing general conclusions and making predictions.

subtraction property of equality (p. 16) Subtracting the same number from both sides of an equation does not affect the solution. If $a = b$, then $a - c = b - c$.

supplementary angles (p. 268) Two angles whose measures have a sum of 180°.

surface area (p. 310) The sum of the areas of all the surfaces of a solid figure.

symmetric property of equality (p. 20) Exchanging the left and right sides of an equation does not affect the solution. If $76 = x + 53$, then $x + 53 = 76$.

symmetry (of a plane figure) (p. 277) The property whereby the figure can be divided into matching parts by folding along a line.

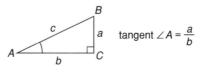

system of equations (p. 354) Two or more equations with the same variables.

tangent of an angle (p. 428) In a right triangle, the ratio of the length of the leg opposite an acute angle to the length of the leg adjacent to that acute angle.

tangent $\angle A = \dfrac{a}{b}$

term (p. 400) One of the monomials a polynomial is composed of.

terminating decimal (p. 212) A decimal fraction, such as 0.5, that is not a repeating decimal.

transversal (p. 272) A line that intersects two or more lines.

Transversal

trapezoid (p. 276) A polygon with four sides and exactly one pair of parallel sides.

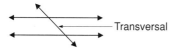

tree diagram (p. 370) A diagram that shows all the possible outcomes of an event.

triangular prism (p. 306) A prism whose bases are triangles.

trinomial (p. 400) A polynomial consisting of three terms.

two-step equation (p. 74) An equation whose solution involves two operations.

variable (p. 2) A symbol, usually a letter, that holds the place for a number.

$$8x + 19 = 23$$
variable

vertex (p. 268) The point at the corner of an angle, plane figure, or solid figure.

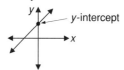
Vertex

volume (p. 316) The amount of space occupied by a space figure.

whole number (p. 154) Any of the numbers 0, 1, 2, 3, 4, and so on.

wholesale price (p. 178) The price a store originally paid for an item it sells.

y-intercept (p. 350) The *y*-coordinate of the point where the graph of an equation crosses the *y*-axis.

y-intercept

SELECTED ANSWERS

Chapter 1 ■ Solving Equations Using Addition and Subtraction

Pages 2–3
1. Batman   2. Steel Magnolias   3. Driving Miss Daisy   4. decreased   5. more   6. Batman   7. $n + 6$
9. $a + 9$ 11. $5 + g$ 13. $b - 8$ 15. $32 - n$
17. $d + 12\frac{1}{2}$ 19. $t + t$ 21. $x + 6$ 23. $x - 2$
25. $x - 2$ 27. B   29. B

Pages 4–5
1. Armstrong   2. Pippen   3. Armstrong   4. King
5. 21   6. Paxson   7. a. 19   b. 5   c. 22.5   d. $20\frac{1}{2}$
9. 3   11. 2   13. 13   15. 0   17. $5\frac{1}{2}$ 19. 7.6
21. 8   23. 20   25. 1   27. 24   29. 4   31. 3   33. 2
35. 6.3   37. 11   39. 7   41. 3   43. 6   45. 13
47. 4   49. Pippen   51. 24   53. King and Nealy
55. King and Nealy

Pages 6–7
1. $11.42   2. 7.625   3. a. 8   b. 7.6   c. 7.63
4. $11   5. 26   7. 0   9. 1   11. 39   13. 54   15. 3.7
17. 8.4   19. 9.2   21. 49.2   23. 6.5   25. 605.4
27. 0.1   29. 64.4   31. 42.11   33. 72.84   35. 72.57
37. 0.04   39. 0.02   41. 0.01   43. 165.79   45. $7
47. $147   49. $28   51. $55   53. $100   55. $18
57. $1.45

Pages 8–9
1. 1.28   2. Lettuce salad   3. tenth   4. Yes
5. Wrong   7. Wrong   9. Right   11. b   13. b
15. 10.4   17. 12.1   19. 8.4   21. 10.2   23. 5.08
25. 28.9   27. 10.8   29. 10.18   31. 9.28   33. a. 1.02
b. 2.675

Pages 10–11
1. 54.29   2. 60   3. whole   4. Yes   5. Right
7. Wrong   9. Wrong   11. b   13. b   15. 6.25
17. 8.2   19. 0.927   21. 51.45   23. 1.94   25. 75.65
27. 2.349   29. 4.566   31. 4.5   33. 15.5   35. 0
37. 7.75   39. 16.75   41. 16.75   43. 4.8 s   45. Dean

Page 14
1. 10.3 mi   3. 8 mi   5. 408 west to 4 south
7. 10.2 mi   9. C

Page 15
1. $y - 4$ 3. $7 + d$ 5. $b + 6$ 7. $5 - s$ 9. $x + 11$
11. $z - 7\frac{1}{6}$ 13. $r + 0.3$ 15. $z - y$ 17. $x - x$
9. 15   21. 12   23. 1   25. 16   27. 1   29. 7
31. 22.73   33. 11   35. 8.2   37. 63.8   39. 78.1
41. 263.7   43. 59.51   45. 270.50   47. 40.96
49. 6.40   51. 13.1   53. 14.98   55. 10.68   57. 20.78
59. 5.1   61. 5.99   63. 13.1   65. 3.11   67. 6.61
69. 1.61

Pages 16–17
1. a. 3   b. Yes   c. $n = 5$ d. 5   2. 3, 5   3. a. 6
b. 16   c. 15.3   4. B: Undo the addition of 8 by subtracting 8, C: Subtract the same number from both sides, D: Subtract 18.5 from 80.2   5. 16   7. 18
9. 8   11. 46   13. 13   15. 14   17. 47   19. 34
21. 27   23. 19   25. 43   27. 38   29. 6   31. 45
33. 3.9   35. 0.92   37. 0.14   39. 0.15   41. 7.98
43. 0   45. $x + 45 = 72$, 27   47. $w + 15 = 43$, 28
49. $a + 18 = 41$, 23   51. $b + 9 = 17.3$, 8.3

Pages 18–19
1. B   2. a. 3, 14   b. 14   3. a. 7   b. 23.6   4. a. 5
b. 6   c. 21.2   5. B: Undo the subtraction of 28 by adding 28, C: Add the same number to both sides, D: Add 17.2 to 60.8   7. 41   9. 32   11. 35   13. 33
15. 24   17. 18   19. 16   21. 0   23. 80   25. 23
27. 49   29. 133   31. 70   33. 25   35. 1.46   37. 0.7
39. 3   41. 1.75   43. 1.55   45. 13.73   47. A, 48
49. B, 72   51. A, 48 min

Pages 20–21
1. a. right   b. left   2. a. 59   b. 80   3. B: Symmetric property of equality, C: Symmetric property of equality, D: Undo the subtraction of 36.7 by adding 36.7   5. 37   7. 27   9. 53   11. 24
13. 67   15. 73   17. 32   19. 0   21. 38   23. 42
25. 10   27. 6.8   29. 7.05   31. 0.93   33. a. 8
b. 15   c. 6   d. 9

Pages 22–23
1. third   2. a. points   b. 26   c. 8   3. Yes   5. B, 41
7. A, 19   11. $g - 14 = 5$, 19   13. $d + 12.75 = 32$, $19.25   15. $39.50 = r - 7.65$, $47.15   17. $34 = f + 10$, 24

Page 24

1. a. Firebird b. 24,825 c. 1986 3. $1240
5. 73,350

Page 25

1. Right 3. Wrong 5. Right 7. Wrong
9. Wrong 11. Right 13. Right 15. Right 17. 14
19. 8 21. 14 23. 0 25. 3.2 27. 1.5 29. 0.19
31. 6.13 33. 149 35. 19 37. 30 39. 36 41. 14
43. 64 45. 9.9 47. 9.7 49. 0.37 51. 91.3
53. 19 55. 0 57. 89 59. 11 61. 41 63. 111
65. 5.5 67. 3.4 69. 0.4 71. 7.25

Chapter 2 ■ Solving Equations Using Multiplication and Division

Pages 30–31

1. Margaret Thatcher 2. Corazon Aquino
3. Nelson Mandela 4. Mikhail Gorbachev 5. times
6. divided 7. $\frac{p}{7}$ 9. $\frac{18}{r}$ 11. $z + 4$ 13. $\frac{a}{b}$
15. $6c - 5$ 17. $\frac{d}{k} - 6$ 19. $0.8y$ 21. $x - \frac{3}{4}$ 23. $5p$
25. $\frac{p}{5}$ 27. $7p + 4$ 29. $29 - p$ 31. A 33. D

Pages 32–33

1. 1 2. 3 3. 2 4. 36 5. 2 7. 3 9. 1 11. 3
13. 2 15. 10 17. 6 19. 20 21. 1 23. 72 25. 5
27. 8 29. 18 31. 36 33. 2 35. 16 37. 3
39. 50 41. 9 43. 5 45. 3 47. 25 49. 13 51. 5
53. 1

Pages 34–35

1. $.59 2. .59 3. $1.42; yes, it assures that you
have enough money to buy the bananas. 4. a. 6, 9,
1, 8 b. 7, 6, 0, 3 5. b, e, f 7. 9 9. 2.604
11. 148 13. 12 15. 22.05 17. 22.776
19. 7.7964 21. 609.9 23. 0.24275 25. 107.8
27. 0.048 29. 5.4 31. 9.7 33. 0.14 35. 1.2
37. 1.9 39. $3.16 42. 38 43. 23 44. 31

Pages 36–37

1. Swimming 2. 2.5×1000 3. 2500 5. 7200
7. 856 9. 8560 11. 860 13. 296 15. 296
17. 51.3 19. 5130 21. 940,000 23. 94,000
25. 8300 27. 83,000 29. 76 31. 7600 33. 9
35. 9 37. 421,000 39. 4210 41. 64,200 43. 642
45. 47.6 47. 47,600 49. Golf 51. 2100
53. Swimming, golf, tennis 55. 800

Pages 38–39

1. 6 2. 64.32 3. 6 4. a. 10.72 s b. The total
number of seconds was reduced for the estimated

quotient. 5. a. 4, 8; 0.05 b. 2, 4, 1; 0.24 7. b
9. a 11. 2.54 13. 4.2 15. 2.74 17. 6.4 19. 3.71
21. 8.05 23. 1.89 25. 0.68 27. 0.70 29. 1.24
31. 7.32 33. 0.51 35. 0.3 37. 2.7 39. 3.6
41. 5.6 43. 2.3 45. 1 47. 3.6 49. Dave
51. 10.15 s

Pages 40–41

1. 26.2188 2. 1.5 3. one 4. 17.4792 5. a. 162,
108 b. 24, 36 7. 15.6 9. 10.5 11. 38 13. 12.7
15. 1.15 17. 0.045 19. 4.6 21. 4.5 23. 6 25. 3
27. 7.9 29. 23.6 31. 4.3 33. B 35. B

Pages 42–43

1. left 2. 2 3. 3 4. a. 3.65 b. 0.063 c. 0.0172
5. 7.6 7. 0.076 9. 21.72 11. 0.2172 13. 0.0681
15. 6.81 17. 8 19. 0.08 21. 4.123 23. 0.04123
25. 0.846 27. 84.6 29. 0.2147 31. 21.47
33. 0.242 35. 0.00242 37. 25 39. 2 41. 5
45. Waikiki

Page 46

1. a. $.27 b. $.92 c. $1.19 3. $1.57 5. $1.81
7. $.28 + .25n$ 9. $.19 + .17n$

Page 47

1. 18 3. 16 5. 13 7. 0 9. 37.9 11. 17.6
13. 2.2 15. 3.81 17. 21 19. 23 21. 29 23. 63
25. 5.11 27. 20.74 29. 6.34 31. 33.7 33. $3w$
35. $z - 6$ 37. $18c$ 39. $s + 15$ 41. $s + t$
43. $r + r$ 45. $y - 4.7$ 47. $0.6 + t$ 49. 3.6 51. 3
53. 6 55. 2.1 57. 9.6 59. 14.88 61. 2.2 63. 2.9
65. 1.48 67. 11.6 69. 12.6 71. 1.44 73. 2.2
75. 0.97

Pages 48–49

1. a. Yes b. $n = 4$ c. 4 2. 3 3. a. 5 b. 0.9
c. 3 4. B: Divide by 7 to undo multiplication by 7,
C: Divide both sides by the same number 5. 5
7. 6 9. 9 11. 17 13. 16 15. 20 17. 4.3
19. 20 21. 7 23. 3 25. 45 27. 17 29. 0
31. 60 33. 107 35. 39 37. 1.5 39. 7.6
41. Yes. It holds 968 pennies, and dimes are smaller
than pennies; so it will likely hold 1000 dimes.

Pages 50–51

1. B 2. a. 5, 75 b. 75 3. a. 3 b. 1.7 c. 5
4. B: Multiply both sides by the same number, C:
Multiply by 0.4 to undo division by 0.4 5. 70 7. 32
9. 56 11. 63 13. 84 15. 9.2 17. 15 19. 0.85
21. 128 23. 75 25. 65 27. 450 29. 42 31. 0.37
33. 18.5 35. 0 37. B, 108 39. D, 2187
41. C, $3

Pages 52–53

1. a. right **b.** left **2. a.** 84 **b.** 17 **c.** 9.4 **3.** C: Symmetric property of equality, D: Divide by 1.2 to undo multiplication by 1.2 **5.** 5 **7.** 19 **9.** 26 **11.** 44 **13.** 66 **15.** 7 **17.** 9 **19.** 150 **21.** 51 **23.** 13 **25.** 0 **27.** 2.6 **29.** 13 **31.** 0.98 **33.** 1.4 **35.** 0.28 **37.** No, because the problem can be solved without Clue 2.

Pages 54–55

1. each **2. a.** cost **b.** 146.25 **c.** 3 **3.** Yes **5.** B, 6 **7.** C, 7.5 **9.** D, 10.5 hours **13.** $\frac{j}{14} = 6.5$, 91 **15.** $n + 17.05 = 46$, 28.95 **17.** $3f = 27$, 9 **19.** $a - 6 = 19$, 25 **21.** $t - 6 = 17$, 23

Page 56

1. $37.33 **3.** 24, slide

Page 57

1. 17 **3.** 25 **5.** 0 **7.** 6 **9.** 15 **11.** 23 **13.** 5.8 **15.** 5.7 **17.** 0 **19.** 7.1 **21.** 7 **23.** 7 **25.** 20 **27.** 1 **29.** 13 **31.** 0 **33.** 0.7 **35.** 7 **37.** 10 **39.** 0 **41.** 42 **43.** 0 **45.** 100 **47.** 28 **49.** 9 **51.** 98 **53.** 0.18 **55.** 3.2 **57.** 1 **59.** 0 **61.** 5 **63.** 8 **65.** 1 **67.** 50 **69.** 0 **71.** 1 **73.** 8 **75.** 0.15 **77.** 230 **79.** 10

Chapter 3 ■ Solving Equations Using Number Properties

Pages 62–63

1. 15 **2. a.** No **b.** Yes **3.** Yes **4.** Because it is within grouping symbols. **5. a.** 3, 11 **b.** 0.6, 0.2 **c.** 1.05, 3.65 **7.** 38 **9.** 1.4 **11.** 3.2 **13.** 99 **15.** 19 **17.** 0 **19.** 192 **21.** 54 **23.** 25 **25.** 215 **27.** 17.6 **29.** 16 **31.** 81 **33.** 0.28 **35.** 4.16 **37.** 0.96 **39. a.** 8 ÷ 4 + 16 = **b.** 4 × 2 + 20 − 9 = **c.** 5.3 × 3.14 = M+ , 22.1 − MR =

Pages 64–65

1. Yes **2.** Yes **3.** Yes **4. a.** x **b.** 7 **c.** 0 **d.** z **e.** r **5.** 59 **7.** 77 **9.** 168 **11.** 45 **13.** 122 **15.** 141 **17.** 134 **19.** 147 **21.** 164 **23.** 15.7 **25.** 17.1 **27.** 1.85 **29.** $160.25 **31.** $200 **33.** $25.75 **35.** $20.50 **37.** $210.25 **39.** $55.75 **41.** $95.50 **43.** $195.50 **45.** $350 **47.** $70.50 **49.** $181 **51.** $96

Pages 66–67

1. Yes **2.** Yes **3.** Yes **4. a.** n **b.** a **c.** t **d.** b **e.** 3 **f.** d **g.** r **h.** 5 **5.** 160 **7.** 280 **9.** 1700 **11.** 720 **13.** 4800 **15.** 6300 **17.** 4900 **19.** 8100 **21.** 36 **23.** d **25.** a **27.** f **29.** 18 **31.** 73 **33.** y **35.** 7 **37.** 27 **39.** 4 **41.** Brent: 2, Dick: 3, Alexa: 4, Sarah: 5

Pages 68–69

1. 54 **2.** B **3. a.** 17 **b.** 26 **c.** 0.2 **4.** C: Commutative property of addition; D: Commutative property of multiplication; E: Symmetric property of equality **5.** 37 **7.** 8 **9.** 20 **11.** 36 **13.** 16 **15.** 16 **17.** 16 **19.** 35 **21.** 76 **23.** 1.875 **25.** 120 **27.** 66 **29.** 0 **31.** 81 **33.** 23.8 **35.** 176 **37.** 1.8 **39.** 17.6 **41.** 9.3 **43.** 13.4 **45. a.** 172 **b.** 312 **c.** 5 **d.** 3 **e.** 51

Page 72

1. b **3.** b **5.** $82.038 = 1.32d$, 62.15 ft

Page 73

1. 4.8 **3.** 9.5 **5.** 8.1 **7.** 29.2 **9.** 5.5 **11.** 406.3 **13.** 0.1 **15.** 34.3 **17.** 11.6 **19.** 15.3 **21.** 0.7 **23.** 2.3 **25.** 16.3 **27.** 26 **29.** 9 **31.** 0 **33.** 7.2 **35.** 4.9 **37.** 26.5 **39.** 12.8 **41.** 14 **43.** 33 **45.** 17 **47.** 93 **49.** 7.2 **51.** 9.1 **53.** 11.3 **55.** 21.4 **57.** 1.1 **59.** 1.3 **61.** 5.5 **63.** 3.8 **65.** 7.18 **67.** 1.4

Pages 74–75

1. $16 **2.** $2 **3.** Number of books ordered **4.** 2, 16 **5.** 3 **7.** 4, 30, 5, 6 **9.** 5 **11.** 9 **13.** 7 **15.** 11 **17.** 7 **19.** 1 **21.** 12 **23.** 15 **25.** 12 **27.** 3 **29.** 1 **31.** 3.1 **33.** 1.9 **35.** 4 **37.** 4.2 **39.** 15.1 **41. a.** 49.85 **b.** 66.35 **c.** 6 **d.** 5 **e.** 4.20 **f.** 3.10 **g.** 148.80 **h.** 164.90 **i.** 5.15 **j.** 13

Pages 76–77

1. a. 19, 7 **b.** 7, yes **2.** B: Symmetric property of equality; C: Divide both sides by the same number **3.** 21 **5.** 3.2 **7.** 0.4 **9.** 2 **11.** 490 **13.** 3.9 **15.** 30 **17.** 3 **19.** A, 32 **21.** C, 2

Pages 78–79

1. Yes **2. a.** 6 **b.** n **c.** 2 **d.** y **3. a.** 7, 6, 42 **b.** 8, 25, 100, 8 **5.** t **7.** 6 **9.** 6 **11.** y **13.** 5 **15.** 80 **17.** 248 **19.** 1414 **21.** 1836 **23.** 40 **25.** 50 **27.** 400 **29.** 700 **31.** $15 **33.** $14 **35.** $11 **37.** $8.25 **39.** $13

Pages 80–81

1. a. 11n **b.** 1 **c.** t **d.** 8 **e.** 1 **f.** 12 **3.** 14n **5.** 7n **7.** 4n **9.** 12b **11.** 16m **13.** 15c

15. $10y + 12$ **17.** $28g + 4$ **19.** $54t$ **21.** $7a$
23. $14r + 1$ **25.** $24c + 3$ **27.** 300   **29.** 2000
31. 600   **33.** 660   **35.** 2400   **37.** 840   **39.** 420
41. 3400   **43.** 1600   **45.** Gary   **47.** $3n + 11$

Pages 82–83
1. a. 10, 10   **b.** \$1.25, yes   **2.** B: Combining like
terms, C: Combining like terms, D: $36 \div 9$ **3.** 4
5. 7   **7.** 20   **9.** 2   **11.** 12   **13.** 3   **15.** 0.8   **17.** 2.2
19. 0.24   **21.** 4.2   **23.** 50   **25.** 9   **27.** 9   **29.** 2
31. 7   **33.** 12   **35.** 3   **37.** 3   **39.** 0.6   **41.** 26.4
43. 3.125   **45.** 0.3   **47.** C, \$1.85   **49.** D, \$3.40

Pages 84–85
1. a. 30   **b.** $4h + 30$ **c.** 30, 98   **2.** Yes   **3. a.** 5
b. 169   **c.** 5   **d.** 169   **e.** 41   **5.** $7n - 9 = 68$, 11
7. $\frac{m}{6} - 5 = 4$, 54   **9.** $7p + 2 = 44$, 6   **11.** $6r + 10 = 64$, 9   **13.** $17 = 3l + 2$, \$5

Page 86
1. B, \$700, 14   **3.** 420   **5.** 560   **7.** B

Page 87
1. 80   **3.** 0.8   **5.** 3420   **7.** 13,600   **9.** 5670
11. 43.7   **13.** 45   **15.** 21.6   **17.** 0.8   **19.** 6
21. 69.8   **23.** 23.4   **25.** 4.2   **27.** 0.086   **29.** 0.3851
31. 0.03851   **33.** 42.1   **35.** 0.01256   **37.** 3.36
39. 8.25   **41.** 0.72   **43.** 0.563   **45.** 9.5   **47.** 24.48
49. 0.96   **51.** 1.666   **53.** 3.84   **55.** 50.9   **57.** 4.9
59. 1.7   **61.** 21.8   **63.** 2.4   **65.** 3   **67.** 19.5   **69.** 5.3
71. 2   **73.** 1.2   **75.** 12.3

Chapter 4 ■ **Integers and Equations**

Pages 92–93
1. $^{+}105°$ **2.** $^{-}19°$ **3. a.** Less than   **b.** Greater than
4. a. 4   **b.** 3   **5.** <   **7.** <   **9.** <   **11.** <   **13.** >
15. <   **17.** >   **19.** >   **21.** >   **23.** >   **25.** 6   **27.** 5
29. 0   **31.** 17   **33.** 19   **35.** positive   **37.** zero
39. less   **41.** California   **43.** Florida   **45.** $^{-}70°$,
$^{-}59°$, $^{-}54°$, $^{-}52°$, $^{-}48°$(two), $^{-}45°$, $^{-}40°$, $^{-}23°$, $^{-}19°$
(two), $^{-}2°$ **47. a.** 0   **b.** 10   **c.** 15   **d.** 77   **e.** 86
f. 212

Pages 94–95
1. opposites   **2.** $^{+}2$ **3. a.** $^{-}2$ **b.** 0   **c.** $^{+}3$ **5.** $^{+}1$
7. $^{+}2$ **9.** 0   **11.** $^{-}1$ **13.** $^{+}2$ **15.** 0   **17.** $^{-}9$
19. $^{+}2$ **21.** $^{+}4$

Pages 96–97
1. $^{+}3$ **2.** $^{-}1$ **3. a.** $^{-}2$ **b.** $^{+}1$ **4.** 0, yes   **5.** $^{-}8$,
$^{+}6$, 0   **7.** $^{+}4$ **9.** $^{-}9$ **11.** $^{+}7$ **13.** $^{-}8$ **15.** 0

17. $^{+}10$ **19.** $^{+}2$ **21.** $^{-}25$ **23.** $^{+}8$ **25.** $^{-}98$
27. $^{-}68$ **29.** 4   **31.** $^{+}2$ **33.** $^{-}10$ **35.** $^{-}42$ **37.** $^{+}32$
39. $^{-}40$ **41.** $^{-}22$ **43.** $^{+}24$ **45.** $^{-}60$ **47.** False
49. True   **51.** False   **53.** True

Pages 98–99
1. $^{+}3$ **2. a.** $^{+}2$ **b.** $^{+}2$ **3.** Effect is the same.
4. a. $^{-}7$ **b.** $^{+}6$ **5.** C and D: To subtract an
integer, add the opposite of the integer; E: $^{-}7 + {}^{+}12$
7. $^{-}3$ **9.** $^{-}5$ **11.** $^{+}4$ **13.** $^{-}13$ **15.** $^{-}6$ **17.** $^{+}9$
19. $^{-}10$ **21.** $^{-}11$ **23.** $^{+}7$ **25.** $^{-}12$ **27.** $^{-}9$
29. $^{-}7$ **31.** $^{-}12$ **33.** $^{+}9$ **35.** $^{-}16$ **37.** $^{+}26$ **39.** 0
41. 0   **43.** $^{-}30$ **45.** $^{-}34$ **47.** $^{+}27$ **49.** $^{-}6$ **51.** $^{+}14$
53. 0   **55.** 18   **57.** 15   **59.** $^{+}13$ **61.** $^{+}11$ **63.** $^{-}15$
65. $^{-}17$ **67.** $^{-}4$ **69.** $^{-}36$ **71.** 13   **73.** 7   **75.** 6
77. $^{+}12$ **79.** $^{+}2$ **81.** $k - {}^{+}8$ **83.** $x + y$ **85.** $r - c$
87. $a + a$ **89.** $d - {}^{-}3$

Pages 100–101
1. 0   **2.** $^{+}4$ **3.** $^{+}4$ **4.** positive   **5.** negative
6. positive   **7.** zero   **9.** $^{-}24$ **11.** $^{-}8$ **13.** $^{-}8$
15. $^{+}64$ **17.** 0   **19.** $^{+}40$ **21.** $^{+}72$ **23.** $^{-}24$ **25.** 0
27. 0   **29.** $^{+}12$ **31.** $^{-}28$ **33.** 15   **35.** 32   **37.** 24
39. $^{+}198$ **41.** $^{-}224$ **43.** $^{-}9$ **45.** $^{-}48$ **47.** 0
49. 18   **51.** $^{-}3$ **53.** $^{+}9$ **55.** $^{+}18$ **57.** $^{+}18$ **59.** $^{-}9$
61. $^{-}11$ **63.** True   **65.** False   **67.** False

Pages 102–103
1. $^{+}5$ **2.** $^{+}9$ **3.** 0   **4.** positive   **5.** positive
6. negative   **7.** negative.   **9.** $^{-}8$ **11.** $^{-}4$ **13.** $^{-}7$
15. $^{+}7$ **17.** $^{-}7$ **19.** 5   **21.** $^{-}6$ **23.** $^{-}4$ **25.** $^{-}4$
27. $^{-}5$ **29.** $^{-}7$ **31.** $^{-}6$ **33.** 7   **35.** 4   **37.** $^{-}108$
39. $^{-}25$ **41.** 78   **43.** $^{-}12$ **45.** 15   **47.** $^{-}2$ **49.** 51
51. $^{-}90$ **53.** $^{-}133$

Page 106
1. \$.08   **3.** \$7.50   **5. a.** 1.20, 1.20   **b.** 1.20   **c.** 5.20
d. 2.5 lb   **7.** $1.80p + 1.89 = 7.29$, 3

Page 107
1. 2.2   **3.** 2.6   **5.** 9.9   **7.** 7.5   **9.** 0.6   **11.** 0.08
13. 6   **15.** 0.96   **17.** 2   **19.** 23   **21.** 14   **23.** 14
25. 7   **27.** 50   **29.** 22   **31.** 17   **33.** 56   **35.** 68
37. 78   **39.** 160   **41.** 34   **43.** 36   **45.** 192   **47.** 2.3
49. 195   **51.** c   **53.** a   **55.** b   **57.** d

Pages 108–109
1. a. 14, $^{-}20$ **b.** $^{-}8$, $^{-}6$ **c.** $^{-}6$ **2. a.** $^{-}4$ **b.** 7
c. $^{-}18$, $^{-}3$ **3.** C: Subtract $^{-}10$ to undo the addition
of $^{-}10$, $^{-}28 + 10$; D: Commutative property of
addition, subtract the same number from both sides,
$^{-}10 + {}^{-}5$ **5.** 14   **7.** $^{-}10$ **9.** 26   **11.** 0   **13.** $^{-}20$
15. $^{-}13$ **17.** $^{-}1$ **19.** 63   **21.** 22   **23.** $^{-}5$ **25.** 6

27. 99 **29.** ⁻10 **31.** ⁻7 **33.** 100 **35.** ⁻5 **37.** ⁻8
39. ⁻27 **41.** B, $50 **43.** C, ⁻50

Pages 110–111

1. a. ⁻3, 5 **b.** ⁻4, ⁻80 **c.** ⁻80 **2. a.** ⁻8 **b.** ⁻3, 15
3. C: Multiply by 7 to undo the division by 7, 7 · ⁻3;
D: Symmetric property of equality, 42 ÷ ⁻6 **5.** ⁻1
7. ⁻8 **9.** ⁻12 **11.** 10 **13.** 0 **15.** 19 **17.** ⁻1
19. ⁻36 **21.** ⁻50 **23.** 66 **25.** 1 **27.** ⁻30 **29.** ⁻3
31. ⁻7 **33.** ⁻18 **35.** 9 **37.** ⁻5 **39.** 18 **41.** ⁻24
43. ⁻6 **45.** 11 **47.** ⁻10 **49. a.** 8 **b.** ⁻4 **c.** 36
d. ⁻18

Pages 112–113

1. ⁻35 feet **2.** Number of minutes of descent
3. ⁻35, ⁻15 **4.** 7, yes **5.** 80, 20, ⁻4, ⁻5 **7.** 8, ⁻32,
4, ⁻128 **9.** 5 **11.** ⁻4 **13.** ⁻12 **15.** ⁻35 **17.** 5
19. ⁻6 **21.** ⁻48 **23.** 35 **25.** ⁻4 **27.** 2 **29.** ⁻2
31. ⁻42 **33.** ⁻5 **35.** ⁻7 **37.** ⁻35 **39.** ⁻36 **41.** 2
43. 0 **45.** ⁻3720

Pages 114–115

1. a. lowest, 187 **b.** $n + 187$, 117 **c.** $n + 187 = 117$
d. 187 **2.** Yes **3. a.** 100 **b.** 100 **c.** ⁻56 **d.** 44°F
5. $d + 9 = ⁻7$, ⁻16 **7.** ⁻12$f = ⁻72$, 6 **9.** $j + ⁻9 =$
⁻12, ⁻3 **11.** $4t - 8 = ⁻32$, ⁻6 **13.** $n - 18 = 7$, 25°F

Page 116

1. 1771 mi **3.** Denver, Los Angeles, 2055 **5.** No
7. Yes

Page 117

1. 5 **3.** 11 **5.** 3 **7.** 0 **9.** 25 **11.** 12 **13.** 1.8
15. 2.9 **17.** 0.1 **19.** 1.7 **21.** 32 **23.** 22 **25.** 7
27. 150 **29.** 4 **31.** 8 **33.** 1 **35.** 0 **37.** 30.1
39. 0 **41.** $8a$ **43.** $4c$ **45.** $13x$ **47.** $17w + 6$
49. $2r + 11$ **51.** $6n + 17$ **53.** $11t$ **55.** $6n + 15$
57. 15 **59.** 13 **61.** 21 **63.** 5 **65.** 13 **67.** 5
69. 0.79 **71.** 0.11 **73.** 6.2 **75.** 2.49

Chapter 5 ■ Algebra Using Exponents and Fractions

Pages 122–123

1. a. 3 **b.** x **c.** 3 **d.** b **e.** 2 **f.** 14 **g.** y **h.** y
i. 3 **3.** 4^3 **5.** $2^3 \cdot 3^2$ **7.** $8^2 \cdot 10^3$ **9.** ab^3 **11.** x^3y
13. x^3y^2 **15.** m^3n^3 **17.** $6a^2$ **19.** $21a^2$ **21.** $12ab^2$
23. ac^2d^2 **25.** ac^3d **27.** $18m^2n^2$ **29.** $12a^2b^2$
31. $24a^3b^2$ **33.** $24r^3s^2$ **35.** $6x^2y^2$ **37. a.** 16 **b.** 64
c. 144 **d.** 256 **e.** 400

Pages 124–125

1. a. Yes **b.** Yes **2.** Yes **3.** Yes **5.** 5 **7.** 6^7
9. 8^8 **11.** 2^5 **13.** d^2 **15.** f^9 **17.** m^7 **19.** t^8
21. 2 **23.** 10^3 **25.** 4^4 **27.** 8^6 **29.** 5^3 **31.** e
33. n^2 **35.** r^4 **37.** ⁻3 **39.** 0 **41.** 3 **43.** 0

Pages 126–127

1. 3.5 **2.** 10^{-3} **3.** 7 **5.** 6 **7.** 7 **9.** ⁻4 **11.** ⁻3
13. ⁻5 **15.** $3.1 \cdot 10^4$ **17.** $5.46 \cdot 10^6$ **19.** $2.36 \cdot 10^5$
21. $5.1 \cdot 10^4$ **23.** $7 \cdot 10^8$ **25.** $1.11 \cdot 10^6$
27. $7.5 \cdot 10^{-4}$ **29.** $6 \cdot 10^{-5}$ **31.** $3.17 \cdot 10^{-2}$
33. $6.49 \cdot 10^{-3}$ **35.** $6 \cdot 10^{12}$ **37.** $8.64 \cdot 10^6$
39. $1.10376 \cdot 10^{14}$ **41.** $5 \cdot 10^9$

Page 129

1. Yes **3.** Yes **5.** Yes **7.** Yes **9.** Yes **11.** Yes
13. Yes **15.** No **17.** Yes **19.** No **21.** Yes **23.** Yes
25. No **27.** No **29.** Yes **31.** Yes **33.** Yes **35.** Yes
37. Yes **39.** Yes **41.** Yes **43.** Yes **45.** No **47.** Yes
49. True **51.** True **53.** True **55.** 96

Pages 130–131

1. 3 **2. a.** 13 **b.** 2 **c.** 5 **3.** 3 **4. a.** c **b.** 3
c. r **5.** $5 \cdot 7$ **7.** $2 \cdot 5$ **9.** $3 \cdot 5$ **11.** $2 \cdot 2 \cdot 3$
13. $2 \cdot 11$ **15.** $2 \cdot 2 \cdot 2$ **17.** $3 \cdot 7$ **19.** $2 \cdot 2 \cdot 3$
21. $2 \cdot 5 \cdot 5$ **23.** $2 \cdot 2 \cdot 11$ **25.** $2 \cdot 19$ **27.** $2 \cdot 2 \cdot 2 \cdot$
2 **29.** $2 \cdot 2 \cdot 2 \cdot 2 \cdot 3$ **31.** $3 \cdot 11$ **33.** $7 \cdot 7$
35. $3 \cdot 3 \cdot 5$ **37.** $2 \cdot 29$ **39.** $2 \cdot 17$ **41.** 5^2 **43.** $2^3 \cdot 5$
45. 2^6 **47.** $2 \cdot 3 \cdot 11$ **49.** $2 \cdot 3^3$ **51.** $2 \cdot 31$
53. $5 \cdot 13$ **55.** $3 \cdot 23$ **57.** $2 \cdot 5 \cdot 7$ **59.** $2 \cdot 2 \cdot$
$x \cdot x \cdot x$ **61.** $2 \cdot 5 \cdot c \cdot c$ **63.** $2 \cdot 2 \cdot 3 \cdot d$ **65.** $3 \cdot 5 \cdot$
$g \cdot g$ **67.** $2 \cdot 2 \cdot 3 \cdot 3 \cdot y$ **69.** $5 \cdot 7 \cdot t \cdot t \cdot t$
71. $2 \cdot 2 \cdot 2 \cdot 2 \cdot 2 \cdot u$ **73.** $3 \cdot 3 \cdot 5 \cdot d \cdot d \cdot d$
75. $11 \cdot g$ **77.** $2 \cdot 2 \cdot 2 \cdot 3 \cdot j \cdot j$ **79.** $2 \cdot 2 \cdot 2 \cdot 2 \cdot$
$2 \cdot a \cdot a \cdot b$ **81.** $7 \cdot a \cdot b \cdot b \cdot b$ **83.** $2 \cdot 3 \cdot 3 \cdot 3 \cdot$
$a \cdot a \cdot b \cdot b$

Pages 132–133

1. 6 **2.** 72 **3.** 4 **5.** 6 **7.** 8 **9.** 12 **11.** 3 **13.** 2
15. 12 **17.** 5 **19.** 3 **21.** 1 **23.** 2 **25.** $3d$ **27.** $2x$
29. x **31.** $5u$ **33.** $3c^2$ **35.** yz **37.** $4a$ **39.** 6
41. 12 **43.** 28 **45.** 24 **47.** 35 **49.** 12 **51.** 33
53. 12 **55.** 60 **57.** 60 **59.** $10w$ **61.** $16x$
63. $45b^2$ **65.** $24d^2e$ **67.** $48x^2y$ **69.** $55x^2y$ **71.** 30

Page 136

1. 1870 **3.** $14.75 **5. a.** .80 **b.** .80 **c.** 8.80
d. $2.00 **7.** $5n + 6.50 = 15.25$, $1.75

Page 137

1. 7 **3.** 7 **5.** 7 **7.** 48 **9.** 18 **11.** 9 **13.** 1.48
15. 3.2 **17.** $12k$ **19.** $2r$ **21.** $9x$ **23.** $9y + 3$
25. $12y$ **27.** $12n$ **29.** 21 **31.** 18 **33.** 15 **35.** 18

37. 1.6   39. 1.05   41. 2.5   43. 29.2   45. <   47. <
49. <   51. <   53. >   55. 6   57. 0   59. 11
61. $^+4$ 63. $^-12$ 65. $^-17$ 67. $^+26$ 69. $^-34$

35. $^-15$ 37. $^+44$ 39. $^+34$ 41. $^-110$ 43. $^-8$
45. $^-150$ 47. $^-12$ 49. 0   51. 19   53. 1   55. $^-7$
57. $^-9$ 59. $^-48$ 61. 2700   63. $^-4$ 65. 3   67. 2
69. $^-77$ 71. $^-28$ 73. 35

Pages 138–139
1. $\frac{1}{4}$ 2. $\frac{2}{3}$ 3. Agree   4. $\frac{15}{24}$ 5. $2b$ 6. a. 6   b. 3,
3   c. $2c$ d. $2t$, $2t$ 7. 8   9. 21   11. 6   13. 10
15. 12   17. 9   19. 42   21. 30   23. 20   25. 3
27. 50   29. 10   31. 40   33. 15   35. 40   37. xy
39. $9t$ 41. $25m$ 43. $4a$ 45. $3z^2$ 47. $10b$ 49. $4cd$
51. $20wx$ 53. 16   55. News anchor, U.S. president

Pages 140–141
1. 3   2. $2x$ 3. divided   4. $2x$ 5. 3   6. $2x$ 7. $\frac{2}{3}$
9. $\frac{2}{5}$ 11. $\frac{5}{12}$ 13. $\frac{1}{3}$ 15. $\frac{2}{3}$ 17. $\frac{2}{3}$ 19. $\frac{2}{3}$ 21. $\frac{11}{5}$
23. $\frac{5}{3}$ 25. $\frac{6}{7}$ 27. $\frac{5}{4}$ 29. $\frac{3a}{2}$ 31. $\frac{2a}{3}$ 33. $\frac{4a}{3}$
35. $\frac{1}{2}$ 37. $\frac{1}{4}$ 39. $\frac{m}{5}$ 41. $\frac{4k}{7}$ 43. $\frac{9}{7q}$ 45. $\frac{11}{24}$
47. $\frac{3}{2m^2}$ 49. $\frac{3m^2}{2n}$ 51. $\frac{3t}{8s}$ 53. $\frac{y^2}{3}$ 55. $\frac{7}{22}$ 57. $\frac{20}{23k}$

59. Utah

Pages 142–143
1. 6, 12   2. b, $3b$, $3b$ 3. a. 24, 24   b. $18n$, $18n$
c. a^2b^2 5. 18   7. 8   9. 24   11. 24   13. 6   15. 24
17. 18   19. 30   21. 30   23. 150   25. $5n$ 27. $6n$
29. $3d$ 31. $4r^2$ 33. a^3 35. $6b$ 37. $6a$ 39. $6a^2$
41. $12t^2$ 43. $18b^2$

Pages 144–145
1. $\frac{4}{9}$ 2. Hamilton   3. $\frac{4}{9}$ and $\frac{2}{9}$ 4. Lincoln   5. >
7. >   9. <   11. <   13. >   15. <   17. >   19. >
21. =   23. >   25. =   27. >   29. >   31. <
33. <   35. <   37. <   39. <   41. <   43. <
45. >   47. =   49. >

Pages 146–147
1. a. 40, n b. $n + n + 40$ c. 2, 400   d. Yes; then
Cora has $180 + 40$, or 220 stamps, and the total is
400.   3. a. 30, 30   b. 124   c. 77   5. $e + e + 9 =$
77, 34   7. $p + p - 10 = 80$, 45¢   9. $t + t - 45 =$
267, 156 lb   11. $d + d - 6 = 128$, 67

Page 148
1. $121.55   3. $166.86   5. 95th

Page 149
1. $^+2$ 3. $^-13$ 5. $^-8$ 7. 0   9. $^-4$ 11. $^+15$
13. $^+28$ 15. $^-7$ 17. $^+144$ 19. $^+160$ 21. $^+4$
23. $^-4$ 25. $^+8$ 27. $^-4$ 29. $^-6$ 31. $^+6$ 33. 0

Chapter 6 ■ Algebra Using Sums and Differences of Fractions

Pages 154–155
1. Yes   2. Yes   3. 3   4. multiply, add   5. $\frac{8}{2}$ 7. $\frac{12}{2}$
9. $\frac{10}{2}$ 11. $\frac{14}{2}$ 13. $\frac{32}{2}$ 15. $\frac{40}{2}$ 17. $\frac{15}{5}$ 19. $\frac{5}{5}$
21. $\frac{35}{5}$ 23. $\frac{60}{5}$ 25. $\frac{55}{5}$ 27. $\frac{45}{5}$ 29. $\frac{3}{2}$ 31. $\frac{4}{3}$
33. $\frac{5}{2}$ 35. $\frac{7}{4}$ 37. $\frac{14}{5}$ 39. $\frac{10}{3}$ 41. $\frac{27}{4}$ 43. $\frac{23}{4}$
45. $\frac{31}{6}$ 47. $\frac{27}{2}$ 49. $\frac{33}{2}$ 51. $\frac{58}{5}$ 53. Stuart Long
55. Nevaro Clark and Edita Sosa   57. Monday

Pages 156–157
1. 1   3. 5   5. 6   7. 4   9. 8   11. 14   13. 15
15. 21   17. 10   19. 3   21. $1\frac{1}{4}$ 23. $1\frac{5}{6}$ 25. $2\frac{3}{5}$
27. $1\frac{2}{9}$ 29. $2\frac{1}{2}$ 31. 4   33. $5\frac{1}{2}$ 35. $2\frac{7}{10}$
37. $3\frac{6}{7}$ 39. 6   41. 7   43. $8\frac{1}{3}$ 45. $17\frac{1}{2}$ 47. $3\frac{7}{10}$
49. 12   51. $6\frac{17}{20}$ 53. 15   55. Less

Pages 158–159
1. Yes   2. Yes   3. $\frac{3}{4}$ 5. $\frac{3}{5}$ 7. $\frac{4}{5}$ 9. $\frac{2}{3}$ 11. $\frac{1}{3}$
13. $\frac{4}{7}$ 15. $2\frac{2}{3}$ 17. $5\frac{5}{6}$ 19. $4\frac{1}{3}$ 21. $6\frac{1}{2}$ 23. $7\frac{2}{3}$
25. $8\frac{4}{5}$ 27. 2   29. $5\frac{2}{3}$ 31. $5\frac{1}{2}$ 33. 5   35. $3\frac{3}{4}$
37. 12   39. 3   41. $1\frac{3}{4}$ 43. $\frac{2}{3}$ 45. $\frac{12a}{b}$ 47. $\frac{4}{3y^2}$
49. $\frac{11r}{8s}$ 51. $\frac{3a}{4b}$ 53. $\frac{4x}{3y}$ 55. 2   57. $\frac{4m}{5}$ 59. $\frac{5}{n^2}$
61. $3y$

Pages 160–161
1. $\frac{7}{21}$ 2. 0.375   3. 0.333   5. 0.4   7. 0.375
9. $0.444\frac{4}{9}$ 11. $0.222\frac{2}{9}$ 13. 0.2   15. $1.666\frac{2}{3}$
17. $0.666\frac{2}{3}$ 19. 0.625   21. $1.166\frac{2}{3}$ 23. $\frac{6}{25}$ 25. $\frac{5}{8}$
27. $\frac{1}{2}$ 29. $\frac{4}{25}$ 31. $1\frac{3}{4}$ 33. $\frac{7}{8}$ 35. $\frac{3}{4}$ 37. $2\frac{1}{4}$
39. $\frac{2}{3}$ 41. $\frac{1}{12}$ 43. 0.313   45. Rivera, Brenner,
Shepard, Carlo, McGrath

1. Swim goggles **3.** Bike lock **5.** Yes **7. a.** .30 **b.** .30 **c.** 1.20 **d.** $.50 **9.** $1.7b + .87 = 16.00$, $8.90

1. 67 **3.** 10.2 **5.** $2\frac{3}{4}$ **7.** 18.2 **9.** 12.8 **11.** 0.1
13. 3 **15.** 12 **17.** 7 **19.** 126 **21.** 108 **23.** 6
25. 3.2 **27.** 6 **29.** 17 **31.** 19 **33.** 14 **35.** 17
37. 1.7 **39.** 1.05 **41.** 3.5 **43.** 33.2 **45.** $^{+}14$
47. $^{-}12$ **49.** $^{+}8$ **51.** $^{+}11$ **53.** $^{-}6$ **55.** $^{+}18$
57. $^{+}60$ **59.** $^{-}48$ **61.** 0 **63.** $^{-}32$ **65.** $^{-}45$
67. $^{+}160$

1. $\frac{3}{8}$ mi **2.** $\frac{5}{8}$ and $\frac{7}{8}$ **3.** $1\frac{1}{2}$ mi **4. a.** 7, 2 **b.** t
c. z, 15 **5.** $\frac{5}{8}$ and $\frac{3}{8}$ **6.** $\frac{1}{4}$ mi **7.** C: $8 \div 2$, D:
Common denominator **9.** $1\frac{2}{7}$ **11.** $1\frac{1}{7}$ **13.** $1\frac{1}{4}$
15. $\frac{2}{5}$ **17.** $\frac{1}{4}$ **19.** $\frac{r+s}{t}$ **21.** $\frac{3x+w}{y}$ **23.** $\frac{12}{a^2}$
25. $\frac{15}{pq}$ **27.** $\frac{32}{a^2b}$ **29.** $\frac{1}{3}$ **31.** $\frac{2}{3}$ **33.** $\frac{3}{4}$ **35.** $\frac{1}{2}$
37. 2 **39.** $\frac{a-c}{d}$ **41.** $\frac{a-2b}{g}$ **43.** $\frac{7}{w^2}$ **45.** $\frac{6}{rs}$ **47.** $\frac{17}{c^2d}$
49. $\frac{3}{4}$ mi **51.** No **53.** Decreases **55.** Increases

1. $\frac{5}{12}$ and $\frac{1}{8}$ **2.** $\frac{13}{24}$ **3.** $\frac{1}{8}$ **4.** $\frac{7}{8}$ **5.** $\frac{5}{6}$ **7.** $1\frac{1}{15}$
9. $1\frac{5}{24}$ **11.** $\frac{4}{5}$ **13.** $\frac{19}{24}$ **15.** $\frac{3a+2b}{6}$ **17.** $\frac{c+2d}{4}$
19. $\frac{5r+4s}{20}$ **21.** $\frac{8+15q}{10}$ **23.** $\frac{8u+25y}{40}$ **25.** $\frac{15b+14c}{18}$
27. $\frac{6g+7a}{10}$ **29.** $\frac{8x+21y}{36}$ **31.** $\frac{1}{4}$ **33.** $\frac{3}{10}$ **35.** $\frac{1}{12}$
37. $\frac{1}{2}$ **39.** $\frac{7}{8}$ **41.** $\frac{6m-5}{30}$ **43.** $\frac{2x-3y}{12}$ **45.** $\frac{8a-3b}{12}$
47. $\frac{15k-4}{10}$ **49.** $\frac{5n-2}{6}$ **51.** $\frac{15x-8y}{20}$
53. $\frac{16m-35n}{40}$ **55.** $\frac{5}{24}$ **57.** $\frac{23}{24}$

1. $2y$ **2.** y **3. a.** $2y$, $2y$ **b.** $3r$, $3r$ **4.** b **5. a.** $2y$,
$2y$ **b.** $5n$, $5n$ **7.** $\frac{a+2b}{ab}$ **9.** $\frac{2+b}{ab}$ **11.** $\frac{6+xy}{3x}$
13. $\frac{xy+3}{y}$ **15.** $\frac{1+2x}{xy}$ **17.** $\frac{3b+2a^2}{ab}$ **19.** $\frac{3+2a}{ab}$
21. $\frac{s+2r}{rs}$ **23.** $\frac{10+r}{2s}$ **25.** $\frac{2+5r}{r^2}$ **27.** $\frac{ab-1}{b}$
29. $\frac{a^2-b^2}{ab}$ **31.** $\frac{10-xy}{5x}$ **33.** $\frac{xy-3}{y}$ **35.** $\frac{y-2}{xy}$

37. $\frac{2b-3a^2}{ab}$ **39.** $\frac{3-2a}{ab}$ **41.** $\frac{r-3s}{rs}$ **43.** $\frac{rs-2}{r}$
45. $\frac{3-2r}{r^2}$ **47.** Numerator and denominator must
be multiplied by the same number. $\frac{7s-6r}{rs}$
49. Squaring numerator and denominator does not
give an equivalent fraction. $\frac{pa+h^2}{a^2}$

1. Add $145\frac{1}{2}$ and $23\frac{1}{4}$. **2.** Subtract $17\frac{3}{8}$ from $38\frac{3}{4}$.
3. $168\frac{3}{4}$ lb, $21\frac{3}{8}$ mi **4. a.** $\frac{3}{5}$ **b.** 6 **c.** $\frac{2}{3}$ **d.** $\frac{3}{12}$
5. $4\frac{5}{6}$ **7.** $5\frac{7}{8}$ **9.** $7\frac{7}{10}$ **11.** $9\frac{13}{24}$ **13.** $16\frac{13}{24}$
15. $4\frac{1}{2}$ **17.** $4\frac{13}{20}$ **19.** $7\frac{11}{12}$ **21.** $2\frac{3}{8}$ **23.** $3\frac{1}{8}$
25. $3\frac{3}{8}$ **27.** $7\frac{1}{4}$ **29.** $9\frac{1}{8}$ **31.** $1\frac{1}{10}$ **33.** $1\frac{3}{8}$
35. $7\frac{11}{24}$ **37.** A, $1\frac{1}{4}$ mi **39.** B, $3\frac{3}{4}$ mi

1. $1\frac{2}{3}$ and $3\frac{1}{2}$ **2.** $6\frac{1}{3}$ and $4\frac{1}{2}$ **3. a.** $5\frac{1}{6}$ **b.** $1\frac{5}{6}$
cups **4. a.** 5, 24 **b.** 10 **c.** 12 **5.** $6\frac{1}{6}$ **7.** $9\frac{3}{8}$
9. $5\frac{1}{2}$ **11.** $9\frac{9}{10}$ **13.** $7\frac{9}{10}$ **15.** $5\frac{1}{24}$ **17.** $5\frac{1}{20}$
19. $8\frac{1}{12}$ **21.** $4\frac{3}{8}$ **23.** $2\frac{3}{4}$ **25.** $5\frac{5}{6}$ **27.** $6\frac{1}{8}$
29. $1\frac{1}{2}$ **31.** $2\frac{7}{8}$ **33.** $\frac{3}{5}$ **35.** $\frac{8}{9}$ **37.** $4\frac{3}{8}$ **39.** $4\frac{7}{12}$
41. 90¢

1. a. Eve is $67\frac{3}{4}$ in. tall. Eve is $4\frac{1}{2}$ in. taller than
Theresa. **b.** Theresa's, $4\frac{1}{2}$ **c.** $t + 4\frac{1}{2}$, $67\frac{3}{4}$
d. $t + 4\frac{1}{2} = 67\frac{3}{4}$ **2.** Yes **3.** A, $62\frac{3}{4}$ in.
5. A, $62\frac{3}{4}$ **7.** A, $62\frac{3}{4}$ **9.** B, $74\frac{1}{4}$ **13.** $n - 3\frac{1}{8} =$
$14\frac{1}{2}$, $17\frac{5}{8}$ **15.** $n + 6\frac{1}{2} = 20\frac{2}{3}$, $14\frac{1}{6}$ **17.** $c + 3\frac{1}{2} =$
$26\frac{1}{4}$, $22\frac{3}{4}$ **19.** $2h + 5 = 43$, 19

1. $213.19 **3.** $751.24 **5.** 39.95, 32.16 **7.** $47.98

1. $^{-}4$ **3.** 18 **5.** 14 **7.** $^{-}12$ **9.** $^{-}10$ **11.** $^{-}2$
13. $^{-}36$ **15.** $^{-}15$ **17.** 4^3 **19.** $6^2 \cdot 8^2$ **21.** x^3y^2
23. $2y^2z^2$ **25.** $6y^3z^2$ **27.** $2 \cdot 7$ **29.** 2^4 **31.** $2 \cdot 3 \cdot 7$

33. $3^2 \cdot 5$ **35.** $3 \cdot 3 \cdot x \cdot x$ **37.** $3 \cdot 5 \cdot d$ **39.** $2 \cdot 2 \cdot 2 \cdot n \cdot n \cdot n$ **41.** $5 \cdot 5 \cdot r \cdot r \cdot s \cdot s$ **43.** $3 \cdot 3 \cdot 3 \cdot y \cdot y \cdot y \cdot z \cdot z$ **45.** 24 **47.** 20 **49.** 15 **51.** $12rs$ **53.** $9ab$ **55.** $16y^2$ **57.** $32f^2g^2$ **59.** > **61.** > **63.** < **65.** < **67.** = **69.** > **71.** < **73.** >

Chapter 7 ■ Algebra Using Products and Quotients of Fractions

Pages 184–185

1. $\frac{1}{2}$ **2.** $\frac{2}{3}$ **3. a.** 1 **b.** 1 **c.** t **d.** a **5.** $\frac{1}{2}$ **7.** $2\frac{1}{4}$ **9.** 6 **11.** $\frac{1}{6}$ **13.** 2 **15.** $\frac{1}{16}$ **17.** $\frac{5}{8}$ **19.** $\frac{1}{4}$ **21.** $\frac{1}{5}$ **23.** $\frac{7}{8}$ **25.** 3 **27.** $\frac{1}{3}$ **29.** $\frac{p}{q}$ **31.** $\frac{2m}{n^2}$ **33.** $\frac{1}{5}$ **35.** $\frac{24}{j}$ **37.** $\frac{k}{2j}$ **39.** $20x$ **41.** $\frac{uv}{y^2}$ **43.** $\frac{v}{wz}$ **45.** 1 **47.** $\frac{d}{6e}$ **49.** $\frac{8}{s}$ **51.** $\frac{b}{15}$ **53.** $\frac{2a}{3b}$ **55.** s **57.** $\frac{21}{e}$ **59.** Orioles: 4, Indians: 3

Pages 186–187

1. $4\frac{1}{2}$ cups **2.** $4\frac{1}{2}$ and $1\frac{1}{2}$ **3.** $6\frac{3}{4}$, yes **4.** B: Canceling, $35 \div 9$; C: Canceling, $23 \cdot 2$; D: $6 \cdot 2 + 5$, $51 \div 2$ **5.** $4\frac{1}{6}$ **7.** $7\frac{7}{8}$ **9.** $6\frac{1}{4}$ **11.** 7 **13.** $22\frac{1}{2}$ **15.** $8\frac{17}{24}$ **17.** $8\frac{2}{5}$ **19.** $15\frac{8}{9}$ **21.** $26\frac{1}{8}$ **23.** $6\frac{3}{5}$ **25.** $9\frac{1}{2}$ **27.** $8\frac{3}{4}$ **29.** $7\frac{5}{16}$ **31.** $10\frac{1}{2}$ **33.** $3\frac{3}{10}$ **35.** $3\frac{1}{2}$ tsp **37.** $11\frac{1}{4}$ cups **39.** No **41.** 12

Pages 188–189

1. a. 10 **b.** 10 **2.** 5 **3.** C: Canceling, $1 \cdot 1$; D: Reciprocal of t is $\frac{1}{t}$, $t \cdot t$ **5.** 8 **7.** $\frac{2}{3}$ **9.** $\frac{5}{3}$ **11.** $1\frac{1}{8}$ **13.** $\frac{9}{16}$ **15.** $1\frac{2}{3}$ **17.** $\frac{4}{5}$ **19.** $12\frac{1}{2}$ **21.** $1\frac{1}{2}$ **23.** $\frac{7}{12}$ **25.** $1\frac{1}{6}$ **27.** $6\frac{2}{5}$ **29.** 4 **31.** $2\frac{4}{5}$ **33.** $\frac{ad}{bc}$ **35.** $\frac{a}{cd}$ **37.** $\frac{r^2}{4s}$ **39.** $\frac{3mq}{2n}$ **41.** $\frac{6d}{ef}$ **43.** $\frac{2q}{p}$ **45.** $\frac{x}{3z}$ **49.** 68 mi

Pages 190–191

1. $3\frac{3}{4}$ **2.** $1\frac{1}{4}$ **3.** $1\frac{1}{4}$ **4.** 3 **5. a.** $\frac{1}{10}$ **b.** $\frac{4}{9}$ **c.** $\frac{7}{4}$ **7.** $4\frac{1}{8}$ **9.** $2\frac{1}{4}$ **11.** $2\frac{7}{16}$ **13.** $3\frac{5}{9}$ **15.** $\frac{7}{9}$ **17.** 3 **19.** $1\frac{1}{3}$ **21.** $1\frac{25}{32}$ **23.** $\frac{6}{25}$ **25.** 4 **27.** $3\frac{1}{3}$ **29.** $3\frac{1}{5}$

31. $1\frac{1}{3}$ **33.** 1 **35.** 1 **37.** 5 **39.** $4\frac{1}{2}$ **41.** $1\frac{1}{4}$ mi **43.** 5 **45.** $1\frac{1}{3}$ **47. a.** $2\frac{2}{5}$ **b.** $2\frac{3}{8}$ **c.** $2\frac{8}{9}$ **d.** $3\frac{1}{3}$ **e.** $7\frac{4}{5}$

Page 194

1. $12\frac{3}{4}$ **3.** 11 **5.** $\frac{3}{5}$ **7.** $2f + \frac{1}{2} = 5\frac{1}{2}$, $2\frac{1}{2}$ **9.** $5s + 2\frac{1}{2} = 12\frac{1}{2}$, 2

Page 195

1. 55 **3.** 161 **5.** 109 **7.** 43 **9.** 19 **11.** 400 **13.** 5.6 **15.** 23.1 **17.** 51 **19.** 21 **21.** 12 **23.** 7 **25.** 1.2 **27.** 90 **29.** 7 **31.** 12 **33.** 37 **35.** 34 **37.** 24 **39.** 10 **41.** 1.5 **43.** 2.6 **45.** ⁻5 **47.** ⁻2 **49.** ⁻30 **51.** ⁻56 **53.** ⁻56 **55.** ⁻9 **57.** $2 \cdot 5 \cdot a \cdot a$ **59.** $3 \cdot b \cdot b \cdot b$ **61.** $7 \cdot g \cdot g \cdot g$ **63.** $2 \cdot 2 \cdot 2 \cdot 3 \cdot w$ **65.** $5 \cdot 5 \cdot n \cdot n \cdot n$ **67.** $3 \cdot 3 \cdot 3 \cdot k$ **69.** $7 \cdot g \cdot g \cdot g \cdot g$ **71.** $7 \cdot r \cdot r \cdot s$ **73.** $2 \cdot 5 \cdot y \cdot z \cdot z$ **75.** $2 \cdot 2 \cdot 3 \cdot u \cdot u \cdot v \cdot v$

Pages 196–197

1. $15 **2.** $\frac{2}{3}$ **3.** $15 **4.** $10 **5.** B: $1 \cdot 15$, $15 \div 4$; C: Canceling, $5 \cdot 6 \div 1$; D: Canceling, $7 \cdot 11 \div 2$ **7.** $6\frac{2}{3}$ **9.** 54 **11.** $3\frac{3}{5}$ **13.** 22 **15.** $7\frac{7}{8}$ **17.** 56 **19.** $19 **21.** $36 **23.** $6 **25.** $28 **27.** $3.20 **29.** $7.50 **31.** $13.13 **33.** $20.83 **35.** $3 **37.** African violets **39. a.** 235.2 **b.** 67.5 **c.** 472.5 **d.** 781.2 **e.** 28.8 **f.** 97.5

Pages 198–199

1. 24 **2.** 6 **3.** Add **4. a.** 30 **b.** Yes, yes **5.** 18 **7.** 28 **9.** 16 **11.** 54 **13.** 81 **15.** 44 **17.** 60 **19.** 80 **21.** 138 **23.** 100 **25.** 10 **27.** 7 **29.** 7 **31.** 22 **33.** 15 **35.** $24d$ **37.** $\frac{s}{60}$ **39.** $12f$ **41.** $36y$ **43.** $2q$ **45.** $\frac{c}{2}$

Pages 200–201

1. $27 **2.** $\frac{3}{4}$ **3.** More **4.** $36, yes **5. a.** 95 **b.** $\frac{3}{2}$ **c.** $\frac{3}{4}$, $7\frac{1}{2}$ **7.** 15 **9.** $10\frac{2}{3}$ **11.** 36 **13.** 25 **15.** 8 **17.** 36 **19.** $11\frac{3}{7}$ **21.** 34 **23.** 90 **25.** 72 **27.** $37\frac{1}{2}$ **29.** 25 **31.** $56 **33.** $20 **35.** C, $64 **37.** B, 30

Pages 202–203

1. $\frac{2}{3}$ of the adults or 54 adults identified all the trademarks. **2.** n, $\frac{2}{3}n$ **3.** $\frac{2}{3}n$, 54; $\frac{2}{3}n = 54$

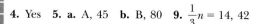

4. Yes **5. a.** A, 45 **b.** B, 80 **9.** $\frac{1}{3}n = 14$, 42

11. $3u + 10 = 55$, 15 **13.** $\frac{2}{3}c = 26$, \$39

15. $\frac{3}{5}a = 1$, $1\frac{2}{3}$

Page 204

1. a. 1.5 **b.** 110 **c.** 346.5 **3.** 1248 **5.** 115.5
7. 9.5

Page 205

1. 5 **3.** 1 **5.** 6 **7.** a **9.** $5w$ **11.** x **13.** $6cd$
15. $5e^2f$ **17.** 36 **19.** 36 **21.** 48 **23.** $24c$ **25.** $9z$
27. $24y$ **29.** $9j$ **31.** $8f^2$ **33.** $35c^2d^2$ **35.** $\frac{2}{3}$ **37.** 3
39. $7\frac{2}{3}$ **41.** 6 **43.** $7\frac{3}{5}$ **45.** 7 **47.** $\frac{3x^2}{4y}$ **49.** 3
51. $\frac{8n^2}{3}$ **53.** $\frac{5}{6}$ **55.** $1\frac{5}{12}$ **57.** $\frac{2m+5n}{10}$ **59.** $\frac{15+4k}{20}$
61. $\frac{15m+8n}{24}$ **63.** $\frac{21r+4s}{36}$ **65.** $\frac{15y+28z}{80}$
67. $\frac{7m+25n}{35}$ **69.** $\frac{1}{6}$ **71.** $\frac{13}{36}$ **73.** $\frac{5m-2n}{10}$
75. $\frac{9a-2b}{12}$

Chapter 8 ■ Algebra Using Rational Numbers

Pages 210–211

1. $\frac{-1}{8}$, $^+1\frac{1}{2}$ **2.** TWA **3.** Greater than **4. a.** Yes
b. Yes **5.** $^-2$ **7.** 3 **9.** $3\frac{1}{3}$ **11.** $\frac{-1}{6}$ **13.** $^-6\frac{1}{3}$
15. 6 **17.** $1\frac{1}{2}$ **19.** $^-4$ **21.** $1\frac{1}{2}$ **23.** $^-5\frac{1}{2}$ **25.** $\frac{-3}{4}$
27. $^-6$ **29.** $^-1$ **31.** 0 **33.** $1\frac{2}{3}$ **35.** < **37.** >
39. < **41.** > **43.** < **45.** > **47.** < **49.** >
51. = **53.** > **55.** < **57.** > **59.** Ford **61.** \$100
63. $^+50$

Pages 212–213

1. Yes **3.** $^-1.75$ **5.** $^-0.8$ **7.** $^-0.\overline{4}$ **9.** 3.5 **11.** 3.75
13. $1.8\overline{3}$ **15.** $0.\overline{45}$ **17.** $1.41\overline{6}$ **19.** $2.08\overline{3}$ **21.** $^-4.\overline{3}$
23. $^-6.8\overline{3}$ **25.** $\frac{-1}{4}$ **27.** $\frac{3}{4}$ **29.** $\frac{4}{5}$ **31.** $\frac{12}{25}$ **33.** $\frac{3}{20}$
35. $\frac{3}{8}$ **37.** $\frac{-2}{5}$ **39.** $2\frac{1}{4}$ **41.** $2\frac{2}{5}$ **43.** $9\frac{7}{20}$ **45.** $3\frac{3}{8}$

Pages 214–215

1. opposite **3.** $\frac{-1}{8}$ **5.** $\frac{-13}{20}$ **7.** $\frac{-11}{20}$ **9.** $1\frac{17}{24}$
11. $\frac{-5}{24}$ **13.** $^-7\frac{3}{4}$ **15.** 1 **17.** $^-2\frac{3}{4}$ **19.** $^-1\frac{19}{20}$

21. $1\frac{23}{24}$ **23.** $\frac{-14}{15}$ **25.** $\frac{-13}{18}$ **27.** $1\frac{17}{24}$ **29.** $6\frac{4}{5}$
31. $^-13\frac{3}{4}$ **33.** $2\frac{17}{24}$ **35.** $^-18\frac{19}{24}$

Pages 216–217

1. a. $\frac{-9}{2}$ **b.** canceling **c.** simplest **3.** $\frac{1}{6}$ **5.** $^-2\frac{2}{15}$
7. $\frac{-5}{8}$ **9.** 1 **11.** $^-6\frac{2}{3}$ **13.** $10\frac{2}{3}$ **15.** $4\frac{5}{18}$
17. $9\frac{13}{18}$ **19.** $^-4\frac{2}{3}$ **21.** $^-11\frac{23}{32}$ **23.** $1\frac{1}{2}$ **25.** $^-1\frac{1}{3}$
27. $1\frac{1}{2}$ **29.** $\frac{-5}{24}$ **31.** $\frac{15}{16}$ **33.** $^-2$ **35.** $\frac{23}{26}$
37. $2\frac{5}{32}$ **39.** $^-1\frac{5}{24}$ **41.** 2 **43. a.** 25 mph
b. 45 mph **c.** 55 mph **d.** 52.5 mph

Page 220

1. \$1.48 **3.** No **5.** $3r + 3.78 = 98.28$, translators
7. $36.95 + 29.95d + 6.28 = 163.03$, 4

Page 221

1. 2 **3.** 0 **5.** $^-63$ **7.** $^-10$ **9.** $^-3$ **11.** $^-36$
13. $1.4 \cdot 10^4$ **15.** $1.2 \cdot 10^5$ **17.** $5.8 \cdot 10^6$ **19.** $1.8 \cdot 10^7$ **21.** $8.3 \cdot 10^{-3}$ **23.** $8.4 \cdot 10^{-4}$ **25.** $1.78 \cdot 10^{-4}$
27. $5.286 \cdot 10^{-3}$ **29.** 12 **31.** 24 **33.** 24 **35.** $2b$
37. $2m^2$ **39.** $6b$ **41.** d^3 **43.** $3f^2$ **45.** $7\frac{1}{2}$ **47.** 3
49. 14 **51.** $\frac{3}{4}$ **53.** $\frac{5f}{2g}$ **55.** $\frac{5}{3w}$ **57.** $\frac{7}{3q}$ **59.** $\frac{y}{3}$
61. $\frac{3d}{4c}$ **63.** $\frac{3s+4r}{rs}$ **65.** $\frac{2n+10}{5n}$ **67.** $\frac{5q-6p}{pq}$
69. $\frac{3b-8}{4b}$ **71.** $\frac{4n-m^2}{mn}$ **73.** $\frac{4r-1}{2t}$ **75.** $\frac{d-c^2}{cd}$

Page 223

1. a. 12, 5 **b.** 5, 6 **c.** $^-10, 6, ^-8$ **3.** 3 **5.** $^-1\frac{5}{7}$
7. 4 **9.** $^-4\frac{2}{3}$ **11.** $1\frac{1}{8}$ **13.** $1\frac{1}{2}$ **15.** $^-1\frac{11}{12}$ **17.** $\frac{6}{7}$
19. $^-5$ **21.** $\frac{2}{15}$ **23.** $\frac{-4}{7}$ **25.** 1 **27.** $^-2$ **29.** 9
31. $^-1\frac{3}{5}$ **33.** 0 **35.** $^-10\frac{1}{2}$ **37.** $1\frac{9}{10}$ **39.** D, $\frac{1}{6}$
41. B, $\frac{2}{5}$

Pages 224–225

1. a. $^-2$ **b.** 0 **c.** $^-1$ **3.** $7n$ **5.** $11n - 6$
7. $8n + 3$ **9.** ^-8x **11.** $5y + 7$ **13.** $9y - 11$
15. $8w - 2$ **17.** $4z - 12$ **19.** $^-9r - 15$ **21.** 8
23. $8a + 1$ **25.** $7c$ **27.** 5 **29.** $17m + 2$ **31.** $^-18$
33. $^-35$ **35.** 0, 48, 64, 48, 0

Pages 226–227

1. combine **2.** \$8, yes **3.** B: Combine like terms,
subtract the same number from both sides, divide

both sides by the same number; C: Commutative property, undo addition of 16 by subtracting 16,

$^-35 \div {}^-2$ **5.** $^-6$ **7.** $\frac{5}{9}$ **9.** $2\frac{3}{7}$ **11.** $^-4\frac{2}{3}$
13. $^-3\frac{1}{5}$ **15.** $\frac{^-1}{8}$ **17.** $1\frac{5}{9}$ **19.** $^-2\frac{1}{2}$ **21.** 2 **23.** 1
25. $\frac{1}{11}$ **27.** $^-1\frac{1}{2}$ **29.** $\frac{^-4}{15}$ **31.** $^-2\frac{1}{7}$ **33.** $\frac{1}{23}$
35. $\frac{21}{23}$ **37.** $\frac{^-10}{17}$ **39.** $4\frac{6}{7}$ **41.** B, \$7 **43.** A, \$6.50

Pages 228–229

1. $2g$, variable **2.** \$4, yes **3. a.** $2x$, 11, $1\frac{5}{6}$ **b.** $2y$,
18, 2 **5.** $\frac{^-1}{2}$ **7.** 4 **9.** $1\frac{3}{5}$ **11.** $^-23$ **13.** 2
15. $^-8$ **17.** $\frac{^-3}{10}$ **19.** $^-1$ **21.** $^-3\frac{1}{2}$ **23.** $1\frac{1}{2}$
25. $^-8$ **27.** $1\frac{1}{11}$ **29.** 3 **31.** $^-2\frac{1}{4}$ **33.** $1\frac{1}{3}$
35. $\frac{^-4}{11}$ **37.** 31

Page 231

1. $4\frac{2}{5}$ **3.** $^-7\frac{1}{2}$ **5.** $4\frac{4}{9}$ **7.** 374 **9.** 140 mi
11. $37\frac{1}{2}$ **13.** 12 km/h

Page 232

1. \$35,000 **3. a.** 1.5 **b.** 112,500 **c.** 3:30–5:30
5. \$900 **7.** Take a loss, \$750

Page 233

1. $1\frac{7}{9}$ **3.** 6 **5.** $\frac{7}{10}$ **7.** $4y$ **9.** $\frac{2}{7df}$ **11.** $\frac{2}{b}$ **13.** $\frac{mn}{3}$
15. $\frac{6}{5d}$ **17.** $1\frac{1}{5}$ **19.** $\frac{7}{16}$ **21.** $\frac{2}{3}$ **23.** $\frac{3j}{2k}$ **25.** $\frac{a}{4b}$
27. $\frac{2q}{3p}$ **29.** $\frac{2}{3x}$ **31.** 15 **33.** 84 **35.** $7\frac{1}{5}$ **37.** $49\frac{1}{2}$
39. 30 **41.** 117 **43.** 160 **45.** 105 **47.** 132 **49.** 5
51. 96 **53.** 90 **55.** $49\frac{1}{2}$ **57.** $57\frac{1}{7}$ **59.** $4\frac{4}{5}$
61. $37\frac{1}{2}$

Chapter 9 ■ Using Algebra in Ratio, Proportion, and Percent

Pages 238–239

1. $\frac{1}{4}$ **2.** $2a$ **3.** equal **4.** $3x$ **5. a.** 3 **b.** 18 **c.** b
d. $6xy$ **7.** $\frac{1}{4}$ **9.** $\frac{4}{3}$ **11.** $\frac{2}{3}$ **13.** $\frac{8}{3}$ **15.** $\frac{2}{3}$ **17.** $\frac{7}{4}$
19. $\frac{2}{3}$ **21.** $\frac{a}{2b}$ **23.** $\frac{d}{c}$ **25.** $\frac{3m}{4n^2}$ **27.** $\frac{3}{2}$ **29.** $\frac{4y}{3z}$
31. $\frac{3m}{5n}$ **33.** $\frac{2cd}{3}$ **35.** $\frac{9h}{5j}$ **37.** 24 **39.** 6 **41.** 36
43. 9 **45.** 110 **47.** 25 **49.** 20 **51.** 21 **53.** mn
55. wz^2 **57.** $10j^2k$ **59.** $25m^2$ **61.** $3cd^2$ **63.** $27ab^3$
65. $35x^2y^2$ **67.** $\frac{6}{5}$ **69.** 9 **71.** $\frac{10}{3}$

Pages 240–241

1. 5 **2.** 5 ft **3.** No **4.** $3\frac{27}{32}$ ft **5. a.** $4\frac{4}{5}$ **b.** 24
c. 7 **d.** 4 **7.** 18 **9.** $2\frac{11}{12}$ **11.** $5\frac{8}{11}$ **13.** 36
15. $6\frac{1}{4}$ **17.** $15\frac{5}{7}$ **19.** $9\frac{3}{8}$ **21.** 1 **23.** $1\frac{3}{8}$ **25.** 1
27. $8\frac{4}{7}$ **29.** 16 **31. a.** $16\frac{2}{5}$ in. **b.** 11 in.
33. $2\frac{1}{82}$ ft

Page 243

1. a. 8.67 **b.** 5 **c.** \$9 **d.** \$17.10 **3. a.** 258.33
b. 335.83 **c.** 15.48 **d.** 10.84 **5. a.** \$150.50
b. 9.30 **7.** Yes

Pages 244–245

1. 25 **2.** $16\frac{2}{3}$ **3.** $\frac{1}{6}$ **4. a.** 75 **b.** 250 **5. a.** 0.47
b. $1.33\frac{1}{3}$ **7.** $1\frac{1}{4}$ **9.** 1 **11.** $\frac{18}{25}$ **13.** $\frac{8}{25}$ **15.** $\frac{33}{50}$
17. $\frac{12}{25}$ **19.** $1\frac{1}{5}$ **21.** 3 **23.** $1\frac{3}{4}$ **25.** $\frac{1}{12}$ **27.** $\frac{2}{3}$
29. $1\frac{1}{6}$ **31.** $\frac{1}{3}$ **33.** $1\frac{1}{16}$ **35.** $2\frac{1}{3}$ **37.** 0.06 **39.** 4
41. 0.037 **43.** 0.75 **45.** 0.0325 **47.** 0.4
49. $0.07\frac{1}{2}$ **51.** $1.66\frac{2}{3}$ **53.** 0.212 **55.** Clothing
57. $\frac{1}{8}$ **59.** $\frac{7}{8}$ **61.** \$48

Pages 246–247

1. 12 **2.** 24 **3.** $\frac{1}{2}$ **4.** $\frac{2}{3}$ **5.** 50, $66\frac{2}{3}$ **6.** Mental
math, proportion **7.** 40% **9.** 90% **11.** 125%
13. 20% **15.** 80% **17.** 250% **19.** $33\frac{1}{3}$%
21. 175% **23.** $83\frac{1}{3}$% **25.** $56\frac{1}{4}$% **27.** $55\frac{5}{9}$%
29. $66\frac{2}{3}$% **31.** 40% **33.** 450% **35.** $33\frac{1}{3}$%
37. 250% **39.** 100% **41.** 6% **43.** $70\frac{5}{6}$
45. $66\frac{2}{3}$ **47.** $16\frac{2}{3}$ **49.** 3

Page 250

1. 507 km **3.** 754 km **5.** 1222 km **7.** 533 km
9. 1131 km **11.** 546 km **13.** About 871 km
15. Pittsburgh

Page 251

1. $^-4\frac{1}{5}$ **3.** $^-52$ **5.** 4 **7.** 0 **9.** $^-24$ **11.** $^-54$
13. $^-2$ **15.** $^-72$ **17.** 10^5 **19.** 5^6 **21.** x^4 **23.** a^8
25. 10^2 **27.** 6^3 **29.** w^3 **31.** c^4 **33.** 3^2 **35.** $2^2 \cdot 3$
37. 3^3 **39.** $2^4 \cdot 3^2$ **41.** $2^4 \cdot 3 \cdot 5$ **43.** $2^3 \cdot 3^2 \cdot 5$
45. $2 \cdot 3 \cdot a \cdot a$ **47.** $2 \cdot 2 \cdot 3 \cdot c \cdot c$ **49.** $2 \cdot 3 \cdot 3 \cdot e$

51. $5 \cdot a \cdot b \cdot b$ **53.** $2 \cdot 2 \cdot 2 \cdot 3 \cdot a \cdot b$ **55.** $2 \cdot 17 \cdot a \cdot a \cdot b \cdot b$ **57.** 1 **59.** 5 **61.** 16 **63.** 1 **65.** y **67.** $8a$ **69.** b **71.** $9f$ **73.** $5wz$ **75.** $8pq^2$ **77.** $\frac{1}{3}$ **79.** $\frac{3}{5}$ **81.** $\frac{2}{3}$ **83.** $\frac{1}{z}$ **85.** $\frac{n}{3m}$ **87.** $\frac{5n}{m}$ **89.** $\frac{3x}{y}$ **91.** $\frac{6}{7q}$ **93.** $\frac{7z^2}{3y}$

Pages 252–253
1. $45 **2.** 20 **3.** $24 **4.** 18 **5. a.** $9, $36 **b.** $4.32, $19.68 **7.** 9 **9.** 5 **11.** 18 **13.** 5 **15.** 49.92 **17.** 128.7 **19.** 40.02 **21.** 1.61 **23.** 0.3 **25.** 21 **27.** 62.4 **29.** 1.68 **31.** 6.93 **33.** 40 **35.** 14 **37.** 13.2 **39.** $9 **41.** $26

Pages 254–255
1. $\frac{1}{5}, \frac{1}{3}$ **2.** $33\frac{1}{3}, 75$ **3. a.** Greater than **b.** Less than **4. a.** 8 **b.** 8 **5.** 9 **7.** 6 **9.** 30 **11.** 30 **13.** 6 **15.** 7 **17.** 5 **19.** 9 **21.** 48 **23.** 100 **25.** 36 **27.** 330 **29.** > **31.** < **33.** >

Pages 256–257
1. 80 **2.** 52 **3.** More **4.** 65 **5. a.** 200 **b.** 630 **c.** 720 **7.** 62.5 **9.** 24 **11.** 800 **13.** 20.3 **15.** 180 **17.** 350 **19.** 7.1 **21. a.** $\frac{3}{4}$ **b.** 75 **23.** 80 **25. a.** 24 **b.** 73.5 **c.** 121.5 **d.** 181.5

Pages 258–259
1. a. $25 **b.** Increase **c.** $50 **2. a.** $175 **b.** Decrease **c.** $50 **3. a.** 14 **b.** 120 **5.** 25% d **7.** 36.2% d **9.** 22.5% d **11.** 19% d **13.** 26.7% d **15.** 40.8% i **17.** 43.8% i **19.** 100% i **21.** 60% d **23.** 46.7% i **25.** $33\frac{1}{3}$% **27. a.** 75 **b.** 75 **c.** No **29.** No **31.** $99

Page 261
1. $672 **3.** $2200 **5.** $800 **7.** 10% **9.** $620 **11.** $372 **13.** $12,000

Page 262
1. $1200, $1350, $1540 **3.** $1450, $1830, $2340 **5.** 6 **7.** $1762.34

Page 263
1. $\frac{1}{9}$ **3.** $\frac{^-5}{12}$ **5.** $\frac{^-7}{10}$ **7.** $\frac{^-13}{32}$ **9.** $1\frac{5}{12}$ **11.** $\frac{^-9}{10}$ **13.** $^-1\frac{17}{24}$ **15.** $\frac{^-1}{24}$ **17.** $\frac{1}{24}$ **19.** $^-2\frac{1}{12}$ **21.** $\frac{7}{10}$ **23.** $^-6$ **25.** $\frac{^-4}{5}$ **27.** $\frac{2}{3}$ **29.** $\frac{^-5}{24}$ **31.** $\frac{^-2}{3}$ **33.** $6x$

35. 0 **37.** $^-3w - 3$ **39.** $^-8r - 8$ **41.** $^-7r - 7$ **43.** t **45.** $^-2\frac{5}{6}$ **47.** 0 **49.** 10 **51.** $^-1$ **53.** $\frac{^-1}{13}$ **55.** $^-1\frac{2}{3}$ **57.** 2 **59.** 2 **61.** $^-1\frac{3}{7}$ **63.** $^-2\frac{1}{3}$ **65.** 0 **67.** 3

Chapter 10 ■ Using Algebra in Geometry

Page 269
7. 25° **9.** 125° **11.** Obtuse **13.** Acute **15.** Acute **17.** *EBF* **19.** *EBH* or *CBF*

Pages 270–271
1. a. Right **b.** Acute **c.** Obtuse **d.** Acute **e.** Obtuse **2.** 180° **3.** Acute, isosceles **5.** Right, isosceles **7.** $n + 2n + 90 = 180$; 30°, 60°, 90° **9.** $n + 2n + 6n = 180$; 20°, 40°, 120° **11.** $n + n + 20 + 90 = 180$; 35°, 55°, 90° **13.** $n + 4n + 90 = 180$; 18°, 72°, 90°

Pages 272–273
1. is parallel to **2.** is perpendicular to **3.** $\angle 1$ **4.** $\angle 7$ **5.** True **7.** False **9.** True **11.** True **13.** False **15.** 40°, 140°, 140° **17.** 120°, 60°, 60° **19.** $6x + 15 = 2x + 39$, 51°

Page 274
1. $\overline{CD}$

Pages 276–277
1. Lot D **2.** Lot A **3.** Lot B **4. a.** Lots C, F **b.** Lot D **c.** Lots D, E **d.** Lots B, D, E **e.** Lot A **f.** Lot F **5.** Quadrilateral, parallelogram, rectangle **7.** Quadrilateral **9.** Hexagon **11.** Quadrilateral, trapezoid **13.** Rectangle **15.** Square **17.** Yes **19.** No **21.** Yes

Pages 278–279
1. m **3.** mm **5.** m **7.** m **9.** km **11.** m **13.** m **15.** m **17.** b **19.** b **21.** b **23.** a **25. a.** 691 km **b.** 321 km **27.** SIXTY-THREE AND FIVE-TENTHS METERS

Pages 280–281
1. a. 100 **b.** 10 **2.** Charles **3. a.** 1, 1.825 **b.** 1000, 9650 **5.** 400 **7.** 0.54 **9.** 25,000 **11.** 3500 **13.** 5800 **15.** 2.5 **17.** 2750 **19.** 7500 **21.** 126 **23.** 0.9 **25.** 84 **27.** 605 **29.** 625 **31.** 5.5 **33.** 0.38 **35.** 8.38 **37.** $100n$ **39.** $1000n$ **41.** $\frac{n}{100}$ **43.** $\frac{n}{10}$

Page 284
1. 25, 5   3. 14, 2   5. 165, 15   7. $2n + 10 = 50$, 20 mph

Page 285
1. $8c$ 3. 0   5. $^-8d + 4$ 7. $^-11s - 8$ 9. $^-10s - 7$
11. $^-4k + 5$ 13. $^-1\frac{4}{5}$ 15. 0   17. 19   19. $\frac{^-3}{4}$
21. 8   23. $\frac{^-1}{2}$ 25. 7   27. $^-1\frac{3}{5}$ 29. $\frac{^-1}{5}$ 31. $^-3$
33. $^-1\frac{5}{9}$ 35. 12   37. 20   39. 8   41. jk 43. $3rs$
45. $36mn$ 47. $5wxy$ 49. $1\frac{3}{7}$ 51. 4   53. 24
55. $12\frac{2}{3}$ 57. 10   59. $2\frac{1}{2}$ 61. $7\frac{41}{49}$ 63. $10\frac{2}{3}$

Pages 286–287
1. $P = 4s$, length of a side   2. $P = 2(l + w)$, length and width   3. 56 m   5. 111 cm
7. 116 km   9. 24 m   11. 12 km   13. 8.8 m
15. 51 m   17. 144 mm   19. 260   21. 15 m

Pages 288–289
1. 3.14   2. $2\pi r$ 3. 36   4. 4   5. 21.98 ft
7. 12.56 ft   9. 9.42 yd   11. 28.26 yd   13. 75.36 in.
15. 251.2 yd   17. 628 in.   19. 1.5 yd
21. a. 81.64 in.   b. 776   c. 2, 168, 144

Pages 290–291
1. 15   2. 9 square centimeters   3. $A = lw$; A is area, l is length, and w is width   4. square   5. 144
7. $6m^2$ 9. 58.5 m^2 11. 810 cm^2 13. 3.9 m^2
15. 84 cm^2 17. 4 km^2 19. 72.25 cm^2 21. 11.56 m^2
23. 0.09 km^2 25. 12m   27. 10 cm   29. Perimeter, 80 ft   31. Area, 8400 ft^2 33. Perimeter, $67.50

Pages 292–293
1. $A = bh$; A is area, b is base, and h is height
2. centimeters   3. 25.2   5. 30.4 m^2 7. 47.46 m^2
9. 450 km^2 11. 120 km^2 13. 125 cm^2 15. 54 m^2
17. $10,000 \text{ m}^2$ 19. 3 m   21. 30 km   23. 46 cm

Pages 294–295
1. height   2. yards   3. 54   4. 60 in.^2 7. 15 ft^2
9. $25\frac{1}{2} \text{ ft}^2$ 11. $55\frac{1}{2} \text{ ft}^2$ 13. 180 yd^2 15. 36 yd^2
17. 10 yd   19. $5(180°) = 900°$, $6(180°) = 1080°$

Pages 296–297
1. a. 36 cm^2 b. 27 cm^2 2. Yes   3. radius   4. r, r
5. centimeters   6. 314   7. 12.56 cm^2 9. 0.785 m^2
11. 113.04 cm^2 13. 1256 cm^2 15. 4.5216 m^2
17. 2.0096 m^2 19. 78.5 cm^2 21. 2826 km^2
23. $20,096 \text{ cm}^2$ 25. 3.14 m^2 27. 37.68 m^2
29. 18.24 m^2

Pages 298–299
1. a. 12   b. $w + w + 12 + w + w + 12$ c. 96
2. 30 cm, yes   3. B, 14 cm   5. A, 20 cm   7. $w + w + 30 + w + w + 30 = 380$, 80 cm   9. $w + 3w + w + 3w = 240$, 30 m   11. $w + 2w + 3 + w + 2w + 3 = 426$, 70 cm   13. $s + s + 3 + s + 4 = 67$, 20 cm

Page 300
1. 260   3. 30-foot square   5. $345, $137.50, $29.50, $512   7. $567.50

Page 301
1. $\frac{9}{25}$ 3. $1\frac{11}{25}$ 5. $\frac{3}{5}$ 7. 3   9. $2\frac{1}{2}$ 11. $\frac{3}{8}$ 13. $\frac{7}{8}$
15. $1\frac{5}{8}$ 17. $2\frac{1}{3}$ 19. 25%   21. 50%   23. 75%
25. 100%   27. $37\frac{1}{2}$%   29. $62\frac{1}{2}$%   31. $433\frac{1}{3}$%
33. $183\frac{1}{3}$%   35. $116\frac{2}{3}$%   37. 21   39. 11
41. 66.04   43. 1.16   45. 4.012   47. 108   49. 32
51. 56   53. 32   55. 96   57. 136   59. 32
61. 100% i   63. 33.3% i   65. 25% i   67. 11.1% i
69. 16.7% d   71. 61.9% i

Chapter 11 ■ Geometry—Surface Area and Volume

Pages 306–307
1. C   2. E   3. a. E   b. B   c. A   d. C, D, F
5. Square pyramid   7. Triangular pyramid
9. Triangular prism   11. Cube
13. Hexagonal prism   15. $E - V + 2$

Pages 308–309
1. a. 5   b. Triangle A, 28¢   c. Square, 90¢
2. $1.18   3. $1.80   5. $3   7. $1.01   9. $1.80
11. $3   13. $1.46   15. $2.08   17. $1.18   19. $2.06
21. $3.58   23. $4.16   25. a. D   b. E   c. C

Pages 310–311
1. Jason   2. back, bottom, right   3. 2   4. 54
5. Multiply the area of one face by 6.   7. 216 in.^2
9. 48 ft^2 11. 54 ft^2 13. 150 yd^2 15. 94   17. 7, 3, 6

Page 314
1. a. D   b. B   c. C

Page 315
1. 1   3. $\frac{19}{24}$ 5. $\frac{8}{15}$ 7. $\frac{3r + 2s}{6}$ 9. $\frac{9m + 4n}{12}$

11. $\dfrac{3p-2q}{12}$ 13. $\dfrac{10a-b}{8}$ 15. $\dfrac{5m-5n}{6}$ 17. $\dfrac{b^2-3}{ab}$

19. $\dfrac{ab-2a}{2b}$ 21. $\dfrac{y^2-2x}{xy}$ 23. $\dfrac{8-y}{4x}$ 25. $\dfrac{5-7y}{y^2}$

27. $2\dfrac{1}{2}$ 29. 1 31. $\dfrac{1}{5}$ 33. $\dfrac{y}{4}$ 35. $1\dfrac{1}{2}$ 37. 1

39. $\dfrac{y}{w}$ 41. 5 43. $\dfrac{1}{4}$ 45. $2\dfrac{2}{3}$ 47. $\dfrac{3c}{b}$ 49. $\dfrac{18}{z}$

51. $\dfrac{1}{3}$ 53. $2d$ 55. $6b$ 57. 30 59. 42 61. 50

63. $10\dfrac{4}{5}$ 65. $68\dfrac{1}{3}$ 67. $\dfrac{3}{4}$

Pages 316–317
1. A 2. B 3. 40 cubic centimeters 4. $V = Bh$ or $V = (lw)h$; V = volume, l = length, w = width, h = height 5. a. 7, 140 b. 4, 4, 64 7. 125 m³
9. 1260 cm³ 11. 238.328 m³ 13. 1440 cm³
15. 780 cm³ 17. 1984 cm³ 19. 26.88 m³
21. Perimeter 23. Area 25. Perimeter 27. 14 cm³

Pages 318–319
1. $V = Bh$ or $V = (\pi r^2)h$; V = volume, π = pi, r = radius, h = height 2. r, r 3. a. 9, 226.08 b. 4, 16, 7, 351.68 5. 452.16 ft³ 7. 769.3 in.³ 9. 28.26 in.³
11. 226.08 in.³ 13. 157 ft³ 15. 6437 in.³
17. 3.14 yd³ 19. 2 in. 21. 7 ft

Pages 320–321
1. $\dfrac{1}{3}$ 2. $\dfrac{1}{3}$ 3. a. 80 b. $V = \dfrac{1}{3}Bh$ or $V = \dfrac{1}{3}(\pi r^2)h$; V = volume, π = pi, r = radius, h = height
5. 105 cm³ 7. 1017.36 cm³ 9. 628 cm³
11. 112 cm³ 13. 160 cm³ 15. 20 m³
17. 136.5 cm³ 19. 235.5 m³ 21. 50.2 m³
23. 303.8 m³ 25. 15.2 m³ 27. a. 34 cm³
b. 10 cm³

Pages 322–323
1. Divide by 1000, multiply by 1000 2. Divide by 1000, multiply by 1000 3. a 5. b 7. b 9. a
11. b 13. a 15. 9000 17. 125,000 19. 2.875
21. 6400 23. 25,750 25. 0.87 27. 7000
29. 28,000 31. 1.25 33. 4600 35. 330 37. 0.575
39. 5 41. 265

Pages 324–325
1. 75, 75 2. 3, 3 3. 8.750, 8.750 5. 4.5
7. 15.625 9. 6.93 11. 1.02 13. 8 15. 1.536
17. 24 cm

Page 327
1. 298 in.² 3. 40.26 in.² 5. 34.4 in.³
7. $A = lw - \pi r^2$, 129.76 ft²
9. $V = \pi r^2 h + \dfrac{1}{3}\pi r^2 h$, $1046\dfrac{2}{3}$ ft³

Page 328
1. Family room 3. 22 ft by 15 ft 5. Bedroom B
7. 192 ft² 9. \$577.50 11. $\dfrac{1}{4}$

Page 329
1. 33 3. 33 5. 56 7. 26 9. $28\dfrac{1}{2}$ 11. $45\dfrac{5}{7}$
13. 45 15. $20\dfrac{5}{9}$ 17. $1\dfrac{3}{8}$ 19. $^-10\dfrac{1}{3}$ 21. $^-1\dfrac{4}{15}$
23. $13\dfrac{1}{2}$ 25. 2 27. $12\dfrac{3}{5}$ 29. $3\dfrac{1}{8}$ 31. $8\dfrac{1}{8}$
33. $9\dfrac{1}{3}$ 35. $13\dfrac{3}{4}$ 37. $45\dfrac{1}{3}$ 39. $2\dfrac{1}{12}$ 41. 8
43. 84 45. 8 47. 150 49. 48 51. 32 53. 25% i
55. 100% i 57. 23.2% i 59. 250% i 61. 57.1% d
63. 50% i 65. 10% d 67. 12.5% d

Chapter 12 ■ Graphing Equations and Inequalities

Pages 334–335
1. Yes, yes, yes 2. Yes, yes, yes 3. No 4. No
5. a. is not b. is 7. b 9. e 11. g 13. f
15. $n \le ^-3$ 17. $n \ge ^-1$ 19. $n \le ^-1\dfrac{1}{2}$ 21. $n < \dfrac{^-3}{4}$
23. $n > ^-3\dfrac{3}{4}$

Pages 336–337
1. multiplied 2. divided 3. a. 5 b. $^-5$ c. $4\dfrac{1}{2}$
5. Yes 7. Yes 9. No 11. No 13. $n \ge ^-11$
15. $n \le ^-8$ 17. $n \le 2\dfrac{1}{2}$ 19. $n < ^-15$ 21. $n < ^-80$
23. $n \le ^-18$ 25. $n < ^-2\dfrac{5}{6}$ 27. $n \ge 8$ 29. $n \ge ^-24$
31. $n < 23$ 33. $y > 2$, c 35. $y < 1$, b 37. $y \ge 1$, a
39. $y < ^-3$, e 41. $n \ge ^-2$ 43. $n \le ^-4$ 45. $n \le 1\dfrac{2}{5}$
47. $n < ^-6$ 49. $n > 0$ 51. $n \le ^-19$

Pages 338–339
1. $^-6$, yes 2. a. 5 b. 4, $\le$ c. 1, $\le$ 3. a. No
b. No c. Yes d. Yes e. No f. Yes g. Yes
h. No i. No j. Yes k. Yes l. Yes 5. $x > ^-3$
7. $x > ^-25$ 9. $x \le 3$ 11. $x \le \dfrac{8}{9}$ 13. $x > 26$
15. $x \le 3$ 17. $x \ge ^-21$ 19. $x \ge 135$ 21. $x \le \dfrac{1}{2}$
23. $x \le ^-2$ 25. $x \ge ^-16$ 27. $x \le 64$ 29. $n \le 2$, d
31. $n \ge \dfrac{1}{2}$, e 33. $a \le 1$ 35. $r < ^-3$ 37. $y < \dfrac{^-2}{3}$
39. $j \ge ^-1$

Page 342
1. Danielle 3. $\dfrac{1}{10}$ 5. Mary 7. 20% 9. Alvita

1. $2^3 \cdot 7$ **3.** 2^4 **5.** 3^4 **7.** $3 \cdot 3 \cdot f$ **9.** $g \cdot h$ **11.** $p \cdot$
$p \cdot p$ **13.** $2 \cdot 2 \cdot 3 \cdot x \cdot x$ **15.** $2 \cdot 19 \cdot y$ **17.** $2 \cdot 2 \cdot 2 \cdot$
$5 \cdot n \cdot n \cdot n$ **19.** 6 **21.** 10 **23.** 10 **25.** 9 **27.** $3c$
29. w **31.** $6n$ **33.** $3b$ **35.** $\frac{1}{2}$ **37.** $1\frac{7}{12}$ **39.** $\frac{5}{24}$
41. $\frac{m+2n}{16}$ **43.** $\frac{3w+4z}{15}$ **45.** $\frac{2x-y}{6}$ **47.** $\frac{2n-9m}{24}$
49. $\frac{10p-9q}{48}$ **51.** $\frac{3}{32}$ **53.** $1\frac{1}{8}$ **55.** $\frac{1}{3}$ **57.** $\frac{y}{6}$
59. $1\frac{1}{2}$ **61.** 1 **63.** $\frac{2y}{w}$ **65.** 5 **67.** $1\frac{1}{6}$ **69.** $6\frac{2}{5}$
71. $\frac{4m}{n}$ **73.** $\frac{12}{z}$ **75.** $\frac{4z}{3x}$ **77.** $3d$ **79.** $6n$

Pages 344–345

1. First coordinate axis **2.** Second coordinate axis
3. Origin **5.** $(5, {}^-2)$ **7.** $({}^-4, {}^-3)$ **9.** $(6, 0)$
11. $(0, 0)$ **13.** $\left(2, {}^-2\frac{1}{2}\right)$ **15.** $\left({}^-2\frac{1}{2}, {}^-1\frac{1}{2}\right)$
17. $\left({}^-6\frac{1}{2}, 1\frac{1}{2}\right)$ **29. c.** No **31.** y-axis, x-axis

Pages 346–347

1. No **2.** Yes **3.** Yes **4.** 1 **5.** $({}^-3, {}^-14)$ and
$\left(\frac{2}{3}, {}^-3\right)$ **6.** ${}^-3$ **7.** ${}^-4, {}^-2, 0, 2, 4$ **9.** ${}^-4, 1, 6, 11,$
16 **19.** $45°\,F$ **21.** $y = 4x$ **23.** $y = {}^-2x$

Pages 348–349

1. y **2.** 4 **3.** Yes, yes **5.** ${}^-5, {}^-4, {}^-3, {}^-2$
7. ${}^-3, {}^-1, 1, 3$ **9.** ${}^-4, {}^-1, 2, 5$ **11.** ${}^-5, {}^-3, {}^-1, 1$
21. $y = 2x - 1$

Pages 350–351

1. ${}^-2$ **2.** $(0, {}^-1)$ **3.** slope **4.** $y = {}^-2x - 1$
5. 2, 1 **7.** 4, 1 **9.** 1, 6 **11.** 3, 0 **13.** $y = 4x + 3$
15. $y = {}^-2x + 4$ **17.** $y = 2x - 1$ **19.** $y = x + 5$
21. 2, ${}^-2$, $y = 2x - 2$ **23.** 3, ${}^-1$, $y = 3x - 1$
25. ${}^-3, 3, y = {}^-3x + 3$ **27.** $y = 2x + 1$
29. $y = {}^-4x$

Pages 352–353

1. a. Yes **b.** Yes **c.** Yes **d.** Yes **2.** Yes **3. a.** Yes
b. Yes **c.** Yes **d.** Yes **4.** Yes **5.** below **6.** above
7. $y \leq x + 2$ **9.** $y \geq \frac{1}{3}x + 1$ **11.** $y \geq \frac{1}{3}x + 2$

Pages 354–355

1. $y = {}^-2x - 6$ **2.** $y = \frac{1}{2}x - 1$ **3.** $({}^-2, {}^-2)$ **5.** Yes
7. Yes **9.** $(1, 1)$ **11.** $({}^-2, 2)$ **13.** $(0, 1)$ **15.** $(2, 2)$
17. $(2, 5)$ **19.** $\left(\frac{{}^-2}{3}, 1\frac{2}{3}\right)$

Pages 356–357

1. a. Cassie, d, $c + d = 12$, $c = 3d$ **b.** 9, 3, 9, 3
2. Yes, yes **3. a.** 20, 8 **b.** 14, 6, 14, 6 **5.** $f + s =$

22, $s + 8 = f$; $f = 15$, $s = 7$ **7.** $f + s = 24$, $3s = f$;
$f = 18$, $s = 6$ **9.** $j + k = 26$, $k + 8 = j$; Jim sold 17
tickets, Kelly sold 9 tickets

Page 358

1. Directly, 3 **3.** Directly, 5 **5.** Direct

Page 359

1. $\frac{2}{5}$ **3.** $\frac{1}{20}$ **5.** $\frac{1}{100}$ **7.** $\frac{9}{50}$ **9.** $1\frac{8}{25}$ **11.** $\frac{16}{25}$
13. 4 **15.** $3\frac{1}{2}$ **17.** $\frac{1}{3}$ **19.** 25% **21.** 60% **23.** 5%
25. 10% **27.** 200% **29.** 150% **31.** 29%
33. $133\frac{1}{3}\%$ **35.** 6% **37.** 16 **39.** 78 **41.** 24.03
43. 1.86 **45.** 0.41 **47.** 125.8 **49.** 85 **51.** 56
53. 32 **55.** 48 **57.** 96 **59.** 43 **61.** $n + n + 18 + 90 = 180$; 36°, 54°, 90° **63.** $n + 2n + 57 = 180$; 41°,
82°, 57°

Chapter 13 ■ Probability and Statistics

Pages 364–365

1. 3 **2.** 3 **3.** 6 **4.** Red Tigers, green Padres **5.** 6
6. Yes **7. a.** 2, 4 **c.** 8 **d.** 4 **e.** 6 **f.** 20 **9.** 30
11. 60

Pages 366–367

1. Candice, Douglas, Aileen **2.** CDA, CAD, ACD,
ADC, DAC, DCA **3.** 6 **4.** 3, 2, and 1 **5. a.** 4
b. 3, 2, 1 **c.** 4, 3, 2, and 1 **d.** 24 **7. a.** 5 **b.** 120
9. 720 **11.** 24 **13.** 48 **15.** 720

Pages 368–369

1. 6 **2.** Yes **3.** $\frac{1}{6}$ **4.** $\frac{1}{3}$ **5.** $\frac{1}{6}$ **7.** $\frac{1}{2}$ **9.** $\frac{2}{3}$ **11.** $\frac{1}{2}$
13. 0 **15.** 1 **17.** $\frac{3}{8}$ **19.** $\frac{1}{8}$ **21.** $\frac{1}{4}$ **23.** $\frac{1}{8}$ **25.** 0
27. $\frac{1}{4}$ **29.** $\frac{5}{8}$ **31.** $\frac{1}{2}$ **33.** $\frac{3}{4}$ **35.** $\frac{1}{2}$ **37.** $\frac{1}{6}$ **39.** $\frac{1}{8}$
41. $\frac{1}{12}$ **43.** $\frac{23}{24}$ **45. a.** No **c.** Point down

Pages 370–371

1. $1.00 off **2.** No **3.** 8 **4.** 2 **5.** $\frac{1}{4}$ **7.** $\frac{3}{8}$ **9.** $\frac{1}{2}$
11. $\frac{3}{4}$ **13. a.** $\frac{1}{8}$ **b.** $\frac{1}{8}$ **c.** $\frac{3}{8}$ **d.** $\frac{3}{8}$ **e.** $\frac{7}{8}$ **f.** 0
15. a. $\frac{1}{36}$ **b.** $\frac{35}{36}$ **c.** $\frac{1}{18}$ **d.** $\frac{1}{6}$ **e.** $\frac{5}{12}$ **f.** $\frac{5}{12}$ **g.** $\frac{1}{6}$
h. $\frac{5}{6}$ **i.** 0

Pages 372–373

1. $\frac{1}{6}$ **2.** $\frac{1}{2}$ **3.** $\frac{1}{12}$ **4.** $\frac{1}{6}, \frac{1}{2}, \frac{1}{12}$ **5.** $\frac{1}{12}$ **7. a.** $\frac{1}{2}$

b. $\frac{1}{2}$ **c.** $\frac{1}{4}$ **9.** $\frac{5}{12}$ **11.** $\frac{1}{48}$ **13.** $\frac{5}{48}$ **15.** $\frac{35}{48}$
17. $\frac{1}{64}$ **19.** $\frac{1}{64}$ **21.** $\frac{3}{32}$ **23.** $\frac{3}{16}$ **25.** $\frac{15}{32}$ **27.** $\frac{1}{12}$
29. $\frac{1}{12}$ **31.** $\frac{2}{9}$ **33.** $\frac{1}{4}$ **35.** $\frac{1}{32}$

Pages 374–375
1. $\frac{1}{10}$ **2.** $\frac{1}{9}$ **3.** multiply **4.** $\frac{1}{90}$ **5.** $\frac{1}{90}$ **7.** $\frac{1}{18}$
9. $\frac{1}{30}$ **11.** $\frac{1}{72}$ **13.** $\frac{1}{9}$ **15.** $\frac{1}{12}$ **17.** $\frac{1}{12}$ **19.** $\frac{1}{4}$
21. $\frac{1}{504}$ **23.** $\frac{1}{3024}$ **25.** $\frac{5}{42}$ **27. a.** $\frac{1}{60,466,176}$
b. $\frac{1}{376,992}$

Page 378
1. $1.20 **3.** 7¢ **5.** $33\frac{1}{3}\%$ **7.** $2c - .55 = 1.65$, $1.10

Page 379
1. 5 **3.** 7 **5.** 21 **7.** 6 **9.** j **11.** $3y$ **13.** $14c$
15. $5xy^3$ **17.** $\frac{7}{6}$ **19.** $\frac{4}{3}$ **21.** $\frac{7}{10}$ **23.** $\frac{4}{3x}$ **25.** $\frac{r}{3}$
27. $\frac{5}{7c^2}$ **29.** $\frac{c^2}{9d}$ **31.** $\frac{19k}{4}$ **33.** $\frac{2}{5z}$ **35.** $\frac{9}{20}$ **37.** $1\frac{17}{24}$
39. $\frac{2a+b}{6}$ **41.** $\frac{10y+3z}{24}$ **43.** $\frac{3a+10b}{35}$ **45.** $\frac{1}{4}$
47. $\frac{5}{36}$ **49.** $\frac{2c-d}{10}$ **51.** $\frac{3j-2k}{24}$ **53.** $\frac{10q-15r}{36}$
55. $\frac{1}{4}$ **57.** 1 **59.** $\frac{1}{4}$ **61.** $\frac{2k}{9}$ **63.** $2\frac{1}{3}$ **65.** 1
67. $1\frac{1}{2}$ **69.** 5 **71.** $1\frac{1}{2}$ **73.** $13\frac{1}{3}$ **75.** $\frac{2b}{a}$ **77.** $\frac{3}{5}$

Pages 380–381
1. $1 **2.** $\frac{1}{8}$ **3.** $6 **4.** $.75 **5.** Less than **6.** Yes
7. $1 or more **9.** $\frac{1}{180}$ **11.** $.67 **13.** A bad deal
15. a. $\frac{1}{2}$ **b.** $3.13 **c.** Yes **17.** Yes **19.** $1.50 for a chance on the watch, $1 for a chance on the radio

Pages 382–383
1. 5, 5 **2.** 3 **3.** 9.8 **5.** 182.6 **7.** 30.7 **9.** 11
11. 10 **13.** 182.5 **15.** 7, 4 **17.** 22, 7 **19. b.** This one, the other one **c.** 46, 57, 60, 60, 63, 64, 67, 72, 78, 88, 90 **d.** 64, 67.$\overline{72}$, 60, 44

Pages 384–385
1. 3 **3.** 15 **5.** 24 **7.** $62\frac{1}{2}\%$ **9.** 16 **11.** 8
13. Edina **15.** 4th **17.** $33\frac{1}{3}\%$ **19.** Elliot, Fred
21. 21 **23.** Ernie **25.** Food, savings **27.** $6
29. $520 **31.** 1986 **33.** 1989

Pages 386–387
1. 3 **2.** 23 **3.** 25 in. **4.** 15 in. **9. a.** 21, 36.5
b. Eastern; western areas are clustered near the top

Pages 388–389
1. Above **2.** 11 **3. a.** 21.6 **b.** 17.7 **c.** 13.1
d. 11.8 **e.** 8.9 **5.** CBS **7. a.** 1987–1988
b. 1987–1988 **c.** 1988–1989 **d.** 1987–1988
e. 1987–1988

Pages 390–391
1. 6 **3.** Yes **5.** There may be many more male drivers than female drivers living in Lake City.
7. 1985: 60, 1986: 61, 1987: 63, 1988: 64, 1989: 65, 1990: 68 **9. a.** 1st **b.** 2nd

Pages 392–393
1. a. 40 **b.** 26 **c.** 65% **2.** 650 **3. a.** 18 **b.** 25
c. 72% **d.** 432 **5.** 375 **7.** 950 **9.** 109

Page 395
1. 13 **3.** 32 **5.** 40 **7.** 35 **9.** $16\frac{1}{5}$ **11.** $52\frac{1}{2}$
13. 16 **15.** 7 **17.** 160 **19.** 48 **21.** 6 **23.** 22
25. 42 **27.** 165 **29.** 56 **31.** 39 **33.** $28\frac{4}{5}$
35. $42\frac{6}{7}$ **37.** $10y$ **39.** 0 **41.** $^-15n + 7$
43. $^-13t - 15$ **45.** $^-12t - 14$ **47.** $^-7a + 2$ **49.** $^-1$
51. $^-5$ **53.** $\frac{2}{3}$ **55.** $1\frac{3}{4}$ **57.** $1\frac{3}{4}$ **59.** $3\frac{1}{2}$

Chapter 14 ■ Polynomials

Pages 400–401
1. a. n^2 **b.** n **2. a.** $2n^2$ **b.** 5 **3. a.** $4n$
b. $2n^2 + 3n$ **c.** $n^2 + 2n + 4$ **5.** $2n^2 + 3n + 2$
7. binomial **9.** binomial **11.** monomial
13. trinomial **15.** trinomial **17.** monomial
19. trinomial **21.** binomial **23.** 3 **25.** 6 **27.** 6
29. $^-8$ **31.** 10 **33.** 0

Pages 402–403
1. a. $n^2 + 2n$ **b.** $3n^2 + 5n$ **2. a.** 7, 7 **b.** 9, 1
3. $6x + 13$ **5.** $8x^2 + 4$ **7.** $9a^2 + 3b$ **9.** $11t + u$
11. $12m + ^-6n$ **13.** $^-3c + 11$ **15.** $^-2n + 5m$
17. $2c + 2d$ **19.** $a + 2c$ **21.** $5a + 4b + 6c$
23. $11r + 5s + 9t$ **25.** $7x^2 + 4$ **27.** $5r + 9, 59$
29. $9r + 8, 98$ **31.** $7s + 7, 707$ **33.** $9s + 7r + 7, 977$
35. $8s + 7r + 9, 879$ **37.** $8x + 28$ **39.** $14x + 14$

Pages 404–405
1. a. $n^2 + 2n$ **b.** $2n^2 + 3n$ **2.** $^-1, 1$ **3.** $n^2 + n$

5. $4n + 1$ **7.** n **9.** $3x + y$ **11.** $3x^2 + 1$
13. $4a^2 + 2b$ **15.** t^2 **17.** $a + 2b$ **19.** $2f$
21. $a + b + c$ **23.** $2x^2 + x + 1$ **25.** Shelby
27. Norris

Page 408
1. $1.83 **3.** No **5.** Kansas City **7.** \$.42 more
9. $p + 20 = 60$, 40 oz

Page 409
1. $\frac{2}{5}$ **3.** $\frac{-3}{8}$ **5.** $\frac{-3}{8}$ **7.** $\frac{18}{25}$ **9.** $\frac{3}{20}$ **11.** $2\frac{1}{4}$
13. $2\frac{2}{5}$ **15.** $4\frac{11}{20}$ **17.** $3\frac{2}{25}$ **19.** $6\frac{2}{5}$ **21.** $\frac{1}{3}$ **23.** $\frac{1}{16}$
25. $^-1\frac{5}{8}$ **27.** $\frac{1}{2}$ **29.** $\frac{-4}{15}$ **31.** $\frac{-3}{20}$ **33.** $^-9\frac{7}{8}$
35. $\frac{19}{24}$ **37.** $\frac{-2}{15}$ **39.** 1 **41.** 1 **43.** $8\frac{5}{8}$ **45.** $^-8$
47. $\frac{-1}{9}$ **49.** $\frac{15}{28}$ **51.** $^-3\frac{5}{13}$ **53.** 2 **55.** $^-1\frac{1}{4}$
57. $1\frac{1}{7}$ **59.** $2\frac{1}{2}$ **61.** $^-1$ **63.** $14y$ **65.** $4v + 12$
67. $^-7j + 9$ **69.** $12v - 6$

Pages 410–411
1. $n + 2$ **2.** Yes **3. a.** 16, 56 **b.** ^-3xy, $^-3y^2$, $12y$
c. $^-6t^2$, $5t$ **d.** $^-14c^2$, ^-7c, 7 **5.** $6n + 15$
7. $4n^2 + 6n$ **9.** $5y^2 + 15y$ **11.** $4z^2 + 8z$
13. $3b^2 + 6b$ **15.** $4t^2 + t$ **17.** $n^2 + {}^-3n$
19. $3a^2 + {}^-6a$ **21.** $4b^2 + 8b$ **23.** $2rt + 2st$
25. $15x^2 + {}^-10x$ **27.** $^-6ab + {}^-8$ **29.** $6y^2 + 18y + 12$
31. $6a^2 + 8ab + 2ac$ **33.** $12c^2 + 4c + {}^-12$
35. $^-8x^2 + 12xy + {}^-8xz$ **37.** 12 **39.** 10 **41.** 6
43. 8 **45.** 10 **47.** 20 **49.** $2n^2 + 6n$ **51.** $6n^2 + 6n$

Pages 412–413
1. 2 **2.** $3n$ **3.** C: $1 \div 1$, D: $(4 \cdot 1 \cdot 1 \cdot c) \div (1 \cdot 1)$
4. a. $6n$ **b.** $2y^2$, y **c.** 3, $3a$, $^-1$ **5.** $2n^2 + 1$
7. $3n + 1$ **9.** $n^2 + n$ **11.** $3x + 1$ **13.** $d^2 + d$
15. $4a^2 + {}^-2a$ **17.** $3a + 6b$ **19.** $x^2 + 2y$
21. $3ab + 1$ **23.** $2xy + 1$ **25.** $5b^2 + 3b + 2$
27. $4x^2 + 3x + 2$ **29.** $3n^2 + 4n + {}^-5$
31. $3n + 6$, $n + 2$, 2

Pages 414–415
1. symbols, like, 84, 4 **2.** 130, 102, 204 **3.** 15, 18,
$^-10$, 5, $^-2$ **5.** 2 **7.** $^-1$ **9.** 0 **11.** 3 **13.** $^-2$
15. 3 **17.** 4 **19.** 2 **21.** C, 110 **23.** A, 117

Pages 416–417
1. a. d, $5(8 - d)$ **b.** $10d + 5(8 - d)$ **3. a.** q,
$10(5 - q)$ **b.** $10(5 - q)$ **c.** 3 **5.** $25q + 10(6 -$
$q) = 90$, 2 **7.** $10d + 5(12 - d) = 90$, 6 **9.** $10d +$
$5(16 - d) = 130$, 10

Page 418
1. $^-5$ **3.** 30 **5.** 10 **7.** $^-81.4°F$ **9.** Above

Page 419
1. 5 **3.** $^-23$ **5.** $^-26$ **7.** $^-4$ **9.** 4 **11.** $\frac{-2}{3}$ **13.** 5
15. $1\frac{1}{3}$ **17.** 3 **19.** 4 **21.** $^-5\frac{1}{3}$ **23.** 1 **25.** 2
27. $\frac{-1}{4}$ **29.** $6\frac{3}{5}$ **31.** $3\frac{15}{16}$ **33.** $13\frac{1}{2}$ **35.** 6 **37.** 6
39. 6 **41.** 24 **43.** 9.6 **45.** 22 **47.** 54.6 **49.** 12.9
51. 36.66 **53.** $n + n + 80 = 180$; 50°, 50°, 80°

Chapter 15 ■ Similar and Right Triangles

Pages 424–425
1. Side ST, side RT **2.** $\angle S$, $\angle T$ **3.** 5 cm
5. 24 mm **7.** 27 mm **9.** 1 cm **11.** 3 cm
13. 3.52 cm **15.** 8.125 m

Pages 426–427
1. 2.0 m **2.** 1.4 m **3.** 4.9 m **4.** 7 m **5.** 15 m
7. 6 m **9.** 9 m **11.** 14.1 m **13.** 3.33 m **15.** 50 m
17. 2.7 m **19.** 18.96 m

Pages 428–429
1. Sides RT and ST, sides WY and XY **2.** 6 units,
9 units **3.** 4 units, 8 units, 12 units **4.** 8 units, 12
units **5.** 3 units, 6 units, 9 units **6.** 0.75 **7. a.** 8
b. 12, 0.75 **c.** $1.\overline{3}$ **d.** 9, $1.\overline{3}$ **8.** Yes, the measure
of the angle **9.** 0.42 **11.** 0.8 **13.** 2.08 **15.** 0.85
17. 0.71 **19.** 1.57 **21.** 0.47 **23.** 0.49

Pages 430–431
1. Side AB **2.** Side RS **3.** 5 ft, 13 ft **4.** 0.6, 0.8
5. C: the length of the hypotenuse is 5 ft, D: the
length of the side opposite $\angle R$ is 5 ft, E: $12 \div 13$
is ≈ 0.92 **6. a.** 0.6 **b.** 13, 0.92 **c.** 5, 0.38 **7.** 1.33
9. 0.8 **11.** 0.6 **13.** 0.42 **15.** 0.38 **17.** 0.92
19. 2.4 **21.** 0.38 **23.** 0.38 **25.** 0.75 **27.** 0.8
29. 0.8 **31.** 1.33 **33.** 0.6 **35.** 0.8 **37.** 0.75
39. 0.8 **41.** 0.6 **43.** $\frac{3}{4}$ **45.** $\frac{19}{12}$ **47.** $\frac{7}{12}$

Page 434
1. $3.06 **3.** \$3.12 **5.** \$1.35 **7.** \$2.86 **9.** $.49h +$
$.95 = 2.42$, 3 **11.** $g + 3g = 12$, 3

Page 435
1. $^-6$ **3.** $^-10$ **5.** 2 **7.** $^-32$ **9.** 24 **11.** $^-7$
13. 21 **15.** 27 **17.** 28 **19.** $8\frac{2}{5}$ **21.** $92\frac{1}{4}$ **23.** $15\frac{5}{9}$
25. 18 **27.** 11 **29.** 60 **31.** 160 **33.** 170 **35.** 7

37. 34 **39.** 72 **41.** 30 **43.** $32\frac{4}{7}$ **45.** $29\frac{1}{6}$
47. $40\frac{4}{5}$ **49.** $9x$ **51.** 0 **53.** $^-6w + 5$
55. $^-10n - 9$ **57.** $^-9n - 7$ **59.** $13y + 7$

Pages 436–437
1. 0.6428 **2.** 0.6428 **3.** 1.0000 **4.** 0.7002, 12.6
5. a. 0.8660, 17.32 **b.** 0.7071, 19.0917 **7.** 68.6 ft
9. 24.6 ft **11.** 18.9 ft **13.** 16.1 ft **15.** 22.9 ft
17. 10.5 ft **19.** 339.6 ft **21.** 300.9 ft

Pages 438–439
1. a. 33° **b.** 39° **c.** 36° **2.** 37° **3.** 0.5446 and
0.5592, 0.5592 **4.** 34° **5. a.** 38° **b.** 35° **7.** 71°
9. 27° **11.** 64° **13.** 51° **15.** 55° **17.** 63°
19. 320.4 ft **21.** 30°

Pages 440–441
1. Yes **2.** 6 and 7 **3.** 6.5 and 7.1 **4.** 6.8 and 6.8
5. 6.8 **7.** 1 **9.** 5 **11.** 8 **13.** 16 **15.** 1.2 **17.** 1.1
19. 6.856 **21.** 9.327 **23.** 11.269 **25.** False
27. 17.3 **29.** 22.5 **31.** 30.3

Pages 442–443
1. 16 square units, 25 square units **2.** 25 square

units **3.** Yes **4.** A, B **5.** 11.4 ft **6.** 13.2 ft
7. 8.6 ft **9.** 12 ft **11.** 8.9 ft **13.** 8 ft **15.** 13.2 ft
17. 5.7 ft **19.** 8.7 in. **21.** No **23.** No **25.** Yes
27. No

Pages 444–445
1. a. 17, 8 **b.** 17, b **c.** 289 **2.** Yes **3.** Diagram A,
12.5 mi **5.** Diagram B, 9.2 ft **7.** 18.0 mi **9.** 127.3
ft **11.** 96.8 ft **13.** 20 ft^2

Page 446
1. 28.3 **3.** Above **5.** 89.4 mi **7.** 37 mi

Page 447
1. 8 **3.** $2\frac{4}{7}$ **5.** $5\frac{1}{3}$ **7.** $^-3$ **9.** $^-3\frac{4}{5}$ **11.** $\frac{9}{19}$
13. $1\frac{1}{3}$ **15.** $1\frac{1}{8}$ **17.** 4 **19.** $^-1\frac{1}{2}$ **21.** $\frac{4}{9}$ **23.** 4
25. 10 **27.** $5\frac{1}{2}$ **29.** $10\frac{2}{15}$ **31.** 40 **33.** $12\frac{4}{5}$ **35.** 7
37. 6 **39.** $13\frac{1}{11}$ **41.** 50 **43.** 34 **45.** 160 **47.** 120
49. 1200 **51.** 60 **53.** 33.3% d **55.** 50% d
57. 44% i **59.** 32.7% i **61.** 25% i **63.** 55.6% d

Index

Graph(s)
  bar, 9, 36, 37, 384, 394
  broken-line, 135, 384, 394
  circle, 141, 144, 244, 385
  of inequalities, 334-335, 337, 339
  of linear equations, 348-351
  of linear inequalities, 352-353
  making, finding data, 394
  making, given data, 37
  of ordered pairs, 344-345
  picture, 385, 394
  of system of equations, 354-357
Greater than, 92, 144, 210, 334
Greatest common factor (GCF), 132, 140, 238
Group projects, 11, 49, 77, 103, 143, 159, 189, 215, 255, 271, 307, 345, 391, 401, 441

Height
  of a cylinder, 318
  of a parallelogram, 292
  of a prism, 311, 316
  of a pyramid and cone, 320
  of a triangle, 294
Hexagon, 276, 295
Hypotenuse, 430, 442

Increase, percent of, 258
Independent events, 372
Indirect variation, 358
Inequality(ies), 334
  graphing, 334-335, 337, 339, 352-353
  linear, 352-353
  solving, 336-339
  two-step, 338-339
Integer(s)
  absolute value of, 92-93
  adding, 96-99
  charged particle model, 94-96, 98, 100
  comparing, 92-93
  dividing, 102-103
  multiplying, 100-101
  negative, 92, 94
  opposite of, 94, 96
  ordering, 92-93
  positive, 92, 94, 103
  subtracting, 98-99
Intercept, 350
Interest, 148
  compound, 148, 262
  simple, 260
Inverse operations, 16, 48, 200, 222
Inverse variation, 358
Isosceles triangle, 270

Kilogram, 322-324
Kilometer, 278-281

Laws of exponents, 124-125
Least common denominator (LCD), 142-145, 168, 170, 214
Least common multiple (LCM), 132-133, 142-143
Leaves, 386
Legs, of right triangle, 428, 442
Length, 198-199, 278-281, 284
  of a prism, 316
  of a rectangle, 286, 290
Less than, 92, 144, 210, 334
Like terms, 80, 224, 226, 402-404
Line(s), 268
  parallel and perpendicular, 272-273
  of symmetry, 277
Line plot, 382-383
Linear equation, 348-351
  graphing, 348-351, 354-355
  system of, 354-355
Linear inequality, 352-353
  graphing, 352-353
Liter, 322-325
Lower quartile, 388
Lowest terms, fraction in, 140-141, 158, 238-239

Markup, 178
Mathematical connections
  linking conceptual and procedural knowledge, 48, 63, 92, 94, 96, 98, 100, 126-127, 145, 197, 277, 292, 296, 314, 337, 339, 348-357, 382, 388, 391, 394, 400-402, 404-405, 410-411, 426-427, 440
  to other curriculum areas, 9, 14, 30, 71-72, 92-96, 98, 100, 103, 114, 116, 127, 138, 143, 157, 164, 202, 204, 210, 215, 225, 232, 250, 279, 322-323, 328, 390-391, 408, 418, 429, 446
Mathematical expression (See *Algebraic expressions.*)
Mean, 382-383, 391
Measurement
  of angles, 268-269
  area, 290-297, 317, 326-328, 407, 410-411, 445
  circumference, 288-289
  converting customary units, 198-199
  converting metric units, 280-281, 322-325
  formulas, 21, 53, 69, 93, 111, 116, 123, 131, 157, 191, 204,

225, 230-231, 257, 286-297, 316-321, 324-327, 347, 418, 429, 446
  group projects, 11, 103, 143, 189, 215, 441
  length, 198-199, 278-281, 284
  perimeter, 286-287, 291, 317
  precision in, 284
  surface area, 310-311
  temperature, 92-93, 347, 418
  volume, 316-321, 324-327
Median, 382-383, 386-389, 391
Mental math, 36, 39, 42, 64-66, 78, 139, 252
Meter, 278-281
Metric-system units
  of area, 290
  of capacity, 322-324
  changing from one to another, 280-281, 322-325
  of length, 278
  of volume, 316, 324
  of weight, 322-323
Milliliter, 322-325
Millimeter, 278-281
Misleading statistics, 390-391
Mixed number(s), 154
  adding and subtracting, 172-175
  and decimals, 160-161
  dividing, 190-191
  and fractions, 154-157
  multiplying, 186-187
    by a whole number, 198
  and percent, 244-245
  quotient as, 156
  in simplest form, 158-159
Mode, 382-383, 386-387, 391
Modeling
  charged particles for integers, 94-96, 98, 100
  tiles for polynomials, 400-402, 404-405, 410
Monomial, 400
  dividing by a, 412-413
  multiplying by a, 410-411
Multiple
  common, 132-133
  least common (LCM), 132-133, 142-143
Multiplication
  of decimals, 34-35
  and division as inverse operations, 48
  estimating products, 34-35
  of fractions, 184-185
  of integers, 100-101
  of mixed numbers, 186-187
  of powers with the same base, 124-125

logical reasoning, 13, 17, 35, 43,
53, 67, 95, 104-105, 113, 129,
133, 155, 185, 219, 229, 248,
282, 340, 376, 406-407
make or complete a chart or
table, 12, 75, 79, 93, 104-105,
111, 134, 135, 175, 191-192,
249, 271, 295, 307, 312, 376,
432
make a model, 192-193, 249,
376
make an organized list, 162-163,
175, 193, 282, 340, 377
more than one solution in, 312-
313
number sense, 7, 9, 13, 39, 44,
49, 69, 125, 129, 135, 163,
167, 192, 202-203, 223, 283,
313, 341, 407
order of operations puzzles, 63,
105, 163
patterns and relationships, 12,
13, 44, 71, 125, 192, 213, 271,
295, 307, 313, 345, 347, 349,
358, 433, 441
solving simpler problems, 134,
192
steps for, 54-55, 84-85, 114-115,
146-147, 176-177, 202-203,
416-417
visual thinking, 12, 44-45, 70,
104, 135, 163, 193, 218, 248,
277, 283, 291, 307-309, 314,
319, 341, 376-377, 410-411,
433, 445
working backward, 312
writing algebraic expressions
for, 2-5, 31-33, 46, 99, 403,
413
writing equations for, 17, 23,
54-55, 72, 84-85, 114-115, 136,
146-147, 164, 176-177, 194,
202-203, 273, 356-357, 378,
416-418
writing formulas, 307, 327
writing problems, 3, 12, 14, 19,
24, 31, 33, 35, 43, 46, 51, 55-
56, 77, 83, 86, 116, 148, 178,
203, 204, 227, 232, 247, 261,
287, 327-328, 342, 393, 418,
446
writing proportions for, 240-243,
250, 258-259, 392-393, 424-
427
Problem-solving applications
for area, perimeter, and volume,
286-297, 310-311, 316-321,
324-328, 341, 445
averages, 38-39, 71, 77, 103,

111, 143, 159-161, 189, 191,
215, 243, 382-383, 391, 432
capacity problems, 9, 13, 174-
175, 186-187, 322-324, 328,
341, 376
consumer math, 6-7, 18, 24, 34-
35, 43, 46, 56, 65, 75, 79, 83-
86, 104, 106, 136, 148, 164,
175, 178, 194, 196-197, 200-
201, 210-211, 213, 220, 227,
232, 242-245, 249, 252-253,
260-262, 300, 328, 342, 378,
408, 434
distances, 14, 45, 69, 72, 81,
105, 112-113, 116, 123, 131,
166-167, 188-191, 217-218,
225, 231, 242-243, 250, 257,
279, 289, 313, 358, 377, 408,
437, 439, 446
evaluating formulas, 21, 53, 69,
75, 86, 93, 111, 123, 131, 157,
163, 191, 204, 217, 225, 231,
257, 260-262, 287, 347, 418,
446
finding errors, 45, 171
geometry–from an algebraic
perspective, 271, 273, 345
geometry–from a synthetic perspective, 12, 44-45, 70, 104,
135, 163, 193, 218, 248, 268-
277, 283, 286-297, 306-314,
316-321, 340-341, 376-377,
407, 411, 424-427, 433, 437,
441, 443
percent, 244-245, 247, 252-253,
255-262, 341, 393, 401
precision in measurement, 284
probability and expectation,
368-375, 380
Pythagorean theorem, 442-445
ratios and rates, 7, 21, 40-41,
53, 69, 71, 103, 111, 116,
189-191, 204, 230-231, 238-
243, 281, 284, 358, 432
scientific notation, 126-127
similar figures, scale drawings,
and models, 240-241, 250,
300, 328, 424-427
temperature, 92-93, 114-115,
135, 347, 418
time, 10-11, 38-39, 46, 69, 135,
159, 177, 215, 232, 243, 262
trigonometry, 430-431, 433,
436-439
weight, 17, 45, 48, 70, 193, 219,
284, 323, 408
Product(s)
cross-products, 240
estimating, 34

Property(ies)
for addition, 64-65, 68-69, 222,
224, 402
distributive, 78-81, 224-225,
410-411
of equality, 16, 18, 20, 48, 50,
52
for multiplication, 66-69, 224-
225, 402-403
Proportion(s), 240
cross-products, 240
and distance from a map, 250
and percent, 246-247, 252-253,
258-259
in rate problems, 242-243
in sampling, 392-393
and similar figures, 424
writing, 240-243, 250, 258-261,
392-393, 424-427
Protractor, 268
Pyramid, 306
volume of, 320
Pythagorean theorem, 442-445
experiment for, 441

Quadrilaterals, 276-277
Quartiles, 388
Quotient(s)
of decimals, 38-41
estimating, 38
mixed number as, 156
of rational numbers, 216-217

Radius of a circle, 288
Range, 382-383, 387
Rate, 242-243, 313, 358
in interest formulas, 260-262
in measurement formulas, 69,
103, 105, 112, 116, 191, 231,
257
pulse rate, 103
Ratio(s), 238-239, 242
equal, 238, 240
rise to run, 429
trigonometric, 428-431, 436-438
Rational number(s), 210
adding and subtracting, 214-215
comparing, 210-211
in decimal form, 212
multiplying and dividing, 216-
217
Rational solutions of equations,
222
Ray, 268
Reciprocals, 188, 216
Rectangle, 276-277, 295, 308
area of, 290-291, 410-411
perimeter of, 286-287, 291

Acknowledgements

Pupil's Edition Design: Traver Design
Cover Design: Design Five Cover Photograph: © The Telegraph Colour Library/FPG International
Page 281, exercise 36: from the *Guinness Book of World Records*, published by Sterling Publishing Co., Inc., New York, NY, © 1984, by Guinness Superlatives Ltd.

Photography Credits

CHAPTER 1: *1:* Jean Kepler/Rainbow; *2:* Shooting Star; *3:* © Rangefinders/Globe Photos; *4:* Focus on Sports; *6:* Donald Dietz/Stock Boston, Eric Roth/The Picture Cube; *7:* Donald Dietz/Stock Boston; *9:* Stephen Frisch/Stock Boston; *10:* Bob Pizaro/Comstock; *11:* Bob Pizaro/Comstock; *12:* Andrew Brilliant; *13:* Tom Di Pace/Sports Illustrated Photos; *14:* Ken O'Donoghue/© D.C. Heath; *16:* Ken O'Donoghue/© D.C. Heath; *17:* Clive Russ/© D.C. Heath; *18:* Ken O'Donoghue/© D.C. Heath; *19:* Ken O'Donoghue/© D.C. Heath; *20:* Bonnie McGrath; *21:* Cary Wolinsky/Stock Boston; *22:* Focus on Sports; *24:* Ulrike Welsch/© D.C. Heath

CHAPTER 2: *29:* Paul Johnson/© D.C. Heath; *30:* Stephanie Hollyman/Picture Group, Alpha/Globe Photos, Dirck Halstead/Gamma-Liaison, Forrest Anderson, Gamma-Liaison; *32:* Melabee Miller/Envision, Bonnie McGrath, Envision, Steve Chambers/The Picture Cube; *34:* Ken O'Donoghue/© D.C. Heath; *35:* M & C Werner/Comstock; *38:* Ken O'Donoghue/© D.C. Heath; *40:* Heinz Kluetmeir/Sports Illustrated Photos; *42:* Eric Cable/Stock Boston; *45:* Gabriella Della Corte, Ken O'Donoghue/© D.C. Heath; *46:* Norman Isaacs/Envision; *48:* Ken O'Donoghue/© D.C. Heath; *50:* Ken O'Donoghue/© D.C. Heath; *53:* Ulrike Welsch/© D.C. Heath; *54:* Jean Marc Giroux/Gamma-Liaison; *56:* Donald Dietz/Stock Boston

CHAPTER 3: *61:* David Lissy/The Picture Cube; *63:* Clive Russ/© D.C. Heath; *64:* Bob Pizaro/Comstock; *65:* Bob Pizaro/Comstock; *67:* Ken O'Donoghue/© D.C. Heath; *69:* Ken O'Donoghue/© D.C. Heath; *70:* Ken O'Donoghue/© D.C. Heath; *71:* Cary Wolinsky/Stock Boston; *74:* Ken O'Donoghue/© D.C. Heath; *76:* M. Grecco/Stock Boston; *78:* Lawrence Migdale/Photo Researchers; *79:* Bettina Cirone/Photo Researchers; *80:* Donald Kaplan/The Picture Cube; *82:* Focus on Sports, Agence de Presse/Photo Researchers; *84:* John Coletti/Stock Boston; *86:* Bonnie DelFavero

CHAPTER 4: *91:* John McDonough/Sports Illustrated Photos; *94:* Stephen Frisch/Stock Boston; *102:* Ken O'Donoghue/© D.C. Heath; *103:* Ken O'Donoghue/© D.C. Heath; *104:* John Iacono/Sports Illustrated Photos; *106:* Steve Hansen/Stock Boston; *108:* Tom Hollyman/Photo Researchers; *111:* Focus on Sports; *112:* Dean Hulse/Rainbow; *114:* Mike Douglas/The Image Works

CHAPTER 5: *121:* Jeff Gnass; *123:* Ellis Herwig/Stock Boston, Focus on Sports; *125:* Ken O'Donoghue/© D.C. Heath; *126:* Dan McCoy/Rainbow; *127:* Hartman-DeWitt/Comstock; *128:* Ken O'Donoghue/© D.C. Heath; *129:* Ken O'Donoghue/© D.C. Heath; *130–131:* Dan McCoy/Rainbow; *133:* Peter Menzel/Stock Boston; *134:* Ken O'Donoghue/© D.C. Heath; *136:* Clive Russ/© D.C. Heath; *138:* Arthur Grace/Stock Boston, Shooting Star, UPPA/Photoreporters, Heinz Kluetmeir/Sports Illustrated Photos; *140:* Ken O'Donoghue/© D.C. Heath; *141:* Kim Massie/Rainbow; *143:* Ken O'Donoghue/© D.C. Heath; *144:* Ken O'Donoghue/© D.C. Heath; *148:* Culver Pictures

CHAPTER 6: *153:* Clyde Smith/Peter Arnold, Inc.; *156:* Ken O'Donoghue/© D.C. Heath; *157:* Comstock; *160:* Rhoda Sidney/Monkmeyer Press; *162:* Ken O'Donoghue/© D.C. Heath, Eastcott/Momatiuk/Image Works; *163:* Cris Corr/Focus on Sports; *166:* Ken O'Donoghue/© D.C. Heath; *168:* Gus Schoenfeld/Berg & Associates; *172:* Mike Douglas/The Image Works; *174:* Russ Kinne/Comstock; *176:* Ken O'Donoghue/© D.C. Heath; *178:* Ken O'Donoghue/© D.C. Heath

CHAPTER 7: *183:* Bob Daemmrich/The Image Works; *184:* Clive Russ/© D.C. Heath; *186:* Clive Russ/© D.C. Heath; *187:* Larry Lawfer/The Picture Cube; *189:* Frank Siteman/The Picture Cube; *190:* The Picture Cube; *192:* Geoffrey Gove/Photo Researchers; *196:* Clive Russ/© D.C. Heath; *198:* Clive Russ/© D.C. Heath; *199:* Clive Russ/© D.C. Heath; *201:* Clive Russ/© D.C. Heath

CHAPTER 8: *209:* Comstock Inc.; *212:* Ken O'Donoghue/© D.C. Heath; *217:* Harriet Gans/The Image Works; *219:* Ken O'Donoghue/© D.C. Heath; *225:* Ken O'Donoghue/© D.C. Heath; *226:* Ken O'Donoghue/© D.C. Heath; *228:* Kirk Schlea/Berg & Associates; *230:* Richard Pasley/Stock Boston; *231:* Bob Daemmrich; *232:* Comstock

CHAPTER 9: *237:* Stacy Pick/Stock Boston; *240:* The Bettmann Archive; *242:* Paul Johnson/© D.C. Heath; *248:* Bruce Henderson/Stock Boston, Ellis Herwig/Stock Boston; *249:* Jerry Cooke/Photo Researchers, Ken O'Donoghue/© D.C. Heath; *252:* Stuart Cohen/Comstock; *254;* Arlene Collins/Monkmeyer Press; *257:* Ken O'Donoghue/© D.C. Heath; *260:* © Pickerell/Comstock; *261:* Ken O'Donoghue/© D.C. Heath

CHAPTER 10: *267:* Michael Stuckey/Comstock; *278:* Miro Vintoniv/Stock Boston; *282:* Julie Habel/Woodfin Camp, Nancy Sheehan; *284:* Ken O'Donoghue/© D.C. Heath; *288:* Ken O'Donoghue/© D.C. Heath; *289:* Larry Kolvoord/The Image Works; *299:* Ken O'Donoghue/© D.C. Heath; *300:* Leonard Lessin/Peter Arnold, Inc.

CHAPTER 11: *305:* Owen Franken/Stock Boston; *306:* Paul Johnson/© D.C. Heath; *310:* Paul Johnson/© D.C. Heath; *312:* Stuart Cohen; *322:* Clive Russ/© D.C. Heath; *326:* Judith Canty/Stock Boston, Ulrike Welsch/Photo Researchers; *328:* Rick Pasley/Stock Boston

Illustration Credits